Treatise on Concrete

# concrete from archeology to invention

## 1700-1769

**Other books by Roberto Gargiani**

*Auguste Perret* (1993), Electra

*Il principio del rivestimento* (1994), with G. Fanelli, Laterza

*Storia dell'architettura contemporanea* (1998), with G. Fanelli, Laterza

*Principi e costruzione nell'architettura italiana del Quattrocento* (2003), Laterza

*Archizoom Associati, 1966-1974* (2007), Electa

Editor of *La colonne. Nouvelle histoire de la construction* (2008), Presses polytechniques et universitaires romandes

*Rem Koolhaas/OMA. The Construction of Merveilles* (2008), EPFL Press

*Superstudio* (2010), with B. Lampariello, Laterza

*Le Corbusier. Béton Brut and Ineffable Space, 1940-1965, Surface Materials and Psychophysiology of Vision* (2011), with A. Rosellini, EPFL Press

Editor of *L'architrave le plancher la plate-forme. Nouvelle histoire de la construction* (2012), Presses polytechniques et universitaires romandes

Treatise on Concrete
Directed by Roberto Gargiani

roberto gargiani

# concrete from archeology to invention 1700-1769

## The Renaissance of Pozzolana and Roman Construction Techniques

Translated from the Italian by Stephen Piccolo

EPFL Press

A Swiss academic publisher
distributed by Routledge

# EPFL Press

The EPFL Press is the English-language imprint of the Foundation of the Presses polytechniques et universitaires romandes (PPUR). The PPUR publishes mainly works of teaching and research of the Ecole polytechnique fédérale de Lausanne (EPFL), of universities and other institutions of higher education.

Presses polytechniques
et universitaires romandes
EPFL – Rolex Learning Center
Post office box 119
CH-1015 Lausanne, Switzerland
E-mail: ppur@epfl.ch
Phone: +41 (0)21 693 2130
Fax: +41 (0)21 693 4027

www.epflpress.org





ISBN 978-2-940222-64-3 (EPFL Press)
ISBN 978-0-415-83346-2 (Routledge)
Printed in Spain





Routledge
Taylor & Francis Group

Taylor & Francis Group Ltd
2 Park Square, Milton Park
Abingdon, Oxford, OX14 4RN, UK

Routledge is an imprint of Taylor & Francis Group, An informa business.

Simultaneously published in the USA and Canada by Routledge,
711 Third Avenue, New York 10017

www.routledge.com

Library of Congress Cataloging-in-Publication Data
A catalog record for this book is available from the Library of Congress.

The graphical layout of the book and the cover design are by Paola Ranzini Pallavicini of Studio Pagina Milan, Italy. The photolithographic services were provided by Walter Bassani, AG-MEDIA, Milan Italy.

This work was translated from the Italian by Stephen Piccolo (www.transiting.eu).

Research for this publication has been financed by the European Research Council grant as part of project ERC2008-AdG, Proposal No. 230394, "The surfaces of cement and reinforced concrete: A history of the formwork and processing of the surface".

ÉCOLE POLYTECHNIQUE
FÉDÉRALE DE LAUSANNE

Series Editors for the publication program in Architecture are Professors Jacques Lucan and Luca Ortelli.

The publisher and authors express their thanks to the Ecole polytechnique fédérale de Lausanne (EPFL) for its generous support towards the publication of this book.

The author is also grateful to Marc Parlange, Dean of the ENAC School of the EPFL, who provided additional support towards the translation.

## Treatise on Concrete

*Directed by Roberto Gargiani*

The volumes of the *Treatise on Concrete* intend to offer a new and documented vision of the evolution of construction techniques and of the creative potential and formal expressions of concrete, from antiquity to the present time. Each book in the series will be based on specific archival research; the analysis of scientific, technical and iconographic original sources; and the investigation into documents made available by companies, engineers, architects and artists. The full spectrum of the use of concrete will be covered, including the building of foundations, ports, moles, bridges, vaults, domes, fortresses, bunkers, houses and monuments, as well as the manufacture of artificial stones and sculptures. In addition, the volumes will address the theoretical issues of *béton brut*, Brutalism and the true nature of matter. All volumes focus on the techniques and materials for the manufacture of formwork, from wood to fabric; on the meaning of the imprint; on the surface processing of concrete by hand or with machine; and on the composition of the mixture.
The series of books aims to provide the scientific community with an unique and updated reference on the historical and contemporary uses of concrete, which will open a new chapter in the knowledge of this crucial material for architectural construction, as well as for masterpieces of engineering and art.

Volumes in preparation:
*Louis I. Kahn, Exposed Concrete and Hollow Stones, 1949-1959*
*Louis I. Kahn, Towards the Zero Degree of Concrete, 1960-1974*
*Tadao Ando, Concrete Spaces*
*Mnemosyne in Concrete*
*The Concrete of Pier Luigi Nervi*
*Research on the Roman Cement, 1770-1779*
*The Origins of Modern Concrete Building in Germany, 1800-1877*
*Concrete Sculpture, from 1950 to Today: Visions of a Material*
*Contemporary Ornament in Concrete*
*Bared Concrete of Natural and Artificial Pozzolanas, 1780-1817*
*Formless: Beyond Formwork*
*The Concrete of Perret and Pupils*

# Table of Contents

Scala
ÆGYPTUS INFERIOR
DELTA
HÆPTANOMIA
ARCADIA
THEBAIS PRIMA
ÆGYPTUS SUPERIOR
THEBAIDIS REGIO
ARABIÆ PETR PARS
Heroopolites Sinus
ARABICI SINUS PARS
MARE ERITHRÆUM
Arabiæ paludes
Arabes Azarei
Thebæ als Diospolis
Mons Psammius
CYNOPOLITES
HERMOPOLITES
LYCOPOLITES
APOLLONOPOLITES
COPTITES
Coptos
Tentyra
Memphis
Heliopolis
Pelusium

# chapter one
# fantastic archaeology and artificial stones

## The mystery of artificial stone in ancient Egypt

The 18th century holds the most unique place in the history of the development of concrete – marked by lively study, experimentation, research and critical advances – since the period of its invention and refinement at the time of the Roman Empire. The techniques used in the Middle Ages derived from the *opus cæmenticium* and the *opus signinum*, and as put into theory and practice in the Renaissance by Brunelleschi, Alberti, Giuliano da Maiano, the Sangallos, Bramante, Vasari and De l'Orme, as well as the experiments and research of Manetti, Michelini, Viviani, De Mari, Barattieri, Bernini, Perrault, Riquet, Vauban, Wren and Shere, are simply steps towards the reinvention of concrete masonry. Thus the techniques developed during the 1700s in Europe for the building of masonry structures (including foundations, embankments, moles, jetties, groynes, bridge piers, canals, cisterns, walls and vaults) find a new, glorious beginning and are destined to orient research for centuries to come. In an historical perspective, the works carried out during the centuries between those of the Romans, but before the inventions of the 18th century appear as brilliant but sporadic attempts to update a technique with extraordinary potential, waiting for a new empire to take form that would substantiate and determine the progressive establishment of that technique, the empire of what would be defined as 19th-to-20th-century capitalism, with its demands for economic construction by non-specialized workers, and speed of execution of magnificent works of engineering for a new dominion of the landscape. Beyond academic paeans in praise of Greek architecture as the model for contemporary monuments, it is no coincidence that the focus of technicians, chemists, inventors and engineers involved in the new, definitive rise of concrete that took place during the course of the 1700s is on Roman construction, with its ruins and the textual sources of Latin authors – from Vitruvius to Pliny – where the secrets of the recipes for making that marvelous artificial material whose solidity had been proven by centuries of wear had been conserved.

Only against the backdrop of the ongoing sequence of inventions and experiments already in progress in the mid-1700s is it possible to grasp the full potential of the drawings of Piranesi and Boullée, and to attribute to them the value that they assume, in certain phases, in the evolution of concrete construction, for which they represent complementary aspects, from the apologia of Piranesi of the Roman era of the invention of concrete to the visions of Boullée of what it is possible to imagine, with that material, in the 18th century; from the impassioned witnessing of the evocative power of the ruins of constructions in *opus cæmenticium*, to the envisioning of synthetic forms for new construction in concrete. Between the extremes represented by the drawings of Piranesi and Boullée, we find constellations of inexorable and fundamental scientific experiments whose protagonists are engineers, architects, masons, chemists, mineralogists and volcanologists, at times taking the form of tables for comparison of numerical data to demonstrate the superiority, with respect to other binding agents, of the concrete made according to the Roman recipe. In all this gained experience, what stands out is a singular use of concrete that emblematically demonstrates, at the same time, its phenomenal constructive potential and the condition of extreme subordination to which it is subjected with respect to traditional materials, from stone to brick: it is poured for foundations, embankments or jetties, remaining invisible. But precisely its potential for construction and fabrication would redeem it from its initial role in which it proved its worth under the ground or under water. The resurrection of concrete from its status as a buried material coincides with the recognition of its endless artistic, as well as technical, potential, which emerges significantly even if it remains limited to an imitation of what was still considered

**1**. Carlo Fontana, "Positioning and overall view of the machines used to raise the Vatican obelisk" (C. Fontana, *Il Tempio Vaticano e sua origine con gli edifici più cospicui antichi, e moderni fatti dentro, e fuori di esso*, Roma, G.F. Buagni, 1694).

the prime parameter of constructive *venustas*: the look of stones and marbles, with their grain and their colors. To understand the multiple technical facets involved in the rise of concrete in the 18th century, we must not overlook the fact that its fabrication in the ancient Roman way is executed according to different techniques, one based on the use of the volcanic product already praised by Vitruvius, pozzolana, and the others based on processes of production of compounds made with other materials. The legendary secret mentioned in the writings of that century regarding the compound of the Romans has to do with these latter ways of making concrete; the former survives with surprising continuity in the main hydraulic constructions along Mediterranean coasts, especially in Italy, which continues to be seen as a repository of technical know-how to be imitated.

During the 18th century, the rise of concrete is marked, especially in the first half of the century, by an extraordinary mixture of scientific research and fantastic interpretation of sources and ruins, in an attempt to date the discovery of concrete and the fabrication of artificial stone to earliest antiquity. At that time, the secret of this discovery is said to have been conserved across the centuries, though it had produced only sporadic, isolated examples of application. Following the exploit of the transport of the Vatican obelisk by Domenico Fontana and his technicians, accompanied by a monumental publication in honor of the invention,[1] awareness of the need to create ingenious machines for the transport of great monolithic blocks contributed to sustain the hypothesis of fabrication on site of the enormous ancient monoliths, by means of *pierre factice* or artificial stone. Precisely such misunderstandings about the presence of an easily moldable concrete compound in the constructive cultures of antiquity are to become the essential stimuli for the creation of the first manifestations of artificial stone in the 1700s.[2] These misunderstandings had already been outlined in previous centuries, regarding Egyptian architecture or monumental

[1] For the reference to Fontana see, Antoine-Yves Goguet, *De l'origine des loix, des arts, et des sciences; et de leurs progrès chez les anciens peuples*, Paris, Desaint & Saillant, 1758, p. 260. *Cf.* Carlo Fontana, *Il Tempio Vaticano e sua origine con gli edifici più cospicui antichi, e moderni fatti dentro, e fuori di esso*, Roma, G.F. Buagni, 1694.

[2] In the area of certain chemistry research on geopolymers conducted at the Geopolymer Institute at Saint Quentin in France, the chemist Joseph Davidovits since the 1970s has formulated the hypothesis that Egyptian stones were artificially produced – though overlooking what had been similarly hypothesized and refuted in the 1600s and 1700s. See, Joseph Davidovits, *The Book of Stone, Volume 1, Alchemy and Pyramids*, Saint Quentin, Geopolymer Institute, 1983, pp. 203-12, and, Joseph Davidovits, *Geopolymer Chemistry and Applications*, Saint Quentin, Geopolymer Institute, 2008[2], pp. 367-86. On the question of the natural or synthetic origin of the stones of the pyramids, debated over the last few decades by chemists, see, among others: Joseph Davidovits, *X-Ray Analysis and X-Ray Diffraction of Casing Stones from the Pyramids of Egypt, and the Limestone of the Associated Quarries*, in, Rosalie A. David, ed., *Science in Egyptology*, Manchester, Manchester University Press, 1986, pp. 511-20; Joseph Davidovits, Margaret Morris, *The Pyramids: An Enigma Solved*, New York, Hippocrene Books, 1988; Margaret Morris, *Pyramid Debate, The Cast-in-Place, Theory of Pyramid Construction*, in, "Concrete International," vol. XIII, August 1991, no. 8, pp. 29, 39-44; Robert Louis Folk, Donald Harvey Campbell, *Are the Pyramids of Egypt Built of Poured Concrete Blocks?*, in, "Journal of Geological Education," vol. XL, 1992, no. 1, pp. 25-34; Kevin D. Ingram, Kenneth E. Daugherty, James L. Marshall, *The Pyramids-Cement or Stone*, in, "Journal of Archaeological Science," vol. XX, 1993, pp. 681-87; Michel W. Barsoum, Adrish Ganguly, Gilles Hug, *Microstructural Evidence of Reconstituted Limestone Blocks in the Great Pyramids of Egypt*, in, "Journal of the American Ceramic Society," vol. LXXXIX, December 2006, no. 12, pp. 3788-96. RMN 29Si, 27Al and 43Ca resonance spectra have been compared of a casing stone of the Bent Pyramid of Dahshur, with those of the calcareous stones of the region. The analyses have established that the casing stone is composed of calcareous grains from the quarry of Tura that are bonded by a calcium silicate substance that was made artificially, perhaps with a silicium additive (*cf.* Kenneth J.D. McKenzie, Mark E. Smith, Alan Wong, John V. Hanna, Bernard Barry, Michel W. Barsoum, *Were the casing stones of Senefru's Bent Pyramid in Dahshour cast or carved? Multinuclear NMR evidence*, in, "Materials Letters," LXV, 2011, no. 1, pp. 350-52). As during the course of the 1700s, contemporary chemists transform their visions of Egyptian artificial stone into hypotheses and experiences to make the science of polymers advance, as well as delving into production of concrete made according to the "secret" of the ancients. What happened in the 1700s and straddling the year 2000, in the chemistry research on Egyptian stones, simply confirms the fascinating hypothesis on the evolutionary processes of science formulated by Paul Feyerabend (*cf.* Paul Feyerabend, *Against Method, Outline of an Anarchistic Theory of Knowledge*, London, Verso Edition, 1975).

complexes like Stonehenge. Even at the end of the 1600s, the authoritative military engineer Vauban spreads word that the columns in the choir of the church of Vézelay, dating back to the 12th century, with monolithic shafts, is made with artificial stone – "recognized as artificial by marshal Vauban."[3] Some of the columns of monuments in Aix-en-Provence, Orange and Lyon are also claimed to be of liquid stone, or *pierre fondue*, at least until men of science insist that they are made, instead, of granite, as the botanist Joseph Pitton de Tournefort (1656-1708) writes.[4] At the start of the 1700s even the columns of the portico of the Pantheon in Rome are said to be made of an artificial material, a *pierre fondue* obtained with an amalgam of marble fragments by means of a cement.[5] The large stone that forms the roof of the mausoleum of Theodoric, in Ravenna, is also considered an artificial creation, due to the "difficulty in understanding how one could have transported and placed it where it is": "Some have suggested that it is pierre fondue made on site; others that one only had to dig underneath it in order to build the Church."[6] But the most heated debate on the existence of artificial stone is undoubtedly the one that continued throughout the 18th century, fed by scholars and travelers, regarding Egyptian architecture and Stonehenge.

The giant blocks that form the pyramids, the columns and obelisks of the Egyptians are seen as artificial products, artificial stones fabricated with molds thanks to a special binding agent whose recipe is said to have been lost. To back up this claim, observations and passages from ancient literary sources are called into play. The composition of the various Egyptian granites of magmatic origin, resulting from solidification of molten silicates mixed with other components, is mistaken for the result of artificial fabrication based on processes similar to those used to cast metal for sculptures. Obelisks, columns, construction blocks and statues are thus said to be the result of the fusion of different stones, creating a compound that could be poured into molds – "pieces [...] melted and poured into molds as one casts a piece of metal."[7]

The reports of Balthasar de Monconys (1611-65), the diplomat, physicist and magistrate from Lyon, regarding the Egyptian *pierre fondue*,[8] are still seen as authoritative support, in the 1700s, of those who believe that granite is an artificial stone, in which red and white fragments are joined by means of a "black material" that "crumbles in the hand, and seems to be a bonding of the other stones that are stronger, and small as if they had been broken up [...]."[9]

"Their opinion," reports the naturalist and French consul in Cairo, Benoît de Maillet (1656-1738), "is based on the fact that this stone is found to be finer in certain works than others, on the fact that it crumbles easily in the hand, and that it separates into red and white pieces, between which one finds a material bright like diamond mixed with a little black; they consider this to be the cement that binds the pebbling. They add that the pebbles are more or

[3] Polycarpe de La Faye, *Recherche sur la préparation que les romains donnaient à la chaux dont ils se servoient pour leurs constructions, & sur la composition & l'emploi de leurs mortiers*, Paris, Imprimerie Royale, 1777, p. 45, note y.

[4] "The columns that pass for pierre fondue are of ordinary granite: those of the Baptistery of Saint Sauver in Aix en Provence, in the hall of Orange, in Lyon in the Abbey of Ainay, are all of the same material, and one can generally affirm that all stone, no matter of which type, can be calcinated in fire, but will not melt" (Joseph Pitton de Tournefort, *Relation d'un voyage du Levant, fait par ordre du roy*, Paris, Imprimerie Royale, 1717, vol. I, p. 308). Also against the opinion of the *pierre fondue*, with which the columns of the "old altar of Lyon" would have been made, see, Claude-François Menestrier, *Histoire civile ou consulaire de la ville de Lyon*, Lyon, J.-B. & N. De Ville, 1696, p. XVII. On the columns discussed by Pitton de Tournefort see, Louis Poinsinet de Sivry, ed., *Histoire Naturelle de Pline, Traduite En François [...]*, Paris, Veuve Desaint, 1778, vol. XI, p. 479, note.

[5] Bernard de Montfaucon, *The travels of the learned Father Montfaucon from Paris thro' Italy*, London, E. Curll, 1712, p. 285.

less large, which is the cause of the fineness or coarseness of the grain that one sees in this stone, and that when the ancient Egyptians wanted columns or obelisks of this composition, they ground the stones to the extent desired, depending on their decision to create a work more or less beautiful."[10]

The hypothesis of concrete parts in the pyramids is advanced by Paul Lucas (1664-1734), the merchant and naturalist, who in 1704 writes that he had found, during a visit to Giza, a "cement" of which the smaller pyramids of the monumental complex has been "coated." "Some people take this coating for marble, and it seems so, but it is just a cement," he specifies.[11] The Benedictine monk Bernard de Montfaucon (1655-1741), a paleographer and expert of antiquity, is aware of the belief that "one had in the past" the existence of the "secret to liquefy stone" with which, according to some, also would have made up the columns of the Temple of Juno in Rome. "Suetonius," he writes, "overturns these opinions when he reports that the columns had been transported to Rome from Thebaid and Egypt."[12]

Certain interpretations of passages from Pliny, Herodotus or Varro on the stones of ancient monuments lend themselves to the spread of the myth of an artificial compound for the fabrication of colossal blocks – thus the so-called Labyrinth of Egypt and the slab covering the tomb of Porsenna at Chiusi are said to be made of artificial stone. What is known as Pompey's Pillar, erected in honor of Diocletian in the place where the Serapeum stood in

**2**. Gilles Robert de Vaugondy, "Veteris Aegypti tabula ex variis auctoribus decerpta, atque itinerariis adornata," Paris, Apud Auctorem, 1743 (Cartes et Plans, GEBB565 (15, 21), Bibliothèque Nationale de France, Paris).

[6] Jean-Etienne Guettard, *Mémoires sur differentes parties des sciences et arts*, Paris, L. Prault, 1768, vol. I, p. 398.
[7] Goguet, *op. cit.*, p. 134.
[8] See, *Journal des voyages de Monsieur de Monconys. [...]. Premiere Partie. Voyage de Portugal, Provence, Italie, Egypte, Syrie, Constantinople, & Natolie*, Lyon, H. Boissat, 1665, pp. 212-13.
[9] Louis Poinsinet de Sivry, ed., *Histoire Naturelle de Pline, Traduite en François [...]*, Paris, Veuve Desaint, 1778, vol. XI, p. 481, note.
[10] Benoît de Maillet, *Description de l'Egypte contenant plusieurs remarques curieuses sur la Geographie ancienne, et moderne de ce Païs [...] Composée sur les Mémoires de Monsieur De Maillet, Ancien Consul de France au Caire, par M. L'Abbé Le Mascrier*, La Haye, I. Beauregard, 1740, vol. II, pp. 137-38.
[11] *Voyage du sieur Paul Lucas au Levant*, Paris, G. Vandive, 1704, vol. I, p. 71.
[12] Bernard de Montfaucon, *Suite de l'extrait du Diarium Italicum*, in, "Memoire pour l'Histoire des Sciences & des beaux Arts," Trévoux, E. Ganeau, July 1703, (pp. 1202-26), p. 1219.

Alexandria, is one of the most often discussed and mentioned objects to demonstrate the Egyptian's knowledge of the technique of artificial stone. Its size (29 meters in height, with a diameter of 3 meters at the base) and the geological composition of its stone, red Aswan granite, is taken as evidence of its on-site construction with artificial stone, precisely to avoid the difficulties of transporting such a huge monolith from a quarry.[13] The columns of the Temple of Artemis at Ephesus, due to their number and size, are also thought to be artificial stone. "These testimonies from ancient history, together with the precious remains of antiquity, would allow one to believe that the columns are of one unique piece and of such an excessive height that they would have had to been liquefied and cast in molds like metal."[14]

The soaring monoliths of obelisks visible not only in Egypt but also in Rome also become a crucial factor in the hypothesis of the existence of a once-flourishing Egyptian production of artificial stone.

"Others have thought that the obelisks were a sort of artificial stone, made of several types of stone pebbles, crushed, bonded & then incorporated by the use of a certain cement hard enough to withstand cutting and polishing. They allege that proof of their feeling is that in the entire world, one cannot find a single quarry where blocks of such volume can be found. In addition, they add, even if one were found, it would be impossible to extract, for example, a piece of the size of the Ramses obelisk, and even more impossible to transport it."[15]

It can be no coincidence that among the first objects to be made with the type of concrete rediscovered in the 1770s, there is precisely an obelisk, as if to restore credibility – through that artificial product – to the hypotheses that circulated at the start of the century regarding Egyptian artificial stone, which in any case, apart from the question of their archaeological accuracy, created the imaginative conditions needed to nurture development of a contemporary technique of artificial stone.

Precisely in the light of what is to happen in the 1770s, in pursuit of the perfect binding agent to make an artificial stone of maximum strength, it is worth looking back at the considerations published in 1758 by Antoine-Yves Goguet (1716-58), an erudite enthusiast of ancient civilizations, to counter the claims of the believers in Egyptian artificial stone, because paradoxically these considerations point to what will become the main objective of the discovery of concrete.

"As for those who believe that the marble of the obelisks is nothing other than a sort of artificial stone," he writes, "an assembly of small stones bonded and incorporated with mastic; the objection is more specious, but it is not more solid. Do they imagine that it is possible to make pieces of such dimension as the obelisks with mastic, with a hardness capable of resisting damage for the many centuries that have passed since the erection of these monuments? We know, it is true, of these types of compounds that can withstand chisels, and that can even be polished. But experience has shown us that we have not at all yet discovered the art of making, with mastic, compounds sufficiently hard and sufficiently solid to resist the action of the sun in our climate, and even more so in Egypt."[16]

[13] De Maillet, *op. cit.*, pp. 137-38.
[14] Gilbert Charles Le Gendre, *Traité historique et critique de l'opinion*, Paris, Braisson, 1741[3], vol. VI, p. 608.
[15] Goguet, *op. cit.*, p. 134. On the belief that the Egyptian obelisks were in "artificially glued stone" see also, Pierre Belon, *Les Observations de plusieurs singularitez & choses mémorables, trouvées en Grece, Asie, Judée, Egypte, Arabie, & autres pays estranges [...]*, Paris, H. de Marnef, 1588, p. 285.
[16] Goguet, *op. cit.*, pp. 134-35.

The question of the existence of a kind of artificial stone in ancient Egypt had already been discussed during the 1600s, among others by Louis Moréri (1643-80), who on the subject of the pyramids – in the dictionary he edits and which continues to be reprinted with success in the 1700s – underlines what a surprising feat it had been to raise such colossal stones in an era in which "most mechanical inventions were unknown." As for the stones themselves, unlike De Monconys, Moréri has no doubt about their nature. "Some do not know where such large stones could be pulled from," one reads in the 1732 edition of his *Dictionnaire*, "and in such large quantity, because one only sees sand thereabout; but they did not notice that under the sand there is solid rock that supplies these stones, in addition to the several mountains not too far away, where there is no lack of stone."[17]

The disputes regarding the material used to make a funerary urn found in France at Saint-Thibéry, near Béziers, reveal the fervent interest in the question of Egyptian artificial stone during the first decades of the 1700s. De Montfaucon stands up against a chorus of voices in favor of the opinion that the urn is in *pierre fondue*, and can therefore be seen as evidence of a technique invented by the Egyptians "had earlier the art of liquefying stone as we liquefy metal," then inherited by the Greeks and, later, the Romans. The artificial stone, in keeping with what De Montfaucon reports, supposed to have been obtained with a *lithocolle* which the 18th-century dictionaries indicate as a binding agent based on old brick and *poix résine* [pitch resin], a "cement that jewelers use to hold in place precious stones that they want to shape on the grindstone."[18]

"Others reject this idea," De Montfaucon writes, "by adopting another which is not better founded. There is marble, they say, made artificially and with a sort of glue, that one calls lithocolla, to bind marble that has been broken and reduced to dust or to other small bits, and to make a compound harder than marble itself. This is how, they continue, the granite marble of Egypt was made: which is manifestly opposed to what the historians, Suetonius and others, say about this marble. They assure us that they were pulled from Egyptian quarries, just as one takes other stones from quarries. Lampridius states it most clearly, when he says of Helagable Ch[apter] 24, that he wanted to make an enormous column, all in one piece, in order to place afterwards the god Helagable on this column [...] but that one could not find an block of marble of sufficient size for that in Thebaid, where one went to look for it. [...]. If these great masses could have been made with a sort of glue or cement, what need would there be to go pull them from the quarries of Thebaid, in order to then transport them to Rome with such difficulty and effort? There would have been nothing so simple as to make them in Rome, on the very pedestal where one wanted to place them."[19]

One of the main arguments in favor of the existence of artificial stone in Egypt, that of the absence of quarries, also discussed by De Monconys and Moréri, already begins to lose ground in the first decades of the 1700s, when the quarries at Mt. Sinai from which the particular "granite marble" of the Egyptian monuments was taken, become known.

[17] Entry *Pyramides*, in, Louis Moréri, *Le Grand Dictionnaire Historique, ou le mélange curieux de l'Histoire Sacrée et profane*, Basle, J. Brandmuller, 1732, p.n.n.

[18] *Cf.* entry *Lithocolle*, in, Joseph-Nicolas Guyot et al, ed., *Le Grand Vocabulaire françois*, Paris, Hôtel de Thou, 1771, vol. XVI, p. 121.

[19] Bernard de Montfaucon, *Supplement au livre de L'antiquité expliquée et representée en figures [...]. Les funerailles*, Paris, Veuve Delaulne, 1724, vol. V, p. 17.

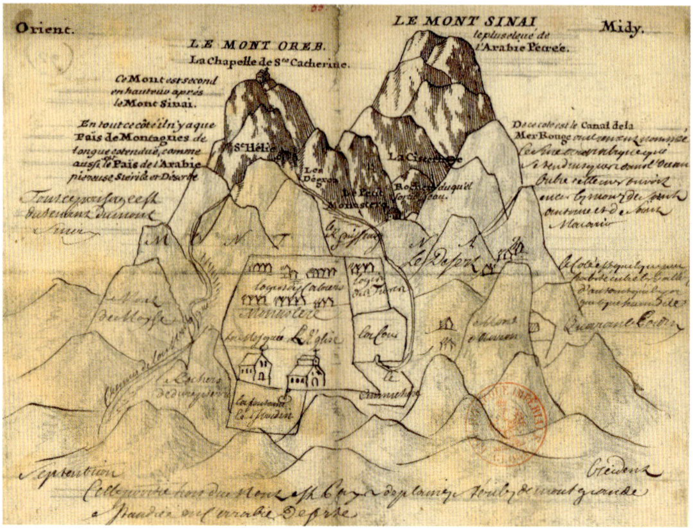


**3**. Anonymous, view of Mount Sinai, late XVII or early XVIII (Cartes et Plans, GED-10559, Bibliothèque Nationale de France, Paris).

Thomas Shaw (1692-1751) is one of the first to reveal the existence of large quarries of "granite marble," which he visits during his travels in Egypt, from 1721 to 1724. The publications of his travel notes, first in English in 1738 and then in French in 1743,contribute to debunk the notion of the existence of artificial stone in ancient Egypt.

"It appears to be a congeries of cawky nodules," he writes on the subject of granite marble, "of different shapes and sizes, beautifully united together: which, from the likeness they bear to a composition of mortar and gravel, might occasion several ingenious travellers to image Pompey's pillar, the obelisks at Heliopolis, Alexandria, and Rome, with other the like extraordinary lumps of this sort of marble, to be factitious, and produced by fusion. That kind of it, which I saw in the neighbourhood of Mount Sinai, and in the midland road from thence to Corondel, is generally of a light grey colour, with little black spots interspersed; though, in some places I have seen it much blacker, and in others of a reddish complexion, like the marble of Syene [...]."[20]

Shaw describes the modes of transport, through special canals and on the Nile, of the large monolithic blocks of the obelisks, for which traces of extraction can also be seen in the quarries: "Today one can still see several quarries where the Egyptians pulled their obelisks and other large blocks of this marble."[21] Instead, he points out how the concrete used in the pyramids has lost its consistency: "Notwithstanding the weight and size of most of these stones, they were all cemented together, but the cement has become friable over the length of time, even though it was certainly hard in the past, because it seems to be of the same sort as that still used today in Barbary."[22]

Another decisive contribution to put an end to the belief in the artificial stone of ancient Egypt can be found in the memoirs of De Maillet, printed in 1740 through initiative of the abbot Le Mascrier. Also relying on information from the Archbishop of Mount Sinai regarding the existence of many marble granite quarries, De Maillet confesses he has changed his mind, and no longer agrees with the "idea of those who only looked at this marble as some kind of compound and an artificial stone."[23]

"It is certain," he writes, "near Mount Sinai there are quarries of this granite marble, that have been considered liquefied marble, and that on the coast itself of the Red Sea one finds entire mountains of the most beautiful white marble that can be extracted from Asia. Certainly nothing more is needed to destroy the opinion of those who hold that this marble granite, which has served in the construction of so many colossal structures in Egypt – sphinxes, columns, superb monuments that will always be admired by posterity – was nothing other than pierre fondue, and composed of stone chippings of different colors."[24]

"[...] the inspection of these quarries," in which it is possible to see "even today the matrices of these famous monuments," suffices to challenge the myth of "a composition whose secret has been lost," flatly states Goguet, who with Shaw and De Maillet contributes to demolish the myth of Egyptian artificial stone.[25]

Yet precisely reasoning like that of De Maillet, intended to demonstrate the non-existence of artificial stone in Egyptian construction, also pointing to the fact that there are no traces of the survival of such techniques in contemporary building, may paradoxically have created the conditions for contemporary research on artificial stone: "In vain to weaken this opinion, one would say, that if these columns and so many other works that we now look upon with astonishment, had been liquefied, one consequently could have liquefied even higher and more massive ones; that it would have even been possible to make entire houses from a single stone, as well as an infinity of other works even more surprising that those that still exist today, and that we quite justifiably admire."[26]

De Maillet's description of the process of geological creation of *marbre granite* contains veritable recipes for artificial mixing of ingredients to create an artificial stone that would be utterly similar to natural stone. That granite is simply a natural blend of "gravel, more or less course or fine," joined by "chance" with "pebbles and other pieces of stone, bigger or smaller," thanks to the "sands & silts" that "probably acted as cement."[27]

De Maillet also reports on certain singular construction practices applied in Egypt, depending on mortars, such as that of building, thanks to a "sort of cement," quick-setting and with a lime base, steps "without the use of scaffolds": "As soon as they have placed one level, they climb on it and set down another; which they continue to do until the work is finished."[28]

Some travelers who cross the territories of Africa facing the Mediterranean do not simply observe the ruins of Egyptian or Roman antiquity, but also report on successive or contemporary construction systems, qualities of mortars and their modes of fabrication, even describing details related to formwork, in the case of poured materials. Shaw's travel notes, again in this case, are full of important information. In different cities, from Tlemcen, near Algiers, to Damascus, he finds a way of constructing walls with concrete poured into formwork made with wooden boards – "molds"[29] or "frames."[30]

"Almost all the walls of Tlem-san," he writes, "are made with a mortar composed of sand, lime and small stones, which one throws into molds; and this mortar well made and well worked has acquired the force and strength of stone. It is possible to see the marks left on the walls by the molds, some of which are one hundred yards long and a fathom in height and thickness."[31]

[20] Thomas Shaw, *Travels, or observations relating to several parts of Barbary and the Levant*, London, A. Millar, W. Sandby, 1757², p. 442. *Cf.*, Thomas Shaw, *Voyages de Monsr. Shaw, M.D., dans plusieurs provinces de la Barbarie et du Levant: contenant des observations géographiques, physiques, philologiques et melées sur les royaumes d'Alger et de Tunis, sur la Syrie, l'Égypte et l'Arabie Pétrée*, La Haye, J. Neaulme, 1743, vol. II, p. 82. For the first English edition, see, Thomas Shaw, *Travels, or observations relating to several parts of Barbary and the Levant*, Oxford, 2 vols., 1738, and supplement, 1746.

[21] Shaw, *Voyages* cit., vol. II, p. 81

[22] *Ibid.*, p. 147.

[23] De Maillet, *op. cit.*, p. 138.

[24] *Ibid.*, p. 137.

[25] Goguet, *op. cit.*, p. 135.

[26] De Maillet, *op. cit.*, p. 138.

[27] *Ibid.*, p. 140.

[28] *Ibid.*, p. 333.

[29] Shaw, *Voyages* cit., vol. II, p. 62.

[30] *Ibid.*, p. 245.

[31] *Ibid.*, p. 62.

Shaw describes remains of works in the sea that indicate the fabrication of hydraulic cements: "the composition likewise is made up of small pebbles and mortar, so well cemented and knit together, that a solid rock cannot be more hard and durable."[32] "It is very probable," he adds, "that, in submarine works of this nature, the Romans might mix and temper this mortar with the earth of Puteoli, which has a surprising property of hardening under water."[33] Shaw also explains how certain peoples, between Algiers and Tunis, prepare a compound, the "cement" that in some cases has the "same consistence and composition with what we meet with in the most ancient fabrics," above all "where any extraordinary compactness or strength is required" (in his notes, he indicates the books of Vitruvius and Pliny).[34]

"[...] they take one part of sand, two parts of wood ashes, and three of lime," he writes, for example, when he is in Carthage, "which, after it is well sifted and mixed together, they beat, for three days and nights incessantly, with wooden mallets, sprinkling them alternately; and, at proper times, with a little oil and water, till they become of a due consistence. This is chiefly used in making arches, cisterns, and the terrasses or tops of their house."[35]

**Artificial stone in contemporary primitive civilizations**

The supposed discovery of artificial stone in Egypt is just one aspect of the curiosity surrounding the making of artificial materials that permeates the technical culture of the early 1700s. Growing interest in the fabrication of artificial stone of different kinds can also be seen in the chronicles of exploration of the various continents, where different practices, techniques and materials are examined for the creation of what is generically indicated as artificial stone in the first half of the 1700s.

Certain travelers focus on the practice among various populations of South America of constructing objects, from wearable adornments to hatchets to furniture, with a particular type of clay mixed with water that dries to take on the hardness of a diamond.

The naturalist Pierre Barrere (1690-1755), professor of medicine, physician at the military hospital of Perpignan and Correspondent of the Académie Royale des Sciences of Paris, describes the pottery the "inhabitants of Brazil" know how to shape with *bol*. "The negroes," he explains, "ordinarily make it for their smoking pipes; it is also called *pipe earth*. This *bol* is colorful, and it thins out easily in water."[36]

The exploration of Peru and the descent down the Amazon River, from 1735 to 1743, allow the geographer Charles-Marie de La Condamine (1701-74) to observe the method of producing utensils and furnishings with what is defined as the "stone of the Amazon," utterly similar to *bol* – his travel memoirs are published in 1744. "It has been quite seriously put forth," he writes, "that this stone is none other than the silt of the river, to which one gives the desired shape by kneading it just after extraction, & that it then acquires in the air that extreme hardness."[37]

[32] Shaw, *Travels* cit., pp. 109-10.
[33] *Ibid.*, p. 110.
[34] *Ibid.*, p. 206.
[35] *Ibid.*
[36] Entry *Bolus*, in, Pierre Barrere, *Essai sur l'histoire naturelle de la France Equinoxiale*, Paris, Veuve Piget, 1749, p. 211.
[37] Charles-Marie de La Condamine, *Relation abrégée d'un voyage fait dans l'intérieur de l'Amérique Méridionale, depuis la Côte de la Mer du Sud, jusqu'aux Côtes du Brésil & da la Guiane, en descendant la riviere des Amazones, lûe à l'Assemblée publique de l'Académie des Sciences, le 28 Avril 1745*, Paris, Veuve Pissot, 1745, pp. 141-42.
[38] René-Antoine Ferchault de Réaumur, *Sur la nature et la formation des cailloux*, in, "Histoire de l'Académie Royale des Sciences. Année MDCCXXI. Avec les Mémoires de Mathématiques & de Physique, pour la même Année," Paris, Imprimerie Royale, 1723, pp. 255-76.
[39] *Ibid.*, pp. 259-60.
[40] *Ibid.*, p. 262.
[41] *Ibid.*
[42] *Ibid.*, p. 270.

### The artificial pebble of Bazin and Geoffroy

Awareness of the artificial stone of ancient Egypt and the artificial stones produced by South American civilizations soon spread beyond the limited circles of archaeologists and curious travelers and collectors. The nascent science of modern chemistry – which develops rapidly in the 1700s – began, from the early decades of that century, to try to imitate the processes of natural formation of rocks and minerals, at times taking the procedures of primitive and ancient peoples as models. The study of natural phenomena and the action of originating agents like water and fire, by naturalists and metallurgists, on the one hand, and the experiments conducted in the laboratories or homes of pharmacists and chemists, on the other, laid the groundwork for the gradual identification of processes for the artificial fabrication of materials. Imitation of calcinated stones or volcanic dusts considered most suitable for the making of concrete, but not easily obtained in the various territories, becomes one of the goals of chemistry as applied to this pursuit of artificial materials. Useful results had already been achieved in the last decades of the 1700s, leading to the production of large quantities of artificial pozzolana. But at the start of the century, chemists advanced in small steps, leading first to the discovery of the production of artificial pebbles about the size of a walnut.

The principle to scientifically create artificial stone dates back to 1721, when the eminent scientist René-Antoine Ferchault de Réaumur (1683-1757) presents his ideas about the origin of stones, offering a glimpse of the possibility of artificial fabrication. Everything seems to begin with the presentation to the members of the Académie Royale des Sciences in Paris of some singular pebbles gathered by Jean-Jacques d'Ortous de Mairan (1678-1771) at Breuilpont, between Anet and Ivry, which when broken are seen to contain, inside, fine earthy substances. Réaumur set out to investigate the origin of these stones – *silex* in Latin and pebble ("*caillou*") or more commonly flint ("*pierre à fusil*") – and in particular of those marked by great hardness, and free of any trace of crystalline structure or fibers of any kind. The observations aimed at discovering the original constituent material – the crystalline material, or "*matière cristalline*," written about by Claude-Joseph Geoffroy (1685-1752), the authoritative pharmacist called on by Réaumur himself.[38] To explain his hypotheses on the origin of the pebble through a process, involving water, of successive petrification of an original stone without always reaching its nucleus, Réaumur provides the example of the limestone ("*pierre à chaux*") or building stone ("*pierre à bâtir*"), a generally spongy stone which, in his view, had been impregnated, to the point of filling all the cavities of the crust, by tiny particles conveyed by water.[39] Therefore he asserts that the "compact earth," such as "marl," can also be transformed into stone through the progressive penetration of "stone sap [*suc pierreux*]".[40] "I have seen physicists who believe that the earth found in the center of these stones was produced there by some sort of calcination, that this earth was the interior of the stone that had been calcinated," he reports, challenging this hypothesis of the genesis of stones through fire.[41] His observations on the nature of pebbles reached the point of considering their calcination in kilns, as practiced in certain regions of France, along the Rhône, for example, where the rocks produce lime considered "excellent for bleaching," but not suitable for masonry.[42]

The hypotheses on the origin of pebbles lays the groundwork for the experience attempted by Gilles-Augustin Bazin (1681-1754), a disciple of Réaumur and a physician from Strasbourg. At the end of 1734, Bazin takes earth used to make pottery, mixed it with water to ensure that only the pulverized earth remained, and then put the mixture into an "bowl of glazed earthenware" which he left on the windowsill in his home, facing north, after adding "iron filings," in order to form a pebble with a metallic color. Over the course of two years, from 1735 to 1736, Bazin adds water to the mixture every day; during the period 1737-38 he leaves the mixture without water, and then finally, in 1739, he observes the results of the process: "a hard body impossible to separate by knife," "a veritable stone the size of a hazelnut, and that can only be broken by hammer applying the same force required to flatten the head of a tack after sinking [...]."[43]

When Bazin's results are made public another experiment is already in progress, conducted with the same goal of exploring the possibility that unspecified "crystalline material" could produce stones, also when manipulated by man. Geoffroy himself sets out to discovery, through laboratory trials, the natural processes of the formation of stone, and already towards 1728 he manages to fabricate, with the quick lime of Melun soaked in distilled vinegar, an artificial silex similar to the flint. The results of the experiments are presented at the Académie Royale des Sciences of Paris in 1746, but will not published until 1751 – the artificial product is left to solidify for about 18 years, to test the degree of hardness that could be obtained.[44]

The words of the French chemist Jean Hellot (1685-1766), future advisor for mines and metallurgy of the Contrôle Général des Finances, spoken in 1739, on the potential of the practical applications of chemistry, offer a glimpse of the fate of artificial stone, starting precisely with the experiments of Bazin. "All these observations," he states, "show the potential utility of chemistry in the discovery of the secret of the production of nature, and whether it would not be a form of injustice to criticize chemistry for some insufficient imitation. The art that can make bitumen, mineral oil, sulfurs, vitriol, alum, cinnabar similar to that found in nature; regenerate salts, revivify metallic limes, imitate thunder, earthquakes, subterranean fires will succeed in making pyrites, petrifying, etc. We just need experimentation and time. We already have the example of the possibility of limited petrification in the experiments of Mr. Bazin, correspondent of this Academy, of which Mr. Réaumur reported to this Company about three months ago."[45]

## The nature of the stones of Stonehenge

The invention and perfecting of a contemporary artificial stone happen against the backdrop of discussions on the nature of stone, not just in ancient Egypt, but also in monumental complexes with a similar accent on the use of colossal stone blocks. The most important of these complexes – also in terms of the impact it had on contemporary techniques through the belief that it had been made from an artificial product – is that of Stonehenge, most ancient and legendary for the origins of English architecture, built with grand blocks of Sarsen stone (quarried at Marlborough Downs) and Bluestone (from the Preseli Hills in Wales). The discussions about Stonehenge resemble those of

[43] *Observations de physique generale*, in, "Histoire de l'Académie Royale des Sciences. Année MDCCXXXIX. Avec les Mémoires de Mathématiques & de Physique, pour la même Année," Paris, Imprimerie Royale, 1741, (pp. 1-3), p. 2.

[44] *Cf. Sur la formation artificielle du silex et sur quelques propriété de la chaux vive*, in, "Histoire de l'Académie Royale des Sciences. Année MDCCXLVI. Avec les Mémoires de Mathématiques & de Physique, pour la même Année," Paris, Imprimerie Royale, 1751, pp. 65-69, and, Claude-Joseph Geoffroy, *Essais sur la formation artificielle du Silex, & observations sur quelques propriété de la chaux vive*, in, "Mémoires de Mathématiques & de Physique," *ibid.*, pp. 284-90.

**4**. Inigo Jones, survey of Stonehenge (*The most notable Antiquity of Great Britain, vulgarly called Stone-Heng on Salisbury plain. Restored by Inigo Jones*, London, J. Flesher, 1655).

European travelers regarding the nature of certain Egyptian stones. In both cases the colossal ancient blocks are described as artificial and are thought to have been made on site, because of the seeming impossibility that quarries and machinery could have existed that allow for the extraction and transport of such large stones.

After the legends narrated by 12th-century observers, in particular by Geoffrey of Monmouth in his *Historia Regium Britanniæ* of circa 1136, according to which the colossal stones of Stonehenge, for which no traces of quarries remained, had been transported by giants from Africa, put first on Mount Killaraus in Ireland and then transported to Salisbury Plain thanks to the magic of Merlin, the hypothesis begins to emerge that those monoliths are not of natural stone, local or transported from faraway lands, but the result of an artificial product made on the site. The homogenous grain, the absence of veins, the extreme strength and enormous size of the blocks convince John Rastell, a writer and publisher from the 1500s, to hypothesize that they were "caste and made in a molde."[46] A contribution seen as authoritative in the 18th-century debate on the stones of Stonehenge is made in 1586 by William Camden, antiquarian, historian and military officer, in his *Chorography of Wiltshire*, claiming that the stones were "artificially made of pure sand, and some gluey und unctuous matter kint [sic] and incorporated together."[47] It is probable that Camden's era marks the beginning of the removal of stone samples from the various megalithic monuments like Stonehenge or Abury, to test the nature of the stones and perhaps to discover their secret.

[45] Jean Hellot, *Sur la liqueur éthérée de M. Frobenius*, 9 May 1739, in, "Histoire de l'Académie Royale des Sciences. Année MDCCXXXIX. Avec les Mémoires de Mathématiques & de Physique, pour la même Année," Paris, Imprimerie Royale, 1741, (pp. 62-83), pp. 82-83.

[46] John Rastell, *The Pastime of People, or, the chronicles of divers realms [...], 1529*, London, F.C. and J. Rivington, 1811, p. 105.

[47] Cit. in, *The most notable Antiquity of Great Britain, vulgarly called Stone-Heng on Salisbury plain. Restored by Inigo Jones*, London, J. Flesher, 1655, p. 32.

The "artificial" stone of Stonehenge becomes a myth that is nevertheless challenged, during the course of the 1600s, by Walter Charleton (1619-1707), president of the College of Physicians and the Royal Society of London, who believes the stones are natural (his hypothesis, formulated against that of Inigo Jones, is that the monument was built by the Danish, not the Romans),[48] and by the antiquarian John Aubrey (1626-97), who notes that by now "the errour runnes from generation to generation concerning Stonehenge, that the stone there are artificiall" (he is the one who indicates the Druids as the creators of the monument).[49] Aubrey also contradicts the belief that the pillars of Salisbury Cathedral are made of the same artificial stone as Stonehenge (they are made of Chilmark limestone): "Tis strange to see how errour hath crept in upon the people, who believe that the pillars of this church were cast, forsooth, as chandlers make candles [...]."[50]

In the first important essay on Stonehenge written by architects, that of Inigo Jones, published by John Webb (1611-72), the question of the artificial nature of the stones proposed by Camden is settled by recalling the capacity of the Romans to transport and raise large stones (Jones mentions the machines of Archimedes, Hero of Alexandria, and also that of Fontana for the transport of the Vatican obelisk, besides his own feat of installing the columns of the portico of St. Paul's Cathedral). Furthermore, according to Jones, the quarries from which the monoliths of Stonehenge were taken exist in its vicinity (pointing to the existence, in the town of Abury, of stones like those of Stonehenge).[51] On the question of artificial stone, Jones demonstrates knowledge of the characteristics of Roman construction and the use of pozzolana, but he doubts that the Romans ever made true artificial stone.

"As for these monuments [in Yorkshire] (for my part)," he writes, "I have not seen, otherwise I would give my sense upon them, and happlly they may be found as far from being artificiall, as those at Stonehenge. [...]. He might as well have told us the rocks in Portland are artificiall. But it's true, this sand of Puteoli, was much used by the Ancients, and it is such a kind of earth, as is very famous for its admirable effects in building, being tempered with the cement of Cuma: for, it not onely yeelds strength to all other buildings, but thereby also, all works made in the sea under water, are most firmly consolidated. Yet, doe I not find, that ever the Ancients made any artificiall stones thereof, or that Vitruvius hath any thing to that purpose [...]. Neverthelesse, whosoever could find out any kind of earth in this island, naturally apt, to make artificiall stones of such greatnesse as these; and, like them so obdurate also, that hardly any too enter them: or, that our auncestors in times of old, did make use of such a cement, and in what manner composed by them. The benefit thereof doubtlesse, would amount so ample to this nation, that records could not but render him deservedly famous to all posterity. In the mean while, as it is most certain those stones at Stonehenge are natural [...]."[52]

[48] *Cf.* Walter Charleton, *Chorea Gigantum, or, The Most Famous Antiquity of Great Britain, Vulgarly Called Stone-Heng standing on Salisbury Plain, restored to the Danes*, London, H. Herringman, 1663 (edition, London, D. Browne, 1725, pp. 44-46).

[49] John Aubrey, *The Natural History of Wiltshire*, London, Wiltshire Topographical Society, 1847, pp. 97-98, cit. in, Christopher Chippindale, *Stonehenge Complete*, London, Thames and Hudson, 2004, p. 28.

[50] *Ibid.*

[51] *The most notable Antiquity of Great Britain, vulgarly called Stone-Heng on Salisbury plain. Restored by Inigo Jones*, London, J. Flesher, 1655, (pp. 33-35), p. 37.

[52] *Ibid.*, pp. 35-36.

[53] John Webb, *Vindication of Stone-Heng Restored In which the Orders and Rules of Architecture Observed by the Ancient Romans, are discussed*, London, G. Conyfrs, 1725$^2$, p. 88.

[54] James Brome, *Travels over England, Scotland and Wales*, London, A. Roper, 1700, p. 45, cit. in, Chippindale, *op. cit.*, p. 28.

[55] William Stukeley, *Stonehenge, A Temple Restor'd to the British Druids*, London, W. Innys, R. Manby, 1740 (reprint, Forgotten Books, 2007, p. 13).

[56] *Ibid.*, p. 47.

[57] *The most notable antiquity of Great Britain: vulgarly called Stone-Heng, on Salisbury Plain, restored by Inigo Jones:*

In 1669, Webb publishes his *Vindication of Stone-Heng Restored*, where the question of natural stone is assumed as a given, though he had been a student of a supporter of the conflicting thesis, John Wilkins (Webb strove above all to challenge the criticisms of Charleton of the thesis of Jones on the Roman origin of Stonehenge). "[...] nature is not accustomed to cast stones in a mold," he adds, to refute those who believed no work had been done on the natural stones erected at Stonehenge.[53] Nevertheless, in 1700 he still thought it a "not irrational conjecture that they are not natural, or had their first growth here, but were artificially cemented into that hard and durable substance from some large congeries of sand, and other unctuous matter mixt together."[54] Among the supporters of this idea we also find Thomas Baker (1656-1740).

The gradual destruction of the monuments under the blows of hatchets and hammers to discover the qualities and secrets of those stones reaches such a point that some believe, in the early decades of the 1700s, that it is the cause of the disappearance of the stones of the complex of Abury, removed because they were considered to be "stones factitious."[55] The Stonehenge complex is similarly damaged: "Not a bit is lost, but what mischievous and silly people knock off with hammers, to see whether, as the wretched vulgar notion would have it, the stones be factitious."[56]

The reprinting of the essays of Jones, Webb and Charleton in 1725 confirms the renewed interest in Stonehenge at the start of the century.[57] The question of the true nature of the stones is still timely when Stonehenge is studied by the Reverend William Stukeley (1687-1765), antiquarian and archaeologist, member of the Royal Society and the Society of Antiquaries of London. From 1721 to 1724 Stukeley makes important surveys and observations (examined in depth and published in 1740),[58] and analyzes samples of stone provided by the astronomer Edmund Halley (1656-1742), and judging it "beyond any controversy" to be natural stone.[59] He also discusses the nature of the stone of Stonehenge with the botanist and theologist Stephen Hales (1677-1761).[60] To reach the conclusion that it is natural stone, Stukeley looks at the samples under a microscope: "[I] find the rock seems to be that of St. Vincents quarry near Bristol, tis a composition of chrystals of different colors red green & white cemented together by natures art with opaque granules or globules of flinty & stony matter [...]."[61] Stukeley also addresses one of the questions that gave rise to the myth of artificial stone: the machines for transporting and raising colossal blocks. He proposes various systems, designing their functioning. In the discussion on machines, he too mentions the case of the transport of the Vatican obelisk by Fontana, which by now has become the standard for comparison in discussions on Egyptian or English artificial stone.[62]

The opinion on the nature of the stones of Stonehenge expressed by a great builder, John Wood (1704-54), coincides with that of those who criticize the belief that they are "factitious": "the marks of nature are manifest and abun-

*to which are added, the Chorea gigantum, or Stone-Heng restored to the Danes, by Doctor Charleton: and Mr. Webb's Vindication of Stone-Heng restored, in answer to Dr. Charleton's reflections, with observations upon the orders and rules of architecture in use among the ancient Romans: before the whole are prefixed, certain memoirs relating to the life of Inigo Jones [...]*, London, D. Browne, 1725.

58 *Cf.* Aubrey Burl, Neil Mortimer, ed., *Stukeley's 'Stonehenge': An Unpublished Manuscript, 1721-1724*, New Haven, London, Yale University Press, 2005.

59 Stukeley, *op. cit.*, p. 13. See also the notes of Stukeley, in, Burl, Mortimer, *op. cit.*, p. 23.

60 Stukeley, *op. cit.*, pp. 13-14.

61 Stukeley, in, Burl, Mortimer, *op. cit.*, p. 23.

62 *Ibid.*, pp. 128-32.

dant in them."[63] The Reverend William Cooke (1711-97), too, in 1754, believes that the stones of Stonehenge and similar monuments are natural – he thinks they had been transported to the site of Stonehenge and other places by a single "great navigator."[64]

But the issue of the nature of the stones is still far from being resolved. One of the most famous makers of microscopes, Benjamin Martin (1705-82), challenges the statements of Stukeley in 1759, continuing to assert the possibility that the large monoliths are made from an artificial cement-like product, not transported from distant quarries by means of complex machines (nor, he observes, are there common stones of the size of those used at Stonehenge).[65] The research on the nature of the stones, also in the case of Martin, is always done by removing fragments from the monoliths to test their strength and consistency. The new factor that emerges from Martin's considerations is that the research on artificial stone conducted by English inventors in the first decades of the 1700s makes it possible to revive the hypothesis of artificial stones for Stonehenge, in spite of the observations of Stukeley.

"That it is possible they might be made by art," Martin asserts, "no man can dispute, who considers, that the substance of common stones, reduced to powder and mixed with proper ingredients, will compose a substance that shall appear like stone, and at the same time be harder and heavier; and a person possessed of this art, to a surprizing degree, was, a few years ago, well known in England, and gave such proof of his skill this way, as prevented his getting a patent for practising his art, lest masonry and other trade should be hurt by it. Of this sort the stones at Stonehenge appear to me; they seem to the eye to be different from common stone, and when I found I was obliged with a hammer to labour hard three quarters of an hour to get but one ounce and half, I was fully convinced, their hardness, or fixity, by much exceeded that of common stone. I was also farther assured by taking the specific gravitey of the pieces [...]."[66]

In the second half of the 1700s the line of those who believe, like Charleton or Stukeley, that the stones are not artificial, begins to prevail. In a lecture held in 1777 at Salisbury, John Warltire (1739-1810), a chemist with a passion for antiquity, asserts that all the great monoliths of the "trilithons," including those of Stonehenge, "are the stone of the country."[67] The writer Samuel Johnson (1709-84) also rejects the opinion that "the materials are not natural stones, but an artificial composition hardened by time."[68] The discovery of dovetail-type connections between uprights and crosspieces is interpreted as further proof of the non-artificial character of the stones (Jones had already noted the system of those connections, attributing it to the Romans):[69] "This proves also the stones not to be factitious; for they that could mould such durable masses could do much more than make mortar, and could have continued the transverse from the upright part with the same paste."[70]

[63] John Wood, *Choir Gaure, Vulgarly called Stonehenge, on Salisbury Plain, Described, Restored, and Explainedin a Letter to The Right Honourable Edward Late [...]*, Oxford, 1747, p. 77.
[64] William Cooke, *An Enquiry into the Patriarchal and Druidical Religion, Temples, &c*, London, L. Davis, 1755², p. 32 (first edition, 1754).
[65] Benjamin Martin, *The Natural History of England; or, A Description of each particular County, in regard to the curious Productions of Nature and Art*, London, W. Owen, vol. I, 1759.
[66] *Ibid.*, pp. 103-04. In 1771 John Smith confirms the natural origin of the stones (John Smith, *Choir Gaur. The Grand Orrery of the Ancient Druids, commonly called Stonehenge, astronomically explained, and proved to be a Temple for observing the Motions of the Heavenly Bodies*, Salisbury, E. Easton, R. Horsfield, J. White, 1771).
[67] Henry Wansey, *New Theory respecting the Origin of Stonehenge*, in, "The Gentleman's Magazine," December 1824,

## Artificial stone of Holt, Ripley, London and Langley

Martin's assertion regarding the notion that Stonehenge had been made with artificial stone is based on awareness of the possibility of making an artificial product similar to stone, like the products invented in England during the first half of the 1700s. Precisely the discussions on the nature of the stones of Stonehenge and the contemporary production of artificial stones demonstrate the extent to which fantastic archaeology and technical innovation can be correlated.

The question of the contemporary production of artificial stone lasts the entire 18th century, involving fabrication techniques based on molds. In England the definition of "artificial stone," for the entire second half of the century and into the first decades of the 19th, includes products derived from the technique of terracotta. Two different technical solutions emerge: the French *pierre factice* that translates, already in the second half of the 1700s, into a cement-based compound, and the English "artificial stone," which from the first decades of the century consists in a sort of terracotta, with the aim of producing a material that is a perfect imitation of stone. Only in the case of English translations of definitions given to compounds invented by the French can we find, in the 1770s, references to "artificial stone" as a synonym for a cement-based material shaped and produced without being fired in furnaces.

Precisely due to the birth of artificial stone from the compounds and fabrication processes of terracotta, the molds used to make the products continue to be the sophisticated sculptural articles developed for working with that material, unlike what happens for *pierre factice*, which instead begins to be shaped inside elementary wooden formwork, to obtain a product to be clad and decorated after its shaping in these molds. Therefore the quality of the shaped design in English "artificial stone" is sculptural, and the production of parts for construction generally has to do with special decorated pieces, like ashlars of arches or balustrades of parapets. The substantial difference in terms of design quality between pieces made in "artificial stone" and those in *pierre factice* has to do precisely with the decoration permitted by the compound and the mold, which is of a sculptural nature in the case of artificial stone, produced as a bronze casting, and is instead more rudimentary in the case of *pierre factice*, so much so that sometimes, when the hardened compound is removed from the mold, it requires covering to conceal the rugged molding marks. The invention of "artificial stone" contributes to guarantee, alongside the techniques of stucco and terracotta, the conservation over the decades of the training of specialized craftsmen and ways of working molds that then combine in the fabrication of true sculptural concrete works, particularly widespread in the 1800s and 1900s – the definition of *pierre factice* gradually vanishes, absorbed by that of artificial stone used to indicate a piece of concrete that is finely shaped, like a sculpted stone.

(pp. 503-05), p. 504.

68 Samuel Johnson, letter to Mrs. Thrale, 9 October 1783, in, James Boswell, John Wilson Croker, *The Life of Samuel Johnson, L.L. D.: including A journal of a tour to the Hebrides*, New York, A. V. Blake, 1843, vol. II, p. 358. *Cf.* Samuel Johnson, *A description of Stonehenge, Abiry, &c. in Wiltshire: with an account of the learning and discipline of the Druids: to which is added, an account of antiquities on Salisbury Plain*, Salisbury, B.C. Collins, S. Crowder, 1783.

69 *Cf. The most notable Antiquity of Great Britain, vulgarly called Stone-Heng on Salisbury plain. Restored by Inigo Jones*, London, J. Flesher, 1655, p. 70.

70 Boswell, Croker, *op. cit.*, p. 358.

The anonymous English inventor, holder of a patent for the making of an artificial stone similar to Stonehenge, described by Martin in the debate on the nature of the Stonehenge monoliths, may have been Richard Holt, who in 1722 filed two patents for "artificial stone," one in collaboration with the architect Thomas Ripley (1682-1758),[71] for a "compound liquid metal, by which artificial stone and marble is made, by casting the same into moulds," and the other in collaboration with Samuel London, for a compound "formed and moulded in a new method."[72]

"Compound Liquid Metal for Making Artificial Stone and Marble" is the title of patent 447 filed by Holt and Ripley. The text of the patent specifies that the operation consists in "casting or running the metall into moulds of any form or figure, great or small, as statues, columns, capitalls, also for house work and garden ornaments, and all other cutt or sculpture work that might be done in natural stone and marble, and which, being petrified or vetrified and finished by strong fire, becomes more durable an harder than stone and marble, insomuch that neither wet or dry, neither frost or snow, heat or cold in the extremity, can do damage to it [...]."[73]

"Mixture for Making White Ware" is the title of patent 448 filed by Holt and London. "[...] they had," we read, regarding their research for the patent, "by many years labour and experience and at great expence and charges, found out and invented a certain new composition or mixture (without any sort of clay) for making of white ware, which is formed and moulded in a method hitherto not known or praticised, and far surpasses the finest of Delf ware, or any other sort made in any other sort made in any part of Europe; and also by their new method of impression make the fabrick of earthen ware of a more equisite shape than the present method of turning could ever perform or arrive to, by which means our subiects will be able to excell all Europe, and not only employ a great many of our own poor to the great benefit of trade and the manufactures of our kingdom, but also prevent the clandestine running of Delf ware, &c from foreign parts into Great Britain."[74]

Holt's invention comes on the wave of the debate on artificial stone, the monuments in Egypt and Stonehenge. Holt himself calls the artificial products of the ancients into play to explain the potential of his invention. He even says that he has conducted experiments to identify the "composition" of the "cement of the Ancients" based on the conviction of the existence of some sort of artificial stone invented in "Asia and African Turkey," and used among "the old Egyptians," to avoid the geological defects of natural stone. Holt states that he has even visited those lands, "to see what I could learn from, and pick up among, the *Turks*, and the race of the *old Egyptians* [...]."[75]

There can be no doubt regarding Holt's position in the context of the hypotheses outlined in the 1600s and 1700s on the possibility of the existence of artificial stone in ancient Egypt.

"The *Egyptians of old* were," he writes, "so sensible of the *use* and *benefit*, and so renown'd for *the make*, of this same artificial stone, that they employed

[71] On Holt see, Alison Kelly, *Mrs Coade's Stone*, Upton-upon-Severn, Self-publishing Association 1990, pp. 31 *et seq.*, and Philip Ward-Jackson, *Public sculpture of the city of London*, Liverpool, Liverpool University Press, 2003, p. 456.

[72] See, Patent Office, 447 and 448, cit. in, *Coade's Artificial Stone Works'*, in, Howard Roberts, Walter H. Godfrey, ed., *Survey of London: volume 23: Lambeth: South Bank and Vauxhall*, London, Survey of London, 1951, pp. 58-61.

[73] *A.D. 1722, N° 447. Ripley and Holt. Compound Liquid Metal for Making Artificial Stone and Marble*, in, *Appendix to reference index of Patent of Invention, containing abstracts from such of the early patents and signet bills as describe the nature of the invention, and which patents have no enrolled specifications*, London, B. Woodcroft, 1855, p. 62.

[74] *A.D. 1722, N° 448. Holt and*

more than six hundred thousand men *in this way* for several Centuries. However our *Interpreters* have been pleas'd to call this sort of *stone*, that the *Israelites* were so employ'd in, by the name of *bricks*, yet I must here observe, that several of the most eminent *writers* of *Egyptian Antiquities*, as well as *Hermes Trismegistus* himself, *Xenophon*, *Herodote*, *Sudas*, and *others*, call it expressly *artificial stone*, a *composition of earths* and *artificial cements*, and as they further describe it, must differ vastly from the *bricks* of our days; and consequently be the same with *my affair in hand*; witness the many considerable remains of *Antiquity*, the numbers of *towers*, the stately *pillars*, the curious *obelisks*, some still standing in and near to *ancient Alexandria*, *Grand Cairo*, &c. The *Egyptian pyramids*, commonly called *King Pharaoh's Hills*, those famous *artificial stone monuments of Antiquity*, which devouring *time* can never consume. The famous *bust*, called *Sphinx*, *all one stone*, [...] can scarce be supposed to be one of our modern *bricks*."[76]

"For these reasons and other defects in this *material*," he continues, referring to natural stone, "the Ancients did not trust their grand *buildings*, at least not the *decorations* and *capital ornaments*, to the work of *natural stone* of any kind; but being well acquainted with the other *products* of nature, were at no loss to find out *an extraordinary tye, a peculiar cement*, for binding of clay together, so as to make an artificial stone far superior, both in strength and beauty, to the best and hardest *natural stone*."[77]

Holt ventures into claims regarding Stonehenge as well, and is categorical about those stones, even though the archaeologists bent on demonstrating the true nature of the stones of that monument never make a direct rebuttal. The stones, Holt says "are the *produce* of the *ancient art, artificial stones*, made upon the spot."[78] He also believes in the artificial character of the stones of other similar complexes, like the "Trophy or Monument of Rollrich Stones" at Witney, in Oxfordshire, that of Boskenna, in Cornwall, and that of Salkeld, in Cumberland. Thus Holt embraces the theses of "famous Camden, Baker, Wilkins, and others,"[79] without informing readers that other antiquarians, like Charleton and Aubrey, have been in disagreement for some time. "All our *Antiquarians*," he writes, "all our *historians* agree in this, that these are *artificial*, not *common stones*."[80]

The experiments conducted by Holt to discover the compound and the production of artificial stone on Holt's part, then, should not be seen only as the result of an activity of technical research and entrepreneurial ability, but also as signs of a particular cultural climate in which technical invention is oriented by antiquarian impulses, giving the contemporary product the value of an archaeological discovery, with an ancient aura much in vogue in the 1700s.

The aim of Holt's invention is to make a product that will improve the natural formation process of stone, avoiding cavities and impurities (like the seashells contained in stones). "[...] the hardest of *natural stone*," he writes, "has no other *cement* or *tye* for the combination of its parts, than what it acquires

*London. Mixture for Making White Ware*, in, *Appendix* cit., p. 62.

[75] Richard Holt, *A short Treatise of Artificial Stone, as 'tis now Made, and Converted into all Manner of Curious Embellishments, and Proper Ornaments of Architecture*, London, S. Austen, 1730, p. 13, note a.

[76] *Ibid.*, pp. 13-14.

[77] *Ibid.*, pp. 10-11.

[78] *Ibid.*, p. 28.

[79] *Ibid.*, pp. 27-28.

[80] *Ibid.*, p. 28. Holt also mentions the monument of John Stow in North Isle of the church of St. Andrew Undershaft in London (*ibid.*); and he discusses the question of the nature of the stones of Stonehenge (*ibid.*, p. 27).

from time, either in the bowels of the earth, or in the bottom of the sea; where the sand and sandy particles, by a sort of natural *trituration*, rarifying and fining, incorporate and unite in a lump, and coalesce and hold together in one solid body." Portland stone, used in small decorations – Holt lists "pines, necks, pots, urns, vases, etc." – gets corroded by atmospheric agents that penetrate into its pores. Holt's descriptions of natural stones, from Portland stone to Bath stone, harp on their imperfection, their "porous body" and the defects encountered in their deployment.[81]

Holt sets up an artificial-stone factory at Lambeth, a borough of London. Its production is publicized in a catalogue and covers a range of architectural parts usually made in stone, adorned with moldings and sculpted features. His product list includes: "columns, pedestals, entablatures, cornices, pediments, ballustrades, statues, rusticks, fascias, coppings of walls and chimneys, chimney-pieces, hearth-stones, architraves, frontispieces of doors, windows, alcoves and grotto's, cascades, obelisques, arches, piazza's, key-stones, steps, pavements, urns, balls, &c."[82] "To all which particulars then," Holt specifies, "I shall only add a few most material articles, such as tomb-stones, monuments, sun-dials, crest for doors, gates, and gateways, statuary of all sorts, pipes of all bores and sizes, for the conveyance of water at all distances, as good and strong as any ever made by the Romans [...]."[83] The production of the pieces is done with molds and purchases are made on the basis of the pieces included in the catalogue, though Holt also indicates that customers can submit different designs.[84]

The catalogue of Holt's artificial stone products includes certain features of ancient architecture, from capitals to balustrades, required for the construction of houses on the model of those of Palladio, starting with the villa built in 1726-29 at Chiswick by Richard Boyle (1695-1752), third Earl of Burlington, or statues, obelisks and urns like those scattered through the garden at Twickenham, property of the poet Alexander Pope, between 1720 and 1736.[85] Artificial stone meets with success in the London suburbs and makes gains in the building materials market precisely in parallel with the spread of the fashion for antiquities led by the circle of artists of Burlington, from Colen Campbell (1676-1729) to William Kent (1685 circa - 1748); that artificial stone becomes the substitute for natural stone, capable of offering economical and durable standard antique elements for the decoration of villas and gardens. Thanks to its nature, it meets with growing success in the second half of the 1700s, when the fashion launched by Burlington becomes the current taste. It is no coincidence that Holt supplies Burlington with pieces made with his compound, to complete the villa and the garden.

Already in the 1730s Holt's production of artificial stone has to compete with initiatives of other businessmen. The most important is Battey Langley (1696-1751), who after prolific activity in the field of making gardens, especially for clients at Twickenham like Thomas Vernon and James Johnston, opens a factory towards 1730, also in Lambeth. Nevertheless, Langley probably orients

[81] *Ibid.*, p. 9.
[82] *Ibid.*, pp. 48-49. *Cf.* Caroline van Eck, ed., *British architectural theory, 1540-1750: an anthology of texts*, Hants, Burlington, Ashgate Publishing Limited, 2003, p. 103.
[83] Holt, *op. cit.*, p. 49.
[84] *Ibid.*, p. 52.
[85] The "grotto" built in 1720-21 by the poet Alexander Pope, on his property in Twickenham, to connect the villa to the garden, passing below the street, becomes a true museum of geology starting in 1739, in which stones from various parts of the world are set, from Peru to Egypt to India.
[86] See, for example, the images of Roman ruins of monuments in *opus cœmenticium* used to illustrate certain pages in, Battey Langley, *New Principles of Gardening, or, the laying out and planting parterres, groves, wildernesses, labyrinths, avenues, parks, & c.*, London, A. Bettesworth and J. Batley, 1728.
[87] Battey and Thomas Langley, *Gothic architecture improved by rules and proportions, In many grand designs of columns,*

his production towards pieces in artificial stone that suit tastes different to those of Holt, if we take account of his lively interest not so much in Roman ruins,[86] but above all in Gothic architecture, to which he devoted the first major English collection of drawings, *Ancient Architecture Restored*, published first in 1742 and later in 1747, under the title *Gothic Architecture, improved by Rules and Proportions*.[87] The subtitle of the 1747 edition lists a range of pieces in competition with those produced by Holt: "columns, doors, windows, chimney-pieces, arcades, colonnades, porticos, umbrellos, temples, and pavilions." This is the beginning of the fabrication of the typical features of Gothic cathedrals in artificial stone, from spires to decorations for rose windows, that become widely available on the market, alongside the articles fabricated in the ancient manner, especially in the second half of the 1700s. Precisely at Twickenham, Horace Walpole (1717-97) makes an entrance gate in artificial stone for his Strawberry Hill House, rented in 1747 and transformed, after purchase, in work that proceeds through successive additions until about 1763, into the most emblematic English example of assemblage of decorative features based on those of Gothic architecture. Walpole's project involves John Chute, Richard Bentley, Thomas Pitt and Robert Adam, and the best-known producers of cement materials capable of imitating fine stones and marbles.

In 1730, precisely when the competition with Langley at Lambeth is starting, Holt publishes an essay, *A short Treatise of Artificial Stone, as 'tis now Made, and Converted into all Manner of Curious Embellishments, and Proper Ornaments of Architecture*, which becomes the manifesto of his artificial stone. The fact that the essay has been written and published to lay claim to the invention of the compound and to defend his production against the competition of Langley is openly declared by Holt himself: "About the beginning of *august* 1729, a certain *pretending architect*, a *meddling busy man* (who shall be nameless, so long as matters stand as they do) thought fit to make use of all manner of little *artifices*, in order to come at, or find out the *secret of my affair*."[88]

Holt dedicates his *Short Treatise of Artificial Stone* to the outstanding figure of English architecture, Burlington, in order to specify the artistic field of application of the products, that of the new fashion for things ancient and Palladian.[89] The fact that Holt intends to tie the success of the cultural project carried on by the circle of artists of Burlington, of an antique-style architecture adorned by precious decorations sculpted in stone, to his own industrial production of pieces in "artificial stone" is also demonstrated by the statement that his stone makes it possible "to build, at least *beautify* and *adorn*, full as *cheap* as evere our *Ancestors* did [...]."[90] From the tone of the dedication, it also appears that Holt's research on artificial stone has taken place under the sponsorship of Burlington.

In *A short Treatise* Holt describes the qualities of his stone, without providing any precise indication of its composition.[91] In the explanation of the invented material, Holt puts the accent on the process of fusion of the stone in a blast furnace, in order to then pour it into molds, as in the casting of metal.

*doors, windows, chimney-pieces, arcadescolonades, porticos, umbrellos, temples and pavillions, & c., with plans, elevations and profiles*, [London], J. Millan, 1747.

88 Holt, *op. cit.*, p. 41.

89 The treatise is dedicated to Burlington, to whom Holt supplies works for the villa and garden at Twickenham (*cf.* Van Eck, *op. cit.*, p. 101).

90 Holt, *op. cit.*, p. 3.

91 For a detailed description of the composition of Holt of the "artificial stone," with "glass," "clay" and "ore," to obtain a product that "in strong fire run almost as flued as water," and of the process of transformation of the "clay," thanks to the addition of "ore," into "stone by its vitrification," see the manuscript, "Richards Holt's Formula for Artificial Stone," dated 3 March 1730, Ms 11394, British Library, published in, Mavis Bimson, *Some Recent Research on Coade Stone*, in, "Transactions of the English Ceramic Circle," vol. XII, 1986, no. 3, pp. 203-05, plates 126-29.

He says he has tested his invention with the fabrication of conduits for aqueducts,[92] after having conducted experiments to identify the compound used by the ancients – referring several times to "a secret," or "grand secret" of the ancients.[93] To demonstrate the strength of his artificial stone he has conducted load tests, comparing its performance with that of Portland stone; he is so convinced of the strength of his artificial stone that he claims to have made garden gates that are stronger than those in iron, and that his stone stands up to fire better than any other artificial or natural product.[94]

"Now as to the nature and quality of what I shall call my secret composition," we read in his treatise, "*it*, briefly then, consists of such *ingredients*, as are, by a strong heat of fire, reduc'd to a state of *fluidity*; melted and run down into *a most compact solid mass*, according to the shape and dimensions of the given *model*. I say, my composition consists of such *ingredients*, as (when hot) have a natural tendency to *fluidity* and *vitrification*; and (when cold) consolidate into the *hardest stone*, and turn into one absolute and perfect *vitreous substance*. Thus, by the means of a few *simple ingredients*, I am able to run down *my clays*, or even the *hardest rock* in *England*, and cast them in the assign'd *moulds*, as fast the *lead-men* do their most *fluid metal*."[95]

Holt's conclusion is a forebear of the pride of certain statements by leading figures on the research on concrete in the 20th century: "in a word [...] *my artificial stone is more strong and durable than the best natural stone.*" "*My composition*," he continues, "in fine, ties down, binds together, and unites the parts of any *given body*, much more closely and strictly, than *nature* does those of *common stone*; and consequently renders whatever is made of *its produce*, infinitely more *compact*, *firm* and *solid*, than any thing made of *natural stone* can possibly be or come up to."[96]

The entrepreneurial activities of Holt and Langley at Lambeth, however, do not meet with the expected success; nevertheless, they represent the indispensable premise for the second phase of the production of that type of stone, starting in the 1770s, still at Lambeth, which would lead to the definitive widespread use of artificial stone in England and the world.

## Venetian "terrazzo" and "scagliola", or the artificial marble

To understand the reasons behind the doubts regarding the true nature of the Egyptian granites in the observations of various travelers at the start of the 1700s, we need to consider the technical ability achieved by certain workshops of artists and craftsmen in the fabrication of artificial multicolored stone composites, which are very similar to *breccias* or to natural marble. The two most famous composites of this type are the so-called *terrazzo*, particularly widespread in Venice to make floors, and the *scagliola* used to make various

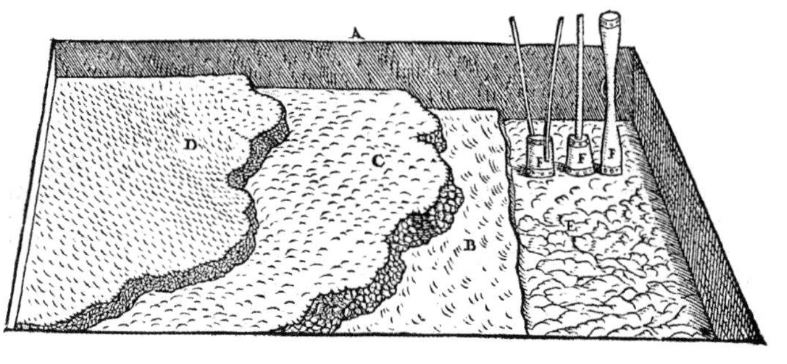


**5**. Explanatory drawing of the *terrazzo* technique (G.A. Rusconi, *Della Architettura*, Venezia, Gioliti, 1590).

**6**. Enrico Hugford, column in *scagliola*, Cappella dei Beati, church of the monastery of Vallombrosa, near Florence, 1757 (photograph by the author, 2010).

sculptural works, from tabletops to columns, whose production centers are Carpi and Florence. In both these composites cement is used as a binding agent, while coloring and marble aggregates create effects that imitate precious natural stones.[97] The evolution of concrete and its possibilities of rivaling stone will also take advantage of these techniques for the making of special cement-based mixtures to imitate the effect of stones with sought-after grains. Some cement-based products, especially starting in the 1770s, are developed precisely to offer an artificial imitation of stone or marble, as already happening with success thanks to *terrazzo* and *scagliola*.

Since the end of the 1600s Venetian *terrazzo* has been seen as an example of a cement-based composite capable of imitating the chromatic qualities, grain and strength of stones – the compound is usually based on *cocciopesto*, a compound of lime mortar and crushed brick or pottery. "One sees oneself everywhere here on the floors, which are made of a certain red cement, hard as stone, & bright as a mirror," writes Jean Du Mont (1667-1727) during his stay in Venice.[98] The perfection of a *terrazzo* is appreciated, as can be seen in the judgment by the masters in the tests for aspiring "terazzeri" helpers, in keeping with the statutes of 1737, for the "quality of the compound," the "perfect smoothing and polishing of the surface" and the "exactness of the level of the surface."[99]

*Scagliola* also has similar technical and finishing characteristics. The knowledge and spread of the Italian *scagliola* technique plays a role in the 18th-century taste for artificial stone. In fact, outside of Italy *scagliola* is often called "artificial marble."[100] Its success is such that in 1760 we see a considerable increase in the number of workers specialized in the fabrication of this type of marble.[101] It is no coincidence that in the treatises of the late 1700s, *scagliola* or artificial marble is discussed in the context of research on concrete. After all, certain recipes for *scagliola* call for the inclusion of pozzolana, mixed in to create a very strong product.

*Scagliola* is often used to give masonry columns the look of marble. The compound used for the columns of the Cappella dei Beati, in the church of the monastery of Vallombrosa near Florence, is made in 1757 by the monk Enrico Hugford (1695-1771) – one of the most famous artists of 18th-century *scagliola* – and makes them seem as if they had been turned from a blue *breccia*.[102] A similar technique is used for the spiral columns of Antonio Ferraboschi for the altar of the church of Santa Maria della Porta, at Guastalla. The architect James Adam (1732-94), during a stay in Florence in 1760-61 – he returns to England in 1764 – meets with the brother of Enrico, Ignazio Enrico Hugford (1703-78), painter and collector,[103] and lists the various uses of *scagliola* in his notebook, from flooring to architectural decorations. It is precisely thanks to

[92] *Ibid.*, pp. 15-16.
[93] *Ibid.*, p. 15.
[94] *Ibid.*, pp. 25-26.
[95] *Ibid.*, pp. 16-17.
[96] *Ibid.*, p. 21.
[97] On imitation and materials see, Reed Benhamou, *Imitation in the Decorative Arts of the Eighteenth Century*, in, "Journal of Design History," vol. IV, 1991, no. 4, pp. 1-13.
[98] *Voyages de Mr. Du Mont, en France, en Italie, en Allemagne, à Malthe et en Turquie*, La Haye, E. Foulque, F. L'Honoré, 1699, vol. IV, p. 275.
[99] Agostino Sagredo, *Sulle consorterie delle arti edificative in Venezia*, Venezia, P. Naratovich, 1856, p. 98. See also Mario Piana, "*Non si vede cosa, per suoli, ne più bella ne più gentile, ne più durabile di questa.*" *I terrazzi e l'edilizia veneziana*, in, Lorenzo Lazzarini, ed., *I pavimenti alla veneziana*, Venezia, Istituto Veneto di Scienze, Lettere ed Arti, 2008, pp. 74-90.
[100] Towards the 1740s, in France there is a major increase of workers specializing in stucco. "It is the material," it is written, regarding stucco "that the Sculptors employ to make ornements & related pieces on the Ceilings and walls, in Churches and in Palaces, in Italy, in Germany and elsewhere. The invention of Stucco, which is very ancient, led to that of artificial marble; but the latter is completely different. Artificial marble or compound marble, called Scajolla in Italy, from the name of its inventor, is also very old." (*Arts. Extrait d'un Mémoire sur le Marbre factice appellé stuc en France*, in, "L'avantcoureur. Feuille Hebdomedaire," 1760, Paris, M. Lambert, pp. 40-41).
[101] *Ibid.*, p. 41.
[102] Anna Maria Massinelli, *Scagliola. L'arte della pietra di luna*, Roma, Editalia, 1997, p. 27.
[103] For the contacts between Adam and Hugford see, John Fleming, *Robert Adam and His Circle in Edinburgh & Rome*, London, J. Murray, 1962, pp. 276-78.

**7**. Charles de Wailly, Augustin Pajou, column with vase, Hôtel d'Argenson (or Hôtel de la Chancellerie d'Orléans), 19 rue des Bons-Enfants, Paris, 1762 (*Charles de Wailly, peintre architecte dans l'Europe des Lumières*, Paris, Caisse Nationale des Monuments Historiques et des Sites, 1979).

the circle of the Adams, and Robert (1728-92) in particular, that *scagliola* finds widespread application in English architecture in the 1760s and 1770s, after its initial success in the second half of the 1600s and the early 1700s, when it is used together with various types of stucco by plasterers like Giovanni Bagutti, Giuseppe Artari, Francesco Vassalli or Francesco Serea, operating under the aegis of John Vanbrugh, James Gibbs, Colen Campbell, Francis Smith or Giacomo Leoni.[104]

Aside from the various types of wall facings, it is in the construction of columns that the techniques based on *scagliola* set the stage for a new, legendary type of 18th-century column. At the Salon de Paris in 1761, in the architecture section, a true column of *marbre factice* is displayed, made by Hermand, based on a design by Charles de Wailly (1730-98). "It is just the model of another porphyry column decorated with different marbles and golden bronze, handled by Mr. Auguste, to serve as a stand for an antique vase in porphyry of the most beautiful form."[105] At this point the question of the imitation of fine stones and marbles using cement-based compounds for artificial stone, to be shaped rather than sculpted, has become a central issue.

The fabrication of columns in compound, after the examples of Roman antiquity and, perhaps, the attempts made in the early 1500s, does not seem to be a solution put into practice in a sufficiently widespread way to offer significant cases or traces. Nevertheless, during the course of the 18th century, dictionaries include the definition of the *colonne moulée*, or molded column. From 1727 to 1759 the definition indicates a type of column made with cement-based multicolored artificial material: "When architects refer to the molded column, it is a column made from a paste consisting of gravel, pebbles of various color, bonded with cement or mastic, and that solidifies perfectly and that can be polished like marble."[106] The description seems to indicate that the molded column is a type obtained through the fabrication process of *scagliola*, therefore resembling columns like those of Hugford. Another definition of a comparable type of column is that of the *colonne fusible*, or fusible column, included in the *Dictionnaire de la langue françoise* of Pierre Richelet. It is probable that for this type of column a mold is required in which to pour the compound, and that behind the *colonne fusible*, then, we can see a reflection of the debate on columns in artificial stone, such as those described by Vauban or in the discussions on ancient Egypt. "With this word, one means," writes Richelet, "columns made of various metals, such as glass and other fusible materials, and those that we believe have been done with *pierre fondue*, whose use has been lost."[107] The definition of the cast column made by pouring a compound inside a mold can also be found in the *Encyclopédie, ou Dictionnaire raisonné des sciences, des arts et des métiers* of Denis Diderot (1713-84) and Jean-Baptiste Le Rond d'Alembert (1717-83), but it is included in order to refute any hypothesis of the artificial origin of the granite and hence also that of

[104] *Cf.* Geoffrey Beard, *Decorative Plasterwork in Great Britain*, London, Phaidon, 1975, pp. 53 *et seq.*
[105] *Supplément à l'article des tableaux exposés au Sallon; par la Société des Amateurs. Ce supplément est tiré de l'Observateur Littéraire de M. l'Abbé De Laporte, ainsi que la Brochure dont nous avons rendu compte dans ce volume, p. 92 & suiv.*, in, "Mercure de France," vol. II, October 1761, (pp. 221-29), p. 228. The column is probably that of the gallery Hôtel d'Argenson, Paris, built by De Wailly in collaboration with the sculptor Augustin Pajou (*cf. Charles de Wailly, peintre architecte des Lumières dans l'Europe*, Paris, Caisse Nationale des Monuments Historiques et des Sites, 1979, p. 53).
[106] Entry *Moulé*, in, Antoine Furetière, *Dictionnaire universel, contenant généralement tous les mots françois, tant vieux que modernes, & les Termes des Sciences & des Arts*, La Haye, P. Husson, 1727, vol. III, p.n.n. "The architects call it a colonne moulée," we read in another dictionary, "a column made with a paste of gravel and stones of different color, bound with cement or mastic that hardens perfectly and that can be polished like marble" (Pierre Richelet, *Dictionnaire de la langue françoise, ancienne et moderne. Nouvelle édition*, Lyon, P. Bruyset-Ponhus, 1759, vol. II, p. 677).
[107] Richelet, *op. cit.*, p. 522.
[108] Entry *Granit*, in, Denis

the fabrication of the Egyptian columns with molds. Thus, under the heading *Granite* the geological origin of the material is described, with indications of the places in which it is quarried Europe: "All columns thought to be of pierre fondue are made of granite from the provinces of this kingdom."[108]

The fantastic columns in artificial stone, the real columns of *scagliola*, and the very process of the transformation of the definitions of the column all the way to the creation of the figure of the shaped or cast column, lay the groundwork for imagining a true process of construction of columns by means of a mold.

Other products already in use in the early 1700s are destined to influence the evolution of the composition of concrete, especially the English culture of construction. They include types of facing stucco with different components, some of which are oily to create a water-repellant surface to protect brick masonry. The main sources for the development of products continue to be the architectural treatises; at times the treatise of Alberti is cited, in the English edition of 1726 edited by Leoni, especially the passages on the composition of cement or stucco.[109] But the observations of travelers like Shaw on the mortar compounds in different parts of the world also influenced the fabrication of English products. Some oriental formulae for protective cement-based coverings for defense against the rain of monsoons met with a certain amount of success in England. One outstanding example is that of the "mortar" prepared in Madras, India, whose recipe is communicated by the governor of the island of Saint Helena, Isaac Pyke, to Halley, to then be published in 1732 by the prestigious journal "Philosophical Transactions" of the Royal Society of London.[110]

The recipes for facing mortars, generically indicated as cements with vaguely waterproof characteristics, circulated through different types of publications, including dictionaries. Recipes like those reported in the literature of the 1730s would influence some of the patents of the most famous English facing cements in the 1760s and 1770s. The entry *Painting on wall* of the *Dictionarium Polygraphicum*, published in 1735, lists the ingredients of hydraulic cement facings, providing an important reference for later technical variations of the material, what would be known as "oil-cement."

"In order to fortify the wall the better against moisture," the *Dictionarium Polygraphicum* reports, "some cover it with a plaister of lime, marble-dust, or a cement made of beaten tiles soak'd with linseed-oil. And lastly, they cover the plaister over with a composition of Greek pitch, mastick and thick varnish boil'd together hot; and when this is dry, they lay on the colours as before. Others make the plaister with lime-mortar, tile, cement and sand; and after this is become thoroughly dry, they apply another of lime, cement and machefer, or seum of iron, all well beaten up together, and incorporated with whites of egge and linseed-oil, which does indeed make an excellent couch or plaister, this being dry: they lay on the colours as before."[111]

Diderot, Jean-Baptiste Le Rond d'Alembert, ed., *Encyclopédie, ou Dictionnaire raisonné des sciences, des arts et des métiers*, Paris, Briasson, David, Le Breton, Durand, 1767, vol. VII, p. 858.

109 *Cf. The architecture of Leon Battista Alberti, in ten books [...] translated into Italian by Cosimo Bartoli, and now first into English by James Leoni*, London, Th. Edlin, 1726.

110 *The metod of making the best mortar at Madrass in East India, described in a letter from the honourable Isaac Pyke, Esq; governor of St. Helena, to Edmund Halley*, in, "Philosophical Transactions," vol. XXXVII, January-March 1732, no. 422, pp. 231-35.

111 Entry *Painting on wall*, in, *Dictionarium Polygraphicum: or, The Whole Body of Arts Regularly Digested*, London, C. Hitch, 1735, vol. II, p.n.n. For information on mortars see also, Richard Neve, *The city and country purchaser's and builder's dictionary: or, the complete builder's guide. Containing an explanation of all the terms of art used by workmen*, London, B. Sprint, D. Browne, J. Osborn, S. Birt, H. Lintot, A. Wilde, 1736. *Cf.* K., *An Appeal to Public on the Right of using oil-cement, or composition for Stucco, &c. containing Provisoes in Letters Patent granted for Invention [...]*, in, "The London Review of English and Foreign Literature," vol. VIII, September 1778, pp. 172-82.

**8**. Agostino Corsini, "Abbondanza dei frutti," attic of the Fountain of Trevi, Rome, 1735 circa (photograph by the author, 2011).

Furthermore, we should not overlook, for an understanding of the origins of technical solutions, like the metal reinforcements used in the making of cement sculptures in the last decades of the 1700s, the practice of sculptures clad in stucco, as reported by the English in the 1730s: "As to the making of figures, the first thing is to form the soul of plaister or lime mortar, and a cement of tile-dust; putting bars of iron into those parts of the figure that stand in need of being borne up. When the soul is form'd, it is cover'd with stuck to work out the figure [...]."[112] The Roman compound of lime and pozzolana is also used to construct the cement nucleus to be clad in stucco to create sculptures, as is demonstrated in an emblematic way by the flowers and fruits of the cornucopia of the *Abbondanza dei frutti* by Agostino Corsini (1688-1772), on the attic of the Trevi Fountain.[113]

## "Cantoni di smalto": in the footsteps of Galileo's pupils

Artificial stone made by pouring a cement compound into molds goes through a crucial period in Italian building techniques. Its origin here is completely independent of the European archaeology of artificial stone and the English production of "artificial stone," and can be traced back to the pioneering experiences in Tuscany during the 17th century, aimed at controlling the flow of rivers. These experiences are conducted by certain pupils of Galileo, first of all by Braccio Manetti and Famiano Michelini, and then by Vincenzo Viviani (1622-1703). The large artificial blocks made in concrete using molds used for this type of project continue to be produced throughout the 1700s and are

[112] Entry *Painting on wall* cit.
[113] Maria Grazia Chilosi, *L'uso delle malte e dei metalli nella costruzione e nella decorazione*, in, Luisa Cardilli, ed., *Fontana di Trevi, la storia, il restauro*, Roma, Carte Segrete, 1991, (pp. 149-55), p. 149.
[114] Vincenzo Viviani, *Discorso al serenissimo Cosimo III Granduca di Toscana, Intorno al difendersi da' Riempimenti, e dalle Corrosioni de' Fiumi, Applicato ad Arno in vicinanza della Citta di Firenze*, Firenze, P. Matini, 1688.
[115] Tommaso Temanza, *Vite dei più celebri architetti, e scultori veneziani*, Venezia, C. Palese, 1778, vol. I, p. 267.
[116] See the list of 1762 of the works done at the Torre del Marzocco; in particular, over the vault of the guard post a "floor" has been made with "*smalto* laden with tufa and pozzolana" (description of the towers and guard posts reporting to Leghorn, in, Giampaolo Trotta, *L'antico Porto Pisano e la Torre del Marzocco a Livorno*, Livorno, Autorità Portuale di Livorno, 2005, (pp. 126-28), p. 127).

known by the same term, *cantoni*, that had appeared in the scientific literature of the previous century, above all thanks to the treatise by Viviani published in 1688, *Discorso al serenissimo Cosimo III Granduca di Toscana, Intorno al difendersi da' Riempimenti, e dalle Corrosioni de' Fiumi, Applicato ad Arno in vicinanza della Citta di Firenze.*[114] This remains the fundamental treatise on Italian hydraulic construction throughout the 1700s.

A vernacular synonym for *cantone*, in 18th-century Tuscany, is *massello*. It is not until the end of the century that the terms *cantone* and *massello* became complementary, only to be replaced by the more generic term of *prisma* that had spread in the scientific literature starting in the early 19th century. All these terms are joined by that of *smalto*, which in Tuscany indicates the cement compound to pour into molds, or *calcestruzzo* (the Italian term for concrete), and which, thanks to the success of the treatise of Viviani, spreads beyond the borders of the Grand Duchy as a synonym for concrete, to indicate the material with which the *cantoni* are made. On the term *smalto* and its circulation, we should recall that Tommaso Temanza (1705-89) describes the composition of Venetian floors as "cement and lime, which therefore is known as smalto, and to us Venetians as terrazzo."[115] In the composition of the *smalto*, at least in certain cases, there is also pozzolana imported to Leghorn from Civitavecchia.[116]

The use of *cantoni di smalto* is not limited to hydraulic works for long rivers. The same *cantoni* turn out to be a practical solution for making solid blocks with which to build the strongest parts of masonry, like the corners – also known as *cantoni* – of the houses of peasants and workers, built in areas far from the stone quarries. So during the 1700s and later, the *cantoni di smalto* go through a dual evolution, in hydraulic works and in country houses. In the discussions on Egyptian artificial stones no one makes mention of the *cantoni di smalto* of the recent tradition of construction in Italy.

The fabrication of *cantoni* is done with molds, or *forme*, in which to pour and shape the compound. These molds can be made in two ways, depending on geographical characteristics. In Tuscany the common practice is to make the formwork with wooden boards; a less frequent and not recommended solution is the use, in the 1700s, of metal panels, because the sides can flex during pouring, producing deformed blocks that are difficult to assemble in geometric structures. In general, the fabrication of *cantoni* with molds is done on the bed of a river, to have a good supply of water, gravel and sand. Once made, the *cantoni* are buried for ageing, for a period of about two or three months. This phase of ageing ensures slow drying of the compound, for greater solidity.

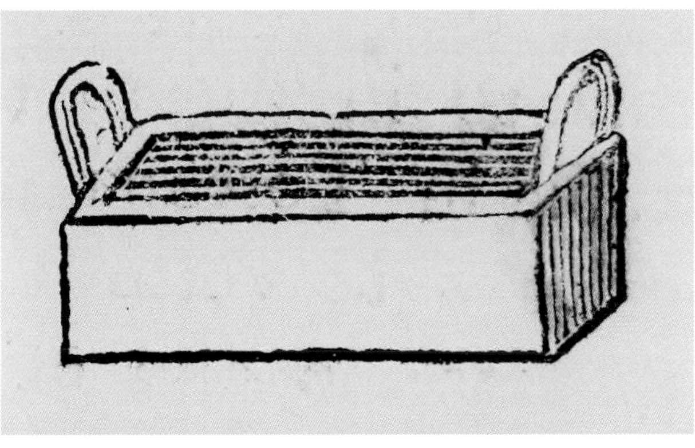

**9**. Metal mould with handles for the fabrication of *cantoni* (G. Del Rosso, *Pratica ed economia dell'arte di fabbricare*, Firenze, J. Grazioli, 1789).

Viviani's treatise contains the most complete outline of the use of concrete for the hydraulic works as constructed by the students of Galileo in Tuscany. The principles indicated by Viviani are the ones followed, with few variations, during the course of the 18th century. Viviani's treatise describes various applications of "poured blocks, of *smalto* or calcestruzzo,"[117] some of which are made "in cases made with boards,"[118] and of "manageable cantoni di smalto."[119] Until the early 1800s, Viviani's treatise remains the text of reference for technicians and engineers involved in the fabrication of this type of artificial stone, while towards the turn of the century Giovan Battista Clemente Nelli (1725-93) still recalls that the discovery of "cantoni, or *smalto* blocks with the form of a parallelepiped placed in a stepped arrangement," to defend the banks of rivers, should be credited to Manetti.[120]

The production of *cantoni di smalto*, according to the technique of the pupils of Galileo, to reinforce riverbanks and control the flow of the water, is confirmed by the initiatives implemented following the flooding of the Arno on 3 December 1740. Some of the massive walls raised to form the substructure of the Strada Regia at Palagina are made with *cantoni di smalto*, fabricated with the method of underground ageing. The blocks found at the start of the 1800s in the valley of the Pesa River, described as "great cantoni with a length of 1 and ½ braccia a width of not quite 1 braccia, and a height of 6 soldi,"[121] are probably part of this series of *cantoni*. "Now these cantoni have become as solid as the very boulders," is the comment written in 1827.[122] Along the course of the Arno the current is controlled at many points by "*masselli di smalto*" – a boat with passengers bound for Figline runs up against a cluster of these blocks in 1843.[123]

Confirmation of the continuity of the constructive tradition of the *cantoni di smalto* can be found in the entry *massello* in the dictionary of Niccolò Tommaseo (1802-74), published in Florence in 1838. Tommaseo, among other things, documents a particular solution of molding in sheet metal: "*Massello* is also the name of an artificial stone block, in the form of a parallelepiped, composed of *smalto* (lime mortar and sand), poured into appropriate sheet-metal molds without bottoms placed in a hole, from which the solidified *smalto* is removed later."[124]

The most widespread use of *cantoni di smalto* happens not only in the towns of Tuscany, but also in Piedmont, where the fabrication of these artificial stones takes place, mainly, in Alessandria and its vicinity. On the plains of Piedmont, besides the option of molds, there is widespread use of direct shaping of the clay-laden soil near the site where the *cantoni* are used. Parallel ditches are dug, with the desired section and depth (the section is usually triangular), with perfectly smooth sides (the earth is moistened and worked with a trowel); to obtain *cantoni* of a given length, wooden boards are in-

[117] Viviani, *op. cit.*, p. 31.
[118] *Ibid.*, p. 56.
[119] *Ibid.*, p. 59.
[120] Giovan Battista Clemente Nelli, *Vita e commercio letterario di Galileo Galilei*, Lausanne, 1793, vol. II, p. 774.
[121] Giovanni Battista Sodi, *Notizie sopra i cantoni di smalto volgarmente detti masselli*, in, "Giornale Agrario Toscano," Firenze, L. Pezzati, 1827, (pp. 176-80), p. 179.
[122] *Ibid.*
[123] See the report of the accident of a boat against these blocks, which took place on 28 February 1843 (*cf.* "Annuario Storico Universale," III, 1843, Firenze, Società Editrice Fiorentina, 1845, p. CCV).
[124] Entry *Massello*, in, Niccolò Tommaseo, *Nuovo Dizionario dei Sinonimi della Lingua Italiana*, Firenze, Gabinetto Vieusseux, 1838, vol. II, p. 593.
[125] *Cf.* Angelo Parrochetti, *Delle pietre artificiali o prismi di smalto per opere idrauliche*, in, "Giornale dell'Ingegnere-Architetto ed Agronomo," vol. VI, 1858, pp. 186-89. From a town in Piedmont, Giuseppe Antonio Borgnis (1781-1863) knew about the tradition of the concrete *cantoni* made in Alessandria, using "concrete of

serted in the ditches at regular intervals – the height of the boards determines that of the pouring of the compound. Once pouring has been done, the ditches are covered with a layer of soil to let the compound dry for a period of two or three years (but if a hydraulic variety of lime has been used, one year can be sufficient).[125]

Near Alessandria ditches are filled from wheelbarrows containing *smalto*, forming a layer of 6-7 centimeters, over which another uniform layer, composed only of pebbles, is poured. The two types of layers are alternated until the height of the boards is reached. It is calculated that one mason with two helpers can produce ten *cantoni* or *prismi* in one day. As evidence of this practice of fabrication of artificial stones, during the excavations for the making of the Carlo Alberto canal, near Alessandria, in 1840, "large slabs" of concrete made at the start of the century are unearthed, at this point with the "solidity of the hardest natural stones."[126]

Other hydraulic works using *cantoni di smalto* are carried out at the end of the 1600s in Piacenza. To protect the walls of the city from the force of the waters of the Po River, after conferring in 1697 with the Jesuit priest and professor of mathematics in Ferrara, Cesare Macrini, the city builds three groynes, or *pennelli*, to deviate the current, with pyramid-shaped *cantoni*. The work is supervised by Domenico Guglielmini (1655-1710), from Bologna, an expert on fluid mechanics, since 1698 professor of mathematics and medicine at the University of Padua, also employed by the Republic of Venice for hydraulic works, carried out in Padua.[127] The report entitled *Riparo alle corrosioni del Po proposto dal Padre Macrini e dal signor Dottore Guglielmini*[128] describes the project of the concrete groynes of Piacenza. For Guglielmini, as for the other leading Italian hydraulic theorists active in the late 1600s and the first decades of the 1700s, the treatise of Viviani becomes the channel of knowledge about the effectiveness of *cantoni di smalto*. The *Riparo alle corrosioni del Po* suggests the mode of fabrication of the concrete barriers: "dig a ditch in the ground for each one, with a width of one braccio, ascending obliquely another braccio, and terminating in a line"; then fill the ditch with "gravel and quick lime of great perfection," and then finally, "leave the mixture covered with earth and buried until it has hardened well."[129] The making of the blocks and the groynes is described in 1765 by the erudite Piacentine Cristoforo Poggiali (1721-1811).

"First he made a large pair of long, great triangular prisms," Poggiali writes on Guglielmini's *cantoni di smalto*, "composed of stones, gravel and lime mortar, with such preparation and mixture as to form a body equally as hard as flint, and superior to any harm from water, air and the seasons; he then connected the prism, at first thrown down in disorder, and later arranged in horizontal layers and closely devised together, thus erecting inside and over the

artificial stones"; these blocks were known as "*prismes*" because "being specially intended for the construction of wall angles, they are given the shape of a triangular prism four feet long and about two feet on the sides"; the ditches in which the prisms are fabricated are "triangular" (Joseph-Antoine Borgnis, *Traité élémentaire de construction appliquée à l'architecture civile*, Paris, Bachelier, 1823, p. 65). See also, Niccola Cavalieri San-Bertolo, *Istituzioni di architettura statica e idraulica*, Firenze, V. Bellini, 1833, vol. I, p. 48.

[126] Parrochetti, *op. cit.*, p. 186.

[127] Pietro Cossali, *Sul corso del fiume Po*, in, *Memorie dell'Imperiale regio istituto del regno lombardo-veneto*, Milano, Regia Stamperia, vol. II, 1821, pp. 103-14. *Cf. Visita fatta da Domenico Guglielmini al Po di Piacenza il giorno 9 aprile 1706*, in, Domenico Guglielmini, *Opuscoli idraulici*, Bologna, Marseille, 1822, pp. 328-36.

[128] *Cf.* Cossali, *op. cit.*, pp. 103-04.

[129] *Ibid.*, pp. 104-05.

waters of the Po River, at a distance of about one thousand paces one from the next, three great, strong protections, or jetties, as you will, which we commonly call pennelli, whose length is in subsextuple proportion to the width of the riverbed, and the width is about twenty-five Piacentine braccia with respect to the base, or the foundation, and of about six of our braccia at the top, which stands out above the water."[130]

Guglielmini also uses the same pyramid-shaped *cantoni* to control the course of the river that threatens to ruin the fortress of Palmanova.[131] His activity in hydraulic construction at the service of Venice and his teaching in Padua mean that precisely Guglielmini can be indicated as one of the fundamental figures for the transmission of the Tuscan constructive practices of the pupils of Galileo, and of their use of *smalto* reported by Viviani, in the constructions of the Republic of Venice. It is Guglielmini who discusses the means of defending the Venetian Lagoon with embankments to be made with rocks larger than those usually employed until then, with a greater height but a gentle slope, as well as the addition of pilings in the form of long groynes.[132] The solution of large rocks would be fundamental for constructed riverbanks during the 1700s.

Viviani's treatise is also well-known to another mathematician, physician and hydraulics expert, Bernardino Zendrini (1679-1747), who is to become one of the main protagonists of the discovery of the potential of pozzolana-base concrete for hydraulic constructions in the territories of the Venetian Republic.[133] Zendrini took a degree at the University of Padua in 1701, where he had the chance to get to know Guglielmini and his ideas on hydrology.[134] Probably thanks to Guglielmini, Zendrini learns of the treatise of Viviani and carries on the constructive tradition of the pupils of Galileo, constructing works along the Po and Adige rivers, where in the 1720s and 1730s he orders the construction of the building, with "excellent success," as he claims, of "cantoni di smalto, as we have been taught by Viviani in the dissertation to defend against the Arno."[135] It is significant that Zendrini also knows about the *cantoni di smalto* constructed by Guglielmini. "Had he taken all those rocks and pebbles of the Trebbia," he writes, "that were used to form the prisms for the jetties of the city of Piacenza, and threw them into the Po at the same site of the jetties, with the idea that this would force that river to stop causing those corrosions, with which it advances towards that city, would have completely wasted his time; but the same pebbles and stones bound with good lime, and reduced to a proper bulk, could be thrown into the current of the Po without danger of being borne away, and have formed three

[130] Cristoforo Poggiali, *Memorie Storiche di Piacenza*, Piacenza, F.G. Giacopazzi, 1765, vol. XII, pp. 184-85.
[131] *Cf.* Francesco Focacci, *Del modo di dirigere e regolare il corso dei fiumi e dei torrenti*, Firenze, N. Carli, 1809, p. 173.
[132] Giovanni Poleni, *Del moto misto dell'acqua*, in, *Raccolta d'autori italiani che trattano del moto dell'acque*, Bologna, J. Marsigli, 1833[4], vol. VI, p. 82.
[133] *Cf.* Bernardino Zendrini, *Leggi e fenomeni, regolazione ed usi delle acque correnti*, Venezia, G. Pasquali, 1741, p. 290.
[134] *Cf.* Angelo Zendrini, *Elogio di Bernardo Zendrini matematico della Repubblica di Venezia*, Venezia, Stamperia Palese, 1807, pp. 9-10.
[135] In "various sites of the Po and the Adige, I have made them with excellent success, and with equal benefit I have put them into practice," he writes about concrete *cantoni* (Zendrini, *Leggi e fenomeni* cit., p. 329).
[136] *Ibid.*, pp. 288-89.
[137] *Ibid.*, p. 330.
[138] *Ibid.*, p. 338.
[139] *Cf.* Antonio Rastovi, letter of 13 May 1763, with drawing, Savi ed Esecutori alle Acque, 782, Archivio di Stato di Venezia. The filler is specified as only loose stones.
[140] *Parere del signor dottore Tommaso Perelli sopra il ponte della Cecinella e sulle cagioni della rovina di esso* in, *Raccolta d'autori che trattano del moto dell'acque, edizione seconda*, Firenze, G. Cambiagi, 1774, vol. IX, (pp. 209-15), p. 211.

jetties of such solidity and strength that all the power of that regal river cannot harm them."[136]

Other solutions of gabions and gravel are described by Zendrini, like the jetties to "form as trilateral pyramids, truncated towards the top, but the section at the top should be oblique to the base, so it should terminate at the bottom in a gentle slope [...],"[137] or the *gorzi* that "are simply large gabions with a truncated conical figure, placed with the larger base in the bed, with a height equal to that of the highest flooding; they are arranged in a straight line, and in two rows, one beside the other, connected and secured by beams" and "they are filled not with earth, but with the largest gravel that can be found in the vicinity."[138] One of the *gorzi* described by Zendrini is designed to repair an embankment along the Piave.[139]

The use of *smalto* in Tuscany is not limited to the fabrication of *cantoni*. The compound is also used for the foundations of the piers and the arches of bridges, and it is probably a current practice, given the fact that it is also used for a bridge like the one on the Arno, known as "ponte della Cecinella," constructed by the master builder Andrea Pesciulli in approximately 1757, which collapses due to a flood in 1769. "Finally, where the work of the masonry of the bridge, of the vaults, of the pavement is concerned, it was found to be of praiseworthy quality, except for the fact that the pavement, instead of resting on the *smalto* poured over the vault in keeping with ordinary practice, rested on a bed of sand, and earth," we read in the expert's report dated 4 March 1769, prepared by Tommaso Perelli (1704-83), who received the degree in physics and medicine from the University of Pisa and is assigned to supervise hydraulic works in the lower Arno valley and at the lake of Bientina.[140]

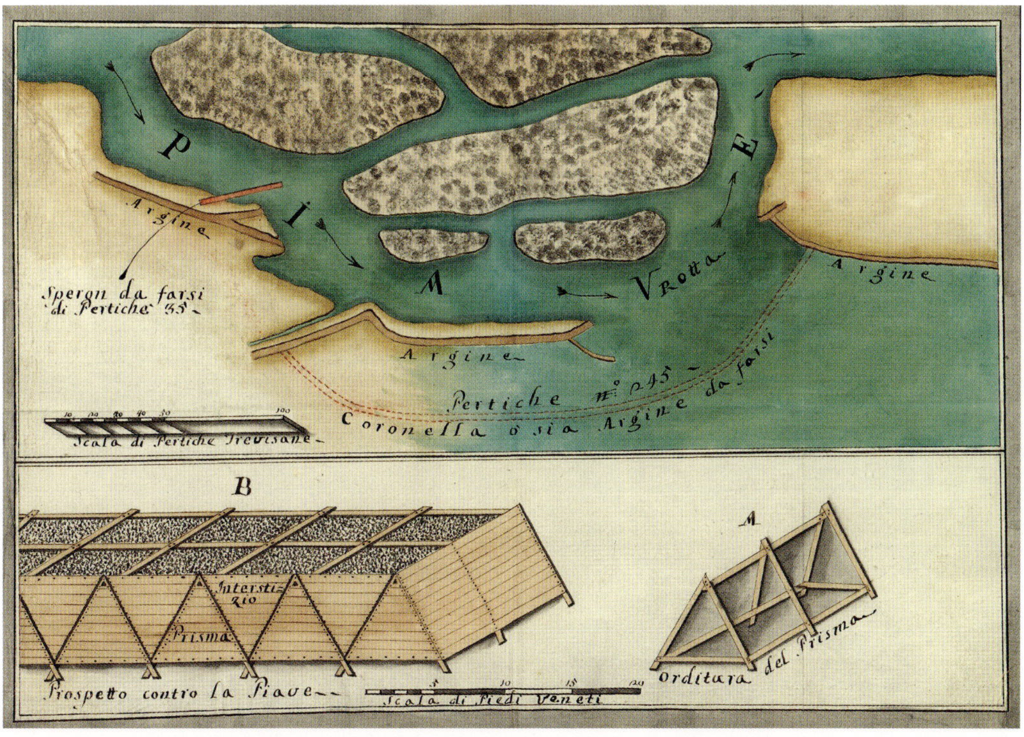


**10**. Antonio Rastovi, project for a "gorzo" to repair a levee along the Piave River, 1763 (Savi ed Esecutori alle Acque, 782, Archivio di Stato di Venezia).

B
D
B

# chapter two
# pozzolana, trass and lime for hydraulic construction

## Exportation of pozzolana from Civitavecchia

The use of concrete with hydraulic properties manufactured in the Roman manner continues during the entire 18th century, without any need of particular inventions or discoveries. Engineers, technicians, masons, experts on architectural treatises and Vitruvius, and scholars of ruins are the main repositories of knowledge of the techniques and materials for fabrication of concrete according to the methods used by the ancient Romans. All the experts on ancient Roman construction acknowledge the excellence of the volcanic rock known as pozzolana for the making of hydraulic cements.

A far-from-secondary corollary of lavic products is the technique of construction of the covering of terrace roofs, or "terrazza," mostly on houses in Naples and vicinity, where the inhabitants go in the evening to enjoy some fresh air.[1] During the course of the 1700s, these roofs become a veritable cult for European travelers, whose descriptions and views play an important part in their international spread.

The 18th century sees a true rise of the subject of cement made by Roman methods, leading those naturalists interested in volcanology – at times the discoverers of extinct volcanoes – to consider how to take advantage of deposits of pozzolana for the making of cement. But in the first decades of the century, until the success of certain grand hydraulic works sets the stage for a general reassessment of the constructive potential of the Roman method in contemporary works, the question of pozzolana remains a mere undercurrent

[1] It is worth remembering the significant mention made, for example, by Charles Thompson, regarding the flat roofs with balustrades of the houses of Naples (*The Travels of the Late Charles Thompson*, Reading, J. Newbery, C. Micklewright, vol. I, 1744, pp. 189-90).

in European construction. Nevertheless information does circulate in the literature regarding the nature of pozzolana and Roman cement, as a legacy of widespread knowledge. It will suffice to look at the entries in dictionaries to observe widespread awareness of the notion of cement as a compound made of lime and sand, or made with pozzolana, as well as the notion of concrete as a compound capable of hardening like stone,[2] and even the definition of *blocage* (rubblework) as an amalgam and, ultimately, a synonym of concrete – *cæmenta*, according to Vitruvius. Members of the Académie Royale d'Architecture of Paris in 1700 reveal a translation error, namely *terre glaise* (clay) for the term *ghiara* (gravel) as used by Palladio, "except that this Italian word means course gravel even crushed stone," which demonstrates awareness of Italian cement construction materials.[3]

Precisely the French builders and engineers are the best informed and most interested regarding pozzolana, importing sizable quantities of the material starting in the second half of the 1600s. "On various occasions, to fortify his squares and Ocean ports, the king of France has sent entire fleets of ships to get it," we find written about pozzolana in 1718.[4] The author of the entry *Pouzzol, ou Pouzzolane* of the famous *Dictionnaire universel françois et latin*, known as the Trévoux, published in 1704 and reprinted many times (the sixth edition is in 1771), lavishes praise on the potential of this Italian product, referring readers to "Vitruvius, Pliny, De Lorme, & other architects." "It is a sort of sand," we read on the subject of pozzolana, "that one finds on the territories of Pouzzol, a city of Italy near Baia, that makes the best cement imaginable. It hardens and petrifies even in water. It penetrates even black pebbles, and bleaches them white. It is used to make moles & other buildings in maritime settings. Agricola believes that its nature is based on alum, bitumen, and sulfur."[5] In a dictionary printed in 1709, under the entry *Pozzolane, poussolane* it is specified that "certain say and write *poussolane*, but incorrectly, it is

**11**. Francesco Serao, view of Mount Vesuvius (F. Serao, *Istoria dell'incendio del Vesuvio, accaduto nel mese di Maggio del' anno MDCCXXXVII. Scritta per l'Accademia delle scienze*, Napoli, N. de Bonis, 1738).

[2] See the entries *Beton*, *Blocage* and *Ciment*, in, François Pomey, *Le Dictionnaire Royale, augmenté de nouveau [...]*, Lyon, Veuve H. Molin, 1708, pp. 104, 111, 178 (*cf.* also the Lyon edition, A. Moulin, 1721). For "cæmenta" see the entry *Blocage*, in, *Dictionnaire d'architecture*, Paris, N. Langlis, 1693, vol. II, p. 27; in the same dictionary see *Ciment* (*ibid.*, p. 52) and *Poussolane* (*ibid.*, p. 211)

[3] Minutes of the meeting of 3 May 1700, in, Henry Lemonnier, *Procès-verbaux de l'Académie Royale d'architecture, 1671-1793*, Paris, E. Champion, 1913, vol. III, p. 96.

[4] Niccolò Madrisio, *Viaggi per l'Italia, Francia e Germania*, Venezia, G.G. Hertz, 1718, p. 549.

[5] *Dictionnaire universel françois et latin [...] Nouvelle Edition*, Paris, Trévoux, 1721, vol. IV, p. 790, entry *Pouzzol, ou Pouzzolane*. In the 1771 edition the entry *Béton* is the same, apart from the appearance of the accent (*Dictionnaire universel françois et latin, vulgairement appelé Dictionnaire de Trévoux [...]*, Paris, Compagnie des Librairies Associés, 1771, vol. I, p. 877).

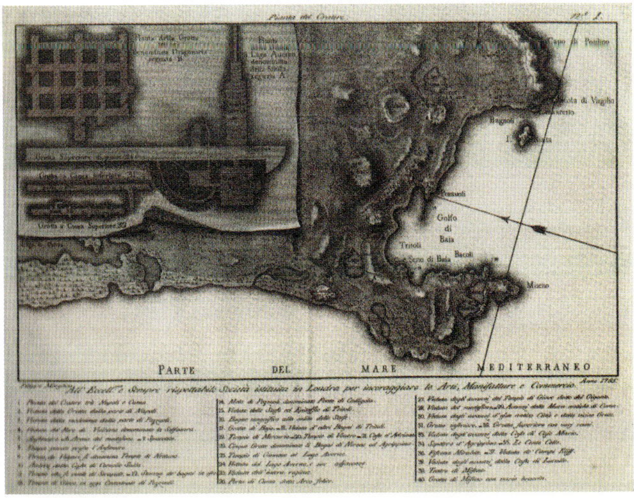


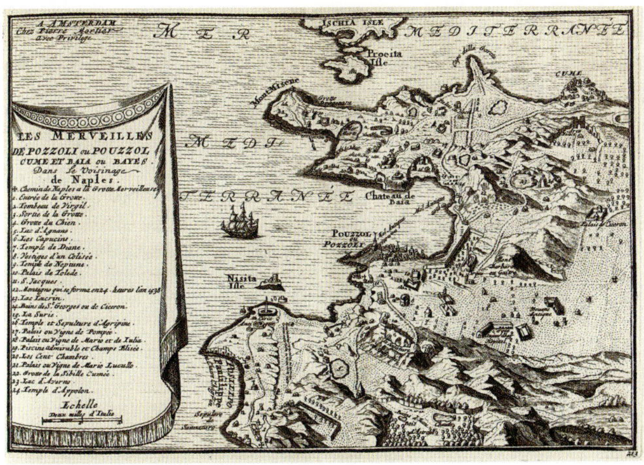


**12**. Pieter Mortier, view of Naples, 1704 (1049B12.081, Atlas Van der Hagen, Koninklijke Bibliotheek, Den Haag, Amsterdam).

**13**. Filippo Morghen, "Plan of the Crater between Naples and Cuma," 1769 (F. Morghen, *Le antichità di Pozzuoli, Baja, e Cuma*, Napoli).

**14**. Pieter Mortier, "The Marvels of Pozzoli or Pouzzol, Cuma and Baia or Bayes, near Naples," 1702 circa (RP-P-OB-83.034-214, Rijksmuseum Amsterdam, Amsterdam).

**15**. Jean-Augustin Renard, "View of the Place of Pouzzoles with the pedestal and statue of Tiberius," 1782
(L. Fino, *Vedutisti e viaggiatori a Pozzuoli, Baia, Cuma e dintorni*, Napoli, Grimaldi & C., 2011).

SETTIMO. 99

*NE' pauimenti, ò laſtriche fatte allo ſcoperto, come s'uſa ſpecialmente à Napoli, per la Sicilia, & à Malta, biſognerà uſar molta diligenza, ſi che poſſano reggere maggiormente all'ingiuria de' tempi: Vuole però Vitruuio, che fatto il primo tauolato, ſe ne faccia un'altro pe'l trauerſo, il quale conficcato con chiodi faccia un'armatura doppia alle trauamēta, dapoi ſia data la terza parte di teſtole peſte al terrazzo nuouo, & due parti di calce riſponda à cinque di eſſo nel mortaio. Fatto il riempiemento, ui ſia poſto il terrazzo, & ben peſto non ſia men groſſo d'un piede, ma indottaui l'anima (come s'è detto di ſopra) ſia fatto il ſuolo, ò pauimento di quadro grande, hauendo in dieci piedi due dita di colmo, come ſi uede nel ſeguente diſſegno.*

*MA*

**16**. "Lastrico" for flat roofs in use in Naples, Sicily and Malta, according to Giovanni Antonio Rusconi (G.A. Rusconi, *Della Architettura*, Venezia, Gioliti, 1590).

believed; pozzolane is the true word."[6] Yet in the range of technical know-how of French architects, at least at the start of the century, pozzolana is not recognized as a product to be used to produce good local cements. The members of the Académie Royale d'Architecture, in the first years of the century, often discuss mortars, the best blends for their composition, the choice of stones most suitable to make lime, processes to make slaked lime, what kind of water to mix with lime and sand, always concluding by returning to the experience of builders and trials to be performed. At times the discussions on mortars happen against the backdrop of the reading of chapters from the books of Palladio and Vitruvius on the subject, or the reading of the minutes of discussions on mortars held in the 1600s at the Académie Royale d'Architecture.[7] The question of pozzolana cement is confronted following the reading, on 19 March 1708, of Perrault's translation of the *De Architectura* of Vitruvius, when the members are considering "what might contribute to making good lime and to choosing appropriate stone."[8] On this occasion they recall the pioneering experience of Bernini on the Louvre worksite, though the members of the Académie Royale d'Architecture reach the same conclusion as their forebears regarding the failure of the results of Bernini's experiment. The surprising final consideration, reported in the minutes, namely that these are scientific questions that do not concern architecture, could not be more different from what was happening at worksites throughout the 1700s, while it confirms the continuing central role for French architects, at the start of the century, of the constructive tradition of stone cutting that found its high point precisely on that Louvre worksite where Bernini tried to introduce pozzolana cement.[9]

The myth of Roman cement also spreads across the northern regions of Europe, and becomes the parameter against which the decline of contemporary construction would be measured, though without putting the accent on pozzolana. Among "Roman Antiquities," the English appreciate precisely the mortar described by Vitruvius and Pliny, making the walls built in Sussex – for example – a model for contemporary construction. "We find indeed, their walls seem to bid fair for eternity; whereas ours, from parsimony and ill management, are scarce able to endure one age," is written in 1717.[10] A sign of the fundamental ignorance regarding pozzolana in these constructive traditions, in spite of the indications supplied by authoritative architects in the 1600s, is that few English travelers in the early 1700s pay any attention to the material. One who does is Joseph Addison (1672-1719), diplomat, writer and politician, who during a trip in 1701-03 briefly describes the constructive characteristics of hardening in the sea of the so-called "Pozzuoli earth" used by the

[6] Pierre Richelet, *Nouveau Dictionnaire François contenant generalement tous les mots anciens et modernes [...]*, Amsterdam, J. Elzevir, 1709, vol. I, p. 254.

[7] The experiences on lime come in the wake of a series of discussions at the Académie Royale d'Architecture of Paris in the 1710s on the best ways of preparing it, starting with observations made on worksites, and the reading of the treatises of Vitruvius, Palladio or Perrault, always underlining the fact that experimentation is needed (see the minutes of the sessions of 12 April 1712, 15 May 1713, 26 June 1713, 14 May 1714 and 11 June 1714, in, Lemonnier, *op. cit.*, vol. IV, 1915, p. 7, pp. 22-23, pp. 24-25, pp. 45-46, pp. 46-47).

[8] *Ibid.*, vol. III, 1913, p. 289.

[9] "Concerning pozzolane, which is amply discussed in the Vitruvius," we read in the report of the session in 1708, "and which is so widely used in Italy, experience shows that here in Paris [in the time of Bernini] this sort of sand does not make good mortar with the best lime used here, either because these are not suitable with each other or that the pozzolana is ruined in transport. The reasoning of Vitruvius and M. Pérault [sic] concerning the nature of pozzolana are better considered physics rather than architecture, which derives more utility in this regard from the experience than from this reasoning, which is based mostly on conjecture" (*ibid.*).

[10] *An accurate Account of a tessellated Pavement, Bath, and other Roman Antiquities, lately discover'd near East Bourne in Sussex. Being part of a letter of January 26. 1717 from the learned Dr. John Tabor of Lewis, to Dr. John Thorpe, R.S.S. and by him communicated to the Royal Society*, in, "Philosophical Transactions," vol. XXX, 1717-19, London, W. and J. Innys, 1720, (pp. 549-63), p. 563.

**17**. Giovanni Battista Piranesi, view of the ancient Appian Way (G.B. Piranesi, *Le Antichità Romane*, Roma, Bouchard and Gravier, 1756, vol. III).

ancient Romans.[11] Even when a description is published in London, in 1745, of the so-called "lavas" of Vesuvius – written by Richard Pococke (1704-65) based on observations made during a journey in 1733-36 – no mention is made of the constructive uses of pozzolana (Pococke planned on descending into the mouth of the volcano to survey it more accurately).[12]

Many other sources confirm that prior to the outbreak of controversy on the superiority of Greek or Roman architecture, in Europe there is a firm conviction that the lasting quality of Roman architecture should be attributed to its cement, and that therefore the discovery of the recipe to make that cement would be essential for the evolution of construction methods. On the durability of Roman cement it is worth reading, as an example, what De Montfaucon writes about the "massif" he had observed along the Appian Way, between Velletri and Sermoneta, where the paving stones had been removed: "the bottom is rough-stone or rubble employed with a very strong cement, and which is very difficult to break: just above is a layer of rubble cemented in the same way, interspersed with small round stones."[13]

An unspoken but effective guarantee of the survival of Roman cement in construction, from the start of the 1700s, is precisely the supply by sea of Italian pozzolana, shipped to the states of the peninsula and to those that can be reached along the main Mediterranean trade routes. The material is used to construct moles, military ports, sluices, banks and canals. Trade was already intense during the 1600s, when Italy supplied pozzolana for important projects, such as that of the Canal du Midi in southern France.[14]

All the trade in pozzolana, until the middle of the 1700s, is affected by the network of alliances among European powers and the patrolling done by English ships to prevent the arrival of goods in France. From the first decades of the century English ships have control of the Mediterranean and its seagoing trade. The Seven Years' War, which broke out in 1756 and ended in 1763, simply sealed the dominion of the sea by the English fleet, and the supply of pozzolana beyond the Italian borders had to submit to this dominion.

The pozzolana shipped across the Mediterranean, at this point, is no longer the famous material from the times of the Romans. The etymology of the name is well known in Europe, though certain misunderstandings can still be found circulating. Precisely the origin of the term makes non-experts suppose that the product on the market still comes from Pozzuoli, or from those areas around Vesuvius – like Cuma and Baia – praised by Vitruvius regarding their *pulvis puteolanus*. Instead, throughout the 1700s the main trade in pozzolana took place through the port of Civitavecchia, in the Papal State, which held a monopoly, exporting from there large quantities of pozzolana quarried in Rome and its vicinity. Only in rare cases was it possible to purchase pozzolana unloaded from boats arriving from Naples at Civitavecchia. There is

[11] Joseph Addison, *Remarks on Several Parts in Italy, etc., in the years 1701, 1702, 1703*, London, J. Tonson, 1705, pp. 424-51. *Cf.* Giovanni Capuano, ed., *Viaggiatori britannici a Napoli nel '700*, Napoli, La Città del Sole, 1999, vol. I, p. 63.

[12] Richard Pococke, *A Description of the East, and some other Countries. Vol. II. Part II. Observations on the Islands of the archipelago, Asia Minor, Thrace, Greece, and some other Parts of Europe*, London, W. Bowyer, 1745, pp. 204-06.

[13] Bernard de Montfaucon, *L'Antiquité expliquée et représentée en figures. [...]. La Guerre, les Voitures, les grands Chemins, les Ponts, les Aqueducs, la Navigation. Premiere Partie*, Paris, F. Delaulne, 1722, vol. IV, p. 179. Roman roads were made with "a cement so hard and so firm that centuries are needed to weaken them" (*ibid.*, p. 80).

[14] *Cf.* Chandra Mukerji, *Tacit Knowledge and Classical Technique in Seventeenth-Century France. Hydraulic Cement as a Living Practice among Masons and Military Engineers*, in, "Technology and Culture. The International Quarterly of the Society for the History of Technlogy," vol. XLVII, 2006, no. 4, pp. 713-33. On the question of pozzolana in the construction of the Louvre see, Roberto Gargiani, *Idee e costruzione del Louvre. Parigi cruciale nella storia dell'architettura moderna europea*, Firenze, Alinea, 1998, p. 95.

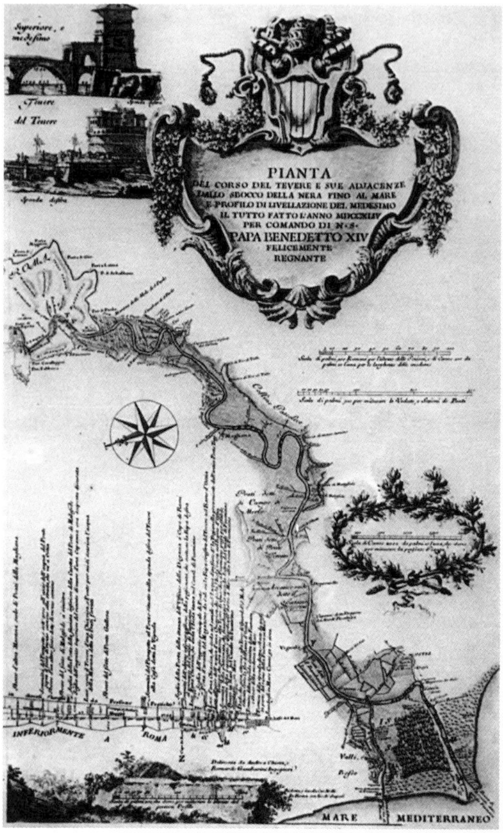


**18**. Andrea Chiesa, Bernardo Gambarini, "Plan of the course of the Tiber and its environs from the mouth of the Nera to the sea," 1744 (Miscellanea Geografica, Collezione Ashby, Biblioteca Apostilica Vaticana, Rome).

**19**. Giovanni Battista Piranesi, "View of the Port of Ripetta," Rome (G.B. Piranesi, *Varie Vedute di Roma Antica e Moderna*, Roma, 1741-48).

**20**. "Elevations and sections of the mouth of Fiumicino with the difficulties of entry of the ships due to the shallow depth there" (Cartes et Plans, Ge CC-1376 (38), Bibliothèque Nationale de France, Paris).

**21**. Mouth of Fiumicino with jetties (Cartes et Plans, Ge CC-1376 (39), Bibliothèque Nationale de France, Paris).

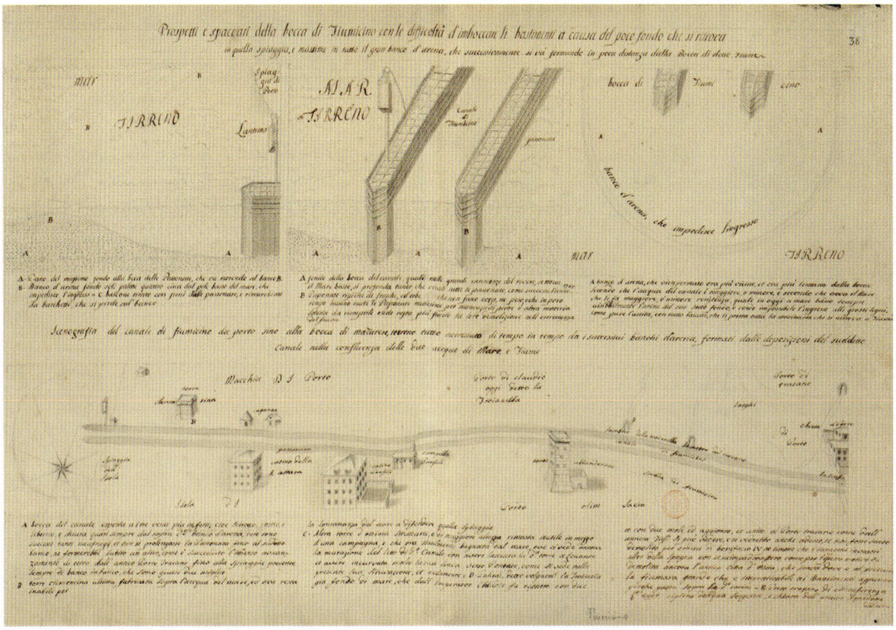
38
MAR
TIRRENO

39
Scala di palmi Romani

**22**. Alfier Giulio Amorini, "Plan of the city, and port of Civitavecchia" (Cartes et Plans, Ge CC-1376 (25), Bibliothèque Nationale de France, Paris).

**23**. Anonymous [Alfier Giulio Amorini ?], View of the port of Civitavecchia (Cartes et Plans, Ge CC-1376 (19), Bibliothèque Nationale de France, Paris).

nothing to back up the claim of Johann Joachim Winckelmann (1717-68) that the pozzolana at Anzio and on the Roman coast was still imported from Naples, "costing less to bring it from there by sea than to transport it by land on wagons from Rome."[15] The port of Naples and above that of Baia, in front of Pozzuoli – and therefore close to the most important and famous historic deposits of pozzolana – were thus no longer able to function as centers of international distribution of pozzolana, though some trade did take place.[16] Shipments left the port of Baia above all for destinations on the peninsula and at Malta, though it had been the most famous port in antiquity precisely for the trade in pozzolana, which was even shipped to Constantinople.

The activity of pozzolana quarrying in Rome and its vicinity remains intense in the first decades of the 1700s, as in the previous century, running the risk of endangering monuments, cemeteries, aqueducts and roads. Tunnel excavations often lead to the destruction of ancient tombs.[17] There are reports of collapses caused by wind currents, and of deaths among the miners of pozzolana.[18] Decrees had been issued, already in the 1600s, to regulate the quarrying of pozzolana, stipulating a minimum safe distance to prevent the quarries from altering the stability of works.[19] The text of reference against the "damage caused by pozzolana quarries" to the antiquities of Rome dates back to 17 May 1580,[20] and decrees in the 1700s confirm and extend these provisions, as in the edict of 5 January 1750.[21] "How many have dug into foundations, simply to obtain pozzolana, making the buildings collapse shortly thereafter!" complains the lawyer Carlo Fea (1753-1836), archaeologist and art collector, commissioner of Roman Antiquities.[22]

*Grottoni* is still the term generally used to indicate pozzolana quarries, which are excavated as tunnels, as in the previous centuries. "The entire territory surrounding Rome is gutted, and undermined by the excavations to mine this

[15] Johann Joachim Winckelmann, *Osservazioni sull'architettura degli antichi*, in, Carlo Fea, ed., *Opere di G. G. Winckelmann, Prima edizione italiana completa*, Prato, F. Giachetti, 1836, vol. VI, p. 59.

[16] The pylons of the port of Pozzuoli are praised, made of "earth known as Pozzolana, or giara, that mixed with lime makes an estimable alloy, as Vitruvius points out, which has been used to make many great Buildings, and brought as far as Constantinople, and not having it the French, as there is peace, load Ships" (Domenico Antonio Parrino, *Nuova Guida de' forastieri per l'antichità curiosissime di Puzzuoli*, Napoli, Parrino, 1727, p. 41).

[17] *Cf.* Giovanni Gaetano Bottari, *Sculture e pitture sagre estratte dai cimiterj di Roma, pubblicate gia dagli autori della della Roma Sotterranea [...]*, Roma, Stamperia Vaticana, 1737, vol. I, p. 18. In the plan of the "Cimiterio superiore, e inferiore de' SS. Marcellino, e Pietro nella via Labicana" there is an indication of the "place where tombs have been ruined to remove pozzolana."

[18] See the chronicles of Rome dated 18 March 1818, in, "Gazzetta di Milano," 29 March 1818, no. 87, p. 374.

[19] Michel Angelo Zaccaria, *Editto contro quelli, che caveranno, e guasteranno edifizj antichi, e loro platee, e pozzolana con burbore*, 9 May 1685, cit. in, Carlo Fea, *Dei diritti del principato sugli antichi edifizj pubblici, sacri e profani*, Roma, Fulgoni, 1806, p. 74.

[20] Carlo Fea, *Relazione di un viaggio ad Ostia e alla villa Plinio della Laurentino*, Roma, A. Fulgoni, 1802, p. 103.

[21] "And because until now by us or by others permits are often given to dig, and to remove treasures, marble, Travertine, statues, pozzolana and other sundry things, the digging of which requires the intervention and assistance of our above-mentioned commissioner, or of the councillor, or another delegated person, to make sure the excavation is not done on a prohibited site, or one that endangers public buildings, roads and cemeteries, as expressed above, we command and order any person of any rank and condition whatsoever, that after the issuing of this edict these permits must be made known to the Office of the herein signed Secretary within ten days, without which notification the permits will be revoked, as was ordered in the above-mentioned Prohibition of 1717, at paragraph 9" (Silvio Valenti, *Proibizione della estrazione delle statue di marmo, o metallo, pitture, antichità, e simili. Silvio del Titolo di S. Calisto Prete Cardinal Valenti della S. Romana Chiesa Camerlengo*, 5 January 1750, Roma, Camera Apostolica, in, Fea, *Relazione di un viaggio* cit., pp. 105-15).

[22] Fea, *Relazione di un viaggio* cit., p. 89.

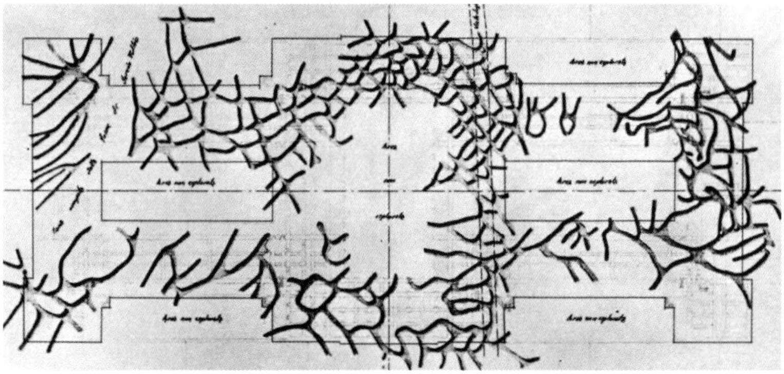

**24**. Survey of the pozzolana quarries along Via XX Settembre, Rome. "Construction of the new Palazzo delle Finanze. Plan of the second order of the ancient subterranean caves for the mining of pozzolana" ("Capitolium," vol. XLV, 1970, no. 12).

**25**. Shaft of the pozzolana quarries on Via Alessandro Cialdi, near the Basilica of St. Paul, Rome (photograph by Luca Cavallari, 2008).

**26**. Pietro Fabris, "View of a tuff quarry on the estate of D. Benedetto Farina in the territory of Castelnuovo" (Private collection, Naples).

earth; and the grottoni have a length of thousands: there were catacombs in them. When work was done on the foundations of the casino of Villa Albani, three of these grottoni were found, one atop the next; so it was necessary to dig further, all the way to the mentioned depth of eighty palms."[23]

The mining activity in the pozzolana quarries is taxed.[24] The use of different grades of pozzolana for certain public works is regulated. In the case of the streets of Rome, the "edict on pozzolana quarries and its qualities," issued on 24 May 1715, prohibits "master builders, masons, architects" from "using black Pozzolana, even of excellent quality, because for the making of these pavements only good red Pozzolana should be used."[25]

Quarriers discover ancient monuments in the course of their work. One of these informs Marcantonio Boldetti (1663-1749), an erudite clergyman and leading expert on Roman underground cemeteries, of the existence of a cemetery with Greek inscriptions along the Via Prenestina; the cemetery had been ruined by the pozzolana quarrying.[26] It is above all in studies or guides of the early Christian cemeteries in Rome and its vicinity that we can find descriptions of the pozzolana quarries (from such sources, for example, we learn that a "very vast cavern made by digging for pozzolana" exists under a convent on the Appian Way).[27]

Pozzolana in Rome is measured by the cartful, the wagonload or the sack. The first two are so-called *a cassa* units, and their measurements are inscribed on a plaque at the Università dei Muratori, in the courtyard of the Palazzo dei Conservatori at Campidoglio – the *barrozza*, or wagonload, corresponds to one cartful (*carrettata*) and a half.[28] A cartful can be divided into 16 quantities known as *scorzi da pozzolana*. A sack contains a quantity of poz-

[23] Carlo Fea, *La calce, ed in ispecie la pozzolana*, in, Joahnn Joachim Winckelmann, *Storia delle Arti del disegno presso gli antichi [...] Tradotta dal tedesco e in questa edizione corretta e aumentata dall'Abate Carlo Fea*, Roma, Stamperia Pagliarini, 1784, vol. III, pp. 26-27.

[24] *Cf. Delle cave di Pozzolana*, in, Pompeo Orsaio, Domenico Orsaio, *Miscellaneum Sacrum et Profanum, Canonicum & Civile*, Typographia Balleoniana, 1726, p. 167.

[25] "Editto sopra le cave di pozzolana e qualità delle medesime," 24 May 1715, Archivio Segreto Vaticano, Miscellanea, cit. in, Daniela Esposito, *Tecniche costruttive murarie medievali. Murature 'a tufelli' in area romana*, Roma, L'Erma di Bretschneider, 1998, p. 86.

[26] Marc'Antonio Boldetti, *Osservazioni sopra i cimiterj de' Santi Martiri ed antichi cristiani di Roma*, Roma, G.M. Salvioni, 1720, p. 567.

[27] *Ibid.*, p. 559.

[28] Feliciano Scarpellini, *Prospetto delle operazioni fatte in Roma per lo stabilimento del nuovo sistema metrico negli Stati Romani dalla commissione de' pesi, e misure*, Roma, M. de Romanis e Figli, 1811, p. 202.

[29] *Ibid.*, p. 203. At the start of the 1800s the difficulty in measuring pozzolana based on the tools presently on the market is emphasized: "The capacity of the Cart with Crates is not a submultiple of any round number, so the calculations for transport are not facilitated as fully as possible" (*ibid.*, p. 202, note

zolana such that eight correspond to one cartful. The sack is four palms long and two and a half wide, and contains two *scorzi.*[29] Information on the cost of pozzolana taken from the quarry of San Paolo indicates that in the first decades of the 1700s one cartful cost 6.65 scutes. This pozzolana would be sent to one of the most prestigious papal worksites of hydraulic construction, at the port of Ancona.

Pozzolana is transported from the Roman quarries to Civitavecchia via the Tiber. But due to the bends, the shallows and the irregular riverbed, only vessels with a flat keel, known as *barchettoni,*[30] are able make the trip from the port of Fiumicino to the docks in Rome, to ship the pozzolana – boats were towed by men or animals, along special trails on the banks. Some boats, at the fork of Capo Due Rami, about two kilometers from Ostia, head rather for Mezzocamino, to load the pozzolana taken from the quarries along the Tiber and on the hills of Rome.[31] Others reach the vicinity of the church of San Paolo, where major trading in pozzolana takes place at the so-called *Scalo della pozzolana*, once known as the port of Grapigliano, near the Ponte Fratto.[32] In the 1700s, the wharf for the shipping of pozzolana on the estate known as "dei Grottoni," six miles from Rome along Via Ostiense, is still active. Ships reaching the port of Ripetta could also transport the pozzolana. The shipping ports above Rome also traded in volcanic materials, including pozzolana. Boats loaded with pozzolana navigated on the Aniene River, a tributary of the Tiber – in this case the pozzolana was loaded at Porto Vecchio, Porto Nuovo and Porto della Cementara.[33]

g). Therefore it is proposed to alter the measurement of the crate so that "3 wagons" contain 100 cubic palms of pozzolana, and 30 one cubic rod (*ibid.*).

[30] *Cf.* Paola C. Scavizzi, *Navigazione e regolazione fluviale nello Stato della Chiesa tra XVI e XVIII secolo*, Roma, Edilstampa, 1991, p. 73.

[31] *Cf.* Anna Alloro, *Roma città di mare*, in, Angela Adriana Cavarra, ed., *Roma, la città dell'acqua*, Roma, De Luca, 1994, (pp. 375-84), p. 383. On the location of the pozzolana quarries at Anzio see, Paola Brandizzi Vittucci, *Antium. Anzio e Nettuno in epoca romana*, Roma, Bardi Editore, 2000, p. 10, note 20.

[32] See the plan where the letter I indicates the deposit of the pozzolana near the church of San Paolo (*cf.* Giovanni Battista Rasi, *Sul Tevere e sua navigazione da Fiumicino a Roma*, Roma, Perego-Salvioni, 1827, p. 52). For the location of the "pozzolana depot" along the Tiber see the letter Y in, Giovanni Battista Rasi, *Sul porto romano di Ostia e di Fiumicino, Osservazioni istoriche*, Roma, L. Contedini, 1826.

[33] *Cf.* Nicoletta Marconi, *I legni e le pietra. Gli approdi per i materiali edili tra XVI e XVIII secolo sul Tevere a Roma*, in, Carlo M. Travaglini, *La città e il fiume (secoli XIII-XIX)*, Roma, Ecole Française de Rome, 2008, (pp. 181-95), p. 186. *Cf.* also, Antonio Scherillo, *Inclusi nelle pozzolane della media valle dell'Aniene*, in, "Periodico di Mineralogia," VI, 1935, pp. 169-78.

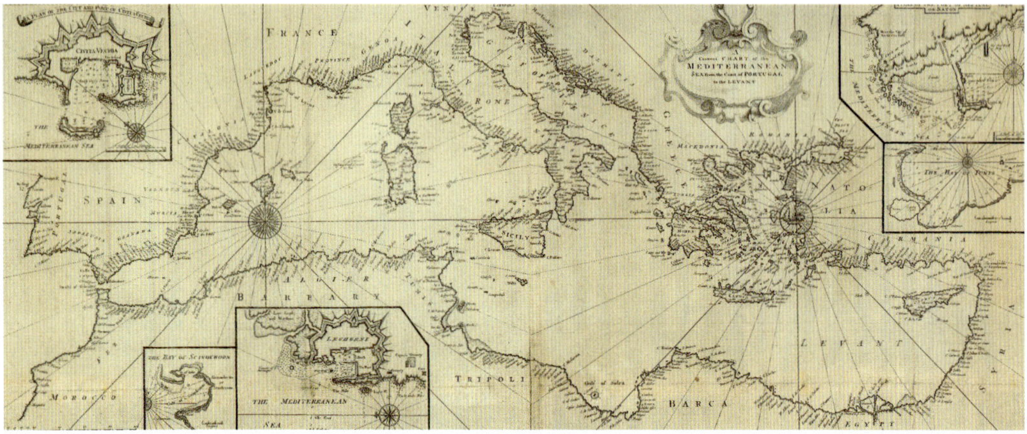


At Fiumicino the pozzolana is unloaded and then placed on larger ships that took it to Civitavecchia, where it is stored awaiting shipment to different ports of the Mediterranean. Some adventurous traders take their ships all the way to Fiumicino, to directly take on pozzolana without paying the duties charged by Civitavecchia. Pozzolana is so easy to sell that in 1757 even what is left over from the work on the houses of the Riminesi, stored at the Forte di Nettuno, is sold.[34]

Thanks to its depth, its piers and the breakwater known as Antemurale Traiano, and to the easy access to supply of the material, the port of Civitavecchia holds sway throughout the 1700s as the port of distribution of pozzolana to all of Europe. Most of the vessels loaded with pozzolana leaving Civitavecchia head for Italian ports like Genoa, Leghorn, Ancona, Pesaro and Rimini; starting in the 1730s they also go to Malamocco, the port of Venice.[35] Trade depends on the works in progress in ports and along rivers and canals, and varies from city to city, depending on the know-how of local builders in the use of pozzolana as a fundamental ingredient of hydraulic mortars. The largest European importer of pozzolana is France, followed by Spain, Portugal and Sweden.[36] Some shipments reached Malta, others go to the coasts of the Middle East. Although the other European states, from England to Holland, send ships to Civitavecchia, they do not import pozzolana, or do so only in exceptional cases. One significant case, that would long remain an isolated one as well, is the load of pozzolana brought by an English merchant to the port of Plymouth, on the Channel, in hopes of selling it in London. The man who would become the most famous English engineer of the 1700s, John Smeaton (1724-92), recalls his experiences with pozzolana

[34] Andrea Fara, *La pesca di mare nei centri costieri del litorale basso-tirrenico dello Stato Pontificio*, in, Valdo D'Arienza, Biagio di Salvia, ed., *Pesci, barche, pescatori nell'area mediterranea dal medioevo all'età contemporanea*, Milano, Franco Angeli, 2010, p. 380.

[35] On the pozzolana trade at the start of the 1700s, see the entry of 15 wagon-loads of pozzolana from Civitavecchia that reach Leghorn with a tartana on 16 October 1714 (*cf.* Jean-Pierre Filippini, *Il porto di Livorno e la Toscana (1676-1814)*, Napoli, Roma, Benevento, Milano, Edizioni Scientifiche Italiane, vol. II, 1998, p. 12, note 9).

[36] For indications on the use of pozzolana-based mortar in Portugal, starting with the Roman tradition, see, Ana Luísa Velosa, Rosário Veiga, João Coroado, Victor M. Ferreira, Fernando Rocha, *Characterization of Ancient Pozzolanic Mortars from Roman Times to the 19th Century*, in, Maria Boştenaru Dan, Richard Přikryl, ed., *Materials, Technolgies and Practice in Historical Heritage Structures*, Bucaresti, Springer, 2010, pp. 235-58.

**27**. "A correct chart of the Mediterranean sea from the coast of Portugal to the Levant," Mount & Page, 1770 (RM. 73670, Institut Cartogràfic de Catalunya, Barcelona).

mortar of his own manufacture, made in the winter of 1756-57: "I very fortunately learnt, that there was a quantity of it then in the hands of a marchand in Plymouth, which had been imported as an adventure from Civita Vecchia during the time Westminster Bridge was building, and which he expected to have disposed of for that work to a good advantage, but failed in his speculation for having found that tarras answered their purpose, neither Commissioners, Engineers, nor Contractors, would trouble themselves to make a trial of it, and therefore refused it."[37]

Documentation provided by the French consul at Civitavecchia, Jean-Baptiste-François Vidau (1721 circa - 1811 circa), makes it possible to reconstruct the background of pozzolana trade across a span of over fifty years, from the start of the 1730s to the end of the 1780s.[38] Starting in the 1730s, the purchasing of pozzolana represents an important part of the trade at the port of Civitavecchia. The ships used for its transport are the *tartanas* used for trade in various types of merchandise; but their hulls had to be treated "in order to take on board pozzolana [la terre possolaine]."[39] The captains and "patrons" of the ships rented for the transport of pozzolana often commit fraud to avoid paying taxes. Some, like Pierre Pericart hailing from Agde, "to load pozzolana at Fiumichino [*sic*]," refuse to pay the rights claimed by the naval Ordonnances, "saying that they are not held to these when only loading earth."[40] Others, like Jourdan de Cannes, report that they are headed for the port of Naples, and leave Civitavecchia without paying duties, docking it instead in Fiumicino, where they board the pozzolana and then "return righteously to Languedoc."[41]

The largest quantities of pozzolana exported from Civitavecchia include shipments to the port of Toulon for the construction of its Arsenal, ordered by the Intendant de la Marine, François-Noël Levasseur de Villeblanche (1648-1759). To bring "the greatest quantity possible of pozzolana [terre possolaine]" to Toulon, in 1746 fake passports and fake contracts of sale are drawn up, with the complicity of the consul of Portugal, indicating that the pozzolana would be going to Lisbon, to avoid the English blockade against the French.[42] Thanks to this ploy, in July-August 1746 over one thousand quintals of pozzolana reaches Toulon (one cartful corresponds to 11 quintal), in multiple shipments (600 and 285 quintals are the amounts of two of the loads) and transported by French ships always carrying "all papers and documents needed in case of meeting an enemy," supplied by the consul of Portugal.[43] One of these concealed shipments to Toulon is intercepted by an English ship and escorted to Leghorn to be sold off. On 8 October two French *tartanas* leave Civitavecchia headed for Toulon with a load of wood and pozzolana (252 quintals of wood, 1600 of pozzolana);[44] a third *tartana* departs a few days later with 184 quintals of pozzolana in its hold (it too is transporting wood),

[37] John Smeaton, *A Narrative of the Building and a Description of the Construction of the Edystone Lighthouse with Stone*, London, G. Nicol, 1791, p. 109.
[38] *Cf.* "Correspondence Consulaire, Civitavecchia, 1732-1743," AE, B1.373, Achives Nationales, Paris; "Correspondence Consulaire, Civitavecchia, 1744-1774," AE, B1.374, Achives Nationales, Paris.
[39] Jean-Baptiste-François Vidau, letter, 20 July 1746, in, "Correspondence Consulaire, Civitavecchia, 1732-1743," AE, B1.373, Achives Nationales, Paris.
[40] Jean-Baptiste-François Vidau, letter, 7 July 1738, *ibid.*
[41] Jean-Baptiste-François Vidau, letter, 7 July 1738, *ibid.*
[42] Jean-Baptiste-François Vidau, letter, 6 June 1746, *ibid.*
[43] Jean-Baptiste-François Vidau, letter, 1 August 1746, in, "Correspondence Consulaire, Civitavecchia, 1744-1774," AE, B1.374. For pozzolana shipments see, Jean-Baptiste-François Vidau, letters, 6 June, 25 July and 1 August 1746, *ibid.*
[44] Jean-Baptiste-François Vidau, letter, 3 October 1746, *ibid.*

again for the destination of Toulon.[45] With the French *tartana* that sailed out of Civitavecchia in October, loaded with wood and 322 quintals of pozzolana, Vidau completes the shipment of construction materials to Toulon as ordered by De Villeblanche. "Up until now, I have sent six ships to Toulon loaded with wood and some pozzolana on each one, I have also sent one fully loaded with the mentioned pozzolana," Vidau writes on 7 November 1746.

Trade in pozzolana with the ports of the oceanic coast, along the route plied by Dutch or Swedish ships that usually called at Leghorn, and not at Civitavecchia, is even more difficult. This is shown by the story of the hiring of a "neutral ship" to bring 6000 quintals of pozzolana to the port of Brest, ordered from Vidau in August 1746 by De Villeblanche.[46] In October Vidau has still not been able to arrange the ship "for the transport of pozzolana to Brest." It is only in November, thanks to the connections of Girolamo Capalti, merchant of Civitavecchia, do negotiations begin to "bring in from Leghorn a number of Dutch or Swedish vessels,"[47] of the proper tonnage for the transport of the cargo. On 30 December a contract is signed in Leghorn between the Dutch captain and the merchant Francesco Morganti for the hire of a vessel that would take on pozzolana at Civitavecchia on behalf of Capalti and transport it to "other convenient places."[48] Given the fact that the ship could transport only 4000 quintals of pozzolana, Vidau wants to hire another ship to bring 3000 more quintals to Brest – which would have been 1000 quintals more than planned. This is because it is hard to find a ship that would carry just 2000 quintals on such a trip.[49] Again, the contract does not precisely indicate the port of destination. On 6 February 1747 the Dutch vessel reaches Civitavecchia, but the captain refuses to sign the new contract, which now indicates a destination under the usual method of a false contract, without guarantees of reimbursement if forced by the English fleet to unload the pozzolana in Amsterdam. The words of the Dutch captain reported by Vidau confirm the fact that the pozzolana would be used above all by the French: "in case the English force us to transport the above-mentioned pozzolana to Amsterdam, saying that if he meets with some sort of crown enemy ship in the waters of Brest, or in the vicinity of this port, and with the English knowing that only France uses this earth, even admitting the false contract [...] he feared that the English would oblige him to transport the earth to where it was intended; therefore, he wanted an obligation by a merchant of Amsterdam for the payment [...] in case that he would be obliged to go there and unload."[50] As the affair with the Dutch captain goes into litigation, Vidau and Capalti resume their efforts to hire a ship to go to Brest.

Still during the course of 1747, in February and March, to respond to the new requests of De Villeblanche for other materials for the Arsenal of Toulon, the French make use of Catalan ships sent to Naples with troops to transport pozzolana and wood on the return trip to Barcelona.[51] "[...] having filled the space between the pieces of wood with pozzolana," we read on the subject of

[45] Jean-Baptiste-François Vidau, letter, 3 October 1746, *ibid.*
[46] Jean-Baptiste-François Vidau, letter, 12 September 1746, *ibid.*
[47] Jean-Baptiste-François Vidau, letter, 7 November 1746, *ibid.* *Cf.* also, Jean-Baptiste-François Vidau, letter, 28 November 1746, *ibid.* From France, in December, a tartana is indicated, *La Diligente*, to send to Civitavecchia to take on part of the pozzolana for Brest (Jean-Baptiste-François Vidau, letter, 16 January 1747, *ibid.*).
[48] Copy of the contract, 30 December 1746, in, "Correspondence Consulaire, Civitavecchia, 1744-1774," AE, B1.374, Archives Nationales, Paris.
[49] Jean-Baptiste-François Vidau, letter, 17 February 1747, *ibid.*
[50] Jean-Baptiste-François Vidau, letter, 8 March 1747, *ibid.*
[51] Jean-Baptiste-François Vidau, letter, 17 February 1747, *ibid.* Vidau reports that on 6 February he has hired "a catalan boat" "to transfer a load of building wood with Pouzzolaine earth for the Toulon Arsenal for a price of 1700 pounds," on the condition that it must be loaded at the expense of the king, and that it will be loaded on its way back from Naples, where it was sent to bring

the loading of one of these ships, completed on 22 March.[52] In advance of the arrival scheduled for May of the ship hired in Leghorn to bring pozzolana to Brest, Vidau, who is concerned "that little of this Earth would be left and perhaps we will be in short supply,"[53] asks for permission to "have some more Pozzolan brought from Rome, to ballast the ships that I will load with wood."[54] Another Catalan ship sets off in June for Toulon with a cargo of wood and 525 tons of pozzolana,[55] while a similar load of wood and pozzolana, also for Toulon, which leaves in August, is sunk by an English ship at the latitude of Viareggio.[56] In 1756 another cargo of pozzolana is also recorded, headed for Rochefort.

One of the most unusual cargos to leave the port of Civitavecchia heads for Lisbon in July 1747; it contains pieces of architecture and sculpture, masonry materials including pozzolana, and the craftsmen for the construction of the Royal Chapel commissioned by the king of Portugal, John V, to Luigi van Wittel (1700-73) – a member of a family of Dutch origin known as the Vanvitelli – to Nicolò Salvi (1697-1751), and to the sculptor Alessandro Giusti (1715-99); Giusti supervises the worksite in Lisbon. "The three laden ships and the new beautiful chapel that the King of Portugal has ordered in Rome," Vidau writes, "are ready to take sail to transport their load to Lisbon, and all three are Venetian, and there all sorts of workers boarded capable of setting up the mentioned chapel, they also loaded pozzolana and other materials for the construction."[57]

During the course of 1748 the shipping of wood and pozzolana from Civitavecchia to Toulon continues with Catalan ships returning from Naples, where they are still transporting troops.[58] In April Vidau tries to hire a ship to bring more pozzolana to Brest: "A Swedish ship was proposed to me that will arrive loaded with wheat from Ancona, for the transport of Pozzolana to Brest, thus I will not fail to take this opportunity to carry out the expedition."[59] The pozzolana for Brest is finally shipped only in September, on a Dutch vessel, at 35 soldi per quintal, paid with a loan provided by Roman bankers, to be paid back to a bank of their Italian partners in Paris.[60] In May 1748 Vidau takes the initiative of sending two loads of pozzolana for a total of 3640 quintals to deliver to De Villeblanche at the port of Toulon or – if the material is no longer needed for the Arsenal – for transport to Marseille where "it will be sold for my account," as Vidau writes, now aware of the possibility of good business in the trade of pozzolana.[61]

Another legal dispute regarding the pozzolana trade, after that of the Dutch captain, has to do with a French captain in charge of a *tartana* hired by the king of Spain, intimidated in January 1751 by two captains and their crews, one from Portofino, the other from Trapani, in the port of Civitavecchia, over a financial matter. The small *tartana* of the captain from Portofino had been hired at Fiumicino by the French captain to transport, via the Tiber, 210 cartfuls of pozzolana from Rome to Civitavecchia, as he had not found enough of the material to be purchased for the king at Civitavecchia.[62]

troops. *Cf.* Jean-Baptiste-François Vidau, letter, 8 March 1747, and letter, 13 March 1747, *ibid.*

52 Jean-Baptiste-François Vidau, letter, 22 March 1747, *ibid.*

53 *Cf.* "Service général," 1756, B66, Archives de la Marine, Paris.

54 Jean-Baptiste-François Vidau, letter, 8 May 1747, in, "Correspondence Consulaire, Civitavecchia, 1744-1774," AE, B1.374, Archives Nationales, Paris.

55 Jean-Baptiste-François Vidau, letter, 29 May 1747, *ibid.*

56 Jean-Baptiste-François Vidau, letter, 30 October 1747, *ibid.*

57 Jean-Baptiste-François Vidau, letter, 31 July 1747, *ibid.*

58 Jean-Baptiste-François Vidau, letter, 19 February 1748, and, letter, 15 April 1748, *ibid.*

59 Jean-Baptiste-François Vidau, letter, 15 April 1748, *ibid.*

60 *Cf.* Jean-Baptiste-François Vidau, letters, 28 July and 2 September 1748, *ibid.*

61 Jean-Baptiste-François Vidau, letter, 8 May 1748, *ibid.*

62 *Cf.* "Extrait du livre des registres de la chancellerie du consulat de France a Civitavecchia," 1 February 1751, *ibid.*

Another important destination for the pozzolana trade is Cartagena, on the Mediterranean coast of Spain, for work on the port planned in the 1730s and consisting in the construction of dry docks for the military port. Towards the end of 1751 Vidau states that by now "the greatest commerce" of the port of Civitavecchia has to do with pozzolana, "that it is a perfect earth for building under water, and of which entrepreneurs now working in the Port of Cartagena in Spain are accumulating large quantities, normally using ships from Agde and Sete in Languedoc in order to transport this earth there."[63]

Confirmations of the Mediterranean constructive tradition using cement also are mentioned during the trip made in 1750-51 by Ferdinando Morozzi (1723-85), who in Tunis reports that the "buildings are covered in smalto to form open terraces."[64] It is no coincidence that during the 1700s trade in pozzolana also takes place along the coasts of Northern Africa, from Algiers to Tunis – trade that is to continue until at least the early 1800s.[65]

### The first non-Italian pozzolanas and the discoveries of Labat

The discovery of pozzolana deposits outside the territories of Rome and Naples, the ministerial decrees to discourage importation of the Italian product in the 1780s, and the invention of processes for the manufacture of alternative artificial products at the end of the century, all fail to reduce the prestige of Italian pozzolana, which continues to be exported throughout the 18th century. A turning point occurs only at the start of the 1800s, and it is no coincidence that its focal point lies in the countries of northern Europe – regions that can be reached by merchant ships bringing pozzolana from Italy only at great expense. Nevertheless, the search for the legendary pozzolana of Roman cement outside of Italy began back in the vicinity of the year 1700.

Awareness of the meaning of the name pozzolana, taken from the town of Pozzuoli, spreads in the early years of the 18th century, when products with the same physical characteristics and mechanical properties are identified beyond the Italian borders.

"It is an error to believe that it exists only in Poussol near Naples," we find written in 1724 regarding *poussolane*, "it is everywhere in the Roman countryside, and there is much in other places where I have been. Maybe the first who used this natural cement found it in Poussol and gave it this name, which was then used by all those who found it in other locations."[66] These words are written by the first discoverer of the existence of pozzolana outside of Italy, the clergyman Jean-Baptiste Labat, or Père Labat (1663-1738). He is the one who recognizes that the "cement or red earth"[67] used in Guadeloupe or in Martinique in the Antilles, the French colony in the Atlantic Ocean where he stays for different periods from 1696 to 1705, is actually a pozzolana.

On his first trip there, Labat observes the "cement or red earth" quarried on the island and used by the natives, and he experiments with its characteristics, rebuilding a canal that powers a mill. Then he has books sent from France to identify the nature of that special red earth. Reading the annotated translation of *De Architectura* of Vitruvius done by Perrault, he realizes that

[63] Jean-Baptiste-François Vidau, letter, 4 October 1751, *ibid.*
[64] Cit. in, Riccardo Francovich, *Materiali per una storia della cartografia toscana: la vita e l'opera di Ferdinando Morozzi (1723-1785)*, in, "Ricerche Storiche," VI, 1976, pp. 445-512.
[65] *Cf. Livorno*, in, "Giornale del Lloyd Austriaco di notizie marittime e commerciali," IV, 8 July 1838, no. 81, p.n.n.
[66] Jean-Baptiste Labat, *Nouveaux voyages aux Isles de l'Amérique, contenant l'histoire naturelle de ces pays*, La Haye, P. Husson, 1724, pp. 386-87.
[67] *Ibid.*, p. 386.

**28**. Johannes van Keulen, "Het Eylandt Martanico of Else de la Martenique," Amsterdam (B.0032(109), 02, 143, Nederlands Scheepvaartmuseum, Amsterdam).

**29**. Gerard van Keulen, "Nieuwe Land en Zee Kaart van de Eylanden Guadaloupe en Mariegalande," Amsterdam (B.0032 (109), 04, 028, Nederlands Scheepvaartmuseum, Amsterdam).

the local earth corresponds to Italian pozzolana.[68] To investigate the efficacy of local mortars and their setting capacities, he conducts elementary tests, such as soaking the compounds in fresh and salt water; he also has a compound prepared with stones embedded in the cement, with which he fills "a cube" he then immerses in water for two or three days, in order to test its strength with blows of the hammer.[69] Instead, the construction of a glacis for a "flour stove" shows that fire rapidly damages the local cement.[70] Following these experiences, Labat informs the engineer Jean-Baptiste de Giou de Cailus (? - 1722), supervisor of the work on the Fort Royal of Martinique and other military works on the island, that the local red earth is a pozzolana.[71] In 1704 Labat replaces the flat roof made with wooden beams and lead sheets of the convent at Martinique, which he built himself in 1698, with a "small terrace" constructed in the Italian way, by means of a "bulk of pumice stones with a good mortar of pozzolana, and well tiled."[72] Labat also offers instructions for the correct use of pozzolana cement in non-hydraulic works: "If we want constructions made of Pozzolana to form a solid body, and last a long time, we must take care to water down the masonry for seven or eight days. Otherwise the lime tends to reignite, consuming the pozzolana, & reducing it to powder."[73] It is possible that precisely thanks to the local "red earth," the natives build the walls of houses with the system of molds and concrete, at least according to the descriptions of the walls in Guadeloupe, made with lavic pebbles gathered on riverbanks and embedded in pozzolana cement.[74]

Also in Martinique, Labat discovers local manufacturing of lime using seashells, "the lime made with these kinds of seashell is excellent, and provides a mortar that hardens like marble."[75]

Back in Europe, Labat studies Roman construction practices during a trip to Italy in 1706, carefully examining ruins of bridges and moles at Baia,[76] and writing down his observations on the use of materials and pozzolana. He also observes the making of pozzolana cement in Sicily, where he sees local pozzolana quarries.[77]

The discovery of various deposits in the Antilles prompts Labat to imagine possible trade in pozzolana to be shipped to France, to avoid the dependency of French worksites on the importation of Italian pozzolana[78] – "we will find ourselves able to construct works which cannot be started without this assistance."

[68] *Ibid.* Probably Labat receives the book by Vitruvius in 1697 (*cf.* Jean-Baptiste Labat, *Nouveaux voyages aux Isles de l'Amérique*, Paris, Th. Le Gras, 1742, vol. IV, p. 231).
[69] Labat, *Nouveaux voyages aux Isles de l'Amérique* cit., 1724, p. 388.
[70] *Ibid.*
[71] *Ibid.*, p. 389.
[72] Labat, *Nouveau voyage du père Labat aux Isles de l'Amérique, contenant l'histoire naturelle de ces pays*, La Haye, P. Husson, 1724, vol. VI, p. 348.
[73] Labat, *Nouveaux voyages aux Isles de l'Amérique* cit., 1724, p. 389.
[74] Félix Longin, *Voyage à la Guadeloupe œuvre posthume*, Le Mans, Monnoyer, 1848, p. 44.
[75] Labat, *Nouveau voyage du père Labat aux Isles de l'Amérique* cit., p. 413.
[76] *Voyages du P. Labat de l'ordre de s.ff.precheurs, en Espagne et en Italie*, Paris, J.-B. Delespine, Ch.-J.-B. Delespine, 1730, vol. V, pp. 89-91.
[77] *Ibid.*, pp. 214-15.
[78] "I have discovered a considerably sized vein at the anchorage of Martinique," Labat writes regarding local pozzolana, "underneath and a bit to the side of the Battery of St. Nicolas. The color was a bit lighter, & the grain finer, but as for the rest it was the same thing. I have used a considerable quantity of it, after convincing myself of its quality using the same tests as those I used to know about that of Guadaloupe" (Labat, *Nouveaux voyages aux Isles de l'Amérique* cit., p. 389). Labat's discoveries also become known thanks to quotations of his passages on pozzolana in, Georges Louis Leclerc de Buffon, *Histoire naturelle des minéraux*, vol. IV, Paris, Imprimerie Royale, 1786, pp. 412-13, note b.
[79] Jean-Baptiste Labat, *Nouveaux voyages aux isles de l'Amérique*, Paris, Th. Le Gras, 1742, vol. IV, p. 276.
[80] On the discoveries of Fréri, the ship's surgeon on the *Glorieux*, at the Isle de Bourbon in 1751, of volcanic materials (pozzolana is not mentioned, but only lava) see, Jean-Etienne Guettard, *Mémoire sur plusieurs morceaux d'histoire naturelle, tirés du cabinet de S.A.S.M. le duc d'Orléans*, in, "Histoire de l'Académie Royale des Sciences, Année MDCCLIII, avec les Mémoires de Mathématique & de Physique, pour la même Année," Paris, Imprimerie Royale, 1757, (pp. 369-400), pp. 376 *et seq. Cf.* also, Fréri, *Excursion au volcan de l'île Bourbon, les 9-16 avril 1751*, in, "Académie des Sciences, Arts et Belles lettres

"[...] we French always have to go get it in Italy," he writes about pozzolana, "to buy it at an expensive price, and often have trouble getting it from foreigners while we have so much of it that we do not know what to do with it. In order not to have shortages in France, we just have to order all the captains of the ships that go out to the Islands to throw their ballast out to the ocean and then to use pozzolana as ballast for the return trip."[79]

After Labat's discoveries, pozzolana is also found in another French holding, the island of Mauritius in the Indian Ocean. Halfway through the century, the overview of the places of quarrying of what is commonly known as pozzolana includes the territories conquered by France. In 1751 the surgeon Fréri, on board the ship *Glorieux*, explores the volcanoes of the Isle de Bourbon, or Mauritius, and brings samples of lavic products back to France,[80] while the surgeon Honoré Lieutaud (? - 1751) has already gathered other samples of volcanic products of Isle de Bourbon, made available to volcanologists so they can make comparisons with the French pozzolanas.[81] Dictionaries of the French language make changes to the definition of pozzolana: "the term refers to a type of sand found around Pouzzol of Italy, near Naples; it is also found in Guadaloupe, in Martinique, and Isle de France [or Mauritius]."[82] But all the discoveries made by the French on faraway ocean islands does not have an impact on national cement production, or challenge the leading role of the port of Civitavecchia, due to transport costs. Exportation to France of pozzolana from Guadeloupe, taken in the 1700s from the many quarries around the capital, Basse-Terre, for use on local worksites, would not be considered economically feasible until the 1800s.[83]

In the villages and cities of South America under Spanish control we can also find construction systems like those introduced at the start of the 1600s based on a constructive tradition of Roman origin: a form of concrete known as *tapia*, composed of lime, sand and shell or stone aggregate mixed with water and poured into wooden molds.[84] The system is then imported by English colonists along the southern coast of North America. In the Spanish colonies of South America, including the Colombian cities of Santa Fede, Antioquia, Popayán, Trujillo and Bogotá, and in Caracas in Venezuela, pozzolana is used to make the concrete that goes into the construction of houses. In certain cases the walls are simply concrete poured into wooden formwork. "These walls are called Tapias by the Spaniards," the Jesuit abbot Filippo Salvatore

de Dijon," 1753, vol. III, pp. 566-73.

[81] On Lieutaud see, Jean-Etienne Guettard, *Mémoire sur plusieurs morceaux d'histoire naturelle, tirés du cabinet de S.A.S.M. le duc d'Orléans*, in, "Histoire de l'Académie Royale des Sciences, Année MDCCLIII, avec les Mémoires de Mathématique & de Physique, pour la même Année," Paris, Imprimerie Royale, 1757, pp. 369-400.

[82] Entry *Pozzolane*, in, Jacques-Christophe Valmont de Bomare, *Dictionnaire raisoné Universel d'Histoire Naturelle*, Paris, Didot le Jeune, vol. IV, 1764, pp. 495-96.

[83] *Cf. Exposition universelle de 1855: Rapports du jury mixte international, publiés sour la direction de S.A.I. le prince Napoléon*, Paris, Imprimerie Impériale, 1851, p. 139.

[84] For the definition of *tapia* still in use in the 1700s, given by the Spaniards to the type of wall made by pouring material into molds described by Pliny, see, *Caii Plinii Secundi Historiae Naturalis Libri XXXVII, quos interpretatione et notis illustravit Joannes Harduinus*, Parisiis Societatits, 1741, vol. II, p. 713, note 1.

**30**. Cotopaxi volcano, near Quito, Ecuador (Ch.-M. de La Condamine, *Journal du voyage fait par ordre du Roi à l'équateur*, Paris, Imprimerie Royale, 1751).

Gilij (1721-89), of Italian origin, reports.[85] He describes certain construction methods practiced in the various cities of Colombia and Venezuela. "Some of them," he writes about the houses of Santa Fede, "are made only of pozzolana, forced into molds suited to the purpose. And it is so sticky and durable that it can hardly be demolished, and it would be very difficult to demolish our walls. In Tungia there are similar very old houses, no longer inhabited, without roofs, and constantly exposed to rain. This shows the durability of the pozzolana that was used to make them."[86] In Antioquia certain houses have walls of only "pozzolana pressed into molds."[87] At Bogotá and Trujillo the houses are also in "gratings of reeds filled with pozzolana."[88]

In Italy the first discoveries are made of extinct volcanoes beyond the confines of Rome and Naples, and beyond the local quarries in Sicily also described by Labat; this raises the possibility of finding new pozzolana deposits. Starting in 1733, the botanist Pier Antonio Micheli (1679-1737) insists on the volcanic origin of Mount Amiata, in Tuscany, though Giovanni Targioni Tozzetti (1712-83) is the one who finally proves, using the notes of his mentor Micheli, that Mount Amiata is an extinct volcano.[89] Nevertheless, it is not until the middle of the 1700s that enough interest develops in extinct Italian volcanoes to lead to hypotheses of trade in pozzolana coming neither from Rome nor from Naples.

### German "tuffstein" and Dutch "trass"

During the same 18th century, the main European center of quarrying and shipping of volcanic material for construction and for the manufacture of hydraulic mortars, after Civitavecchia, is in Germany, along the Rhine between Koblenz and Bonn, the site of a volcanic crater that had become a lake, the Laacher See. Below the crater, along the river, stand the village of Brohl and the town of Andernach, with their small trading ports and quarries. The various types of volcanic rocks found in the vicinity of the Laacher See had been used locally at least since the time of the Romans, and the Germans had already set up some mills, in Brohl and other villages, at least by the 1600s, to make the powders used to prepare mortars.

The quarries active during the 1700s, usually dug on the surface of the land, are in the villages around the Laacher See. Plaidt, Kretz, Kruft, Tönisstein, Mendig and Weibern are some of the main zones of quarrying. German workers use different names to indicate different qualities of rock. The best of them, exported for the making of mortar, is called "tuffstein," where *tuff* is an adaptation of the Italian term *tufo*, meaning a volcanic product.[90] The rocks are removed in pieces and transported with carts to Andernach or Brohl, for river shipping. In the case of rocks for transformation into powders, the product is marketed in two forms, depending on the destination.

The product for the German market and that of the French territories near the border, towards Metz, reached along the Rhine, is usually processed

[85] Filippo Salvatore Gilij, *Saggio di Storia americana, o sia storia naturale, civile, e sacra de' regni, e delle provincie Spagnuole di Terra-ferma nell'America meridionale*, Roma, Perego Salvione, vol. IV, 1784, p. 474, note 1.
[86] *Ibid.*, p. 363.
[87] *Ibid.*, p. 451.
[88] *Ibid.*, p. 266.
[89] Giovanni Targioni Tozzetti, *Relazioni d'alcuni viaggi fatti in diverse parti della Toscana*, Firenze, Cambiagi, 1776, vol. IX, p. 363.
[90] *Cf.* Barthélemy Faujas de Saint-Fond, *Mémoire sur le Trass ou Tuffa volcanique des environs d'Andernach*, in, "Annales du Muséum National d'Histoire Naturelle," 1802, vol. I, pp. 15-26. Faujas de Saint-Fond publishes the drawing of the quarry at Plaidt (*ibid.*, plate II). *Cf.* also, Christian Ludwig Ziegler, *Von dem Terras, der Puzzolana und deren substitutionsarten*, in, "Hannoverisches Magazin," XI, 19 January 1773, no. 6, coll. 81-94; Fournet, "Rapport sur le trass ou la Pozzolane de Broll," 4 February 1808, Ecole des Ponts et Chaussées, Paris; Gustav Haegermann, *Von Cæmentum zum Zement*, in, Günter Huberti, ed., *Von Cæmentum zum Spannbeton. Beiträge zur Geschichte des Betons*, Wiesbaden, Berlin, Bauverlag, 1964, pp. 30-34.

**31**. Lorenz Janscha, Johann Ziegler, View of Andernach, 1798 circa (KAR0501734, Österreichische Nationalbibliothek, Wien).

**32**. Trass quarry at Plaidt, near Andernach (A. Hambloch, *Le Trass et son emploi pratique dans les constructions*, Andernach, Chez L'Auteur, 1908).

DE STAT DORDRECHT
DEN AMER FLU.
DEN POLDER
VAN WIEL
DRECHT
DEN ZUYT POLDER
HET OUDE LANT VAN DUBBELDAM
DEN NOORT POLDER
Den Biesbos
De Merwede
DEN ALBLASSER WAERT
DEN SWINDRECHTSEN WAERT

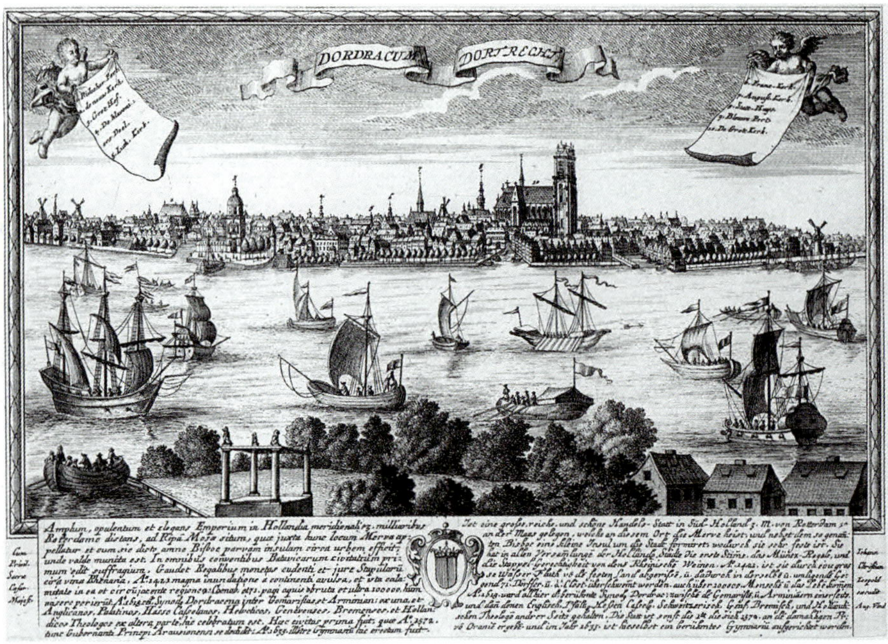
DORDRACUM
DORTRECHT

**33**. Mattheus van Nispen, plan of the city and territory of Dordrecht, 1673 (1049B11.085, Atlas Van der Hagen, Koninklijke Bibliotheek, Den Haag).

**34**. Johann Christian Leopold, Friedrich Bernhard Werner, view of Dordrecht, 1725 circa (P323N435, Universiteitsbibliotheek Leiden).

prior to river shipping, ground up in the mills of Andernach and Brohl and then sifted. "[...] most often the pieces are mere shapeless chips, intended to be pulverized under the pestles of a mill," writes an observer, on the subject of the materials taken from the quarries; instead, the material removed from the quarries already in the form of powder, called *wild* by the workers, and of inferior quality, is immediately put aside for the making of mortar.[91]

The unit of measure of the volcanic stones sold at Andernach and Brohl is the *wagen*, or cart, containing two wooden crates, for a total weight of 60 quintals; once milled, the content produces 14 tons of powder.[92]

The quantity of volcanic ash taken from the quarries and that of the powders obtained with the mills, used in German construction, is less than the production of volcanic powders launched on a large scale by the Dutch in the 1700s – the importing in Holland of that material dates back at least to the 1500s and is already important in the 1600s, when the *tuffstein* is shipped mostly to Utrecht.[93] The fact that in 1712 a Dutch *tuffstein* merchant, Van Nispen, built his own house in Brohl, shows the importance of the transport of volcanic material for construction to Holland.

As the largest importer of *tuffstein*, Holland purchases the product as it is brought by the quarrymen to Andernach and Brohl. The stones are loaded onto boats and transported to the Dutch city of Dordrecht, on the Rhine-Meuse delta, and the port of arrival for all goods coming from Switzerland and Germany. It is at Dordrecht, no longer at Utrecht, that the greatest quantities of *tuffstein* in the 1700s, renamed as *duif steen* by the Dutch, are processed and traded.[94] Precisely at Dordrecht, the site of one of the largest concentrations of windmills for water drainage, the *duif steen* is ground up by the mills. A certain amount of *tuffstein* also reaches the cities of Rotterdam, Utrecht, Zutphen and Haarlem, where the German product is processed, again by mills.

Ongoing evolution of the mechanics of mills leads, precisely in the context of the processing of *duif steen*, to the adaptation of the gears of one type of mill, for the grinding into powder of quarried materials from Germany. To transform the mill into a continuously functioning machine for the production of volcanic powders, crates and conduits are applied (and the gears are also modified) so that the mill can continuously receive material to be ground, while, thanks to sieves, the finished product is automatically selected in special differentiated containers, depending on particle size. Overly large particles fall onto a board and are channeled back through the mill. The mills for the production of *tuffstein* powders have different devices in different provinces, and though they are usually powered by wind, others are driven by horses, as in Haarlem, or with water wheels, as at Zutphen. In 1777 accurate drawings are made of one of the windmills of Dordrecht, with the aim of spreading the Dutch machine for the production of volcanic powders.[95] At the end of the 1700s at Dordrecht there are six mills for the grinding of *tuffstein*, powered by wind or horses.[96]

[91] Nicolas de Pigage, "Mémoire sur quelques pierres singulieres propres à la construction des bâtimens, qui se trouvent en Allemagne, sur les bords du Rhin, et dans son voisinage," Mannheim, 28 December 1777, B21, Archives de l'Académie Royale d'Architecture, Institut de France, Paris.

[92] *Ibid.*

[93] *Cf.* Anton Jospeh Wolf, *Vom Eifeler Tuffsteinhandel im 17. u. 18. Jahrhundert*, Euskirchen, Verlag des Eifelvereins, 1923.

[94] *Cf.* Smeaton, *A Narrative of the Building* cit., p. 111.

[95] Nicolas Desmarest, *Extrait d'un mémoire sur la détermination de quelques époques de la nature par les produits des Volcans, & sur l'usage de ces époques dans l'etude des Volcans*, in, "Observations sur la Physique, sur l'Histoire naturelle et sur les Arts," Paris, Cuchet, vol. XIII, February 1779, pp. 115-26, pp. 198-203, and plates I-III.

[96] Gratien Le Père, *Deuxième recueil de divers mémoires sur les pouzzolanes naturelles et artificielles, Paris*, Imprimerie de Cellot, 1807, p. 33. *Cf.* Salvatore Ciriacono, *Acque e agricoltura. Venezia, l'Olanda e la bonifica europea in età moderna*, Milano, FrancoAngeli, 1994.

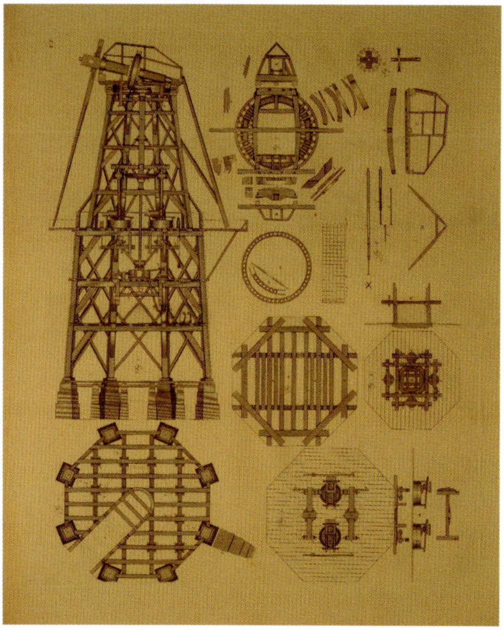

**35**. Dutch windmill (T. van der Horst, *Theatrum machinarum universale; of keurige verzameling van versheide grote en zeer fraaie waterwerken, schutsluizen, waterkeringen, ophaalen draaibruggen*, Amsterdam, P. Schenk, vol. I, 1736).

**36**. Jehan-Guillaume Ecrevisse, "Plan and elevation of the cement and Andernach-stone mill built at Dordrecht in Holland" (B21, Archives de l'Académie Royale d'Architecture, Institut de France, Paris).

**37**. Mill for grinding stones from Andernach, Dordrecht ("Observations sur la Physique, sur l'Histoire naturelle et sur les Arts," Paris, Cuchet, vol. XIII, February 1779).

Pl. I.
F. 1.
50
40
30
20
1 2 3 4 5 6 7 8 9 10 11 12
Maro 1779.

**38**. Simon Fokke, Ary van Wanum, "Gezicht van het Biesbosch en den Noordendyk by de stad Dordrecht," Dordrecht, A. Blussé & Zoon, 1770 circa (RP-P-1926-410, Rijksmuseum, Amsterdam).

It is probable that the type of mill surveyed in 1777 is not the type that had been widespread at the middle of the century, given the fact that Smeaton, during a stay in Holland in 1755, observes that once ground into still rather large particles, the product goes through further processing, this time by hand: "It is beat by iron-headed stampers upon an iron bed, till it passes through a sieve of certain fineness, equivalent to one of ours, having about eight wires in an inch, it is ready for use [...]."[97]

At this point in the work, when it has been reduced to powder, the *tuffstein* takes on the definitive name of *tras*, or more commonly *trass*, or *tarras*[98] – a contraction of *terras*, which probably comes from the Italian *terrazzo*. The fact that the term *trass* contains an echo of the Italian word that indicates both a flat roof for buildings and also a cement-base product used for the construction of its roof coating – which in Naples and vicinity is of volcanic origin – demonstrates the desire of the Dutch to introduce a product on the market that can compete with the most famous volcanic material in Europe for hydraulic mortars, the pozzolana.[99] Only if we consider the various terms used in Europe, starting with the Neapolitan *terrazzo*, meaning a cement-based compound for durable surfaces, can we perhaps understand the reason behind the name *terrazzo* of the Venetian product for floors.

The process of manufacture of Dutch *trass* includes phases of quality control supervised by official technicians. At Dordrecht the controls imposed by the municipality on the type of *tuffstein* to be imported and the *trass* milled in the city and stocked for sale are so rigorous that the reputation for quality

[97] Smeaton, *A Narrative of the Building* cit., p. 111.
[98] For the meaning of trass see, Barthélemy Faujas de Saint-Fond, *Mémoire sur le Trass ou Tuffa volcanique des environs d'Andernach*, in, "Annales du Muséum National d'Histoire Naturelle," 1802, vol. I, pp. 15-26, p. 15. *Cf.* also, Rudolf August Vogel, *De pulvere cæmentariis nonnullis usitato, tras vulgo dicto, commentatio chemical*, in, "Novi Commentarii Societatis Regiae Scientiarum Gottingensis," vol. III, 1773, (pp. 50-59), pp. 50-51. During the 1700s and later, various opinions circulate in Europe on the meaning of *trass* – some believe it means the material quarried in the German provinces, others think it indicates a type of cement made by mixing *tuffstein* with quick lime (*cf. Gabinetto mineralogico del Collegio Nazareno*, Roma, Lazzarini, 1792, vol. II, p. 296). Desmarest contributes to the misunderstanding, so much so that *trass* is often used to indicate the German stone, not

spreads throughout Europe. The controls are guaranteed by commissioners appointed to oversee the operations. Prior to the unloading of boats from Andernach or Brohl, the commissioners check on the quality of the rocks imported from Germany, to see if they are *tuffstein* or substitutes. The commissioners themselves examine the product, after pulverization in the mills of Dordrecht, and before it is put onto the Dutch market, to guard against fraudulent imitations. The quality of the *trass* is checked by constructing masonry samples, and certain tests are invented to avoid putting altered products into circulation, given the fact that *trass* powder can easily be cut with powders of the same color, such as wood or peat ash. "[…] one makes a vase with the powder, combined with lime, which is then filled with water, and if, after three days, no apparent filtration has appeared, we take it as good quality, otherwise it is rejected and can not be brought into Holland."[100] The *trass* is conserved in special warehouses, under constant surveillance.[101] The same commissioners have the job of making periodic checks on the state of maintenance of the mills. Nevertheless, the problem of counterfeit versions of the German stones and *trass* continues throughout the 1700s. On the same note, there are also suspicions that the Dutch themselves may take advantage of German *wild* purchased at Andernach to dilute the *trass* obtained through milling.[102]

*Trass* becomes an essential material for the composition of hydraulic mortars, and its trade by sea by the Dutch enters into competition with Italian pozzolana. The *trass* made by the Dutch is not used only for their hydraulic works, but also becomes a lucrative export to the nations facing the northern seas. It makes inroads on the English market, where it is called *tarras*, and in the provinces of Germany closest to Holland. It easily reaches French worksites, especially in the provinces bordering Flanders, where it is harder to bring in shipments of Italian pozzolana at a reasonable cost – *terrasse de Hollande* is the French term. On ships, *trass* is transported in such a way that it is protected – it "is generally sent from Holland in casks."[103]

Like the English, the Dutch do not take part in the seagoing trade of pozzolana, and only a few of their ships that ply the route of the Mediterranean contribute to ensure, throughout the 1700s, the distribution of loads of pozzolana from Civitavecchia along the oceanic coasts and the North Sea.

In Holland, with *trass*, two types of cement of different quality are manufactured, known as *starker trass* and *bastard*. The *bastard* compound is preferred for masonry exposed to the open air, as it offers savings with respect to the use of the *starker trass*, which under such conditions does not set as well as in hydraulic works.

Confirmation of the spread of the Dutch practices can be found in the writings of Friedrich Ludwig Aster (1732-1804), who knows the Dutch technical literature and has traveled throughout Holland to study the processes of fabrication of hydraulic mortars. Aster reports on the main compounds of

the result of its working in Holland (*cf.* Desmarest, *Extrait d'un mémoire* cit., pp. 115-26).

[99] For the apotheosis of the flat roof made possible by construction materials and techniques in Naples see the project of Mario Gioffredo for a palazzo identified as that of Caserta (*cf.* Cesare de Seta, ed., *Luigi Vanvitelli e la sua cerchia*, Napoli, Electa Napoli, 2000, pp. 227-28).

[100] Gratien Le Père, *Deuxième recueil de divers mémoires sur les pouzzolanes naturelles et artificielles*, Paris, Imprimerie de Cellot, 1807, p. 33.

[101] *Cf.* Friedrich Ludwig Aster, *Gesammelte Nachrichten von dem in den Vereinigten Niederländischen Provinzen gebräuchlichen Cement aus Trassen oder gemahlnen Cöllnschen und Andernachschen Tuffsteine*, 1791, p. 11, (first edition 1773).

[102] "The Dutch buy little wild trass, yet enough to fear that on the market they do mix it with the fine trasse" (Nicolas de Pigage, "Mémoire sur quelques pierres singulieres propres à la construction des bâtimens, qui se trouvent en Allemagne, sur les bords du Rhin, et dans son voisinage," Mannheim, 28 December 1777, B21, Archives de l'Académie Royale d'Architecture, Institut de France, Paris).

[103] Smeaton, *A Narrative of the Building* cit., p. 111.

Dutch *trass* cement, and also on certain local practices, which differ from city to city: "Even in Dutch provinces have I seen the following: there are two main mixtures of trass and lime in cement or water mortars: one the Dutch call starker trass or cement, and they use it for walls under water and waterproof works; the other is called bastard or false trass, also slop or weak bastard, and they use it for walls above water. The first is made of lime and trass, the other out of lime, trass and sand. For common mortars, they take, as we do, only lime and sand. For starker trass or true water cement, one takes: in Grave and Vlissingen, 3 parts lime and two parts ground trass; in Masterdam and Zütphen 1 part lime and 1 part trass."[104]

Outside of Holland, in the early part of the 1700s, *trass* is used for the mortars of certain major hydraulic works that would become outstanding examples of hydraulic engineering. One of them is the lock on the Aa River, at Gravelines, in French territory at the border with Flanders. Designed by Vauban in 1699, the lock is built at the start of the 1700s under the supervision of the engineer De La Guerre de Charbise, using hydraulic mortar made with Dutch *trass* – a "cement made of two thirds quicklime from Boulogne and one third Dutch *terrasse*, of highest quality, well ground and sifted."[105] Dutch *trass* is also used to prepare the mortar for the masonry of the piers of one of the most important bridges of the early 1700s, Westminster Bridge in London.

Throughout the century the Dutch hold the largest share of seagoing trade in volcanic powders in the northern seas. Their ships bring *trass* to England, France and Denmark. The quality and cost are competitive, also with respect to the products of Andernach, meaning that Dutch *trass* also reaches the German markets of Hanover, Bremen or Hamburg. Among the terms used to indicate *trass*, we find, significantly, "Holland cement."

"This fine trass powder, manufactured on site, as well as the wild trass, which is made separately, are intended for consumption from the upper to the lower Rhine, which is little compared to the production of the Dutch, who sell it in the North, in the Netherlands, England and France. It is because of this trade, entirely in the hands of the Dutch, that trass powder, ground in their own mills, has taken the name of Holland cement."[106]

The exportation of Dutch *trass* continues into the early years of the 1800s, though on a smaller scale, due to the international spread of new artificial products.

Manuals of hydraulic engineering call for the making of hydraulic mortars with *trass*. One emblematic example about *terrasse de Hollande* is written in one of the most widely used and important manuals of the middle of the 1700s, the one on fortifications by Bernard Forest de Bélidor (1698-1761); it discusses, among other things, the practice of fraudulent adulteration of the material.

"The other type of powder," he writes, after having explained the characteristics of pozzolana, "is made of an earth that is found close to the low Rhine in Germany and in the region of Cologne, it is fired like plaster, and then it is

[104] Aster, *op. cit.*, p. 18.
[105] Bernard Forest de Bélidor, *Architecture Hydraulique, ou L'Art de conduire, d'élever et de ménager les eaux, pour les différens besoins de la vie. Premiere Partie, Tome Premier*, Paris, Ch.-A. Jombert, 1737, p. 320. De La Guerre de De Charbise does not describe the mortar with *trass* when he explains the *ciment* to use for the new Picardy canal, in the estimate of 1731 (*cf.* De La Guerre de De Charbise, *Canal de Picardie. Devis des ouvrages nécessaires à faire pour la construction du nouveau canal*, Paris, P. Simon, 1732, p. 58). "The cement is made with fragments of old tile, especially hard, without brick in the mixture, well ground and seived."
[106] De Pigage, "Mémoire sur quelques pierres" cit.

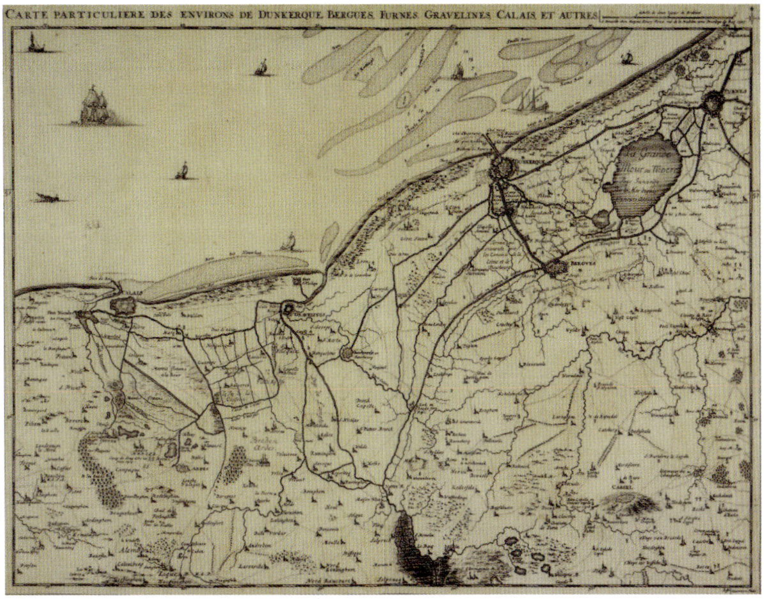


**39**. "Map of the region of Dunkerque, Bergues, Furnes, Gravelines, Calais, and others," Bruxelles, E.-H. Friex, 1707 (2606.29, Sammlung Ryhiner, Stadt- und Univeristätsbibliothek, Bern).

**40**. Philippe Buache, "Partial map of Gravelines [...] for the understanding for the project for a channel, canal and sluices," 1733 (Cartes et Plans, GE SH 18 PF 34 DIV 3 P 2 D, Bibliothèque Nationale de France, Paris).

**41**. "Map of Gravelines with its new canal to the sea," 1741 (Cartes et Plans, GE D-14657, Bibliothèque Nationale de France, Paris).

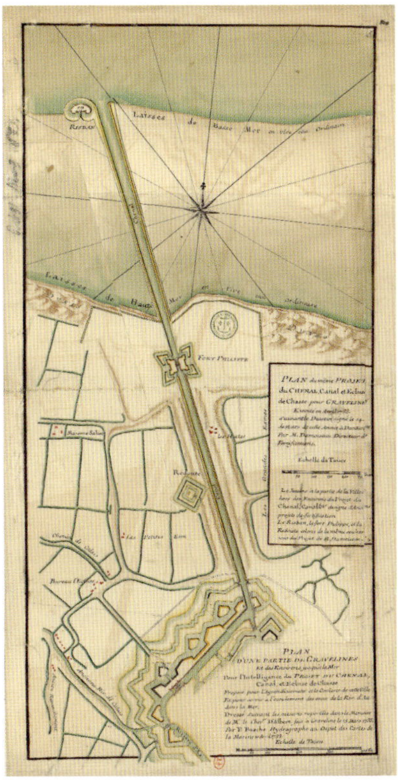


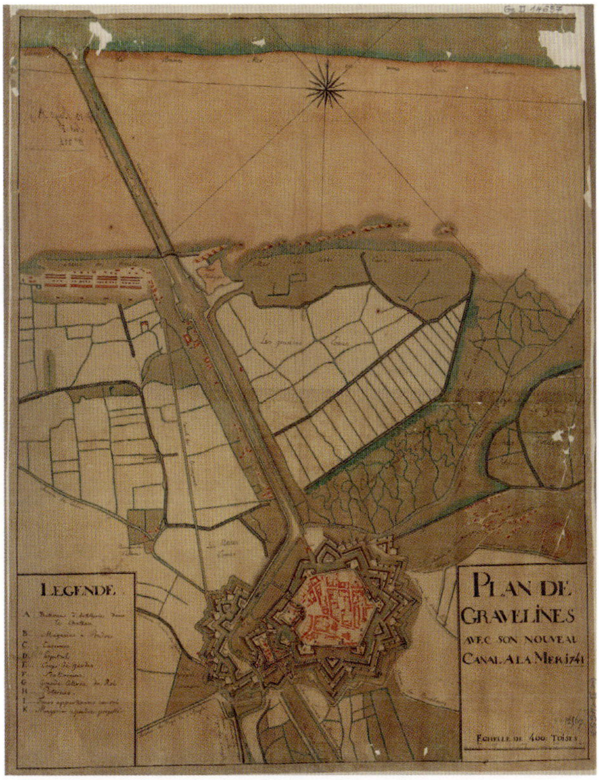

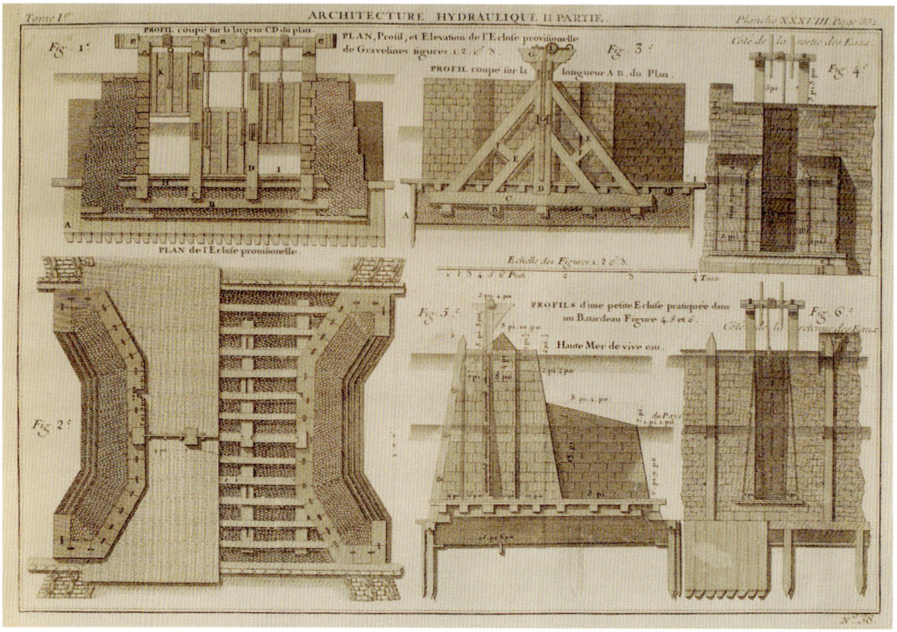
ARCHITECTURE HYDRAULIQUE II PARTIE
PLAN, Profil, et Elevation de l'Ecluse provisionelle de Gravelines figures 1. 2. & 3.
PROFIL coupé sur la largeur CD du plan
PROFIL coupé sur la longueur A B du Plan
PLAN de l'Ecluse provisionelle
Echelle des Figures 1. 2. & 3.
PROFILS d'une petite Ecluse pratiquée dans un Batardeau Figure 4. 5. et 6.
Haute Mer de vive eau
Fig. 1.
Fig. 2.
Fig. 3.
Fig. 4.
Fig. 5.
Fig. 6.

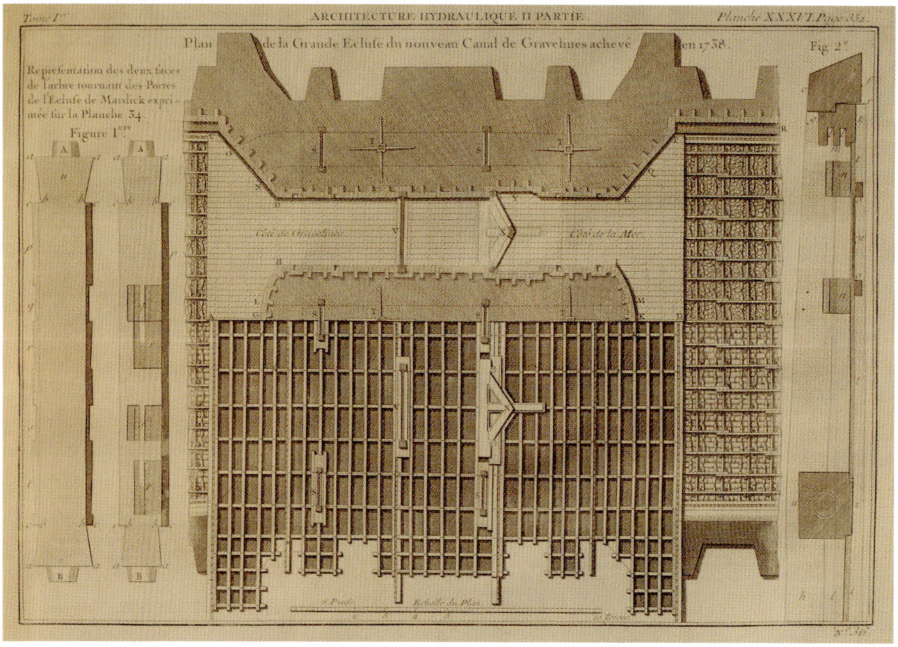
ARCHITECTURE HYDRAULIQUE II PARTIE
Plan de la Grande Ecluse du nouveau Canal de Gravelines achevé en 1738.
Representation des deux faces de l'arbre tournant des Portes de l'Ecluse de Mardick exprimée sur la Planche 34.
Figure I.
Fig. 2.
Côté de Gravelines
Côté de la Mer
Echelle du Plan

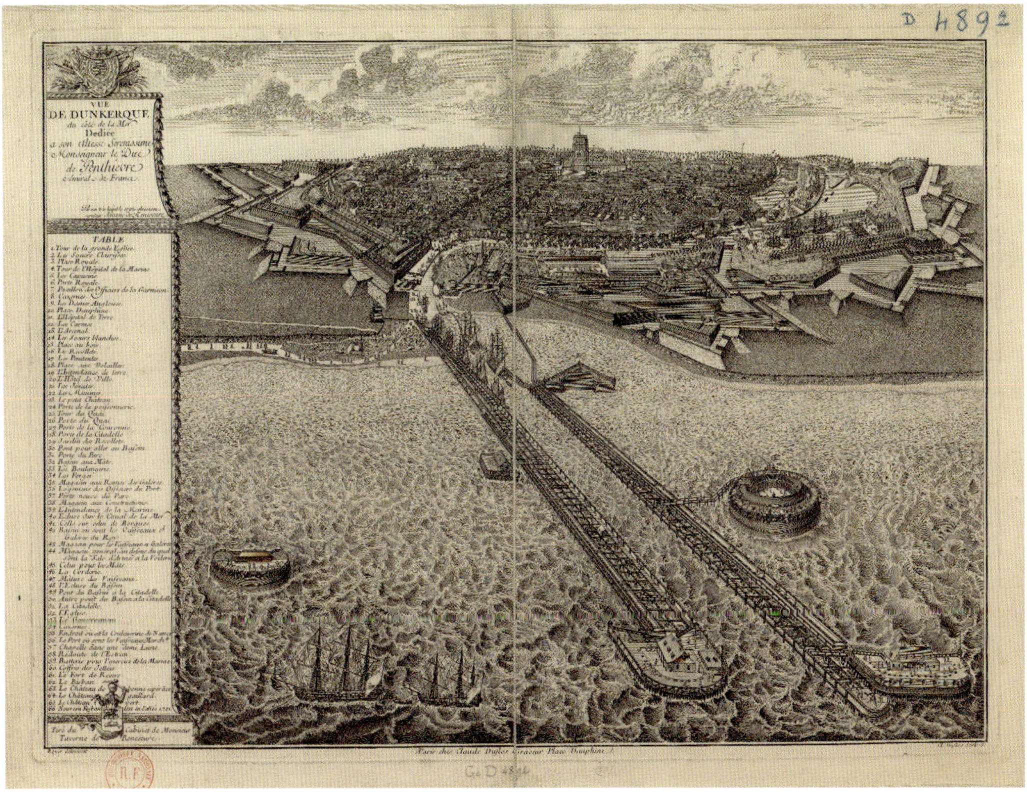


**42-43**. Lock at Gravelines (B.F. de Bélidor, *Architecture Hydraulique. Seconde Partie, qui comprend l'art de diriger les eaux de la mer & des rivieres à l'avantage de la défense des Places, du Commerce & de l'Agricolture*, Paris, Ch.-A. Jombert, vol. I, 1750).

**44**. "View of Dunkerque from the sea," early XVIII century (Cartes et Plans, GE D-4892, Bibliothèque Nationale de France, Paris).

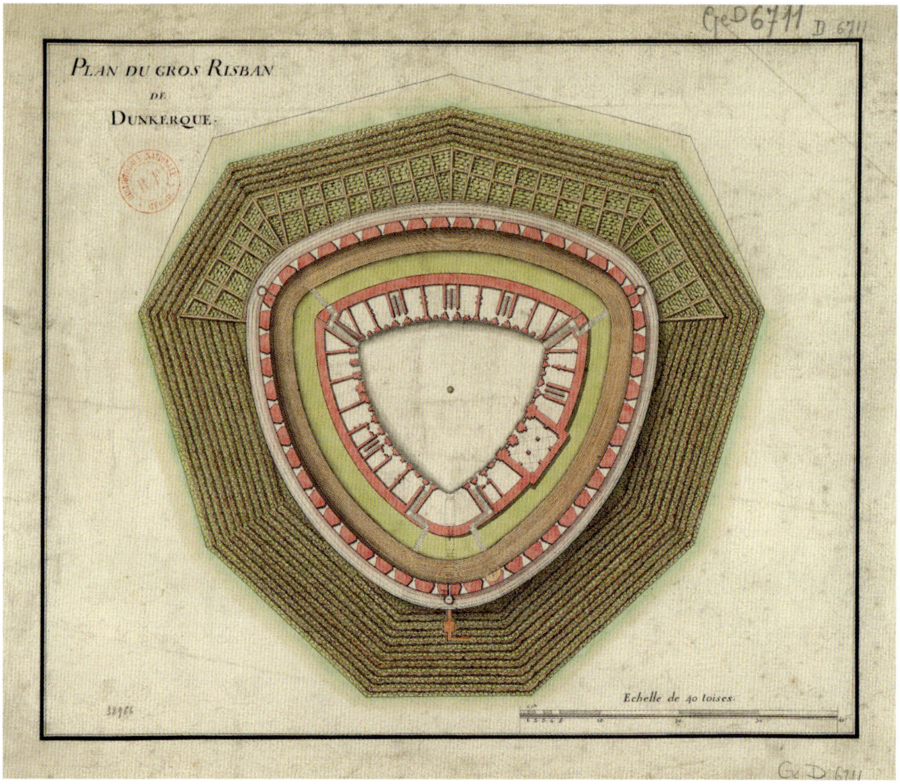
Plan du gros Risban de Dunkerque.
Echelle de 40 toises.

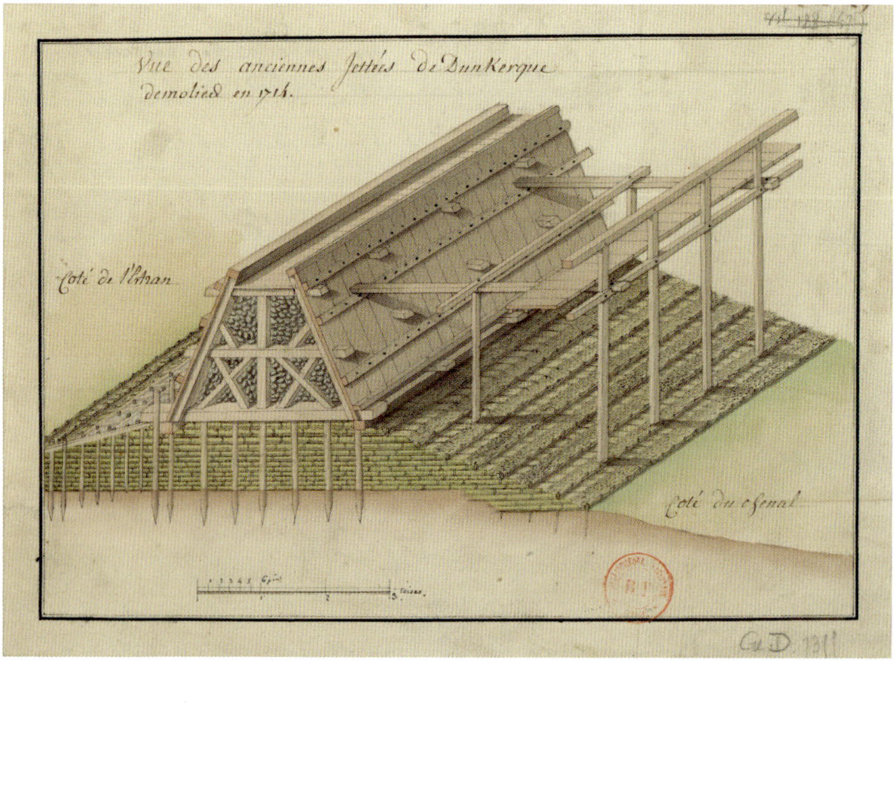
Vue des anciennes Jettées de Dunkerque demolies en 1714.
Coté de l'Estran
Coté du chenal

**45**. "Map of greater Risban," Dunkerque, early XVIII century (Cartes et Plans, GE D-6711, Bibliothèque Nationale de France, Paris).

**46**. "View of the old piers of Dunkerque destroyed in 1714" (Cartes et Plans, GE D-7355, Bibliothèque Nationale de France, Paris).

ground with millstones to reduce it to powder, it is so common in the Netherlands that is has taken the name of Holland cement [*terrasse de Hollande*], it is gray in color, and when it is pure and has not been faked, which is rare, it is excellent for works that sit in water, and it also resists the wear of changing seasons, humidity and drought do not change it, it holds together stones and other materials with force and steadfastness, which is why it is used in France and in the Netherlands for the construction of aquatic works, because of the difficulty of obtaining pozzolana at a fair price [...]."[107]

It should be noted that Bélidor makes an error – taken up by other authors during the course of the 1700s – when he states that the German volcanic rocks are fired in furnaces. Only much later would the Dutch begin to produce an artificial *trass* made with clay taken from the sea and fired as bricks and then crushed into powder for use in hydraulic mortars.[108] One of the most accurate reports on the working the German rocks comes from Smeaton, who during a visit to Holland investigates products for hydraulic mortar. "It is brought to Holland," he writes about the German stone, "in the same state as it is taken out of the earth, and the only art that is there employed in preparing it for use is, to reduce it to a coarse powder, by means of mills for that purpose, which I saw."[109]

The same volcanic product sold at Andernach and Brohl is worked with in different ways by Dutch and German masons to make cement, obtaining utterly different results. The quality of Dutch cement is exemplary, which is a guarantee for export of *trass* to other nations. The recipe for what would become "Holland cement," considered "admirable against water and humidity," calls for putting into a large wooden vessel (five by five feet, height of one foot) two parts of *trass* and one of unslaked quick lime, mixing constantly with additions of water for about 4-5 hours, until the compound has reached a certain consistency, "that is, until it forms a cement that is thick and smooth, which can be kept for several days if it is well covered." "When one takes some for use, the entire mass must first be blended, in which a bit of water has been added if it is too compact, or a bit of trass or quick lime if it is too liquid. Sometimes we add some ground and sieved coal [*charbon de pierre*] into the mixture, sometimes crushed glass, iron filings: but in this case, it is necessary to increase proportionately the amounts of cement and quick lime, as well as the work required for blending."[110]

The cement is used in Holland for masonry works along rivers and canals, for bridges, sluices, cisterns and all kinds of works in contact with water, like fountain basins, tubs and watering troughs for livestock and birds, cellars and other underground parts of buildings.[111] There is also documentation of the use of *trass* cement for roofs, "perfectly impervious to the rain," in the boathouses for the *treckschuyt* or *trekschuit* – the typical Dutch barge used to transport passengers, pulled by horses or men.[112]

[107] Bernard Forest de Bélidor, *La Science des Ingénieurs dans la conduite des travaux de fortification et d'architecture civile. Livre troisième*, La Haye, P. Gosse, 1754, p. 12.
[108] For this type of "trass artificiel" see, Emiland-Marie Gauthey, *Œuvres de M. Gauthey. Traité de la construction des ponts*, vol. II, Paris, Firmin Didot, 1813, p. 280. *Cf.* also, Joseph-Mathieu Sganzin, *Programmes ou résumés des leçons d'un cours de construction*, Paris, V. e Courcier, 1821, p. 51
[109] *Cf.* Smeaton, *A Narrative of the Building* cit., p. 111.
[110] De Pigage, "Mémoire sur quelques pierres" cit.
[111] *Ibid.*
[112] *Cf.* John Carr, *A tour through Holland, along the right and left banks of the Rhine, to the south of Germany, in the summer and autumn of 1806*, Philadelphia, C. and A. Conrand and Co., 1807, p. 263.

One of the leading international experts on volcanology, Nicolas Desmarest (1725-1815), who visits Germany and Holland specifically to learn about the volcanic products used in construction, reports in 1768 on some of the selection criteria used by the Dutch when working with materials imported from Germany.

"The Dutch identify two types of volcanic rubble," he writes, "depending on the quality of the cement that it yields after it has been reduced to powder. The rubble of Andernack, which is softer, easier to crush, and whiter; it provides a cement that, once mixed with lime, gives a less solid mortar, and one that is less capable of stopping all flowing of water. The rubble of Boul [Brohl], on the other hand, is generally more compact, harder to crush, and is a dark gray color; it provides a cement that, once mixed with an equal quantity of lime, forms a very solid mortar, one that is impenetrable to water: thus, one uses this in the preparation of important dikes and underground dwellings, where it is of greatest interest to prevent the infiltration of water. It is used, mixed with the first type, when one does not have to meet the same demands. These materials, used either alone or mixed to different proportions, provide the basis for almost all the mortars used in Holland, and it is general practice to use equal portions of lime and trass."[113]

German building practices differ from those of the Dutch in terms of the fabrication of cement. In Germany the goods are purchased directly at Andernach or Brohl, permitting control over quality and cost only in the case of important shipments and public acquisitions. Private buyers, instead, have to purchase the goods directly along the Rhine, when the boats pass, paying as much as four times the original price. These dealers also adulterate the product with the material of lower quality known as *wild*.[114] The German masons come up short, above all, in the choice of the ingredients to mix with the volcanic product to make cement, and in the mode of the working of the compound. "The Germans do not take enough precaution in the preparation of their cement: they even use rather often slaked lime, instead of quick lime, because the latter is hard to obtain on time. They do not mix the materials together with care, they do not crush enough, which results in a cement that is inferior to that of the Dutch."[115]

### Powders for cement, from the "cendrée de Tournai" to seashell lime

The term cement indicates, throughout the 1700s, a vast range of compounds all marked by hydraulic properties, each of which is made by mixing lime, sand and water with a particular substance that gives the mortar a character similar to what can be made with pozzolana or *trass*. The treatises, starting with *De Architectura* by Vitruvius, offer indications on how to make cement that are still timely and useful in the 1700s, explaining how to mix pulverized brick or roof tiles with lime, which continues to be a widespread practice.

113 *Lettre sur les différentes sortes de pozzolanes et particulièrement sur celles qu'on peut tirer de l'Auvergne, à M. l'Abbé Bossut, de l'Académie des Sciences [...]. Extraite du Journal de Physique, mois de mars 1779*, Paris, Clousier, 1779, p. 15. For a description of the quarries and the German products dating back to 1778 see, Jean-André de Luc, *Lettres Physiques et morales sur l'histoire de la terre et de l'homme*, Paris, Duchesne, 1779, vol. IV, pp. 404-16.

114 De Pigage, "Mémoire sur quelques pierres" cit.

115 *Ibid.* De Pigage recalls having seen at Brihl, at the country residence of the Electeur de Cologne, two leagues away from Cologne, "a small pavilion, called la Maison de l'Escargot, which was located in the middle of a rather deep body of water." "[...] the cellar under the pavilion," he explains, "which served as the kitchen was sunk under the water which was at the level of three vents that provided daylight: this cellar, built with trass cement, formed [...] a vessel of a single piece into which the water was not able to enter." With certain modifications to the positioning of the openings for light, "it would have been an underwater room similar to the one Mr. Chambrez [sic: Chambers] of London (in his book on Chinese gardens) claims to have seen in China, in which one could see fish swimming all around the head

Some advise using only crushed roof tiles, because the clay has been fired at a higher temperature than that of bricks.[116] "It is a compound of lime, of crushed roof tile, and water. [...]. The pavers use it to pave," we read under the heading *Ciment* in the dictionary of Richelet.[117]

The cement compounds indicated in the manuals can be very different, depending on the materials available near the worksite. One of the recipes to make "ordinary concrete" calls for three equal parts of finely sifted powder, of "the hardest possible crushed roofing tile," "chips of ashlar," and "slag or cinders from the foundries where iron is poured," to mix using proven procedures with quick lime and water.[118] The so-called *ciment de fontaine*, or *perpétuel*, is well known, obtained by mixing pulverized fragments of "ceramic pots," "stoneware vases" and of "slag" from the "coal burnt in the forges," to add to "millstone" and lime.[119] In certain zones a powder for cement is obtained from the calcination of river stones, achieving properties comparable to those of Dutch *trass*.

Under the entry *Ciment* of Trévoux in the 1721 edition we read: "Mortar appropriate for bonding stones in buildings. [...]. The best cement in the world is pozzolana. Bitumen is the cement that was used on the walls of Babylon. In France, cement from crushed roof tile or brick is made, and it is mixed with lime. Fountain basins are made with lime and cement. Cement is well employed for works located in water. [...]. Eternal cement can also be made from crushed bricks, glass, coal, well washed course sand, iron flakes that break off under the hammer, with well-crushed quicklime, one that dissolves in wine or in common water."[120]

A generic definition in the leading dictionary of the Italian language, the *Vocabolario degli Accademici della Crusca*, is provided by the entry *Calcestruzzo*, but without listing its ingredients, as is the case in dictionaries in France and England: "mixture of slaked lime with other materials, to increase durability."[121] In subsequent editions of the *Vocabolario* a remark is added on the technique of the pupils of Galileo for the fabrication of *cantoni*: "commonly known as poured cantoni, of cement, or of concrete."[122]

But the best known ingredient for making a hydraulic mortar, when volcanic products like pozzolana and *trass* are not available, is the so-called *cendrée de Tournai*, from the name of the town in Flanders where the powder is produced. On hydraulic worksites in Flanders, the Netherlands and France extensive use is made of this *cendrée de Tournai*. Certain 18th-century experts on cement even assert that the Dutch prefer to use this product for the construction of dikes, sluices and foundations.[123] The stone calcined in furnaces at Tournai is fired with coal, or *charbon de terre*, and the process produces three substances: a compound of "lime & ashes," as well as pure lime and pure ash. The latter substance is used to make cement, after having been treated to slake the remaining particles of lime, and repeatedly beaten at different times,

of the spectator."

116 *Cf.* Pierre Patte, *Mémoires sur les objets les plus importans de l'architecture*, Paris, B. Rozet, 1769, p. 118.

117 Pierre Richelet, *Dictionnaire de la langue françoise,ancienne et moderne*, Lyon, Frères Bruyset, 1728, vol. I, p. 458.

118 Bernard Forest de Bélidor, *Architecture Hydraulique. Seconde Partie, qui comprend l'art de diriger les eaux de la mer & des rivieres à l'avantage de la défense des Places, du Commerce & de l'Agricolture*, Paris, Ch.-A. Jombert, vol. I, 1750, p. 194.

119 Jacques-Raymond Lucote, *Maçonnerie*, in, Denis Diderot, Jean-Baptiste Le Rond d'Alembert, ed., *Encyclopédie ou Dictionnaire raisonné des Sciences, des Arts et des Métiers*, Neufchastel, S. Faulche & Compagnie, 1765, vol. IX, p. 825.

120 *Dictionnaire universel françois et latin [...] Nouvelle Edition*, Paris, Trévoux, 1721, vol. I, p. 1811.

121 *Vocabolario degli Accademici della Crusca*, Venezia, L. Basegio, 1724, vol. I, p. 194.

122 *Vocabolario degli Accademici della Crusca*, Venezia, F. Pitteri, 1741, vol. I, p. 358.

123 Gabriel Jars, Carrey, *Différens procédés pour employer le charbon de terre*, Paris, Imprimerie Royale, 1770, pp. 24 *et seq*. See, André Guillerme, *Bâtir la ville: révolutions industrielles dans les matériaux de construction*, Editions Champ Vallon, 1995.

about ten in all, in keeping with a complex and costly process.[124] Masons in different regions, from Lille to Arras or Douai, to the towns of Flanders, apply different criteria to the preparation of mortar with *cendrée de Tournai*. The *cendrée de Tournai* is also indicated in hydraulic architecture manuals as a fundamental substance for cement.[125]

Certain limes continue to play a fundamental role for the manufacture of local hydraulic cements, as alternatives for various imported products of volcanic or non-volcanic origin. One of the limes still in use in the 1700s, having spread thanks to hydraulic works in Lyon and in the Dauphiné, in particular for the piers of bridges, is obtained by firing stones in furnaces fueled with coal, giving it characteristics similar to those of *cendrée de Tournai*. This type of lime is also produced in furnaces built starting in 1738 near Fontainebleau by Martin and Pluyette, entrepreneurs of the coal mines, and it is widely used for the construction of the sewers of Paris.[126] The range of hydraulic limes is quite vast, and engineers are aware of the types and qualities, relying on these limes to construct foundations in water without the use of pozzolana, *trass* or *cendrée de Tournai*. For example, there is also "a mortar that quickly hardens in water," obtained from the lime of Senonches, in the Généralités d'Alençon.[127] A lime particularly renowned for its hydraulic properties, recommended as an alternative to *cendrée de Tournai*, is made in Boulogne-sur-Mer by firing pebbles. The lime of Boulogne-sur-Mer is also exported to England, and becomes a model, at the start of the 1800s, for research on artificial pozzolana and Roman cement.[128] In Italy, among others, we should mention the hydraulic lime obtained from the stones of the area of Padua, already praised by Palladio for its excellent performance and still widely used on local worksites in the 1700s. "Certain flaky stones are quarried in the hills of Padua, whose lime is excellent in works made out of doors, and in water: because it sets quickly, and is very durable," we read in the chapter of the *Quattro Libri dell'Architettura* dedicated "on lime, and the way to mix it," still a timely work in the 1700s, and not only for Italian builders.[129]

Among the various products for making cements, one widespread type comes from the calcination of seashells. In Europe, in the 1700s, the Dutch continue to be the main producers of this, which is widely utilized for making mortars on a local level. Their claddings also gain renown, which appear "hard as stone" and "very delicate."[130] We cannot rule out the idea that the Dutch

[124] Carrey, *Procédé de la cendrée de Tournai*, in, "Observations sur la Physique, sur l'Histoire naturelle, et sur les Arts," vol. I, November 1771, pp. 370-74. See also, Jars, Carrey, *op. cit*. In the book, Carrey describes the qualities of the cement made "from lime and coal ash" – with properties for strengthening with water, and after several years becoming harder than the stones that it bonds together" (*ibid*., pp. 24 *et seq*.).

[125] Bernard Forest de Bélidor, *La Science des Ingénieurs dans la conduite des travaux de fortification et d'architecture civile*, Paris, C. Jombert, 1729, vol. I, *Livre III*, p. 79.

[126] *Cf*. "Brique, chaux et plastre cuite avec charbon de terre," O.1.1293, Archives Nationales, Paris. On the hydraulic mortar obtained by firing with coal, and on its various applications in France, see, "Copie des certificats de M. De Faylle sur le mortier de composition," 15 October 1741, O.1.1293,176, Archives Nationales, Paris.

[127] "Expériences sur la force du mortier composé de cendrée et de chaux de Tournay comparée avec celle de plusieurs autres mortiers ûsités sans la maçonnerie," Ms 137 fol., Ecole des Ponts et Chaussées, Paris.

[128] *Cf. Rapport fait par M. Mallet, ingénieur en chef des Ponts et Chaussées, à la Société d'Encouragement; au nom d'une commission spéciale, sur le ciment découvert par M. Lacordaire, ingénieur des Ponts et Chaussées, à Pouilly, en Auxois, Departement de Saône-et-Loire*, in, "Journal du génie civil, des sciences et des arts," vol. VI, 1830, (pp. 575-606), pp. 582-84; Gratien Le Père, *Recueil des rapport et observations sur les expériences faites a Cherbourg pour remplacer la pouzzolane dans les constructions hydrauliques [...]. Seconde édition, revue et augeméntée, suivi du Mémoire de M. Guyton-Morveau, Membre de l'Institut national, sur les Cimens et Pouzzolanes*, Paris, Cœury, 1805, p. 2; Pierre-Charles Lesage, *Rapport fait à la Société d'agriculture, de commerce et des arts, de Boulogne-sur-Mer*, Boulogne, Dolet, year X (1801-02).

[129] Andrea Palladio, *I quattro Libri dell'Architettura*, Venezia, D. de' Franceschi, 1570, p. 8.

[130] This is the impression described by James Thornhill during a visit in 1711 to the house of Jacques Meijers, in Rotterdam (*cf*. Geoffrey Beard, *Decorative Plasterwork in Great Britain*, London, Phaidon, 1975, p. 72).

[131] On Dutch imports of seashells from India see, *A compendious system of natural*

learned the practice of calcinating seashell from the methods of natives in their colonies, thanks to trade with America or India[131] – here the lime from seashells is widespread along the coasts, also for stucco. After all, the Dutch obtained a similar product from the calcination of seashells in Manhattan, making use of the great heaps of shells on the beaches of Long Island, after having observed the practices of the natives.[132] During the course of the 1700s we find documentation on the construction of the walls of the homes of Dutch colonists using shell lime along Atlantic Coast.[133] In the colonies of South Africa the Dutch also made use of shell lime.[134]

Certain outstanding personalities of French construction are aware of the way of mixing mortars and stuccos based on seashells, according to Indian methods.[135] In dictionaries of the French language, under the heading *Chaux*, we find information on shell lime, as is written in 1721: "The foundation walls are made of lime & sand. The Siamese have a lime that lasts one and two hundred years, with which they make statues & mausoleums. There is a lot of limestone near Malmo in Sweden. There are two types, one of which is taken from the ground, and the other which they take along rivers or at the ocean shore, the latter type is better. [...]. In the Indies, the lime is usually made with sea shells; that which is made of snail shells is used to whiten the houses, and that from limestone is used for chewing with betel leaves."[136]

In the experiments conducted by chemists on the firing of natural products, seashells are often utilized for calcination trials. A great chemist of the early 1700s, Caspar Neumann (1682-1737), states that "the Dutch have no other lime for building, than calcined sea-shells, particularly those of muscles," and that "when tempered into mortar, [it] proves more durable and more compact than the limes made from mineral stones or earths."[137] Not until the scientific tests of Smeaton do doubts arise regarding the effectiveness of Dutch lime for foundations in the sea. But since they are determined, like the Danes, the French and the English, to find an alternative to Italian pozzolana and to other imported volcanic materials for the fabrication of hydraulic mortar, the Dutch continue in any case, at the end of the century, when artificial substances are being developed everywhere for the making of hydraulic mortars, to make improvements to the process of calcination of seashells.

The Dutch are the leading exporters not only of *trass*, but also of their shell lime. The list of importers of this sort of lime includes English builders,

*history*, in, "The Universal Magazine of Knowledge and Pleasure," vol. XX, January 1757, pp. 13-15.

132 *Cf.* Charles Rau, *Artificial shell-deposits in New Jersey*, in, *The miscellaneous documents of the house of representatives, printed during the second session of the thirty-eighth congress 1864-65*, Washington, Government Printing Office, 1865, (pp. 370-74), p. 373. On the use, since the 1500s, of this type of lime for the homes of colonists, especially the Dutch, see, Marian Card Donnelly, Leland M. Roth, *Architecture in colonial America*, University of Oregon Press, 2003, pp. 24, 41, and Leland M. Roth, *American architecture, a history*, Westview Press, 2001.

133 *Cf.* Helen Wilkinson Reynolds, *Dutch Houses in the Hudson Valley before 1776*, New York, Dover Publications, 1929.

134 *Cf.* Clarence Van Riet Lowe, ed., *The Monuments of South Africa. Including an account of the work of the Commission for the Preservation of natural and Historical Monuments, Relics and Antiques during the period 23rd April, 1935, to 22nd April, 1940*, Pretoria, L. S. Gray & Co., 1949, p. 22.

135 Soufflot is known as an expert on mortar, as demonstrated by the sending to his attention, on the part of Perronet, of a report of the chevalier Jean-Ignace Courtin, written on 17 July 1763, on the composition also of the "lime made of river shells" ("Stuc des Indes Orientales, le 23 juillet 1763 le S. Perronet a envoyé l'original de cette recette à M. Soufflot," Ms 2179, Ecole des Ponts et Chaussées, Paris).

136 *Dictionnaire universel françois et latin [...] Nouvelle Edition*, Paris, Trévoux, 1721, vol. I, p. 1710. A particular cement compound for cladding is mentioned by an anonymous traveler in 1743, along the road from Campengpet to Tian-Tong, where there is a type of tree "from which the Siamese harvest fruit to make cement." "It is mixed with lime," he observes, "and is used to whiten walls, to which it gives a luster that differs little from that of marble" (Joseph de la Porte, *Le voyageur françois, ou La connoissance de l'ancien et du nouveau monde, mis au jour par M. l'Abbé Delaporte*, Paris, Vincent, vol. IV, 1765, p. 343).

137 *The chemical works of Caspar Neumann, M.D.*, London, J. and F. Rivington, vol. II, 1773², p. 390.

who have made use of it at least since the 1600s, especially for claddings. In 1716 John Vanbrugh (1664-1726), an architect and descendent of a Dutch or Flemish family, explains that all the decoration of the Blenheim Palace, at Woodstock, built in 1704 based on his design, is done with facings "after the Dutch manner," specifying that this is a "manner which nobody is got into the practice of here in England; but it is well worth what it costs more than our manner, being as much difference as between fine paper and coarse, besides it being much more lasting."[138]

Confirmation of the English pursuit of various types of stucco comes from news of the invention, in 1744, by Charles Neville, of a cement in the Dutch manner, obtained through calcination of seashells: "lime, stucco plaster, mortar, cement and manure of and from cockle, oyster, and other sea shells, by grinding, pounding, or burning the same."[139] The characteristics of the mortar compound based on calcined shells become known in 1750, thanks to the book *Parentalia, or, Memoirs of the family of the Wrens*, on the life of Christopher Wren, written by his son, where it is explained that the "spandrels" of St. Paul's cathedral are "of found brick invested with stucco of cockle-shell lime, which becomes as hard as Portland stone."[140]

Certain particular recipes for products classified as cement have to do with the composition of the mortar for filling masonry joins immersed in river or sea water. "[...] for it we use," we find written, for example, in one of these recipes, "oil, egg whites, flayed slugs, cement and slag ground and sifted: each in appropriate amounts."[141]

In England the use of shell lime is part of a wider range of cement facings for which the ideal compound has been sought since the start of the 1700s. The *Cyclopædia, or, an Universal Dictionary of Arts and Sciences*, right from its first edition in 1728, represents the most popular dictionary of reference for construction terminology in the world of English building. From this source, we learn that for the English "cement" is a synonym for bitumen and asphalt, substances capable of cementing the blocks of the walls of Babylon, evoked as a famous example of cement construction.[142] The entry *Mortar* offers a description of the legendary cement of the ancients: "The Antients had a kind of mortar so very hard and binding, that, after so long a duration, 'tis next to impossible to separate the parts of some of their buildings; tho' there are some who ascribe that excessive strength to time, and the influences of certain properties in the air, which is found to harden some bodies very surprisingly. De Lorme observes, that the best mortar is that made of Puzzuoli [...]."[143] The authors of the *Cyclopædia* provide certain pithy indications regarding the amounts of the ingredients for mixing "mortar," starting with the technical literature – Vitruvius, Palladio and Félibien are mentioned. Apart from the *Cyclopædia*, in other English dictionaries generic definitions, like that of "mortar" as a compound of "lime, sand, etc. mixed up together with water, for a cement in building," or that of "cement" as a "compound of pitch, brick-dust, plaister of Paris, etc.,"[144] seem to prevail. For more qualified technical information on the various combinations of mortars and also of stuccoes, at the

[138] John Vanbrugh, letter to the Duchess of Marlborough, 19 June 1716, in, Geoffrey Webb, ed., *The complete works of Sir John Vanbrugh. The Letters*, London, Nonesuch Press, vol. IV, 1928.

[139] *The sessional papers printed by order of the House of Lords, or presented by royal command, in the session 1845, vol. XXVIII. Sixth report of the deputy keeper of the publics records*, London, W. Clowes and Sons, 1845, p. 121.

[140] Christopher Wren, *Parentalia, or, Memoirs of the family of the Wrens*, London, T. Osborn, 1750, p. 291.

[141] "Expériences sur la force du mortier composé de cendrée et de chaux de Tournay" cit.

[142] Ephraim Chambers, ed., *Cyclopædia, or, An universal dictionary of arts and sciences*, London, J. and J. Knapton, 1728, vol. I, pp. 176-77.

middle of the century preference still falls to the expanded versions of the *Cyclopædia, A New and Universal Directory of Arts and Sciences*, in the editions of 1751 and 1764.

As in other nations, in England limes with hydraulic properties represent a fundamental complement of cement. The hydraulic limes in use in England, from that of Barrow to that of Aberthaw, are subjected to research and analysis in the 1750s, in a period that sees some of the most important works of English hydraulic construction.

## Masonry prototypes for the levees of Venice, from Padua lime to Vicenza pozzolana

In the early decades of the 1700s pozzolana and *trass* were not used for the construction of two of the most outstanding European hydraulic works: the levees of the Venice Lagoon, and of the polders in the Netherlands. In both cases the construction practices of the levees are still traditional, though subject to refinements, based on the use of mud, wooden posts, rocks and cultivation of vegetation to consolidate the structures. The first important modification becomes necessary following the spread in the northern seas and the Mediterranean of the so-called shipworms – especially the *teredinidae*, bivalve mollusks with wormlike bodies and small shells – that devour any kind of wood immersed in water, from the hulls of ships to wooden piles and posts. To avoid costly periodic replacement of the posts, greater use is made of stones that are imported, in the case of Holland, from northern countries like Norway, and in the case of the Republic of Venice from the Veneto inland and Istria. In this process of adding stone to the levees, the change set in motion by the Venetian engineers, known as *proti*, in charge of the shoreline, leads in the first years of the century to the invention of a powerful masonry structure that gradually – through the contributions of technicians, architects and mathematicians – forms a solid masonry levee, which is perfected during the course of the 1730s. The relevant characteristic that eventually brings this levee international renown is precisely the nature of the binder of the stones, a pozzolana-based compound: Roman cement. The phases of development of the pozzolana levee point to the discovery of the role of hydraulic mortars in the construction of sea barriers and mark a technical breakthrough, which spreads across Europe, in the use of loose stones, an advance that will have major international repercussions.

After a range of projects prepared during the course of the 1600s and following the proposal of Guglielmini, around 1700, of using large stones, the levee structure considered most feasible in the first years of the 1700s to defend the shores of Venice is the one conceived by the *proto* Carlo Guberni and implemented at the Pellestrina waterfront in 1701. Guberni's levee is composed of loose stones placed on an earthen escarpment and held in place by four parallel rows of poles, or *tolpi*, known as a *graticolato*, or latticing – the poles get gradually taller towards the top of the levee.[145] This framework of poles becomes the binding force of the stones when suitable mortars are lacking, while also functioning as a breakwater. Also applied by the *proto* Domenico Margutti

[143] *Ibid.*, vol. II, p. 582.
[144] Entries *Mortar* and *Cement*, in, Nathan Bailey, George Gordon, Philip Miller, Thomas Lediard, ed., *Dictionarium Britannicum: or, A more compleat universal etymological English Dictionary*, London, T. Cox, 1736, p.n.n.
[145] See for the printed drawing, *Dissegno Geografico, & forma dell'opperatione Raccordata da me Carlo Guberni per la riparatione de Publici Lidi, & pratticata l'anno 1701 sul Lido di Pelestrina fra le Palade num 45:46.*

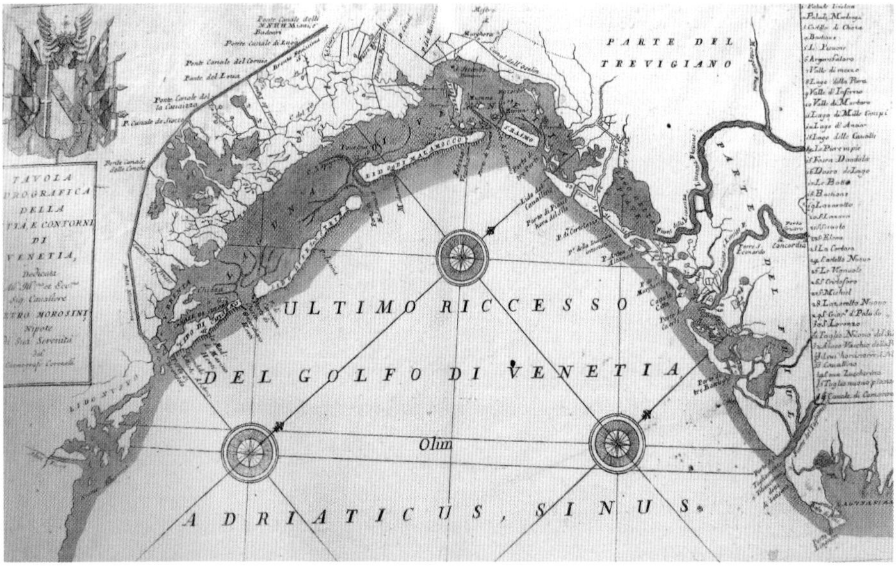


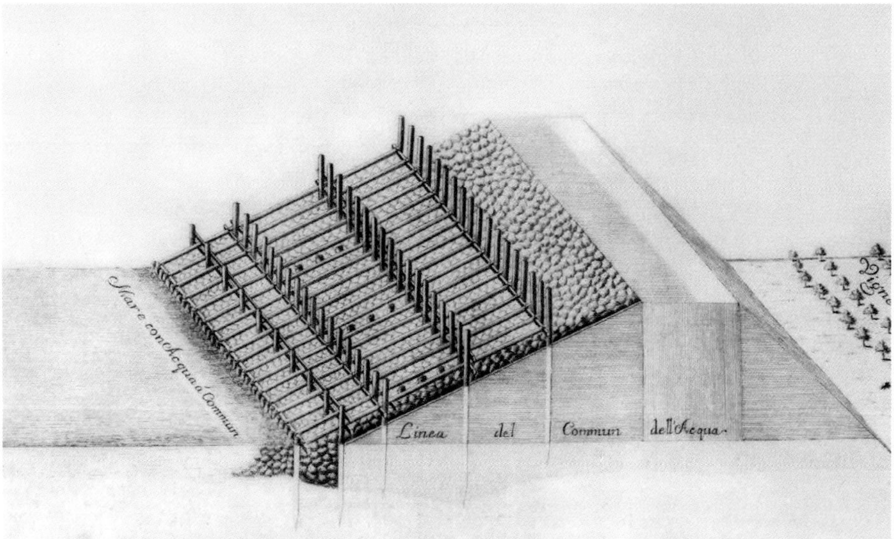


**47**. Vincenzo Maria Coronelli, "Hydrographic table of the city and vicinity of Venice" (V.M. Coronelli, *Corso geografico universale, o sia la terra divisa nelle sue parti e subdistinta ne' suoi gran regni*, Venezia, 1692).

**48**. Carlo Guberni, drawing of the levee at the shoreline of Pellestrina, Venice (*Dissegno Geografico, & forma dell'opperatione Raccordata da me Carlo Guberni per la riparatione de Publici Lidi, & pratticata l'anno 1701 sul Lido di Pelestrina fra le Palade num 45:46*).

**49-50**. Andrea Tirali, masonry levee design, Venice, 1715. The pencil corrections on the section drawing are probably made by Angelo Minorelli (Savi ed Esecutori alle Acque, 74, Archivio di Stato di Venezia).

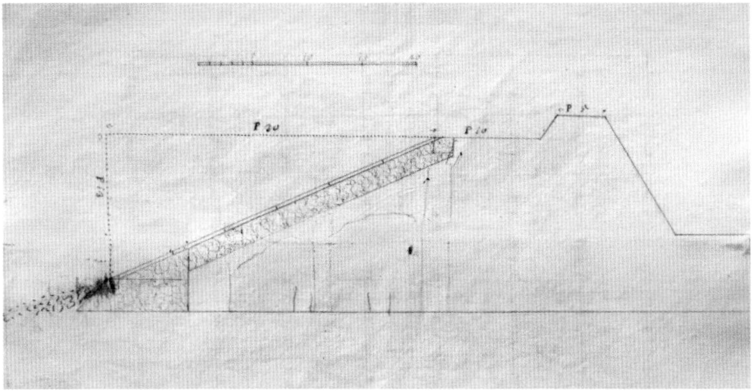

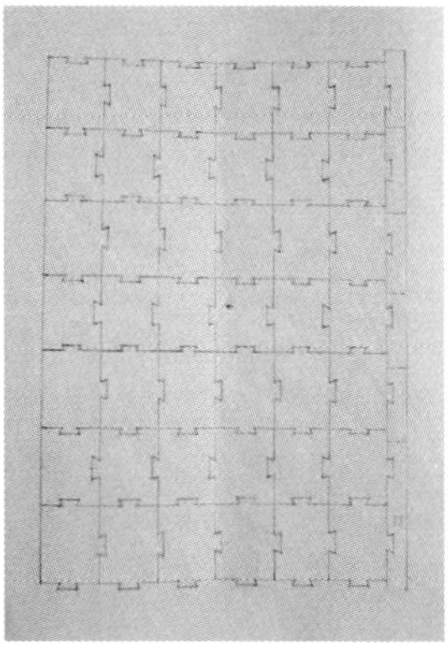

(1659-1721), Guberni's latticing proves to be effective, but requires costly on-going maintenance (the poles have to be replaced every five years or so). And the levee is still subject to breakage during heavy storms.[146]

The first important transformation of the waterfront protection systems used along the shores of the Venice Lagoon, in which the drastic reduction of the use of *tolpi* for the binding of stones begins, relying instead on mortar or on the size of the stones themselves, dates back to the middle of the 1710s. To respond to the problem of a shortage of poles and deforestation, while seeking a waterfront defense structure that would be more durable, in 1715 Domenico Gritti and Andrea Tirali, or Tiralli (1657 or 1660 circa -1737) – mason and stonecutter, student of Andrea Musalo, and *proto* of the shoreline as of 1700[147] – propose two different solutions for masonry levees, also using stones embedded in mortar.

According to Gritti, the idea is to build a levee with stone "columns" protruding from the ground to break the force of the waves – "stones standing in place of poles above the surface of the levee," as the proposal of Angelo Minorelli, water engineer for rivers, was correctly interpreted.[148] Instead, Tirali suggests spreading, over the surface of the earthen levee, a layer of stones of the size of the loose ones,[149] set in place with lime mortar, and protected against the water by means of a covering of thin, square stone slabs. The thickness of the stone and lime mortar masonry increases at the base of the levee, where the masonry forms a sort of socle to replace the row of poles, or *palisella*, to combat the erosion of the foundation caused by the movement of the waves. To prevent the covering from being dislodged, the stone slabs are held together by means of dovetail joints – a detail Tirali may have seen in the steps of staircases in Venice. "[…] stones covered with large slabs of exposed stone that are so tightly joined together that they cannot be rent apart by the sea" is Tirali's definition.[150]

[146] On the various projects for improvement of the levees see also the proposal of Lorenzo Boschetti that consists, among other things, of a sizeable increase in the bulk of the stones: "That the stones should be of greater size than usual and of such obliged connection and union that they cannot be easily carried away by the sea" (Lorenzo Boschetti, *Dimostrazione scenografica et ortografica d'un nuovo riparo per li publici lidi veneti, Venezia*, G. Albrizzi, 1707). *Cf.* Maria Francesca Tiepolo, ed., *Laguna, lidi, fiumi. Cinque secoli di gestione delle acque. Mostra Documentaria, 10 giugno-2 ottobre 1983*, Venezia, Archivio di Stato di Venezia, 1983, p. 129.

[147] *Cf.* Andrea Tirali, report, 20 July 1733, Savi ed Esecutori alle Acque, 75, Archivio di Stato di Venezia. For the proposals on masonry levees formulated during the 1600s see that of Sebastiano Roccatagliata and Giuseppe Benoni in 1659 (*cf.* Susanna Grillo, *Venezia, le difese a mare. Profilo architettonico delle opere di difesa idraulica nei litorali di Venezia*, Venezia, Arsenale Editrice, 1989, p. 27).

[148] Angelo Minorelli, report, 18 May 1716, Savi ed Esecutori alle Acque, 74, Archivio di Stato di Venezia.

[149] The measurement can also be deduced from Tirali's drawing.

[150] Andrea Tirali, report, 23 August 1715, Savi ed Esecutori alle Acque, 74, Archivio di Stato di Venezia.

In December 1715 a commission of technicians reporting to the Magistrato alle Acque, composed of Margutti, Minorelli, Andrea Cornello, Domenico Piccoli and Lorenzo Boschetti, expresses its views on the two proposals. While it is suggested to give a pyramidal form to Gritti's columns – "giving them the figure almost of pyramids," it is written[151] – substantial changes are also indicated for the structure beneath the "pavement" of Tirali, to prevent yielding of the foundation, replacing the earthen levee with reinforcement walls and brick vaults – "vaults of terracotta bordered with stones, i.e. filled at the sides of the vaults themselves with stones in mortar."[152] The definition of "stones in mortar" offers a glimpse of a masonry-like concrete, such as the one proposed by Minorelli shortly after the opinion expressed by the commission of which he was a member.

It is probable that the observations of the commission also caused Tirali to reconsider his orientation, since in March 1716 he proposes a levee conceived as an improvement and apotheosis of Guberni's system, though without altering the materials – clusters of poles, loose stones and earth. The only variation that would be implemented in the stone levees is a less accentuated slope of the surface.[153] Margutti and Minorelli are ordered to evaluate the Tirali's new proposal and the works under construction (Margutti is assisted by Cornello, Piccoli and Boschetti). Margutti states that he favors Guberni's method and, in the case of a levee covered with squared stones, that of Gritti.[154]

Minorelli's observations are fundamental for the rise of poured concrete construction. Minorelli criticizes not just the recent project of Tirali but also, more generally, any system based on loose stones held in place by latticing, because they require constant maintenance, and also have to be built under the direct supervision of the water engineers according to the instructions of Guberni. In his reasoning against the idea of loose stones, Minorelli tellingly makes reference to the advice of Guglielmini on the size of the stones to utilize in the levees to create a strong structure, and then gets to the issue that interests him most: the possibility of transforming loose stones into a strong conglomerate thanks to lime mortar, without having to resort to costly, large Istria stones, while eliminating the need for poles. The solution of stones held in place by a lime mortar binder already appears in Tirali's proposal in 1715; but Minorelli significantly transforms the system, no longer specifying loose stones, as indicated by Tirali, but proposing true aggregates, of the size of gravel and pebbles, to form a conglomerate.

"These small stones could be employed just as well as larger ones," Minorelli writes, "if they are mounded and held in place with some artful method, and united, and joined with some mixture of lime mortar, and other ingredients, making them into a large, heavy figure that could be used in the same way and with the same effect as large stones."[155]

After the proposals of Gritti and Tirali, and following the levee breaches at Pellestrina and Malamocco, in June 1716 other proposals for masonry levees are advanced by Minorelli and Piccoli, the latter in collaboration with Cornello and Boschetti. Minorelli returns to the solution previously proposed by

[151] Domenico Margutti, Andrea Tirali, Angelo Minorelli, Andrea Cornello, Domenico Piccoli, Lorenzo Boschetti, report, 10 December 1715, *ibid.*
[152] *Ibid.*
[153] Andrea Tirali, report, 26 March 1716, *ibid.* See what Minarelli writes on the non-innovative character of Tirali's proposal, a method practiced since 1670 as he states (Angelo Minorelli, report, 27 March 1716, *ibid.*).
[154] Domenico Margutti, with Andrea Cornello, Domenico Piccoli and Lorenzo Boschetti, 12 May 1716, *ibid.*
[155] Angelo Minorelli, report, 18 May 1716, *ibid.*

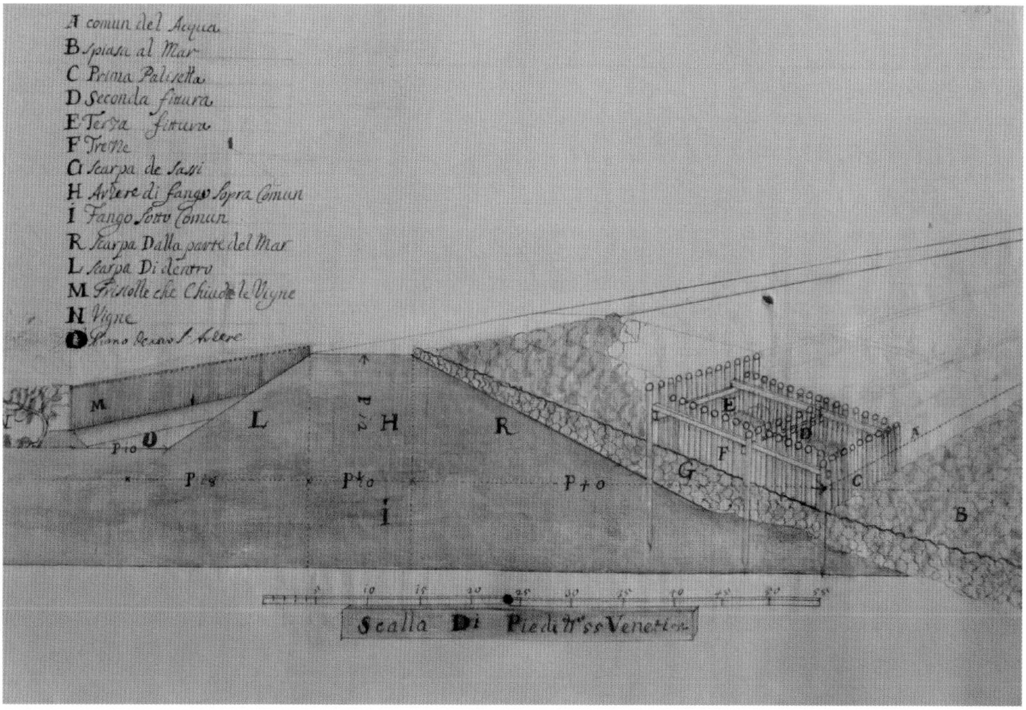

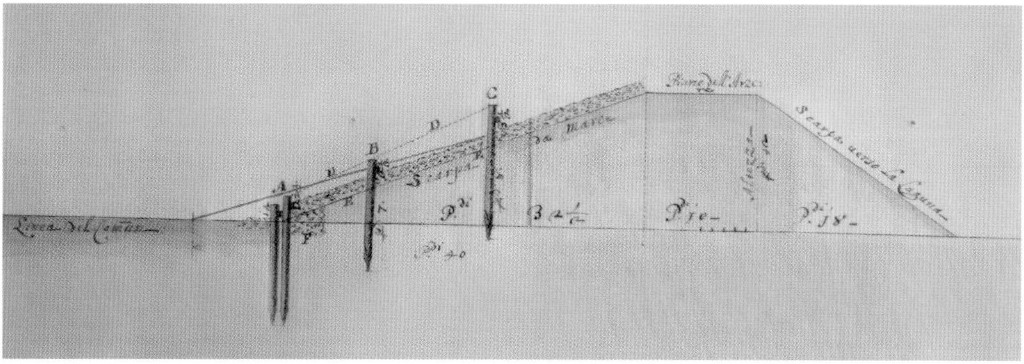

**51**. Andrea Tirali, project for a levee with piles and loose stones, Venice, 1717 (Savi ed Esecutori alle Acque, 74, Archivio di Stato di Venezia).

**52**. Angelo Minorelli, project for a levee in masonry and piles, Venice, 1716 (Savi ed Esecutori alle Acque, 74, Archivio di Stato di Venezia).

Tirali of masonry made with stones and lime mortar; but he eliminates the cladding in stone slabs, while recouping from the traditional system its rows of poles at the levee toe and along the embankment – the *palisella* and the *graticola*. The foundation of the embankment can return to the usual pressed mud, rather than the complex system of masonry vaults suggested during the course of the assessment of Tirali's project. The fact that the masonry of the levee proposed by Minorelli is stripped of the protective stone cladding envisioned by Tirali explains why Minorelli concentrates on the technical details of the composition of the lime mortar of what he calls the *massiccio*, or massif, (in his drawing the aggregates are shown with smaller sizes). The lime recommended by Minorelli to form the *massiccio*, comes from stones quarried near Padua, already indicated by Palladio. In the complex phases of development of the project for the masonry levees of the shores of Venice the idea of using Padua lime has already appeared, as Minorelli himself recalls in the report to explain the model of his levee, when after having underlined the economic advantages of the lime mortar *massiccio*, he states that it is a method "suggested by some in other times, but not heeded, though it can still be considered worthy and even extremely perfect."

"[...] it would be the binding of these stones with a mixture of good Padua lime mortar," he explains on the topic of the massif, "a soft mixture laden with good river sand; these stones are crushed and pressed together so that the embankment is covered with a massif for its full height, at least of 5 ounces, except from a height of 3 feet upward on the embankment from the main row of piles, which should be a massiccio entirely of stones at a height of 3 ½ feet, like the letter F in profile. This work should be done in the spring or the fall, in good weather, not in the winter, and done quickly, particularly when close to the sea, to avoid the spray until it has set; otherwise it could be covered with seaweed."[156]

The masonry structure with a base of Padua lime would consolidate and strengthen, as Minorelli writes, "the rows of poles for the containment of the stones with this mortar in the openings between the poles," and the stones, "united and walled up will remain stronger, without the appearing of stones that move, are lighter or looser, as happens at present."

The fact that he has set aside the large Istria stones that are nevertheless emerging, in the statements of Guglielmini and the proposals of Piccoli, as the ideal solution to stand up to the destructive force of the waves, makes Minorelli's project a particularly clear indication of the orientation of certain water engineers towards a conglomerate-type masonry – *maniera riempiuta* or *a cassa* ("coffer-work") according to Palladio's definitions[157] – in which the job of providing strength is assigned to the mortar, not the stones, in keeping with the constructive logic of Roman cement.

The proposal made by Piccoli, Cornello and Boschetti moves in the direction of an earthen levee covered with a layer of loose Lispida or Istria stones, held in place by means of rows of poles and latticing; but the system is revised to take into account the suggestion, which emerges around 1700, of increasing the weight of the individual stones to obtain better resistance to the

[156] Angelo Minorelli, report, 3 June 1716, *ibid*.
[157] Palladio, *op. cit.*, p. 13. For the english translation, by Isaac Ware, see *The four books of Andrea Palladio's Architecture*, London, I. Ware, 1738, p. 9.
[158] Domenico Piccoli, Andrea Cornello, Lorenzo Boschetti, report, 4 June 1716, Savi ed Esecutori alle Acque, 74, Archivio di Stato di Venezia.
[159] According to Piccoli, on 16 July 1716 it is decided to make a prototype of the levee at the Lido di San Pietro, "towards Pelestrina" (Domenico Piccoli, report, 4 June 1720, Savi ed Esecutori alle Acque, 27, Archivio di Stato di Venezia).
[160] "Proposte del padre Coronelli di ripparare i veneti lidi," March 1716, Inquisitori

disruptive force of the sea. Thus in the levee of Piccoli, Cornello and Boschetti, in the part of the embankment towards the sea, large squared Istria stones appear to form a sort of socle – "Istria marble regularly cut into the suitable form."[158] Though the conclusion reached on 16 July 1716 recommends experimentation with Piccoli's levee, work on the trial does not begin.[159]

So two opposing methods of resistance of the levee against the force of the waves of the sea emerge in the proposals of Minorelli and of Piccoli, Cornello and Boschetti, respectively: relying on the strength of the mortar, or on the weight of the stones.

Precisely as the water engineers are identifying a form of levee protected by masonry of large Istria stones and stone and lime mortar, the Franciscan monk Vincenzo Maria Coronelli (1670-1718), cosmographer of the Republic of Venice, writes to the Savi ed Esecutori alle Acque, on March 1716, to explain his views on the defense of the Lagoon – "Proposals of Father Coronelli on the protections of the Venetian shores."[160] Coronelli imagines earthen levees with different structures in the materials that cover the embankments: wooden planks attached to poles (a variation of the stone paving of Tirali); slabs of marble held together by means of metal anchors instead of dovetail joints, considered as a fragile solution (confirmation of Coronelli's criticism of Tirali); and, finally, iron sheets. In all three cases Coronelli calls for sealing of the joints and protection of the metals with a "mixture" of his own invention that is water-resistant and composed of "ammonium salt," "verdigris," "common salt" and "the strongest vinegar."[161] The stones are to be assembled as masonry without lime mortar. "The protection I propose," he writes, "is that instead of poles that do not last more than five years, the levee be made of blocks of Istria marble, left raw and rugged, because *without putting them in lime mortar* they can sit better, and remain one atop the next."[162] Certain solutions proposed by Coronelli to protect the wood of the levees are based on those in use since the second half of the 1600s to keep shipworms from damaging ships (since the 1600s hulls were protected with tinplate or lead, with a dense covering of nails or with cow hair, ashes, lime, foam, or as in the case of the Portuguese, through burning of the surface to create an impervious crust).[163] Coronelli also proposes groynes with an iron structure, again protected against rust thanks to his "mixture" (without this protection he calculates that they can last at least fifteen years, like the metal parts of ship rudders immersed in water). But the process of development of the strongest possible Venetian levee continues along the path of the experiments initiated by the water engineers, without being distracted by the proposals of Coronelli.

The decisive evolution of the waterfront defense structure takes place in the period from about 1719 to 1721, when the masonry levee system proposed starting in 1715 is finally translated into a prototype built at sea. The prototype makes it possible to take important corrective measures, leading to the final phase of the design of the definitive structure. The orientation of the prototype towards a structure announcing the final phase is still based on the projects proposed by Tirali, the development of which are still being discussed in partular by Minorelli, Piccoli and Boschetti, along with Zendrini.

di Stato, 902, Archivio di Stato di Venezia, in, Tiepolo, *op. cit.*, p. 129. *Cf.* Grillo, *op. cit.*, pp. 44-51.

[161] Cit. in, Tiepolo, *op. cit.*, p. 129.

[162] *Venezia tra Sette e Novecento. Brevi interrelazioni*, in, *Scienze e tecniche agrarie nel veneto dell'Ottocento. Atti del secondo seminario di storia delle scienze e delle tecniche nell'Ottocento veneto, Venezia, 14 e 15 dicembre 1990*, Venezia, Istituto Veneto di Scienze, Lettere ed Arti, 1992, (pp. 209-43), p. 214.

[163] Pierre Massuet, *Recherches interessantes sur l'origine, la formation, le developement, la structure* [...], Amsterdam, F. Changuion, 1733, pp. 62-63.

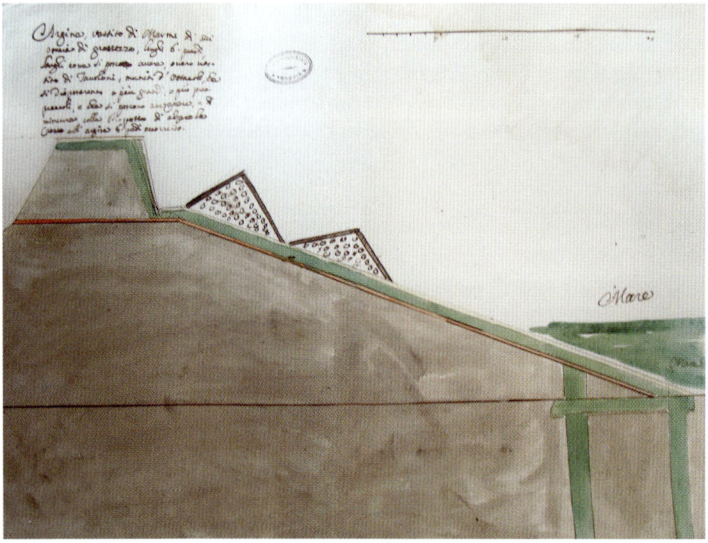

**53-55**. Vincenzo Maria Coronelli, levee designs, Venice, 1716 (Inquisitori di Stato, 902, Archivio di Stato di Venezia).

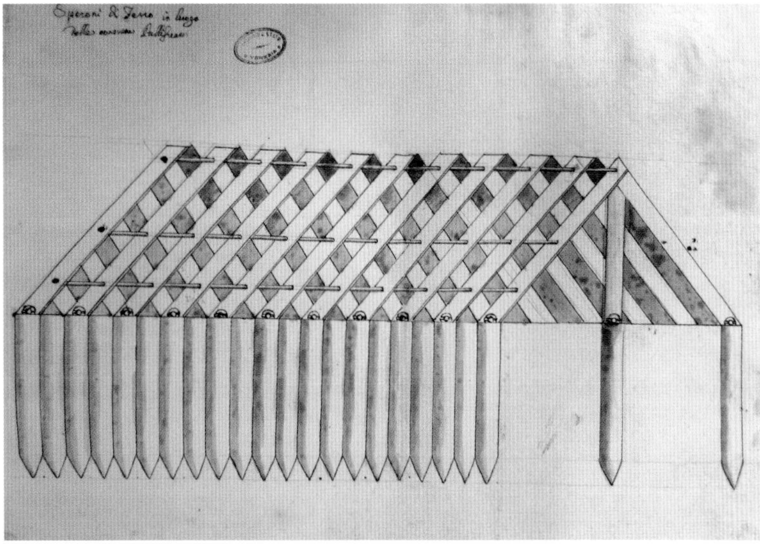

In October 1719, Tirali works on three variations based on the insertion – to protect the part of the earthen embankment facing the sea – of colossal Istria stones to be arranged in steps, resting on a masonry base (with poles at the levee toe). The difference between the variations consists in the degree of extension of the stone structure with respect to the earthen embankment behind it, and in the masonry to support the surface on which to place the stepped Istria stones. For the support masonry, Tirali calls for three solutions: the use of a poured layer of only pebbles and lime mortar; the arrangement of large squared Istria stones concentrated at the foot of the levee toe, to create a very strong socle to resist the force of the waves – the rest of the support masonry of the surface is in stone and lime mortar; finally, there is masonry entirely made with large blocks of Istria stone. The three types of masonry suggest an implied hierarchy of construction systems that, according to Tirali, ranges from large Istria stones, considered the structure of greatest strength, to the pouring of stones with lime mortar, which represents a fallback solution for economic reasons. A fourth levee variation, defined by Tirali as "embankments, built for defense of the coast," involves implementation of a system derived from that of Guberni.[164]

The set of solutions does not only attempt to identify the structure of greatest strength. Probably Tirali intends to suggest a range of works to implement based on exposure to the force of the sea, to be varied along the coastline at different portions of the levees.

The Magistrato alle Acque is determined to continue developing the idea of large stone levees and orders the construction, at Pellestrina, on the site known as Caroman, of a prototype of Tirali's stepped system, "with large slabs of hewn Istria stone."[165] After inspection of the worksite in April 1720, to view the experimental efforts following study of the three variations of Tirali,[166]

[164] See drawings "A," "B," "C" and "D" of Tirali and related expenses in, Savi ed Esecutori alle Acque, 27, Archivio di Stato di Venezia.

[165] Domenico Piccoli, report, 4 June 1720, *ibid.*

[166] *Cf.* Andrea Tirali, letter, 1 June 1720, *ibid.*

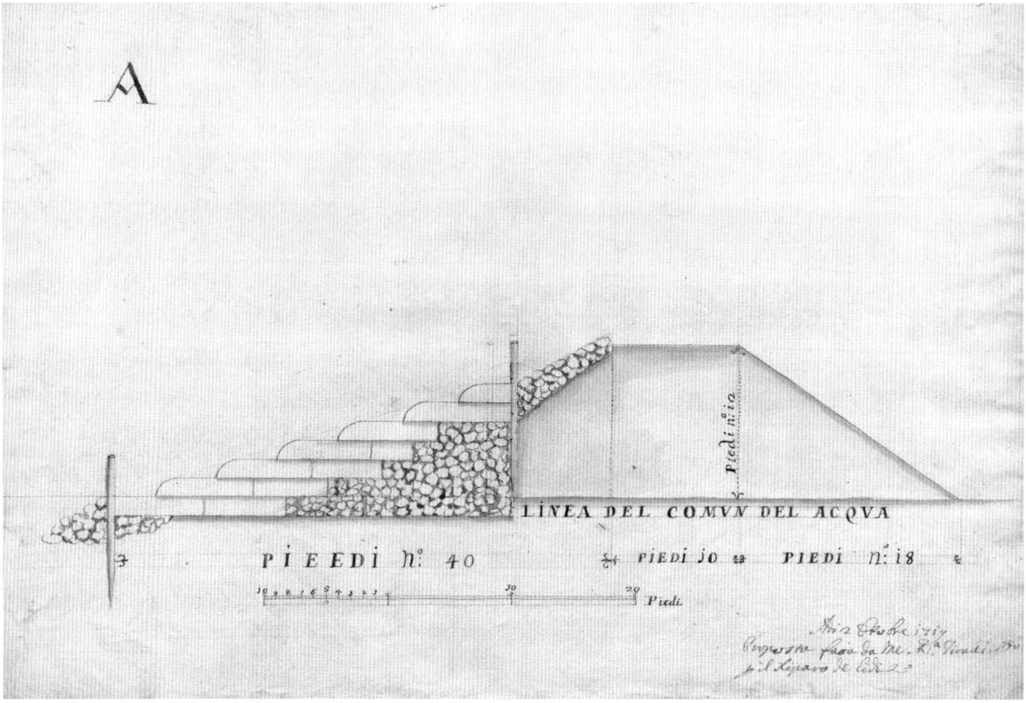


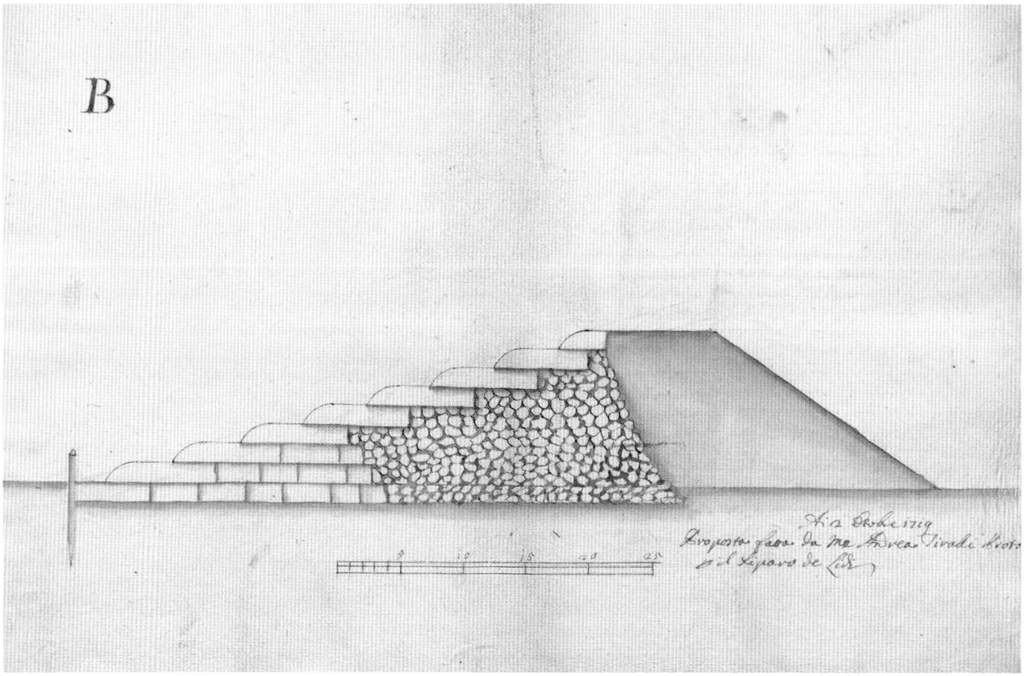


**56-59**. Andrea Tirali, masonry levee designs, Venice, 1719 (Savi ed Esecutori alle Acque, 27, Archivio di Stato di Venezia).

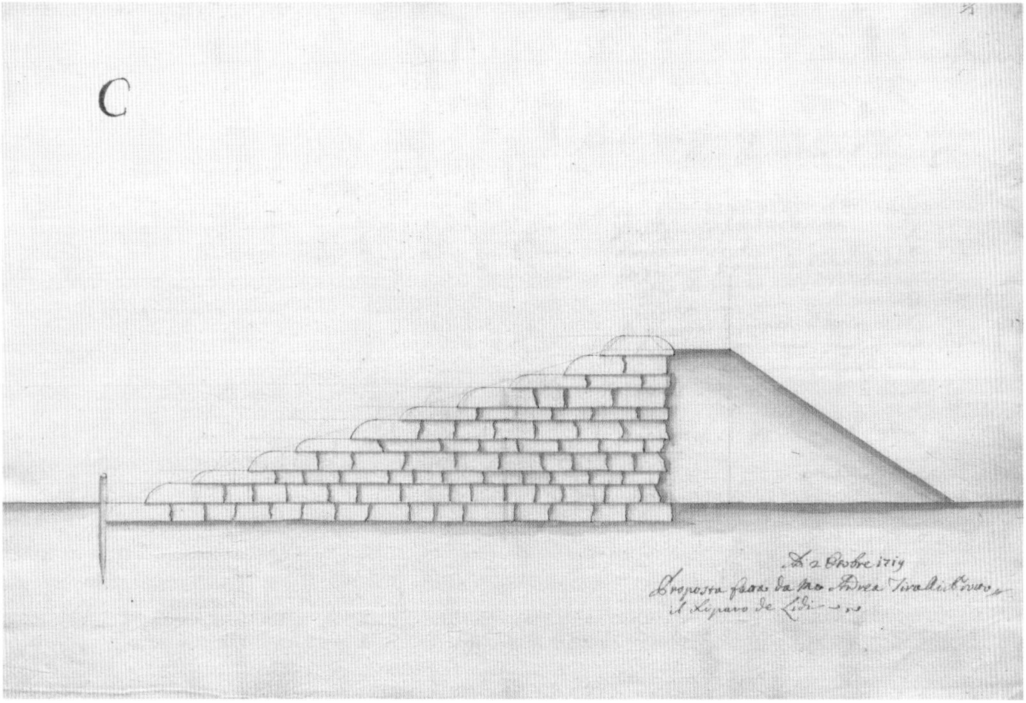
C

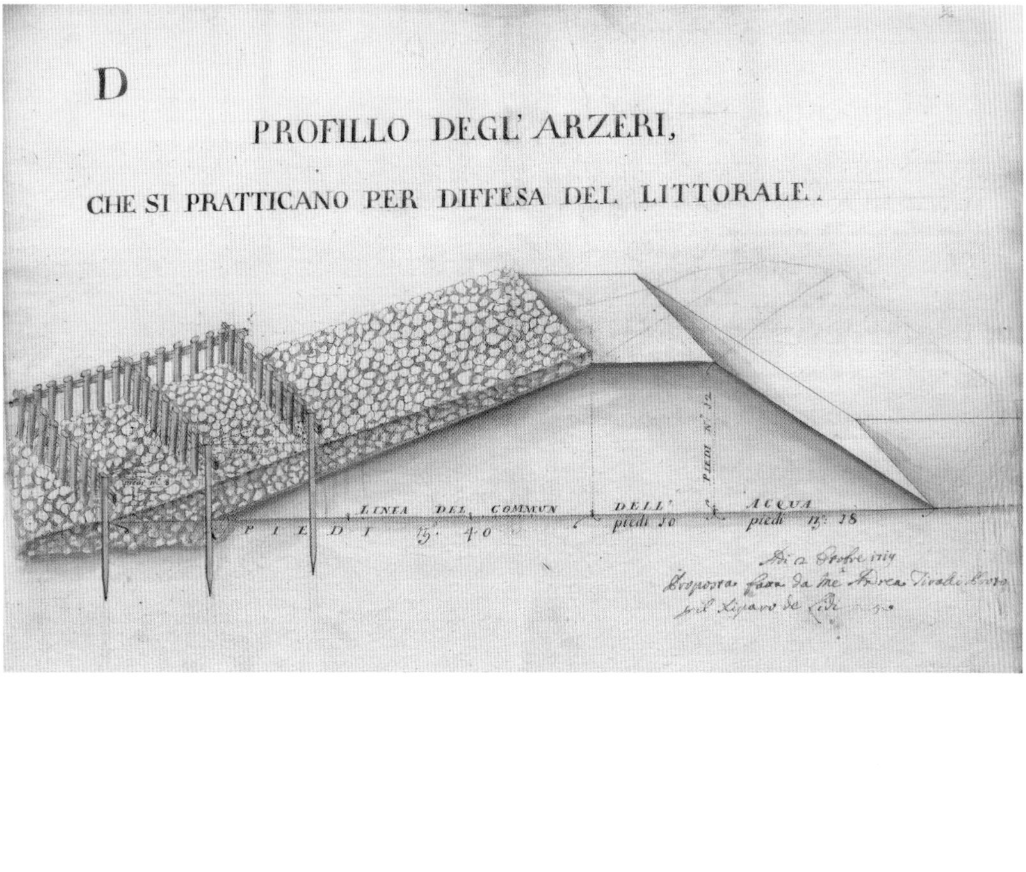
D
PROFILLO DEGL' ARZERI,
CHE SI PRATTICANO PER DIFFESA DEL LITTORALE.
LINEA DEL COMMUN DELL' ACQUA
PIEDI 15. 40
piedi 30
piedi 15: 18
PIEDI N° 12

**60**. Domenico Piccoli and Lorenzo Boschetti, masonry levee design, Venice, 1720 (Savi ed Esecutori alle Acque, 27, Archivio di Stato di Venezia).

Piccoli and Minorelli report back to the Magistrato with an initial opinion on 4 June. Piccoli recommends, for the segments exposed to the greatest force of the sea, a levee of large stones, also in the masonry at the "foot," but positioned along the slope of the embankment, without the steps proposed by Tirali – for the masonry supporting the surface, he indicates that of Tirali, which he calls "massif of boulders."[167] In this circumstance Piccoli lays claim to the invention, in 1716, of a "mixed construction" in stone slabs, pebbles and wood, and complains that it has not been put into practice.[168] Minorelli, who uses the definition of "stone protections," sees Tirali's proposals as a detailed system for the different portions of the shoreline, commending its logic of construction and its reliance, for the strength of the levee, on the "natural resistance of the material," on its "absolute weight" and the "good arrangement and concatenation of the pieces when arranged in steps," though he also corrects certain details of the variations – among other things, he suggests modifying the slope of the steps.[169]

Given the differing conclusions of Piccoli and Minorelli, on 8 June the Magistrato sets up a commission to find the best construction system for the stone levee, starting with the work in progress and the proposals of Tirali. The commission is composed of Minorelli and Piccoli, but also Boschetti and the outstanding mathematician and hydraulics expert Zendrini.[170] What happens next represents a crucial step in the definition of a construction system capable of holding the stones together. It is precisely at this phase that the question of lime becomes fundamental, obtained by the firing of stones to give them hydraulic properties, or the lime from a local volcanic product, a pozzolana that can give the mortar the same characteristics.

Piccoli and Minorelli propose alternative solutions to those of Tirali. Not convinced about the system of steps, Piccoli, supported by Boschetti, suggests

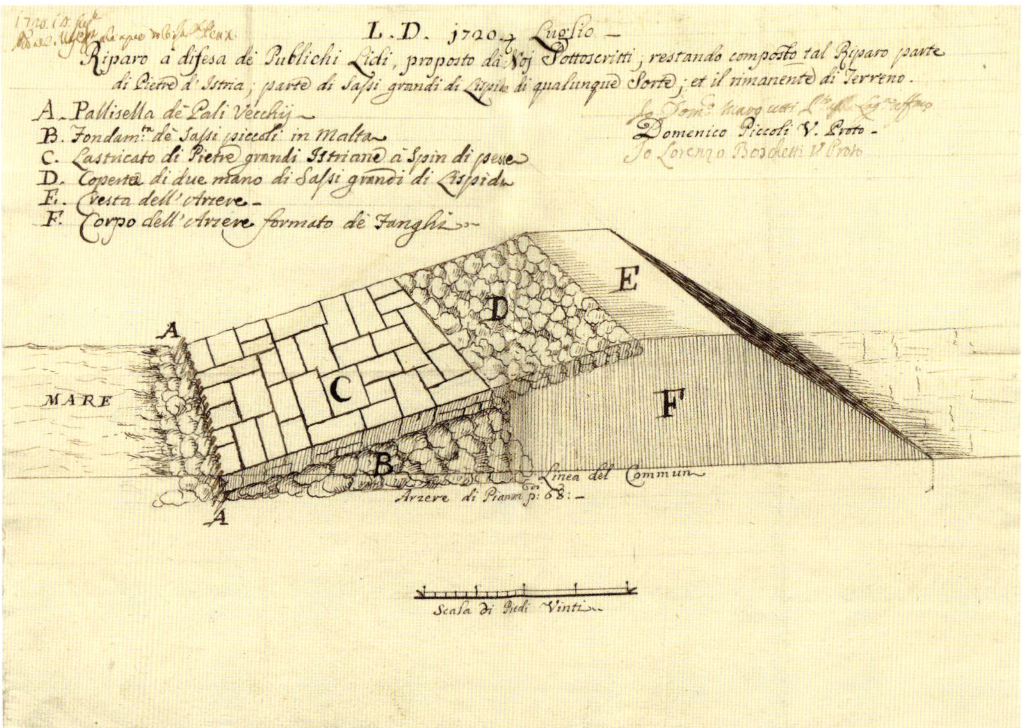

that the mantle of large blocks of Istria stone be smooth and arranged in a herringbone pattern to rest on a "foundation of small stones set in mortar"; the rest of the levee would be in pressed earth covered with loose Lispida stones.[171] At the levee toe, he recommends only poles and loose stones. The masonry structure suggested by Minorelli is the most complex to date, and stands out for the major role assigned to Padua lime for a cement to be poured, whose composition goes beyond what was envisioned by Tirali or Piccoli.[172]

With respect to 1716, when he suggested use of Padua lime to form a masonry crust to spread over the embankment in pressed earth, Minorelli now invents a levee that has been transformed into a powerful structural machine equipped with a long embankment in the sea, entirely in large stones set up to form steps, a sector behind it, to be poured, organized with vaults and pillars, and an embankment on land, facing the Lagoon. Minorelli proposes two different versions of this structure, varying the size of the stepped stone embankment and the form of the vault to be made in the pouring. The project is influenced by the solution that took form following the critique of Tirali's project in 1715, when the proposal was made to reinforce the foundation of the masonry of the surface with a structure of vaults and pillars (Minorelli was part of the commission that reached that conclusion). In the drawings of the two versions, Minorelli demonstrates, with graphic precision, the innovative technical part of his proposal, namely the lime mortar conglomerate.

By now Minorelli is using for his structure the term *murazzo* (or masonry wall), and he does not miss the chance to invoke the example of ancient Roman construction to justify some of its technical particulars. Not even Zendrini manages to construct a machine in the Roman manner, like the one proposed by Minorelli. To connect the large Istria stones of the stepped surface, Minorelli suggests the use not of mortar, but of iron stirrups, or *pironi di metallo*, because "this was the practice of our ancestors in real works." This is far from a negligible detail to understand the evolution of the masonry technique of the *murazzo* applied at the end of the 1730s with large stones and mortar, as well as the criticisms voiced in the 1750s regarding the definitive solution. Minorelli's explanation of the method of constructing the conglomerate of the vaults and pillars – what he defines as "regarding the fabrication of the murazzi, piles and embankment walls to be made with small stones in mortar" – also demonstrates a focus on the principles of Roman construction, given that this is a pouring of stone material and river pebbles embedded in Padua lime.[173] "To fabricate, then, these murazzi," Minorelli writes, "one must attempt to use the best fresh, soft Padua lime mortar, mixed with good large sand, which will become incredibly hard when it sets."[174]

Minorelli goes into detail on the making of the conglomerate and of the support structures. The materials of the compound can be rocks found locally – "the old large and small stones found scattered on the coasts," the rubble from quarries – "the shards and bits of the same pits," and river stones gathered along the Brenta and transported to the Lagoon with barges.[175] The centering for the construction of the vaults is prepared directly on the worksite,

167 Domenico Piccoli, report, 4 June 1720, *ibid.*
168 *Ibid.*
169 Angelo Minorelli, report, 4 June 1720, *ibid.*
170 Alvise Zusto, Nicolò Erizzo, Andrea Soranzo, Lunardo Donado, report, 8 June 1720, *ibid.*
171 Domenico Piccoli, Lorenzo Boschetti, report, 5 July 1720, *ibid.*
172 Angelo Minorelli, report, 25 June 1720, *ibid.*
173 *Ibid.*
174 *Ibid.*
175 *Ibid.*

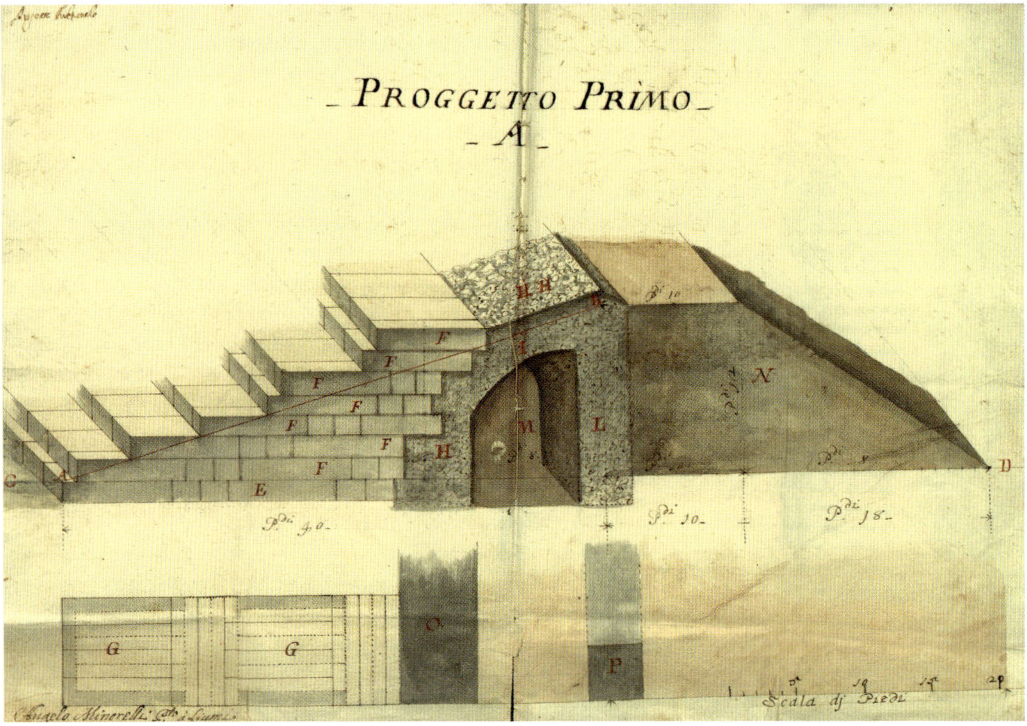
PROGGETTO PRIMO
A
HH
F
F
F
F
F
F
F
E
H
M
L
N
I
C
A
D
G
G
O
P
Scala di Piedi

PROGGETTO SEGONDO
B
HH
B
F
I
M
L
N
C
A
F
E
H
D
G
G
O
P

**61-62**. Angelo Minorelli, masonry levee design, Venice, 1720 (Savi ed Esecutori alle Acque, 27, Archivio di Stato di Venezia).

making use of the old technique of earthen banks: the earth itself is shaped in the form of a vault, or as Minorelli rights, of "firm terrain, and well pressed, that is filled, and all the place between the Murazzi of the pillars, and vaults, and arranged in such a way that is serves as an arch, or forms a profile over which to pour said vaults."[176] In the successive proposals over the course of the centuries for the levees of the coasts of Venice, no one prior to Minorelli had introduced a similar embankment in the ancient manner, with a core of conglomerate and a powerful embankment wall of large squared stones.

But Minorelli's proposal is fated to remain on the drawing board, at least until the concept of a poured conglomerate with hydraulic mortar resurfaces thanks to Zendrini in the second half of the 1730s, though formulated on a different technical basis, in keeping with a more direct relationship with the compound made by the Romans, rather than that of Palladio.

Zendrini's position in the evaluation in June 1720 goes in the direction indicated by Piccoli's project, regarding the masonry structure of the levee embankment wall. His report opens with praise of the emerging idea, for the protection of the Venice shoreline, of a structure of large stones: "It will be a work that can stand up to the centuries, and will generate amazement in the more educated nations, seeing the judicious assemblage of so much marble and stones for the construction of a very strong obstacle to the furious attack of the siroccos, which beating at almost a right angle against the short quickly ruin any of the ordinary protections built until now."[177] To urge the view that it is necessary to make use of large stones, he refers not to the statements of Guglielmini about the coasts, but to the direct source from which this principle has spread among hydraulic experts, the treatise of Viviani *Intorno al difendersi da' Riempimenti, e dalle Corrosioni de' Fiumi*. The principle that guides Zendrini in expressing his opinion on the coastline is that of the strength of masonry based either on the weight of the single stones, or on the binding force that holds them together "like a single solid" against the "force of the waves" (an expression that in the project of the levees for the coasts had been used by Minorelli). The proposals of the water engineers make him familiar with the role of lime mortar in the construction of stone levees, given that they call for – as Zendrini writes – either a "levee of marble in large steps, filled as a whole with stones of all kinds worked with lime mortar" (Minorelli's project), or an embankment wall covered with stones arranged in a herringbone pattern, but also "connected" with "lime mortar" (Piccoli's project).[178] Zendrini is already so convinced of the crucial role of the connection between the stones that he believes, like Piccoli before him, that the "stepped" levee would be subject to dislocations caused by the power of the waves, because the large stones, in spite of their size, would not be sufficiently stacked one atop the other. It is the logical technique of connection that makes him favor the sophisticated herringbone masonry arrangement, above all – he suggests – if "the blocks of marble as are large as possible" and the seams between "marble and marble" are made "imperceptible." To back up his suggestions, Zendrini cites the model of ancient dams, "masterpieces of the Romans," renowned for their "excellent and truly very diligent connecting of the boulders of which they are composed."[179]

Following the deliberations, Tirali and Piccoli are assigned by the Magistrato alle Acque the task of developing a masonry structure starting with the

[176] "Spiegat.[io]ne che serve per li due dissegni in profilo, e pianta d'un nuovo riparo de lidi suggerito dal P.to Angelo Minorelli," [25 June 1720], *ibid.*
[177] Bernardino Zendrini, report, 18 July 1720, *ibid.*
[178] *Ibid.*
[179] *Ibid.*

herringbone solution, taking Zendrini's observations into account. In their project the earthen levee is reduced to the sector facing the Lagoon; the levee facing the sea is formed by stones embedded in lime mortar ("foundation of small stones in mortar"), by a cladding of the embankment wall in large Istria stones laid in a herringbone patter, with at the top – the levee crown, or *coronella* – a banquette of stone, and at the foot an even larger stone base, laid in two rows (and known as *fondamento*), protected by poles and stones. In the list of materials for the construction of the "paved herringbone" levee, Tirali and Piccoli include "lime mortar" and "stone chips" for the "massif of stones" with which to form the foundation.[180] No reliable information exists on the nature of this lime mortar, though it is probably from Padua. This leads to the second embankment prototype, slated for construction at the Pellestrina coast, on the site known as the "Magre," for the stretch identified by the numbers 44-46.[181] In July the supply modes of the stones are indicated, both "large Istria stone," and "old stones" to place "in lime mortar" to form the "screed entirely of small stones," and the criteria for the fabrication of the mortar and the way of juxtaposing the large stones so that there are "no cracks or wedges," with each stone worked in such a way that "it fits perfectly in the fishbone pattern."[182]

All the considerations of specialists like Guglielmini regarding the need to get way from the practice of loose stones in favor of the use of large blocks, or like Tirali and Minorelli regarding the need to connect the stones with mortar, find confirmation in the experiment of Tirali and Piccoli, which marks a decisive passage in the evolutionary process of the constructed form of the levee of the Lagoon. And it is significant that in this experience lime mortar is also used to connect the stones in a herringbone pattern in the sectors of the embankment wall in closest contact with the sea and the impact of the waves.

By 1720 the term *murazzo* is already circulating among the technicians and water engineers of the Savi ed Esecutori alle Acque, to indicate the masonry structure of the new levees. Halfway through the 1740s the two prototypes made under Tirali's supervision[183] are called "old murazzi of large stones but made without any cement."[184] In 1746 Zendrini writes about *murazzo* at the Magre formed by "marble bound by lime mortar, but without pozzolana that, back then, was not yet in use, and what is worse without foundations, so that its base, though formed of large Istria marble blocks, threatens to yield with every passing day."[185]

Confirmation that the experiment of Tirali and Piccoli contains the premises for what was to become the murazzo of Zendrini can be seen from the start of 1721. From analysis of the prototype, and in the wake of the discussions with the workmen on the site – the "practitioners" – an ingredient is identified that turns out to be fundamental for the definition murazzo solution: pozzolana. The setting for this discovery is the coastline at the "Magre," where in January Zendrini goes with Tirali to examine the condition of the structure in large blocks of Istria stone after a storm – *antemurale*, or breakwater, is the term used by Zendrini for the work, evoking the ancient hydraulic constructions

[180] Andrea Tirali, Domenico Piccoli, letter, 15 July 1720, *ibid.*
[181] Report, 23 July 1720, *ibid.* For the description of the work see, Andrea Tirali, report, 13 September 1723, Savi ed Esecutori alle Acque, 74, Archivio di Stato di Venezia.
[182] Andrea Rizzo, copy of deliberation, 23 July 1720, Savi ed Esecutori alle Acque, 73, Archivio di Stato di Venezia. For the progress of the work in 1722 and the defects in its construction, some of which can be attributed to the builders, see Vincenzo Gradenigo, copy taken from the "Terminazioni," 29 July 1722, Savi ed Esecutori alle Acque, 2, Archivio di Stato di Venezia.
[183] Bernardino Zendrini, Tommaso Temanza, report, 5 June 1745, Savi ed Esecutori alle Acque, 75, Archivio di Stato di Venezia. Both write of "massi" or "murazzi" regarding the levees of Tirali.
[184] Matteo Lucchese, report, 14 November 1746, *ibid.*
[185] Bernardino Zendrini, report, 24 September 1746, *ibid.*

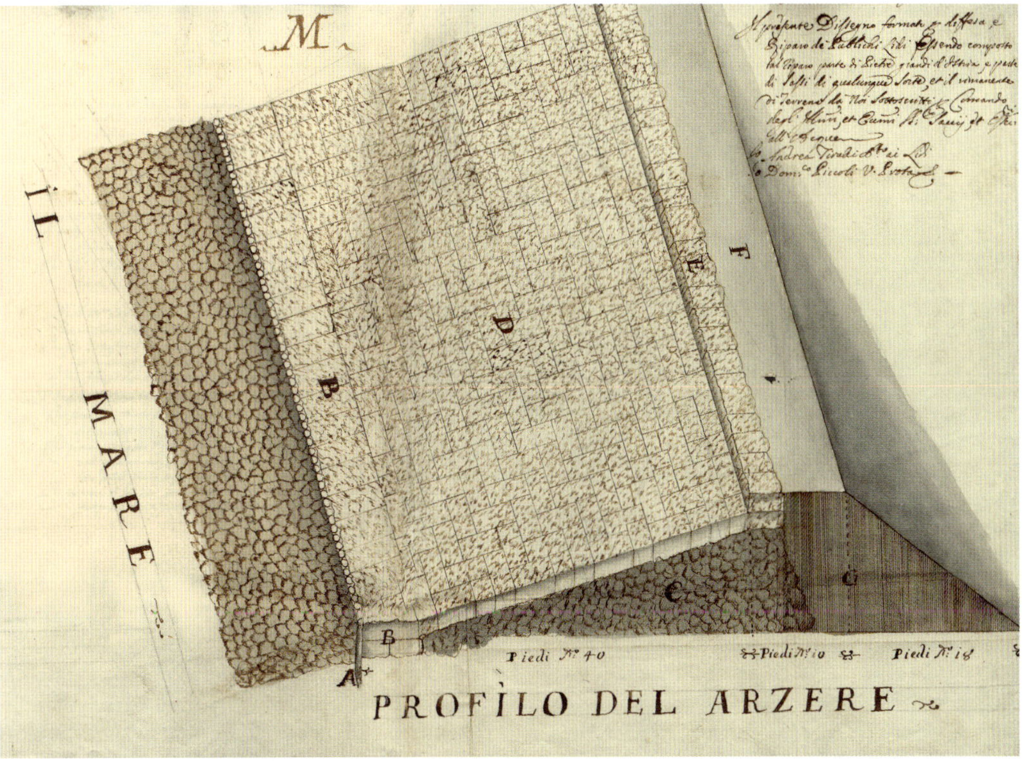


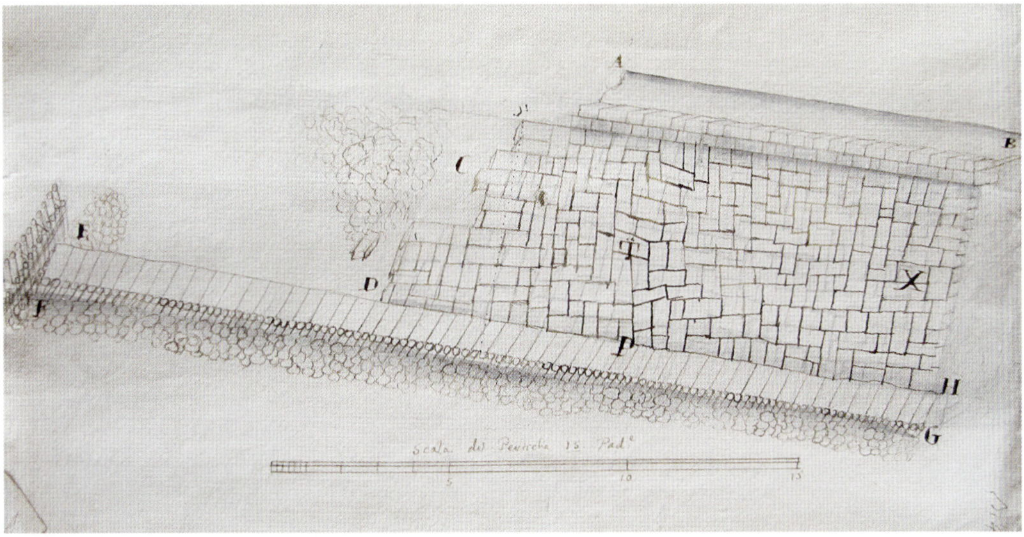


**63**. Andrea Tirali and Domenico Piccoli, masonry levee design, Venice, 1720 (Savi ed Esecutori alle Acque, 27, Archivio di Stato di Venezia).

**64**. Bernardino Zendrini, survey of the masonry levee of Tirali and Piccoli at the Pellestrina shoreline, at the site known as Le Magre, Venice, 1721 (Savi ed Esecutori alle Acque, 74, Archivio di Stato di Venezia).

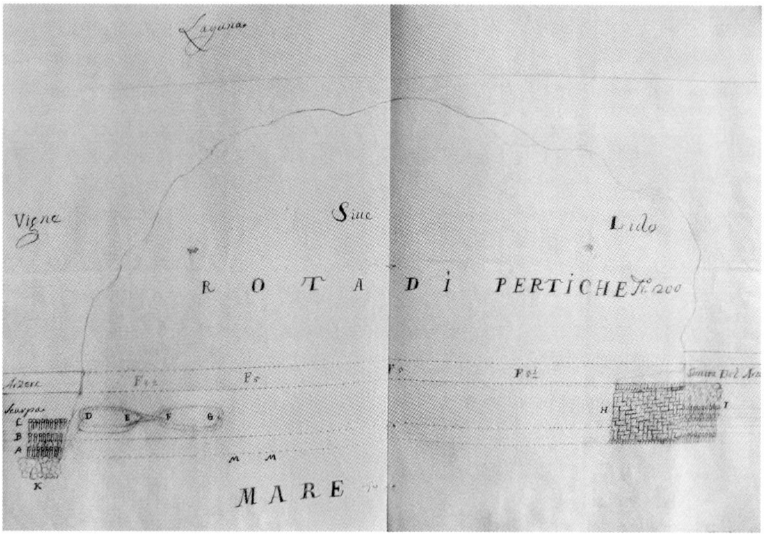


**65**. Bernardino Zendrini, levee design, with ships to sink filled with stones, for the Pellestrina shoreline at the Le Magre site, Venice, 1722 (Savi ed Esecutori alle Acque, 74, Archivio di Stato di Venezia).

made by the Romans to protect their ports.[186] During the inspection Zendrini talks with Tirali about the principles applied up to that point in the construction of the levees of the Lagoon, and makes a survey of that "levee of Istria stone." The constructive issue that worries Zendrini, related to the binding between the stones, can be seen clearly in the drawing where he precisely indicates, with the letters T and X, the level up to which, starting from the foot of the embankment wall, lime mortar has been used to bond the stones – "large Istria stones, welded with lime mortar," as Zendrini reports.[187] But he does not agree about the type of lime mortar used, and in its place he recommends "cement."[188] The shifts of the blocks of the barrier are attributed mostly to the fact that the work was interrupted after construction of just 11 rods and 4 feet of the 25 planned rods, leaving the side of the levee exposed – the sector indicated with the letters ADP in Zendrini's drawing, the one that was most damaged by the storm. The levee has a solid base of large rectangular prisms of Istria stone – the *gran marmi*[189] – installed pointing at the sea to constitute "the large embankment wall of the levee," as Zendrini writes (rows of poles and stones protect this large embankment wall).

During the inspection Zendrini and Tirali also decide on the way to repair the breach at the Magre, to be done – as Zendrini himself reports – following the indications of Tirali and the fixed principles of the antemural. He simply corrects the slope of the embankment, as a mathematician taking into account the observations of practitioners: "I deduced this not just based on reason, but much later also having understood, from the practitioners, the way in which the waves impact the shore and the protections." It is at this point in the report that Zendrini introduces the hypothesis of replacing the lime mortar used by Tirali with a type of earth quarried in the area of Vicenza, which

186 Bernardino Zendrini, report, 21 January 1721, Savi ed Esecutori alle Acque, 74, Archivio di Stato di Venezia.

187 *Ibid.*

188 "The lime mortar," Zendrini writes, "has not been placed as necessary, nor is it of the necessary quality to grip the marble, so the cement must be replaced in this breakwater, where required, to prevent the possible penetration of water in the cracks that remain between one piece of marble and the next, so that the incessant beating of the waves will not disconnect an operation that must stand up to the test of the centuries" (*ibid.*).

he calls "a sort of pozzolana." Probably this is a lime mortar made with volcanic earth, similar to the "Lavezzara."[190] The fact that Zendrini formulates the proposal in close contact with Tirali and with the workmen leaves the question open of whether the technical solution is discussed with those who know the best way to make lime mortar using local products (therefore Tirali himself), though the words used by Zendrini in his report leave little doubt about the fact that this is his own, personal intuition.

"As far as the base to make with Istria stones," he writes, "once the materials are ready, as long as the blocks of marble are nicely squared, connected and topped off with lime mortar that sets and grips well, it would be incomparably better than simply setting the levee on the naked ground, even more so if it is the desire of Your Excellencies to continue the work of the stone levee, for which the greatest, most real and most eternal safety ever achieved on the public shores will be the result of this principle, as fabricated above the same measurements; and since the duration of this work depends on the perfect connection of the marble blocks, I do not think it is out of the question to try a certain mineral quarried in the area of Vicenza, which is a sort of Pozzolana, which may be of service to achieve this effect."[191]

After the long series of designs and trials, and once it has been ascertained through experience that the levee of Tirali and Piccoli constitutes the prototype to improve in order to achieve a durable barrier against the sea, for Zendrini the task is not just to specify the details of that antemural, from the slope of the embankment to its profile, the rows of poles and their heights: also, and above all, he senses that the crucial question lies in the nature of the binder of the stones. His discussion with Tirali and examination of the experimental masonry levee mark the symbolic entry in the history of the shoreline of the man who would bring about the last, decisive evolution in the invention of this construction system. By now it is clear that the design process leading towards the project for the barriers is a true system of interchange among multiple forms of knowledge and expertise: those of the water engineers, the masons and the mathematicians.

The works envisioned by Tirali and Piccoli slated for the Magre site are underway in 1722, when Zendrini makes his proposal for the experimental construction of a segment of levees along the same breach; he suggests sinking two flat-keel ships, or *arsili*, to fill with mud and stones, as a way of speeding up the work.[192] In September 1723 the leading experts on the shoreline are again called on by the Savi ed Esecutori alle Acque for other opinions on the way to intervene to consolidate the "stone work" at the Magre, when about 40 rods have been completed, given the fact that the connections between the stones have loosened, leaving gaps that are "too wide and dilated all over."[193] Minorelli, Cornello, Boschetti, Zendrini, Tirali and Giovanni Filippini each suggest a different remedy. Minorelli offers an overview of the Lagoon defense system, which he feels should not rely on the strength of a single element, but on a detailed system of levees, *guardiani* (or groynes) and *palade* (or framework of poles), organized in such a way as to create a sort of skeleton extended

189 *Ibid.*
190 *Ibid.* Giovanni Maironi da Ponte, *Ricerche sopra alcune argille e sopra la terra vulcanica della provincia bergamasca*, Bergamo, Locatelli 1791, p. 28. For previous hypotheses of the use of pozzolana in Venetian construction see, *Conterminazione lagunare: storia, ingegneria, politica e diritto nella laguna di Venezia, atti del Convegno di studio nel bicentenario della conterminazione lagunare, Venezia, 14-16 marzo 1991*, Venezia, Istituto Veneto di Scienza, Lettere ed Arti, 1992, p. 53.
191 Bernardino Zendrini, report, 21 January 1721, Savi ed Esecutori alle Acque, 74, Archivio di Stato di Venezia.
192 Bernardino Zendrini, report, 23 July 1722, *ibid.*
193 Bernardino Zendrini, 1 September 1723, *ibid.*

**66**. Jan Rokusz Douw, Agge Roskam Kool, Johannes Keulen en Zonen, "Hollandse Wal," coast of Holland from Maas to Texel, 1773 circa (B.0032 (109), 02, 020, Nederlands Scheepvaartmuseum, Amsterdam).

**67**. Johannes Keulen en Zoonen, "Nieuwe Pas-Kaart van Texel en den Horst," 1771 (B.0032 (109), 01, 097, Nederlands Scheepvaartmuseum Amsterdam).

into the sea, capable of gradually forming an artificial beach thanks to the movement of the waves.[194] Filippini, Cornello and Boschetti share this view, and recommend an increase of the structures capable of attracting sand.[195] Zendrini, who had already emphasized the critical point of the work in stone – the joining of the stones – proposes filling the larger cracks with stone fragments and lime mortar, to keep the waves from breaking up the masonry – "filling with lime mortar and fragments of marble should be considered."[196] But once again it is Tirali who provides the technical suggestions that point in the direction of what becomes the definitive solution for the levees of the Lagoon. Probably remembering his discussions with Zendrini in 1721, Tirali proposes the use of pozzolana together with a special cutting of the stones, perhaps the dovetail design he had already envisioned in 1715. "[...] it would be possible to also work the same stones connecting them one to the next, and it would be worthwhile to put them with pozzolana lime, and in this form many stones would be converted into one single stone," are his significant words. For the second time, with Tirali directly involved, the use of pozzolana is urged for the shoreline.[197] By now it is clear to the Venetian water engineers that the goal to achieve is that of "many stones" held together to form "a single stone," and that there is only one material that can make this possible, the same material that had made ancient Roman construction and its cement famous: pozzolana.

## Seafront barriers in Flanders and the Netherlands

The process of development of the dikes for the shores of Flanders and the polders in the Netherlands does not follow an evolution similar to that of the levees of the Lagoon in Venice, involving masonry with large stones laid with lime mortar. In spite of the production of *trass* and its use for foundations and different hydraulic constructions, the Dutch engineers do not set off along the path of masonry to make the barriers to protect the land from the sea. The construction of dikes, or "Hollandse Wal," until the middle of the 1700s, continues with conventional methods, using sand, mud, vegetation, poles and loose stones, without any mortar.

Dikes of Flanders, like those near Ostend, are composed of sand banks formed by a natural process of accumulation during the spring tides (although the city is protected by structures made of poles and stones). Smeaton indicates this type of natural defense of the shore, where the sand, free of any "bituminous [substance] to cement together," is shaped and shifted by the wind.[198]

The protection of certain segments of the Netherlands coast is done with wooden palisades, with poles organized in rows and inclined at different angles, based on a system developed towards the end of the 1600s – the *krebbingen*. The palisade is often protected at the base by a cliff, with loose stones held in place by a row of poles. The poles of the palisade are held firm against

[194] Angelo Minorelli, report, 9 September 1722, *ibid.*

[195] Giovanni Filippini, report, 9 September 1723, and, Andrea Cornello, Lorenzo Boschetti, report, 7 September 1723, Savi ed Esecutori alle Acque, 74, Archivio di Stato di Venezia.

[196] Bernardino Zendrini, 1 September 1723, Savi ed Esecutori alle Acque, 74, Archivio di Stato di Venezia.

[197] Andrea Tirali, report, 13 September 1723, *ibid.*

[198] *Cf.* Arthur Titley, *John Smeaton's Diary of his journey to the Low Countries, 1755*, The Newcomen Society, 1938, p. 9. See, Otto S. Knottnerus, *Die Angst vor dem Meer. Der Wandel kultureller Muster an der niederländischen und deutschen Nordseeküste (1500-1800)*, in, Ludwig Fischer, ed., *Kulturlandshaft Nordseemarschen*, Bredstedt, Westerhever, Nordfriisk Instituut, 1997, pp. 145-74, and, Dirk van de Vijver, *Ingenieurs en architecten op de drempel van een nieuwe tijd, 1750-1830. Essays en catalogus. Vorwort und Endredaktion, Krista de Junge*, Leuven, Presses Universitaires de Louvain, 2003.

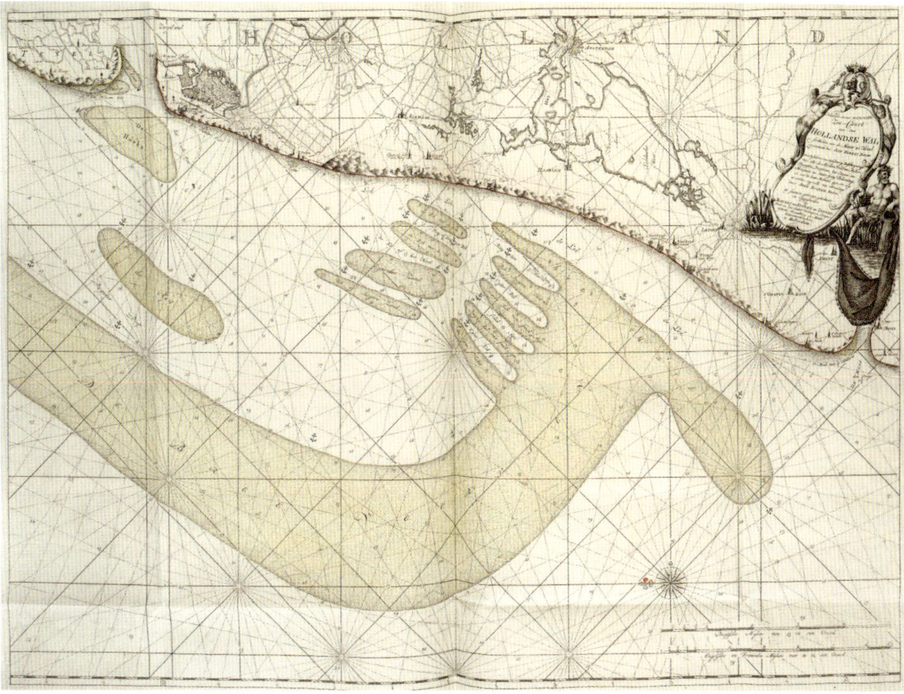
H O L L A N D
HOLLANDSE WAL

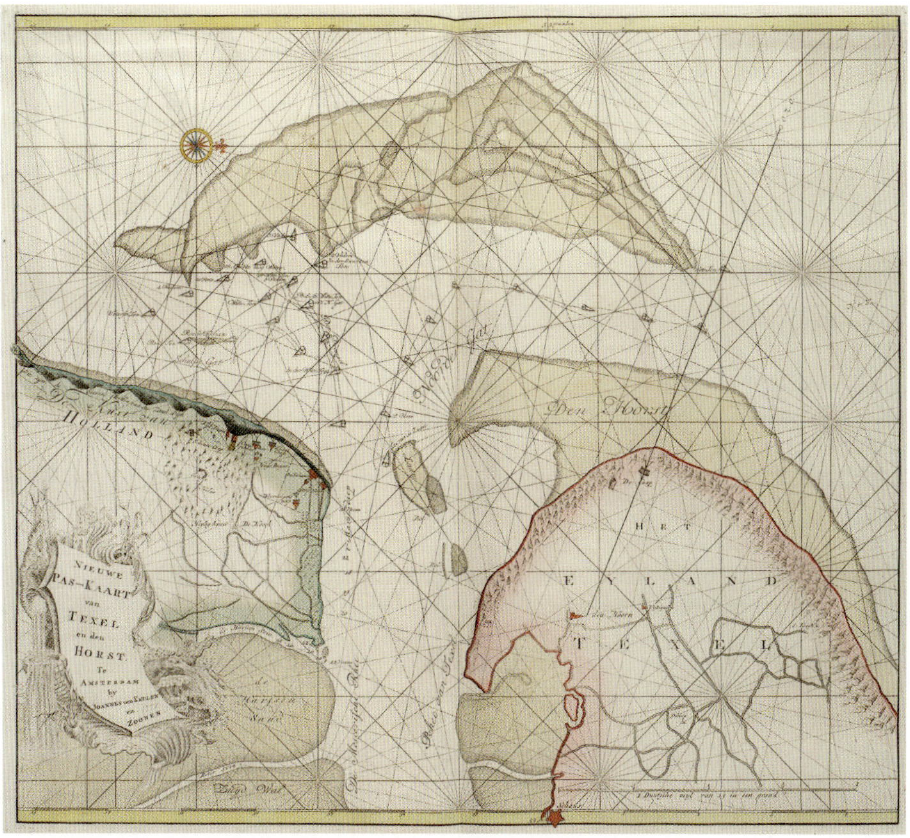
Nieuwe
Pas-Kaart
van
Texel
en den
Horst.
Te
Amsterdam
by
Joannes van Keulen
en
Zoonen
Den Horst
HOLLAND
H E T
E Y L A N D
T E X E L
Noorder Gat

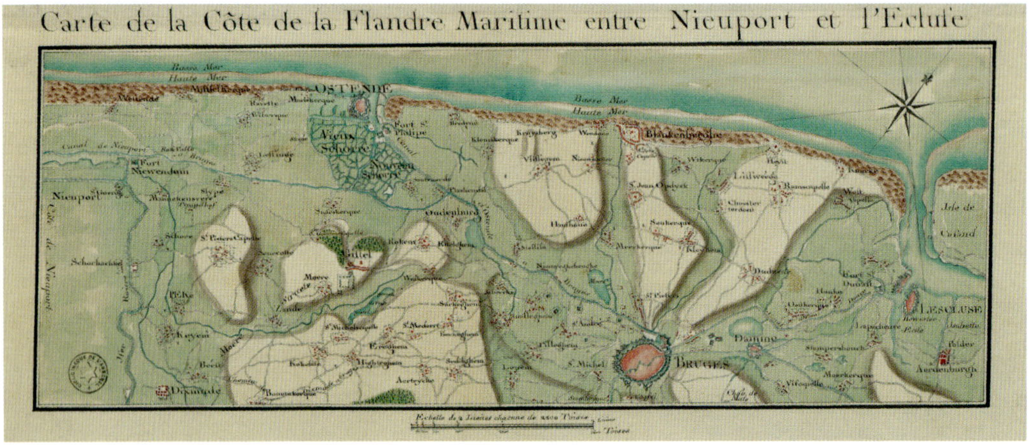


**68**. "Map of the maritime coast of Flanders between Nieuport and the locks," [before 1758] (Ms 6461(645), Bibliothèque de l'Arsenal, Paris).

**69**. Plan of Ostend with the shoreline defense in wood and stone, "Grondtekening det stad Ostende," [1760 circa] (I. Tirion, *Beknopte Atlas, van omtrent honderd Platte Gronden der Voornaamste Vestingen, Kasteelen en Steden van Europa, en byzonder van de Nederlanden en eenige Kolonien [...]*, Amsterdam, Leiden, Dordrech, Harlingen, Compagnie Van Boekverkopers).

**70**. Stretch of levee between Amsterdam and Muyden. "2d Figúúr, Vertoonende den teegenswoordigen toestandt, van de Paalen-Dyk, op twee plaatzen na 't Leeven afgeteekent, in Mey 1703" (P323N325, Universiteitsbibliotheek Leiden).

**71**. "The city of Muyden, with its Castle, as seen from Zee-dyk," wooden levees, near Muyden (P331N448, Universiteitsbibliotheek Leiden).

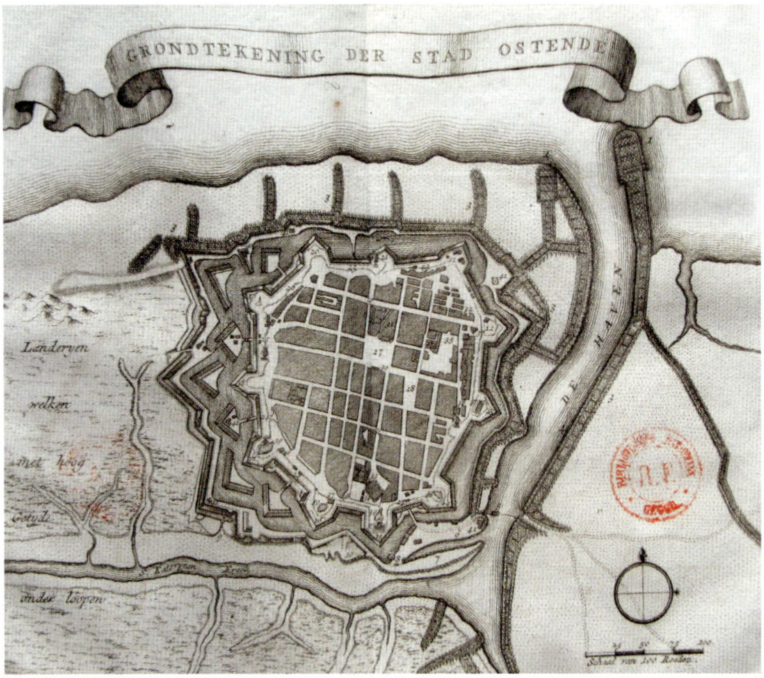

2e Figuur. Vertoonende den tegenwoordigen toestandt van de Paalen Dyk, op twee plaatzen na 't Leeven afgeteekent in May 1703.
R. UNIV. BIBLIOTHEEK LEIDEN
R. UNIV. BIBLIOTHEEK LEIDEN

R. UNIV. BIBLIOTHEEK LEIDEN
De Stad MUYDEN en haer Slot; van de Zee-dyk af te sien.
La Ville de MUYDEN, avec son Chateau, qui se Regarde du Zee-dyk
98

**72**. Matteo Alberti, palisade in Holland (V.M. Coronelli, *Teatro delle città e porti principali dell'Europa, in pianta, in profilo, ed in elevatione [...]. Parte III*, Venezia, Accademia degli Argonauti, 1697).

**73**. Jan Belkmeer, plan of the new canal and port of Enkhuisen, 1720 (Atlas der Neederlanden (Band V, 62), Special Collections, University of Amsterdam).

the impact of the waves by a connection at the top attached to struts fastened in the ground (with cross trestles at the end in the terrain). The space between the palisade and the shore, or the earthen embankment, can be filled with seaweed and grasses to avoid infiltration of water through the poles. One embankment stretch protected by palisades extends from Amsterdam to Muyden, as is documented in views painted in 1703. This palisade system that tends to save the use of stones attracts the attention of a Venetian technician, Matteo Alberti (1660 circa - 1716), who after working for the Magistrato alle Acque practiced architecture in Düsseldorf. Alberti compares the "shore as done in Holland" with that "as done in Venice" in his "Dimostrazione scenografica et ortografica de' ripari che si fanno sopra i liti del mare all'uso di Venetia e d'Holanda" presented in 1692 to the Savi ed Esecutori alle Acque.[199]

On the background of the proliferation of essays on hydraulic construction in the German and Dutch technical literature in the first decades of the 1700s, the catastrophic flooding on 24-25 December 1717 impacts vast territories of northern Europe, from Norway to Denmark, Prussia to Holland, facing the North Sea (over 11,000 persons lost their lives). But an initial radical transformation of the Dutch waterfront defenses comes only in the wake of the deterioration of the wooden poles. Though the Dutch were aware, since the end of the 1500s, that shipworms would eat into the poles of the dikes, the problem does not become urgent until September-October 1730, when inspectors notice the damage caused to the dike of West-Kappel; the problem spreads in September 1731, when the inspectors of the dikes of Drechterland, Texel, Helder and West-Frise, and of the *krebbingen* from Broekerhaven to Venhuysen, report on the threat of the shipworm infestation, discovering that poles inserted from 1718 to 1729 (especially those in fir wood, as opposed to the more resistant oak) had been decayed and destroyed by worms of different sizes - the *teredo navalis*. The report of 12 January 1732 reveals an alarming situation and the urgent need for intervention on the dikes and the *krebbingen* to avoid the danger of breaches in West-Frise, that would have had disastrous consequences for the whole country. The repairs suggested have to do with

[199] *Cf.* "Relazioni Periti", 23 September 1692, Savi ed Esecutori alle Acque, 72, Archivio di Stato di Venezia (*cf.* Grillo, *op. cit.*, p. 30).

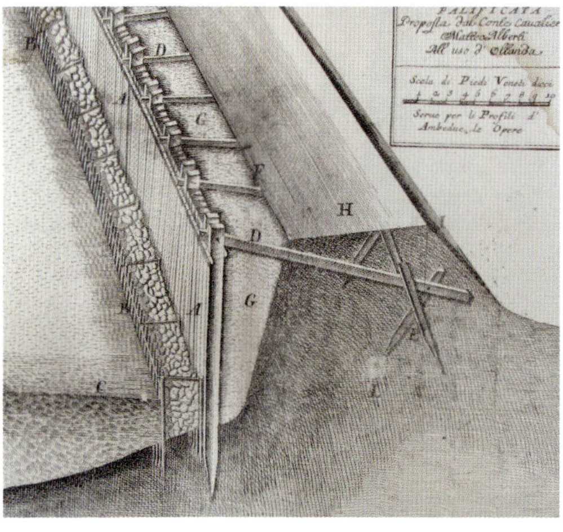


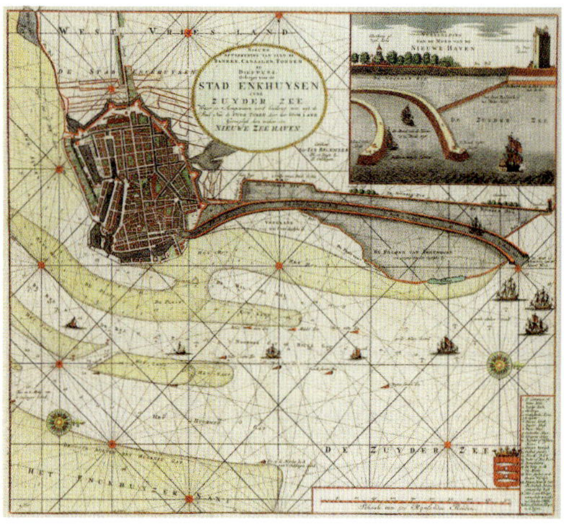

**74**. Abraham Allard, "Hindeloopen a city on the southwestern coast of Friesland 11/2 hours north from Staveren" (P327N290, Universiteitsbibliotheek Leiden).

**75**. F. Patton, "A View of the New Boat Sluices in Amstel/Veuë des Nouvelles Ecluses a Bateaux a Amstel" (P320N410, Universiteitsbibliotheek Leiden).

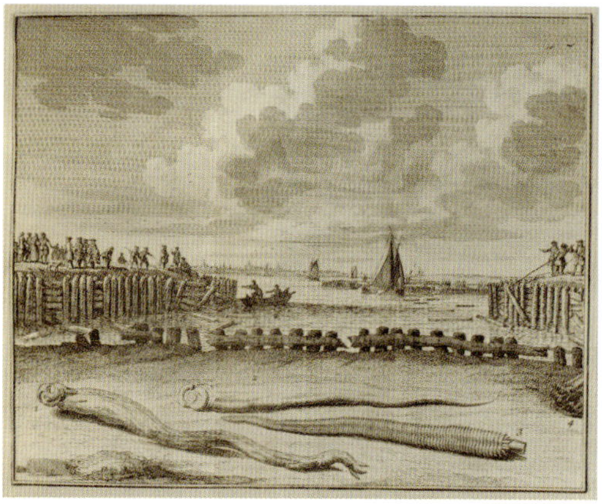

**76-78**. The "teredo navalis" and the wood of the Dutch levees, or "krebbingen," destroyed by worms (E. Baeck, called Heldenmuth, "Abildung deren höchst schädliche unbekandten See-Würmer welche aus West-Indien Zuerst nach Ost-Friesland, in den Texel, und Amsterdam gekomen, und alldorten unbeschreibliche Schaden verursachet," Augsburg, 1732; P. Massuet, *Recherches interessantes sur l'origine, la formation, le developement, la structure, &c. des diverses especes de vers à tuyau, qui infestent les vaisseaux, les digues, &c. de quelques-unes des Provinces-Unies*, Amsterdam, F. Changuion, 1733; J. Rousset de Missy, *Aanmerkingen over den oorsprong, gesteltheit, en aard der zee-wormen die de schepen en paal-werken doorboren*, Leiden, G. Langerak, 1733).

AANMERKINGEN
Over den
OORSPRONG, GESTELTHEIT,
en AARD
der
ZEE-WORMEN.
Die de Schepen en Paal-werken
doorboren
door
Den H^r. ROUSSET,
Lid van 't Koninglyk Genootschap
der Wetenschappen van Berlin.
*Met Kopere Platen.*
DEN TWEDEN DRUK.
Overgezien, verbetert en vermeerdert met verscheide *Aanmerkingen*; als mede met eene *Beschryvinge der Dyken*, en 't Onderzoek van verscheide *Hulpmiddelen* tegen de Wormen.

*TE LEYDEN,*
By GYSBERT LANGERAK,
Boekverkoper in de Pieters Choorsteeg. 1733.

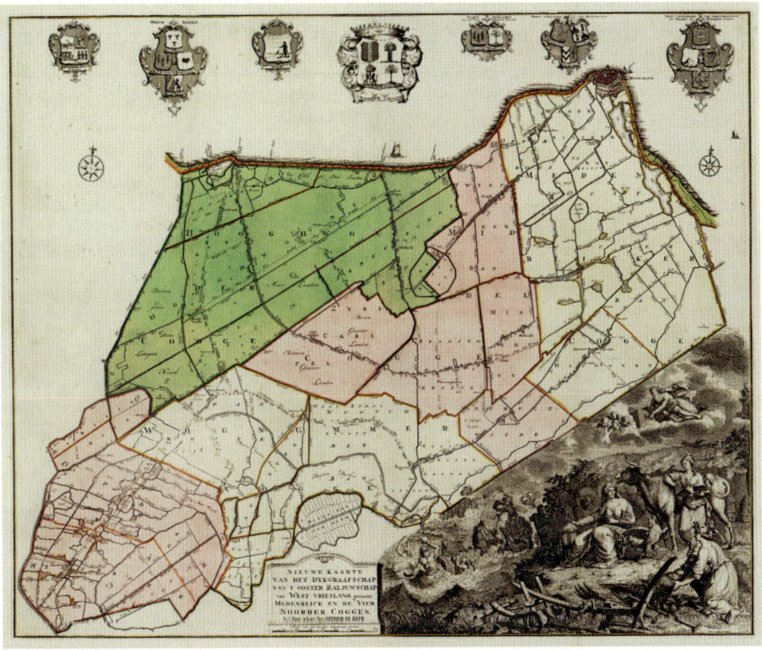

**79**. Hendrik de Leth, "New map of dike country east of the Bailiwick of West-Vriesland, called Medenblick and the Four Noorder Coggen," 1730 circa (21550, Stichting Rijksmuseum het Zuiderzeemuseum, Enkhuizen).

additional supplies of wood, machines and workers, for systematic maintenance and monitoring of the various stretches of the dikes and *krebbingen*,[200] to gain time while waiting for the invention of another kind of seafront defense system that would stand up to the shipworms. In February of that same year the hypothesis arises of considering dike maintenance "like a barrier or fortress that protects all of Holland from a dangerous enemy" in order to cover the expenses under the administration of the fortresses.[201] In March the inspectors make an official request that the dike of West-Frise should become part of the public domain and be considered, "as it faces the ocean, the state's most advanced and exposed fortress."[202] In the meantime, the first proposals for technical remedies are advanced, ranging from the creation of a protective barrier of oak poles for insertion in front of the fir structures, to the idea of charring the surface of the poles, as the Portuguese had done to protect the hulls of their ships. The first scientific research on marine worms, including analysis under the microscope, is published precisely when the extent of infestation is discovered; the paper is written by Pierre Massuet (1698-1776), a physician educated at Leiden and working in Amsterdam. In Amsterdam a fad even develops, that of worms conserved in containers, to be used as gifts or sent abroad as a curiosity.[203]

The Dutch method of waterfront construction is modified by turning to a defense system using stones, which spreads above all when the major hydraulic works fall under the supervision of the superintendent of rivers and of

[200] Pierre Massuet, *Recherches interessantes sur l'origine, la formation, le developement, la structure, &c. des diverses especes de vers à tuyau, qui infestent les vaisseaux, les digues, &c. de quelques-unes des Provinces-Unies*, Amsterdam, F. Changuion, 1733, pp. 175 *et seq.* See also, Jean Rousset de Missy, *Aanmerkingen over den oorsprong, gesteltheit, en aard der zee-wormen die de schepen en paal-werken doorboren*, Leiden, G. Langerak, 1733.

[201] Massuet, *op. cit.*, pp. 193-94.

[202] *Ibid.*, p. 216.

[203] *Ibid.*, p. 2. In the attached drawing the worms are shown in the foreground, while the scene is that of the dikes and levees with wooden poles, some of which have been corroded and destroyed.

Pag. 9.
Roeden
1 2 3 4
Aarden Dyk.
Wier
Schaale van 4 Rhynlandsche Roeden.
1 2 3 4
Nieuwe Glooyinge van Steen.

DE
SUYDER
ZEE
De Veght

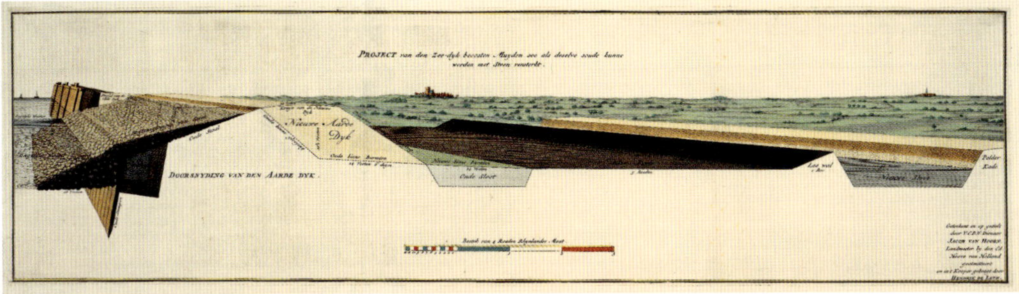


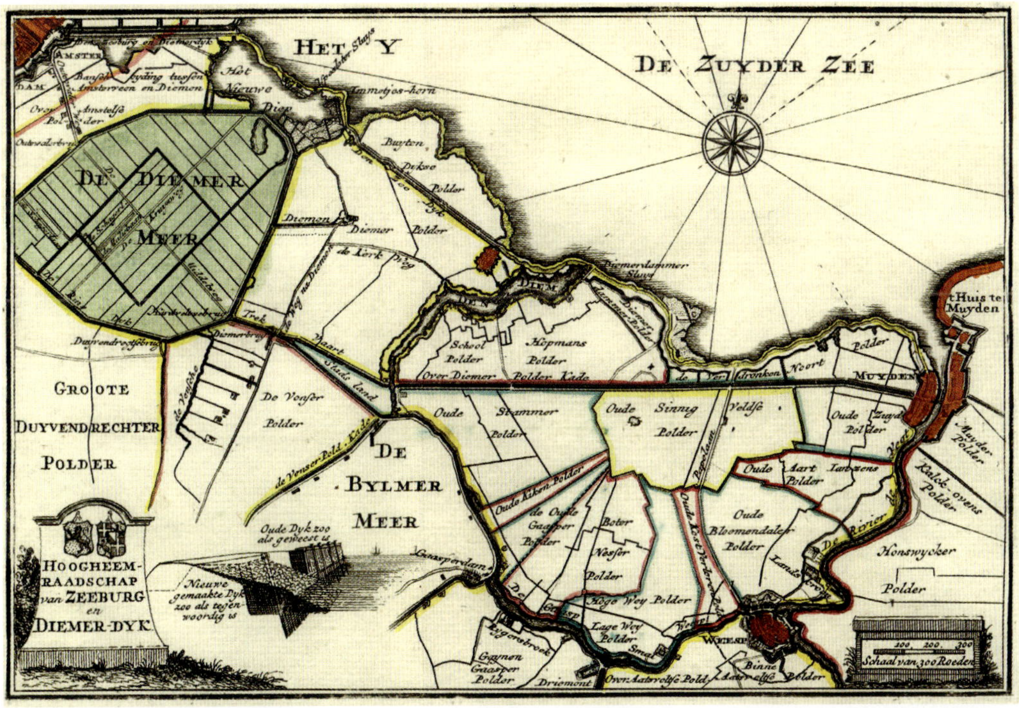


**80**. Pieter van Der Deure, Pieter Straat, levee design (P. van Der Deure, P. Straat, *Ontwerp tot een minst kostbaare, zeekerste en schielykste herstelling van de zorgelyke toestand der Westfriesche zeedyken, zonder dat het voortknagend Zeegewormte daar aan eenige hindernisse kan veroorsaken* [...], Amsterdam, J. Oosterwyk, 1735).

**81**. Jacob van Hoorn, "Caarte van Den Nieuw vermaakte Zee-diyk be Oosten Muyden," plan with sections of the construction for the new dike of stones at Muyden, drawing by Hendrik de Leth, 1737 (001076A, Stichting Rijksmuseum het Zuiderzeemuseum, Enkhuizen).

**82**. Jacob van Hoorn, "Project van de Zee-dyk beoosten Muyden soo als deslve soude kunne werden met Steen versterkt," drawing by Hendrik de Leth, 1735 (Atlas der Neederlanden (Band IV, 85), Special Collections, University of Amsterdam).

**83**. Dike of the Zuiderzee, at Diemen, with indication of the traditional system of piles and the new system with levees covered with stones (H. de Leth, "Hoogheem-raadschap van Zeeburg en Diemer-Dyck," 1740, Roberto Gargiani Collection).

**84**. Jan Wandelaar, view of the levee at Diemen, 1749 (Atlas der Neederlanden (Band IV, 81-84), Special Collections, University of Amsterdam).

**85**. Govert Oostwoud, Joannes Pool, Teunis Pool, Jacob Groes, "Aan de Wel Edele Gestrenge Heeren de Heeren Dykgraven Mitsgaders aan de Heeren Heemraaden en Secretaris van Dregterlandt [...]," 1775. Detail of a levee in wood and stones (022013A, Stichting Rijksmuseum het Zuiderzeemuseum, Enkhuizen).

western Friesland, Johannes Lulofs (1711-68), professor of mathematics, astronomy and philosophy at the University of Leiden. The cause of the transformation of the protection system from palisades to rocks is precisely the shipworm infestation. On the coast of western Friesland, where the Zuiderzee had formed, Pieter Straat and Pieter van der Deure design dikes with large stones similar to those of Venice, though with different masonry techniques and without the use of *trass* (but the main difficulty is precisely to find stones similar to those that were used in Venice)[204]. Other barriers are installed along the coast of the Noorderkwartier. One of the systems developed as a result of the infestation of the shipworms is to reinforce the slope of the embankments facing the sea with stones, as in Venetian examples. These types of projects are made for example by Jacob van Hoorn for the levees at Muyden, in 1735. Another variant includes the creation of wooden coffers with poles and compartments made of boards or woven branches, to be filled with large loose stones, while windmills are used to drain the water remaining in the polder. The stones for the new Dutch waterfront barriers are imported mainly from Norway: "All Dutch ships that trade in the North are obliged to load a certain number of these stones as ballast when returning."[205]

Unlike what happens in the Venetian Lagoon, the authorities responsible for the Dutch dikes hesitate to initiate a process of transformation of the construction system of their waterfront defenses based on elimination of wooden structures. Halfway through the century, in spite of different technical measures, the Dutch system still seems antiquated in the eyes of a French observer: "This work has been praised as the most surprising marvel of Holland; but I have to admit that I was impressed only by its immense expanse;

[204] *Cf.* Pieter van Der Deure, Pieter Straat, *Ontwerp tot een minst kostbaare, zeekerste en schielykste herstelling van de zorgelyke toestand der Westfriesche zeedyken, zonder dat het voortknagend Zeegewormte daar aan eenige hindernisse kan veroorsaken* [...], Amsterdam, J. Oosterwyk, 1735. See also, Seger Lakenman, *Ontwerp van een onkostbaar en zeker middel, om de Westvriesche Zeedyken [...]*, Amsterdam, J. Ratelband, 1733; Max Dendermonde, Herman Anne Marie Cato Dibbits, *De dijken. Een nationale uitgave*, Amsterdam, De Bezige Bij, 1954, pp. 73-75, and Jan Albert Bakker, *Protection, Acquisition, Restoration, and Maintenance of the Dutch Hunebeds since 1734: an active and often exemplary policy in Drenthe*, in "Berichten van de Rijksdienst voor het Oudheidkundig Bodemonderzoek," vol. XXIX, 1979, pp. 143-83.

[205] Joseph de Laporte, Louis-Abel de Bonafons abbé de Fontenay, Louis Domayron, *Le voyageur françois, ou La connoissance de l'Ancien et du Nouveau Monde, Mise au jour par M. l'Abbé Delaporte. Nouvelle Edition*, Paris, L. Cellot, vol. XX, 1778, p. 146. On the visit to the seafront dikes in 1756 we read: "Along the shore, piles are sunk in the sand; they were covered with planks or interlaced with tree branches to form a sort of encasement, into which were poured large stones brought from Norway. All Dutch ships that trade in the North are obliged to load a certain number of these stones as ballast when returning. These encasements, on the shore side, are overlooked by a parapet of fifteen to eighteen feet high, and seven or eight in width, composed only of seaweed, with no further fashioning. These plants collapse under their own weight, are thus converted into earth, and are a pretty solid mass, behind which is located the real dike, elevated about two fathoms above the highest sea, but four or five feet lower than the parapet. Numbered stakes separated one from the next by forty to fifty fathoms, signal to those taking care of the dikes the part entrusted to their care. The dike is made uniquely of sand, it can be damaged by the wind, to prevent this trouble, rushes & reeds are grown there that hold them together and harden them. This part, covered with grass on the shore side, has a embankment which, by a gentle slope, slowly drops to the level of the countryside, over which the parapet dominates by more than twenty feet" (*ibid.*, pp. 146-47).

**86**. Anonymous Dutch, view of the port and city of Jakarta, 1724-26 (B.0032 (109), 06, 120, Nederlands Scheepvaartmuseum Amsterdam).

**87**. Wooden levees ([G. Walter], *Brücken-bau oder anweisung wie allerlei arten von brücken sowohl von holz als steinen nach dem desten regeln der zimmerkunst dauerhaft anzelegen sind [...]*, Augsburg, Beith, 1766).

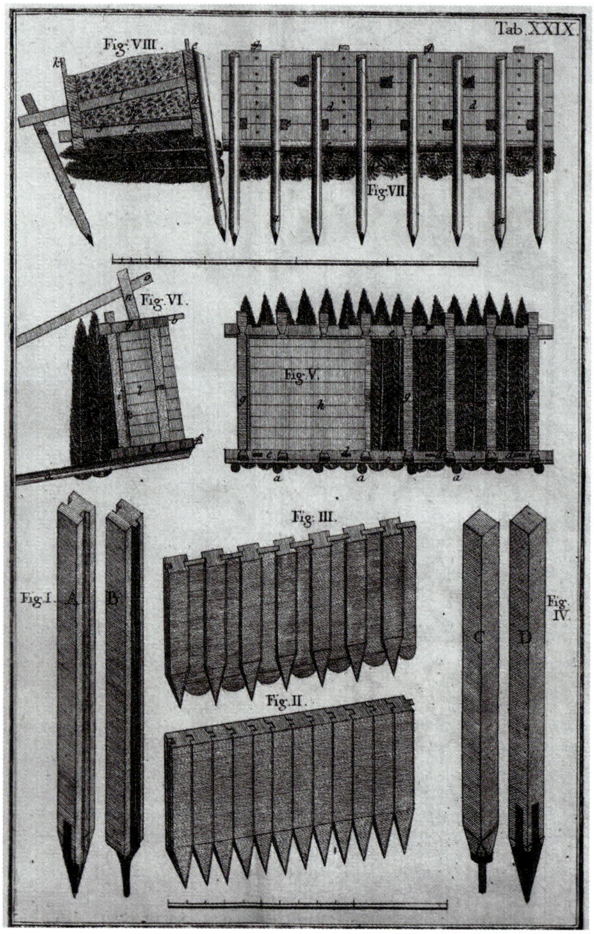

because, even if there was some wind, the ocean produced some small foamless waves that rolled up to the first stones. The battlements of Saint-Malo, the jetty at Havre-de-Grace, and the one at Dunkerque oppose their efforts with, for quite some time now, a sea much more terrible than the one at Zuyderzee. In a word, I saw the result of a lot of work, but rather simple: I had the idea of a marvel."[206]

## Cofferdams for the foundations of hydraulic works

The most important field of experimentation for the processes of construction with cement – where complex carpentry structures are developed – is still that of the construction of the foundations of bridges and piers in the first decades of the 1700s. The references that recur most often are still the recommendations found in the treatises of Vitruvius, Alberti and Scamozzi. The carpentry for making underwater works is fundamental for the success of the construction. Its varying degree of complexity depends precisely on the type of masonry, the quality of the mortars and the possibility – or lack of it – of using pozzolana.

For the foundation of sluices or the piers of bridges in shallow water use continues to be made, as in the previous centuries, of temporary watertight enclosures, known as cofferdams, "*passonate*" and "*batardeaux*," to create zones on the bed of a river, a lake or the sea, where once the water trapped inside during their installation has been removed, it is possible to proceed with construction of the pile foundations and masonry. The making of these enclosures represents a crucial chapter for the future development of formwork for concrete, due to the methods of assembly of wooden poles and boards, which are gradually perfected across the 18th century (nevertheless, the cofferdam can also be a dam made only with packed earth).

As in the previous centuries, in the 1700s to make a carpentry cofferdam it is necessary to use a technique similar to that of adobe construction in which earth is packed into wooden coffer molds: first two parallel sides composed of wooden boards are assembled, arranged horizontally and held in place by posts set at regular intervals and connected at the top by pairs of small beams; then clay-rich soil is poured and pressed into the space between the sides, creating a solid watertight enclosure.[207]

"We plant two rows of piles," explains Bélidor, "one parallel to the other, placed at a distance proportional to the height of the water, held together with ribs and spacers, and then sheet piling is driven into the interior of the batardeau along these piles, forming a box that is filled with clay or other binding soil, or even *crayon* [a sandy clay soil], which is as strong as the clay when it is wrought. [...]. To properly use the clay, it is first reduced into egg-sized pieces, so that it can be sliced open to see it is devoid of sand or small gravel. Then we water it, to beat & knead with feet on a board, which is done only the day after it was moistened, taking care that it neither too much nor too little moistened. We then make loaves of it that are thrown to the bottom of the cofferdam, where the water emerges as it is filled, the workers beat it layer by layer with a rammer until we have reached about two feet above the water level outside, and even higher, if it is in the sea, fearing that when agitated it might pass over.

[206] *Ibid.*, pp. 147-48.
[207] See the chapter *De la construction des batardeaux servant à faciliter l'exécution des ouvrages hydrauliques*, in, Bélidor, *Architecture Hydraulique. Seconde Partie* cit., vol. I, 1750, pp. 123 *et seq.*

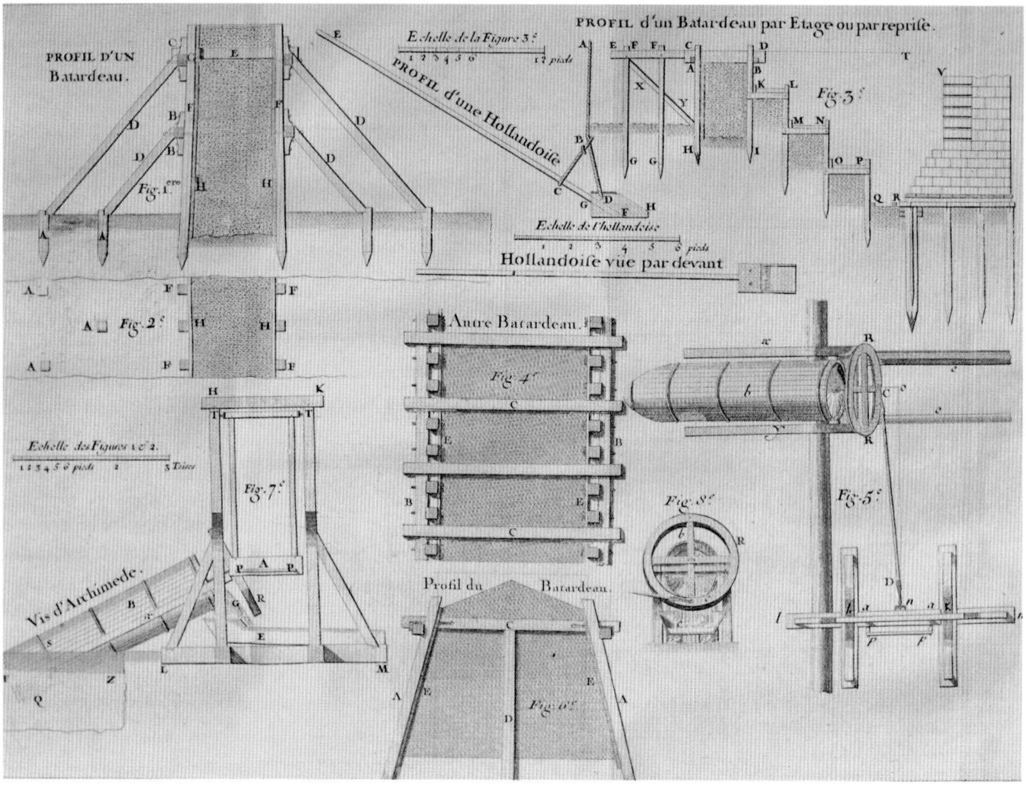

**88**. Bernard Forest de Bélidor, drawing of a cofferdam (B.F. de Bélidor, *Architecture Hydraulique. Seconde Partie, qui comprend l'art de diriger les eaux de la mer & des rivieres à l'avantage de la défense des Places, du Commerce & de l'Agricolture*, Paris, Ch.-A. Jombert, vol. I, 1750).

**89**. Draining by hand of a foundation excavation (B.F. de Bélidor, *Architecture Hydraulique, ou L'Art de conduire, d'élever et de ménager les eaux, pour les différens besoins de la vie. Premiere Partie, Tome Premier*, Paris, Ch.-A. Jombert, 1737).

**90**. Tileman van der Horst, cofferdam drawing (T. van der Horst, *Theatrum machinarum universale; of keurige verzameling van versheide grote en zeer fraaie waterwerken, schutsluizen, waterkeringen, ophaalen draaibruggen*, Amsterdam, P. Schenk, vol. II, 1737).

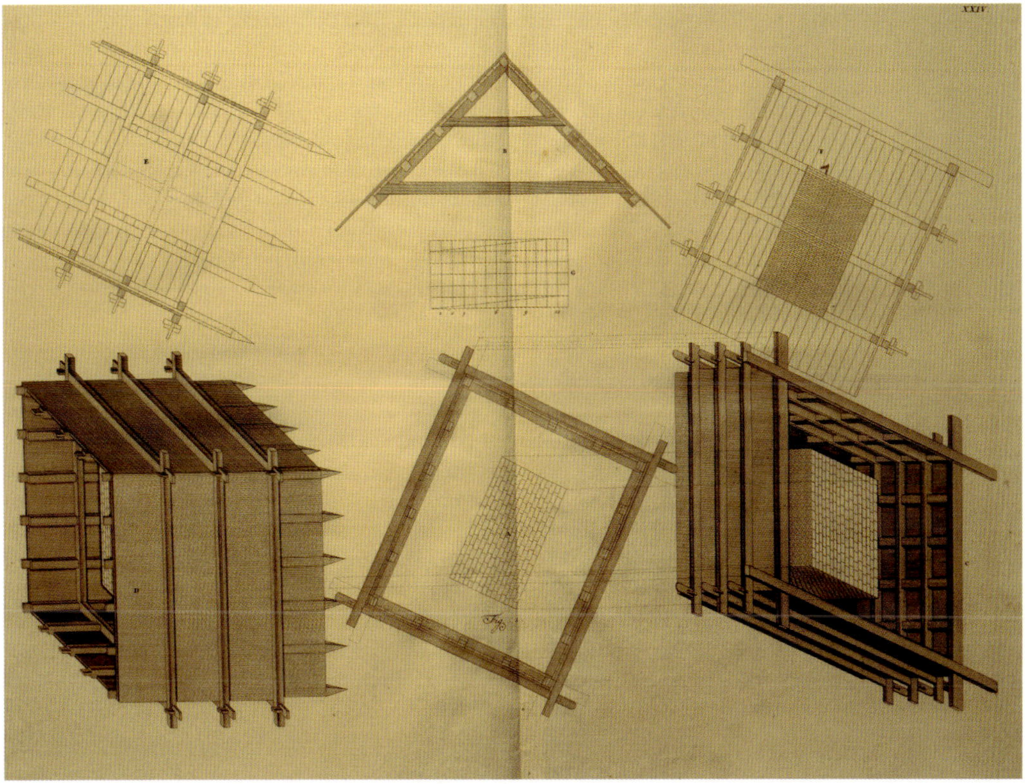

When cofferdams done by encasement and with good earth, it is not necessary to give them as much thickness as for the other case [...]."[208]

Once the water has been removed from this chamber by machines or buckets, the large space bordered by the cofferdam becomes the site of the work on the pile foundations and masonry construction (the cofferdam is dismantled when the height of the masonry reaches a level above water). Emblematic 18th-century examples of the use of the wooden cofferdam, often mentioned in treatises on hydraulic construction, include the lock on the Aa River at Gravelines,[209] and the bridge on the Loire at Blois, built in 1716-24 based on a design by Jacques-Jules Gabriel (1667-1742) – with the collaboration of Robert Pitrou (1684-1750) and Jean-Baptiste de Règemorte (? - 1725). One of the first detailed technical drawings for the construction of a *batardeau* to use for masonry in water was published in a work by Tileman van der Horst.[210]

[208] *Ibid.*, pp. 125-26. The cofferdam is described not only in the treatises that specialize in hydraulic works, but also in the *Encyclopédie* of Diderot and D'Alembert that documents their main features (*cf. Recueil de planches, sur les sciences, les arts libéraux et les arts méchaniques, avec leur explication, seconde livraison, en deux parties, Première partie*, Paris, Briasson, David, Le Breton, Durand, 1763, plate XIX).

[209] Bélidor describes the cofferdam used for the locks at Gravelines: "The interior of the cofferdam was then cased with E boards 2 inches thick; moreover all is well attached with harpoons, bolts & iron pins. This chest was filled with earth pounded [à la demoiselle], raised up to 5 or 6 feet above the highest spring tides, to prevent waves from passing over" (Bélidor, *Architecture Hydraulique. Seconde partie* cit., vol. I, 1750, p. 127).

[210] Tileman van der Horst, *Theatrum machinarum universale; of keurige verzameling van versheide grote en zeer fraaie waterwerken, schutsluizen, waterkeringen, ophaalen draaibruggen*, Amsterdam, P. Schenk, vol. II, 1737.

### Foundations of jetties and moles with coffers or caissons, from Civitavecchia to Genoa

In the ports of the Mediterranean, where it is not possible to take advantage of the low tides of the Ocean, and where pozzolana is available, a carpentry enclosure continues to be generally used to build underwater foundations, known as a caisson or a coffer, depending on the size.

The caisson has a prismatic form. It is constructed on dry ground and caulked so it can float; then it is put into the water and anchored with ropes at the site selected for the foundation. When it is filled with masonry, with stones and pebbles embedded in pozzolana mortar, it sinks. During the course of the 1600s the construction of foundations using caissons is discussed in manuals, and practiced with different variations from Genoa to Savona to Palermo (some suggest a truncated pyramid form to improve the connections between the sunken caissons).[211]

In the early decades of the 1700s, the construction of jetties and moles in Italian ports continues to be done with caissons and pozzolana-based cement. Outstanding examples include Civitavecchia and Viareggio. The project to reinforce the antemural in the port of Civitavecchia, developed between 1700 circa and 1708 by Carlo Fontana (1638-1714) and Giovanni Battista Contini (1642-1723), calls for the sinking of seven caissons to form a new barrier in the sea.[212] The construction of jetties in the port of Viareggio is done with caissons – construction is undertaken in the early 1720s. The drawing of one of these caissons, sunk in 1723, shows the carpentry structure ("main skeleton"), the boards and the iron reinforcements, while in the list of required supplies we find pozzolana.[213] The operation of sinking the last caisson in the port of Viareggio does not go as planned, because it "slipped a few yards towards the canal," blocking the passage of boats, as Zendrini writes.[214]

A variation of the caisson carpentry method that goes through major evolutions and meets with international success starting in the 1730s as an alternative to the cofferdam, features special anchors that make it possible to disassemble the sides of the caisson from the bottom, after it has been sunk and after the masonry has been constructed; the sides are then reassembled on another base to form a new caisson. To make a solid barrier against the pressure of the water with only the wooden sides, and to permit their disassembly,

[211] *Cf.* Gianfranco Faina, *Ingegneria portuale genovese del Seicento*, Firenze, Giunti Barbèra, 1969.

[212] *Cf.* "Profilo per la dimostrazione dell'esperienza," "profilo per la dimostrazione dell'operazione proposta," "pianta dell'antimurale del Porto di Civita Vecchia," drawing from the start of the 18th century (Giovanna Curcio, Paola Zampa, *Il porto di Civitavecchia dal XV al XVIII secolo*, in, Giorgio Simoncini, ed., *Sopra i porti di mare. IV. Lo Stato Pontificio*, Firenze, Olschki, 1995, fig. 44, Camerale III, busta 845, Archivio di Sato di Roma). See also, *Civitavecchia, survey of the 'antimurale' (1708 and earlier)*, in, Allan Braham, Hellmut Hager, *Carlo Fontana. The Drawings at Windsor Castle*, London, A. Zwemmer, 1977, pp. 167-70.

[213] "Disegno, e pianta, col quale si dimostra l'ordine di ciascheduna delle parti che compongono la fabbrica del Cascione, che dall'Ill. Offizio sopra la Foce è stato fatto porre in mare alla Fossa di Viareggio dalla parte di ponente in quest'anno 1723," in, Corrado Benzio, *Viareggio. Storia di un territorio. Le Marine lucchesi tra il XV e il XIX secolo*, Pisa, Pacini Editore, 1986, p. 148.

[214] Bernardino Zendrini, report, May 1735, cit. in, Flavio Serafini, *La flotta scomparsa. Storia degli armamenti velici viareggini. I bastimenti*, Milano, Gribaudo, 2010, p. 10.

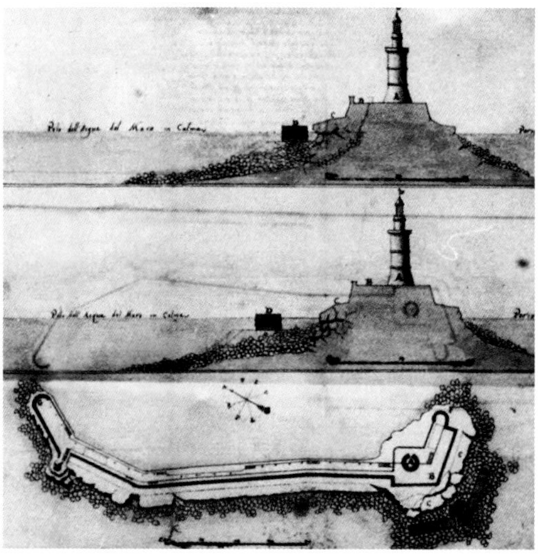

**91**. Carlo Fontana, Giovanni Battista Contini, project to reinforce the breakwater of the port of Civitavecchia, 1700-08 circa (G. Simoncini, ed., *Sopra i porti di mare. IV. Lo Stato Pontificio*, Firenze, Olschki, 1995).

**92**. Project of a caisson for the construction of the jetty of the port of Viareggio, 1723 (C. Benzio, *Viareggio. Storia di un territorio. Le Marine lucchesi tra il XV e il XIX secolo*, Pisa, Pacini Editore, 1986).

**93**. Plans and section of a caisson used for the construction of the mole of the port of Nice, and section of the walls in the port of Toulon (B.F. de Bélidor, *Architecture Hydraulique. Seconde Partie, qui comprend l'Art de diriger les eaux de la Mer & des Rivieres à l'avantage de la défense des Places, du Commerce, & de l'Agriculture*, Paris, Ch.-A. Jombert, vol. II, 1753).

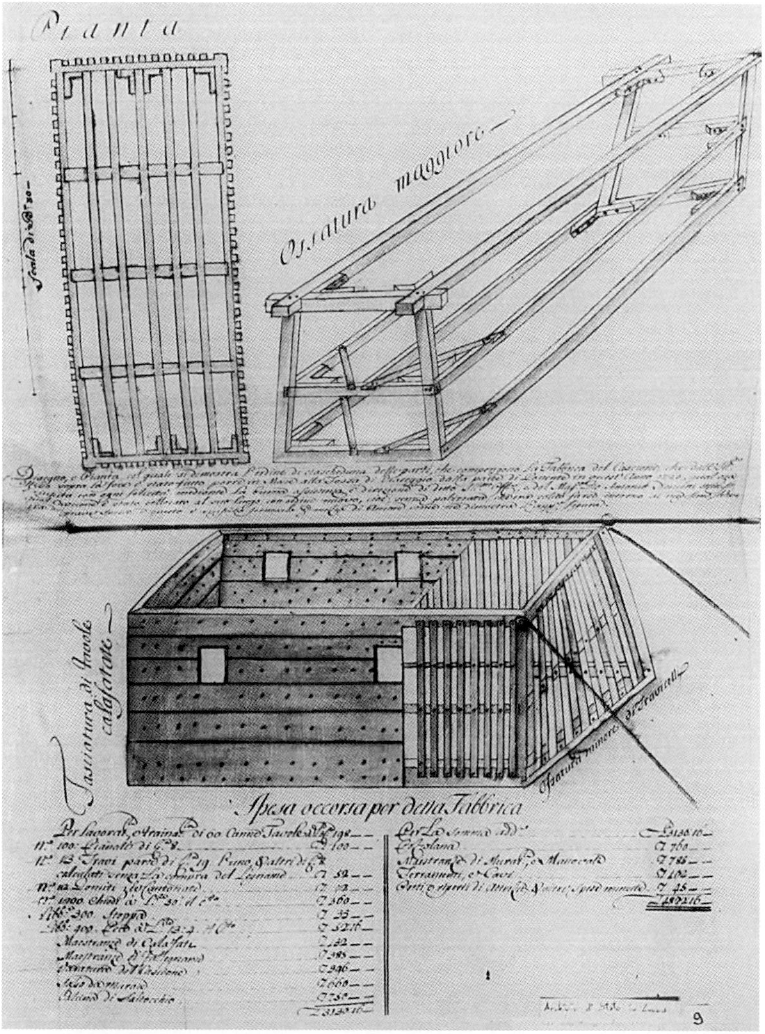


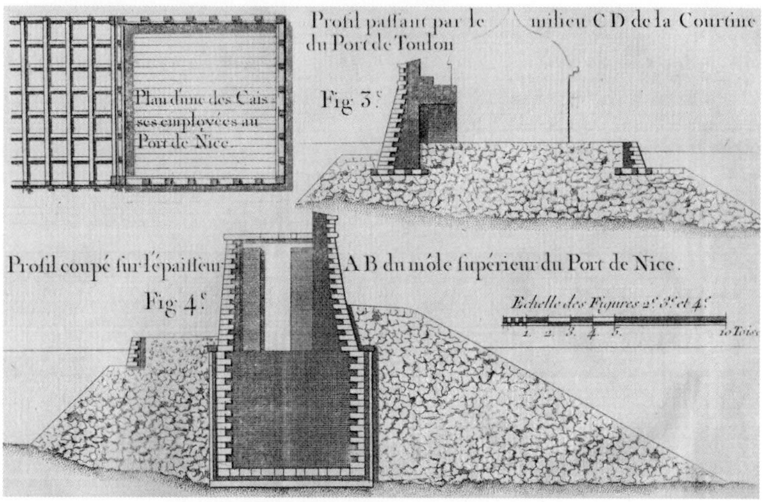

GENOA.

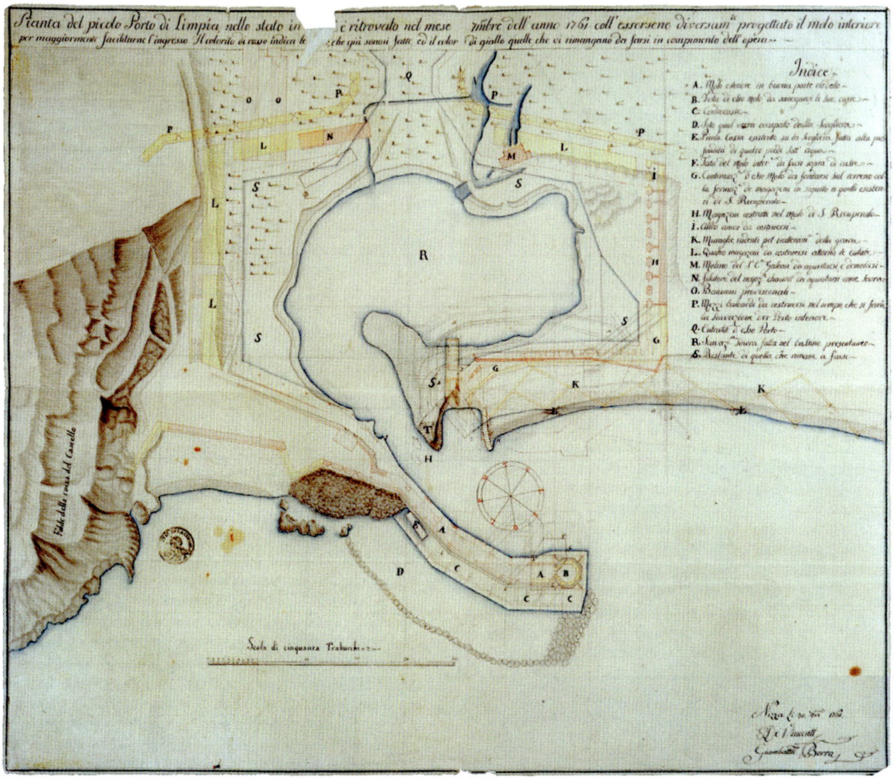
Pianta del picolo Porto di Limpia nello stato in ... ritrovato nel mese ...mbre dell'anno 1761 coll'essersene diversam.te progettato il molo interiore
per maggiormente faciltarne l'ingresso. Il colorito di rosso indica le ... che già sonosi fatte, ed il color di giallo quelle che vi rimangono da farsi in compimento dell'opera.
Indice
Scala di cinquanta Trabucchi

**94**. Pieter Mortier, view of Genoa, Amsterdam, 1704 (1049B12.062, Atlas Van der Hagen, Koninklijke Bibliotheek, Den Haag).

**95**. Giovanni Battista Borra, plan of the port of Lympia, 1761. The letter E indicates a "small chest existing on the rocks" (Ph. Graff, *L'exception urbaine. Nice: de la Renaissance au Consiglio d'Ornato*, Marseille, Parenthèses, 2000).

the carpentry technique is refined with sophisticated solutions comparable to those applied in naval shipyards. This advanced type of floating caisson is used as an alternative to the cofferdam, in the construction of the wall of the quai of Place de l'Hotel de Ville in Toulon (the caisson has a base of 60 x 12 feet, with removable sides 23 feet in height), and for the foundation of the upper mole of the port of Nice (the base of this caisson measures 42 by 42 feet, while the sides reach a height of 33 feet).[215]

Similar caisson solutions are proposed for the port of Toulon starting in the 1600s, for the floating drydocks used for ship maintenance, an idea the architect Jean Pomet puts back on the agenda in 1720 – it is also discussed at the Académie Royale des Sciences.[216]

Another solution, again relying on the use of pozzolana concrete, continues to be applied for the construction of jetties or moles foundations. Rather than making the grand carpentry of the caisson, coffers are fabricated to fill with pozzolana concrete with small aggregates. The coffers are then lowered into the sea. This type of masonry is more similar to Roman concrete, given the fact that the aggregates are smaller than those used in caissons, which reduces the time needed for the mortar to set. Nevertheless, it is not always possible to distinguish a coffer from a caisson, since the measurements of the structures vary continuously, even on the same worksite.

The construction system involving the sinking of coffers filled with pozzolana concrete is applied in the port of Genoa starting in the 1600s by Ansaldo de Mari. From 1705-06, for the expansion of the old mole, the system is examined by technicians and eminent scientists like the mathematician and astronomer Giovanni Domenico Cassini (1625-1712) – work begins only at the end of the 1720s, under the supervision, among others, of Niccolò Delle Piane, Giovanni Battista Storace and Benedetto Fernando.[217]

At the port of Genoa a technique is proposed for filling of the gaps between the blocks, different from the manual operation done by workers in the water, in some cases aided by a bell of copper or lead, with an air tube to let them breathe under water. For the new pier Bartolomeo Rossi, in 1728, suggests filling "with pozzolana and fresh lime mortar" the gaps between the "boulders" poured to form the foundation. To explain the effectiveness of his proposal, Rossi cites the example of what is done on Greek islands: "something the Greeks in the East put into practice, in filling damages to the moles and making large blocks under water, as in Crete [...]."[218]

[215] *Cf.* Jean-Rodolphe Perronet, *Mémoire sur les différentes méthodes qui ont été employées pour fonder les ouvrages de Mâçonnerie dans l'eau, & principalement sur celles qui tendent à supprimer les batardeaux & épuisemens dans la construction des Ponts*, in, "Histoire de l'Académie Royale des Sciences. Année MDCCLXVI. Avec les Mémoires de Mathématiques & de Physique, pour la même Année," Paris, Imprimerie Royale, 1769, (pp. 139-48), p. 143. The *Mémoire* is dated 13 November 1765.

[216] *Cf.* Vincent Félix Brun, *Guerres maritimes de la France. Port de Toulon, ses armements, son administration, depuis son origine jusqu'à nos jours*, Paris, H. Pilon, 1861, vol. I, p. 251. Among the various foundation solutions, see that of Gabriel, proposed in 1731, for an "*encaissement*" (encasing or coffer-work) to fill with "masonry with lime and cement in the water" (minutes of the meeting of 9 April 1731, in, Lemonnier, *op. cit.*, vol. V, 1918, p. 73).

[217] *Cf.* Francesco Podestà, *Il Porto di Genova, dalle origini fino alla caduta della repubblica genovese (1797)*, Genova, E. Spiotti, 1913, pp. 141 *et seq. Cf.* Anna Boato, Tiziano Mannoni, *Materiali e tecniche nella Genova portuale: i calcestruzzi 'alla pozzolana' dall'età moderna alla rivoluzione industriale*, in, *Atti del Congresso Calcestruzzi antichi e moderni: storia cultura e tecnologia*, Padova, Libreria Pogetto, 1993, pp. 11-20.

[218] Cit. in, Podestà, *op. cit.*, p. 209. On the port of Candia and the construction of its piers with coffers filled with pozzolana masonry, in keeping with a practice applied since the period of Venetian domination, between 1299 and 1363, see, Ruthi Gertwagen, *The Venetian Port of Candia, Crete (1299-1363). Construction and Maintenance*, in, Irad Malkin, Robert L. Hohlfelder, ed., *Mediterranean cities: historical perspectives*, London, F. Cass and Co., 1988, pp. 148-52.

The system of coffers filled with pozzolana cement is used in other ports in Italy halfway through the century to build foundations in the sea. In the inner harbor of the port of Carrara, in 1749-50, moles are built by sinking coffers of concrete. The fact that the composition of the masonry in the coffers is different from that of the caissons is confirmed by the observations of Lazzaro Spallanzani (1729-99), who describes the "calcistruzzo" used to fill the coffers in the port of Carrara as a "compound of small stones, lime mortar and pozzolana [...]."[219]

In the same period as the work in Carrara, the jetty of Port Lympia in Nice is built, designed in 1749 probably by the military engineer Antonio Felice de Vincenti (1690-1778), reporting to Charles Emmanuel III of Sardinia. At Lympia jetty foundations are created by sinking large coffers of caulked wood filled with pozzolana cement (work begins in 1750 and is still in progress in 1761). The first coffer, about 20 by 12 fathoms, is sunk at the end of the jetty, and the stones begin to be lowered. In October 1750 a storm damages the work. "The entire jetty was turned over and the stones transported onto the banks behind the coffer," an observer reports. "All the planks that covered the coffer were knocked off down to the stones covered in iron that were at the angles of the coffer. The masonry, however, held up against the action of the sea."[220]

### Archaeological research of Fontana on the Roman concrete moles of the port of Anzio

At the start of the century, there is no construction treatise that describes technical methods of underwater foundations using coffers or caissons. Even later, when the descriptions become more precise, there are no explanations of the details of the deployment of coffers or caissons to construct the foundation of a pier. So one of the most useful sources on this score continues to be a "description of the manner of construction of sea moles, according to practice in several places in Italy" presented at the Académie Royale d'Architecture of Paris in September 1699 by Jean-François Félibien (1658-1733), son of the famous André.[221] Only the eminent papal builder of hydraulic works, Fontana, provides a fundamental essay for knowledge of Roman concrete construction processes, and for the development of such techniques in the 1700s: *Antio e sue antichità*, in which we find useful considerations on the system of coffers and caissons. The essay is published in 1710, at the completion of work on the moles of the port of Anzio, done precisely with coffers, caissons and pozzolana cement masonry – the work was begun in 1697 by Alessando Zinaghi (the coffers are done by Sebastiano Morazzo and designed based on those used at Viareggio, measured for the purpose by Giovanni Battista Ricamonte), continued by Leonardo Forlini and finally completed by the Saxon engineer Johann Escher in collaboration with Carlo's grandson, Francesco Fontana.[222]

[219] Lazzaro Spallanzani, letter, 15 October 1783, in, *Lettere di vari illustri italiani e stranieri*, Reggio, Torreggiani e Compagno, 1843, vol. II, pp. 176-77.
[220] Anon., letter, 28 November 1750, in, Philippe Graff, *L'exception urbaine. Nice, de la Renaissance au Consiglio d'Ornato*, Marseille, Parenthèses, 2000, p. 112.
[221] Jean-François Félibien, "Description de la manière de construire des moles avancez dans la mer, suivant ce qui se pratique en plusieurs endroits de l'Italie," in, Lemonnier, *op. cit.*, vol. III, 1913, pp. 381-83. Félibien's report is presented at the session of 22 September 1699. The construction process described by Félibien should correspond to what was implemented for the old mole of Genoa, which is one of the cases cited precisely for construction with coffers, when the members of the Académie Royale d'Architecture discuss foundations in deep water, on 2 December 1709, with Félibien himself, not coincidentally, in attendance. At the session of 1709 the members of the Académie Royale d'Architecture attempt to evaluate major masonry works done with loose stones, or *pierres perdues*, in the sea or in deep rivers. In this context, mention is made of the "large coffers or even flat ships, that one fills with masonry and stones" and with

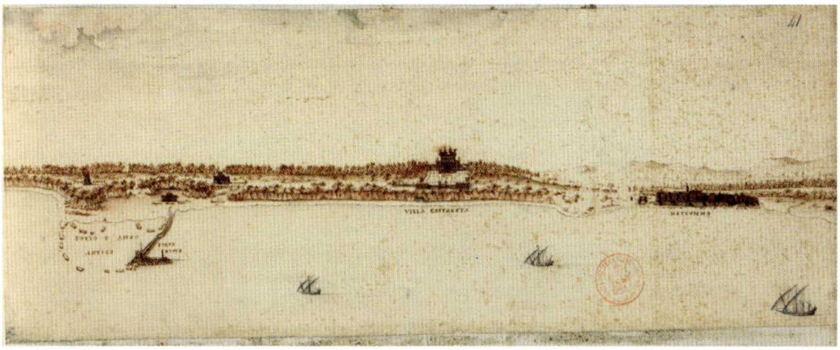

**96**. Anonymous, view of the coast with the port of Anzio and Neptune (Cartes et Plans, Ge CC-1376 (41), Bibliothèque Nationale de France, Paris).

In *Antio e sue antichità* Fontana narrates the observations he made during visits to the ancient Roman moles of Anzio in the company of Francesco and a group of expert master masons and seamen, with the aim of gaining detailed knowledge of the Roman works and of basing his own design for the port, an alternative to that of Zinaghi, his own project is based on those very structures. Precisely this archaeological study permits Fontana to discover a process of construction in water of a masonry made without coffers or caissons.[223]

The first observation that emerges from the surveys has to do with the state of conservation of the masonry of the remains of the "antemurals," which Fontana attributes to the "good binding of the very consolidated Cement"[224] and to the quality of the construction done with "abundant Lime, and perfect Cements."[225] Then he examines the imprints still visible on the mass of concrete, left behind by the materials used to contain the poured concrete, from boards to iron brackets. The circular holes of irregular size, some crossing the masonry and others extending deep into the terrain, are attributed to the use of wooden trunks to reinforce the containers into which the concrete was poured.[226] Thanks to all these marks, the group of experts can outline the system of temporary works developed by the Romans for the pouring of concrete. According to Fontana, the Romans used a system of molds made with two parallel and continuous banks of boards, or coffer, held in place with tree trunks, some driven into the ground, and others as crossbars, positioned at various heights and equipped with metal braces. Confirmation of this hypothesis, according to Fontana and the group of experts, comes from the absence of rocks at the foot of the pier, which would have been placed there in the case of construction using coffers because of their unstable bases, and from the presence of a series of diagonal openings at the base of the masonry, for the outflow of water, which could not have been made except inside molds.[227] Fontana calls the continuous molds for

which piers continue to be built in Italy, starting from Roman practices – the examples are the ancient port of Ostia, the recent old mole of Genoa, as well as the French case of the port of Sète (minutes of the assembly of 2 December 1709, in, Lemonnier, *op. cit.*, vol. III, 1913, p. 329).

[222] The construction work on the new moles of Anzio are assigned to Alessandro Zinaghi, begin in 1698 and are completed in about 1700, using the system of coffers filled with concrete and arranged in rows. The captain Leonardo Forlini, in 1699, seals with "small materials, & excellent lime" the gaps between the coffers, while by order of the pope he prepares a large caisson, 90 palms square, for the construction of a platform for cannons – but the caisson breaks when it is sunk, once it has been filled with masonry, and the platform is not completed until the first years of the next century, thanks to the intervention of the Saxon engineer Johann Escher in collaboration with Francesco Fontana (Carlo Fontana, *Antio e sue antichità descritte dalla Punta S. Giovanni ai Volaci in vicinanza del nuovo Porto*, Roma, G. F. Buagni, 1710, p. 31).

[223] On the methods used for the construction of the moles of Anzio at the time of the Romans see, Enrico Felici, *Osservazioni su porto neroniano di Anzio e sulla tecnica romana delle costruzioni portuali in calcestruzzo*, in, "Archeologia subacquea. Studi, ricerche e documenti," I, 1993, pp. 71-104.

[224] Fontana, *Antio e sue antichità* cit., p. 13.

[225] *Ibid.*, p. 23.

[226] *Ibid.*, p. 24.

[227] *Ibid.*, p. 27.

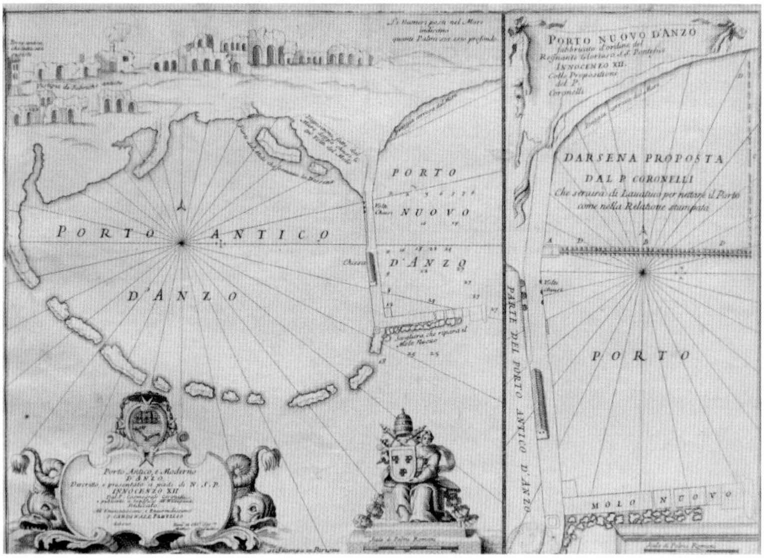


**97**. Vincenzo Maria Coronelli, "Antique and Modern Port of Anzio," 1699 (Cartes et Plans, Ge CC-1376 (43), Bibliothèque Nationale de France, Paris).

the pouring of the conglomerate prepared for the moles of Anzio *vacui arginati*, or dammed up voids. One of the characteristics he assigns to these *vacui arginati* is speed of deployment, with respect to the methods of pouring concrete into coffers or caissons.[228] The absence of fillers between one masonry sector and the next, typical of constructions using coffers or caissons, is seen as further confirmation that the concrete of the moles of Anzio was poured into *vacui arginati*, or coffer-work like that documented by Palladio.

In his reasoning on the *vacui arginati*, Fontana also offers readers a succinct description of the method of construction of the conglomerate with coffers whose sides are boards, with crosswise reinforcements, filled with masonry and sunk; he thus brings out the defects of this system when compared to the system of the moles of Anzio – the usual gaps in the masonry between one coffer and the next. Though the painstaking and knowledgeable observations of Fontana find their way to only a very limited audience, they offer evidence of the expertise of Italian architects in working with concrete, using knowledge acquired by studying ruins. Precisely the technique of the *vacui arginati* becomes famous in the works done in Toulon and narrated in one of the most famous construction manuals of the 18th century.

"Modern coffers," writes Fontana, "are made to contain masonry, and composed of wood with strong reinforcement, with crosswise pieces both above and in the middle to make a good connection. These coffers are placed, empty, at the selected location in the sea; then they are filled with masonry, and its weight makes them sink into the water. Instead the barriers that are built for defense against all kinds of water and to keep dry sites selected for the construction of edifices for ports, or for other uses, that can be seen in different parts of the world, are constructed of strong and well-connected wood, and filled with a material that is not dense, and is therefore suitable to break up the unity of the sea's waves. Then there are other breakwaters, also in wood, but

[228] *Ibid*.

woven in a denser fashion and filled with a denser material, to prevent penetration by the water and to make the voids, in which the structures are built, dry so that the workers can operate in total freedom. [...]. The walls of the mole that can be seen towards the Fortress of Neptune are well joined and reveal no openings; this makes it possible to state that that mole was not made with coffers, given the fact that there are none of the interspaces generated by the disappearance in time, due to corrosion, of the boards of the ends of the coffers, placed one against the next. This observation leads us to believe that the mole was not built with coffers, but laid in a continuous way inside those openings thus well defended against the water. Furthermore, various wall parts have been found that display a union that would not be possible with the use of coffers, given the fact that the coffers, when they are placed one beside the next, cannot produce such continuity, for the reasons already explained. Moreover, the use of coffers produces irregular and winding wall shapes [...]. All this leads us to believe that the said mole was constructed through excavation on site, defended by barriers. Also, the above-mentioned visible parts of the mole display, in their masonry mass, lines that are higher here, and lower there, which indicate the process of fabrication through gaps, which can be lower or higher, depending on the needs of the construction. Now, if the mole had been constructed with coffers, its form would be different, in the grafts and in the heights, and this fact too leads me to believe that the mole was made by means of vacui arginati [dammed up voids]. Given that perpendicular holes are found here and there in the body of the above-mentioned walls, without there being a regular distance between them, it is possible to glimpse, in those holes, the imprints of the irregular pieces of wood that were inserted in those excavations made to act as barriers. The function of these pieces of wood can only be that of supporting the decks to allow the workers to have close at hand the materials ready to be poured into the large openings. Once the wall has been built, the wood remains buried. This too is another clue as to the fact that the pier was built with formwork compartments, since the wood of the reinforcement of coffers would have had quite a different position and form. In like

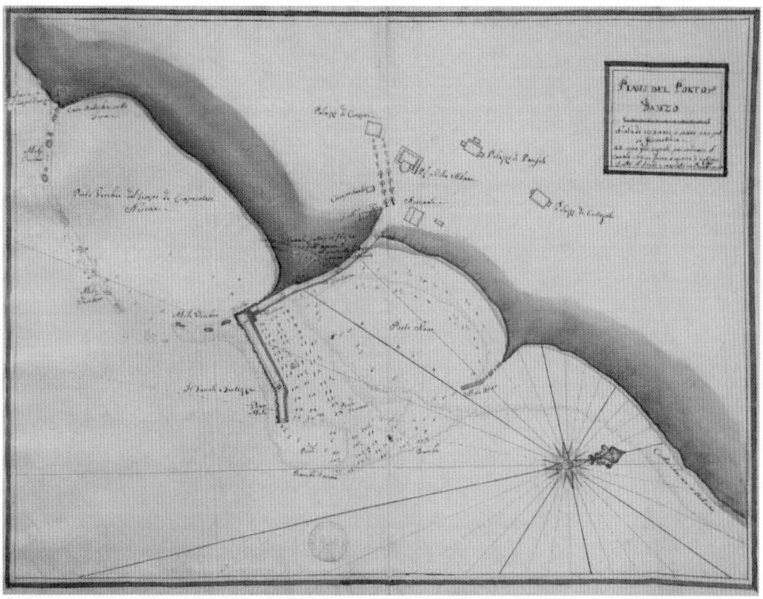

**98**. Anonymous, "Plan of the Port of Anzio," 1700 circa (Cartes et Plans, Ge CC-1376 (42), Bibliothèque Nationale de France, Paris).

manner, crosswise holes have been found, without any regular distance, corresponding to the perpendicular holes. From their imprint, we can understand that those holes were left by the pieces of wood placed there to form the work deck, for the supply of the construction materials. Once again, this demonstrates that the mole was made with dammed up voids, and not with coffers, given that the crosswise pieces of the coffers would have had another form. Furthermore, I have found between the lateral walls of the port, under water, some of them made with matched and level faces, with orderly, well-laid stones, and without any imprint of wood that would indicate some reinforcement, or the surface of the coffers. And this is an evident reason to suppose that that mole was made with dry, dammed up voids, and not otherwise, given the fact that similar work cannot be done in the water."[229]

The port of Anzio with the recent work on its piers also becomes the pretext for another description of the caisson system. After having discovered pozzolana in different French holdings, Labat narrates – in the sixth book of the series *Voyages en Espagne et en Italie*, published in 1730 – the main technical methods for foundations in the sea used at Anzio, where he traveled in the summer of 1711.

"Caissons are used to make the moles," he writes, "these are wooden cases of dimension 18 by 20 feet squared, well nailed and caulked; we bring together as well as possible the stones carefully tossed to the bottom of the sea, on which the caisson will be placed, and the caisson being held with ropes and anchors on this spot, it is filled with good masonry, into which stones are placed in a bathing of lime mortar and pozzolana. The caisson sinks as it is loaded with masonry, and, as soon as it touches the bottom and it is good and firm, and we attach it with iron rings to those right next to it, and then we throw large stones and quarter boulders in the sea in order to support it on the side that the sea comes, and we thus cover it with an antemural against which the waves break and dissipate before their force can be felt by the masonry work. Depending on the width desired for the mole, we put two or three caissons in front; they are attached to one another with strips of iron, and as the wood in their constructions eventually rots, the moles receive no harm because the large blocks of masonry constitute a single body, even if composed of several parts, and are so well closed up and so well supported by the antemural, that they are in the condition to resist the fury of the most agitated sea, or, if they are damaged, they are easy to repair, as I have observed elsewhere, the antemural of Civitia-Vechia [*sic*]."[230]

The work on the antemural to which Labat refers may be the one done with a sort of *passonate*, as described in 18th-century documents regarding the port of Civitavecchia: "coffers with vertical posts, lined inside with straw and canvas," into which the masonry is poured.[231]

[229] *Ibid.*, pp. 25-26.
[230] Jean-Baptiste Labat, *Voyages en Espagne et en Italie*, Paris, J.-B. Delespine, 1730, vol. VI, pp. 53-54.
[231] "Relazione dello stato in cui si trova l'Antemurale del Porto di Civitavecchia, e de' provedimenti applicabili al mantenimenti, e restaurazione," 1778, Camerale III, busta 845, Archivio di Stato, Roma.
[232] Henri Gautier, *Traité des ponts, où il est parlé de ceux des Romains & de ceux des Modernes*, Paris, A. Cailleau, 1979, p. 117.
[233] *Ibid.*, p. 15.
[234] *Ibid.*, 155.
[235] *Cf.* Y.H. Piet, M. Champion, *La réparation de la huitième arche du pont Jacques Gabriel sur la Loire à Blois*, in, "Revue générale des routes et des aerodromes," LIII, July-August 1716, no. 555, pp. 7-22. About French engineers see, Antoine Picon, *Architectes et ingénieurs au siècle des lumières*, Marseille, Parenthèses, 1988.

## Masonry and "encaissements" according to Gautier

The construction criteria of bridges and the questions regarding their materials, as outlined during the course of the centuries, are summed up in a treatise destined to become a reference point for European engineers and architects, that of Henri Gautier (1660-1737), *Traité des ponts*, published in 1716, expanded in the 1728 edition and reprinted many times until 1768, when it is replaced by the treatise on bridges written by Émiland-Marie Gauthey (1732-1806).

The *Traité des ponts* is particularly important to assess knowledge on the subject of hydraulic mortars, and it allows us to better appreciate the fundamental change that takes place in the science of engineering, little more than one decade after the first publication of the treatise.

The concept of masonry for the piers and arches of a bridge outlined by Gautier corresponds to the logic of the stone cutting, or *coupe des pierres*, calling for a type of stone masonry, for the nucleus or the filler, generally not different from that of the cladding, with courses leveled at the same height. "Indeed," Gautier writes in a crucial passage of the treatise, "stone cutting is the soul of all vaults and all stone bridges, and must be considered as the main basis of their construction."[232] The masonry model for the piers is the one described by Palladio and summed up by Gautier as follows: "their structure must be made of large quarters of good stone well tied together with iron clamps, or other metal, so that with this manner of chaining them together they are as if one single stone."[233] A different type of masonry, closer to Roman poured concrete (though this reference is not indicated), is described for the reinforcement of piers with cribs, that is through "tossed material, either loose stones, which will fill the space underneath the piers by rolling on top of eachother, or otherwise loose masonry that flows everywhere, and where it takes hold when it is made on the spot, with the lime slaked instantly and well enclosed."[234] The fact that the use of special limes makes it possible, in the construction of bridges, to fabricate mortars that are similar in solidity to Roman cements, to guarantee the construction of a true pouring of concrete, is borne out not only by certain sporadic excerpts of Gautier's text, but also by the poured masonry in the arches of works like the Gabriel bridge on the Loire, at Blois.[235] It is worth noting, on the subject of poured materials, that Gautier acknowledges the existence of a process of geological creation of a natural conglomerate similar to the one discussed in the debate on ancient Egypt, when he describes the nature of the terrain on which the piers of the Cesse aqueduct (Pont-Canal) stand, begun by Vauban and completed by Antoine Niquet (1636-1726): "The Cesse bridge that is under the Canal Royal of Languedoc, has only a founda-

**99**. Jacques-Jules Gabriel, bridge on the Loire, Blois, 1716-24. Photograph of the restoration work on the eighth arch ("Revue générale des routes et des aerodromes," LIII, July-August 1979, n. 555).

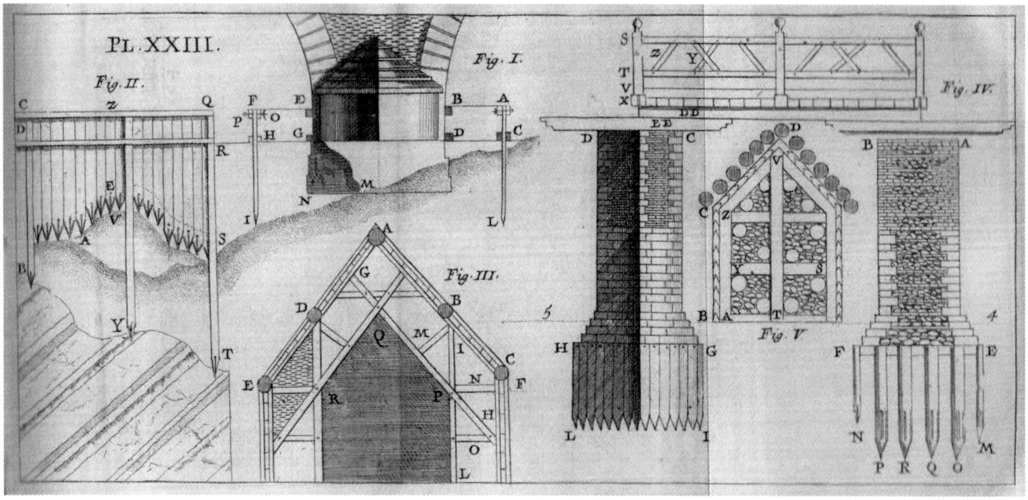


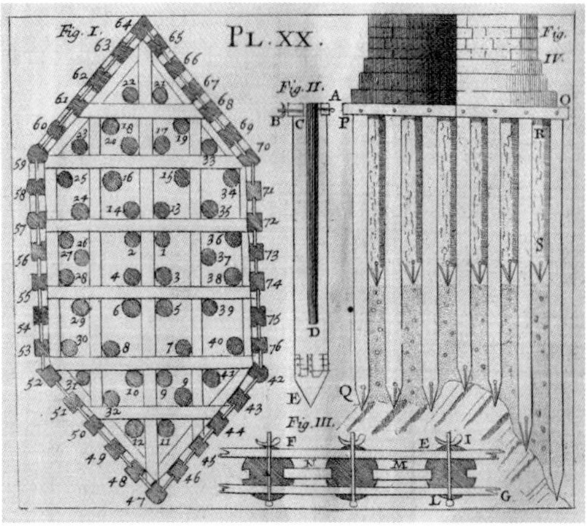


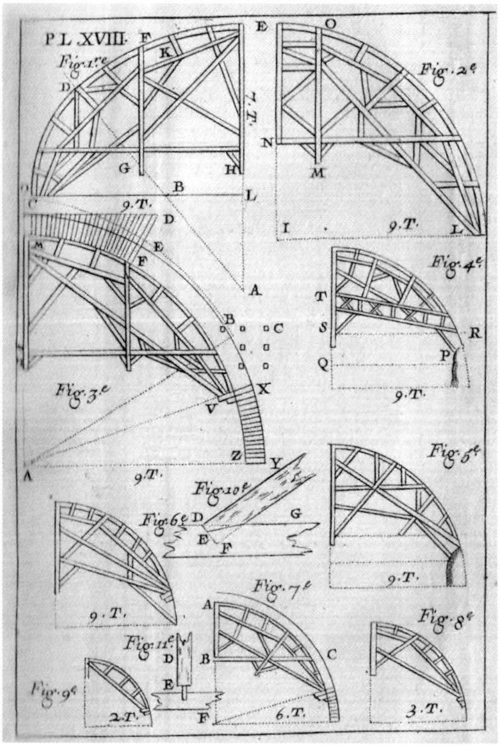


**100-102**. Henri Gautier, foundation systems for bridge piers and reinforcements for centering of bridge arches (H. Gautier, *Traité des ponts, où il est parlé de ceux des Romains, & de ceux des Modernes*, Paris, A. Cailleau, 1716).

tion of large stones, of which several are massed together, frozen together by a petrifying substance, that has united them in the form of a bank, and that as the river hollowed out its bed and banks, caused it to crumble at the top at the level of the plain, where it is completely strewn."[236]

The *Traité des ponts* summarizes the main systems of pier foundations, starting with those described by Scamozzi, all based on a type of masonry work protected from water, ranging from the construction of cofferdams with earth pressed into an interspace, to the deviation, where possible, of the course of the river until "the masonry has been able to grip."[237] One of the foundation systems described by Gautier is composed of floating wooden decks and known as "grids, or rafts." Precisely in the explanation on the masonry to install on the rafts, Gautier makes the sole reference to pozzolana in his whole treatise, though without offering any technical information about its characteristics, confirming the fact that at the start of the 1700s, away from the Mediterranean ports, masonry with the Roman method was certainly not practiced by the leading European construction specialists.

"The second," he explains, speaking of the foundation of the piers, "is made by building the foundation on grids, or rafts made of good oakwood, strong and well tied together, supported on the water's surface with cables, or mechanically, building them up with large quarter boulders clasped together, with joints of good lime mortar, or of pozzolana, or cement; then letting then descend with the use of the same cables and mechanics, slowly, and straight down to the bottom of the water, as was done, [Scamozzi] says, during the times of the emperor Claudius at the port of Ostia and as Draguet Reys did a century ago in Constantinople, for the beautiful mosque he built in the sea."[238]

From the technical literature on construction in the sea, Gautier also reports on methods of masonry structure in deep water, either through piers, that is to say large stones sunk and then connected "with stone chains constructed during the lowest of tides, with masonry cladding – solid block structures on which lighthouses, storerooms, gun batteries, docks, etc., were built [...]"[239], or through different types of molds, like the caisson, or *encaissement*, though here Gautier provides only pithy descriptions: "This is any carpentry work in which one loosely pours masonry, dry stones [...]."[240] In the explanation of this type of work, the type applied in the Mediterranean ports where pozzolana is used, Gautier does not specify the particular nature of the masonry. "We make a base with cases and boats built specially for the task," he explains, "in which material is placed, and which we sink differently, depending on the need: and whose blocks are tied together in different ways, thus one builds dock walls, piers, storerooms, and piles in rivers [...]."[241]

Until about midway through the 1700s, the main task of engineers – especially where it is hard to get supplies of pozzolana, *trass* and other products with similar hydraulic properties – lies in perfecting cofferdams and caissons to be able to continue to use a type of stone masonry like that described by Gautier, where the mortar plays a secondary part and the stones are connected by metal spikes. The rise of Roman cement alters the logic of European construction sites, especially starting with two Italian hydraulic works, in Ancona and Venice, that are destined to bring pozzolana back to the center of attention.

[236] Gautier, *Traité des ponts* cit., p. 104.
[237] *Ibid.*, p. 87.
[238] *Ibid.*
[239] *Ibid.*, p. 100.
[240] *Ibid.*, p. 152.
[241] *Ibid.*, p. 103.

## Cements, coffers and foundations in the water: the recommendations of Bélidor

Readers of architectural treatises possess an important source of technical information on Roman cement and pozzolana in ancient times, namely the *De Architectura* of Vitruvius, enhanced by annotated editions and printed in many variations starting in the 1500s, though the authors that carry on this theoretical tradition in the first decades of the 1700s lack the practical and scientific knowledge needed to rival the technical investigations of the various mortars described by De l'Orme or Perrault.

The technical information gathered by Giovanni Biagio Amico (1684-1754) on the questions of mortars and concrete is too succinct for a book that claims to be a manual, *L'architetto pratico*, published in 1726. Certain constructive details refer to ways of making foundations and walls, some by pouring concrete into ditches, into stone casings and even into a special wooden mold, known as a *cassa*, or coffer, with horizontal wooden boards nailed to poles, for which a drawing and a pithy description are supplied (the drawing published by Palladio on molds for poured materials, on which Amico's is based, was of quite another force and importance).[242] "[...] the cut boards are placed at a distance from each other," Amico writes, "to correspond to the thickness of the wall; they are then filled with mortar, and stones of any kind, one next to the other, course after course."[243] The term *cassa* is also used for the mold formed by surfaces of "square stones" with "the space filled with stones, and earth, or lime mortar."[244] The compound for foundations is described as being composed of "small stones" that "are mixed with lime mortar in a kind of mash."[245] A comparable compound is identified in the "walls of Turin made of river stones."[246] "In Sicily," Amico reports, talking about his own homeland, "we mostly use masonry in stones of different sizes and forms, hewn in a rustic way, filling the spaces with pebbles and lime mortar packed well with hammers."[247]

Perhaps the work that best demonstrates the difficulties of making progress with the theoretical system of the treatise, after the achievement of Perrault, is the edition of *L'Architettura di M. Vitruvio Pollione* edited by Berardo Galiani (1724-74) and printed in 1758 in Naples, which contains only a few additional bits of information on the fabrication of cement with pozzolana (unlike the revised edition of 1790). If anything, it should be noted that the studies that aim more at exploring particular aspects of Roman construction, like roads and ports, published during the years straddling 1700, are the ones that bring new considerations on the art of Roman construction and the quality of its cement.[248] Yet precisely in the first decades of the 1700s, an epochal breakthrough is taking place in construction publications, and one of the central issues of this breakthrough is the question of the composition of mortars and concrete.

Practical rules for the fabrication of mortars, observations on the various compounds and their properties, and the criteria of the deployment of concrete, are explained for the first time – without simply repeating the contents of the treatises dating back to *De Architectura* – by a military man who became

[242] Palladio, *op. cit.*, p. 13.
[243] Giovanni Biagio Amico, *L'architetto pratico*, Palermo, G.B. Aiccardo, 1726, vol. I, p. 63. See the drawing, fig. 11, no. 7.
[244] *Ibid.*, p. 63.
[245] *Ibid.*, p. 61.
[246] *Ibid.*, p. 63.
[247] *Ibid.*
[248] See, for example, Nicolas Bergier, *Histoire des grands chemins de l'empire romain*, Bruxelles, J. Leonard, 1728 (first edition, 1622), or, Giovanni Pietro Lucatelli, *Del porto di Ostia e della maniera usata da' romani nel fabbricare i porti nel mediterraneo*, Roma, Stamperia di Pallade, 1750.

**103-104**. Giovanni Biagio Amico, systems of masonry and foundations (G.B. Amico, *L'architetto pratico*, Palermo, G.B. Aiccardo, 1726, vol. I).

a construction engineer for defensive structures and battlefields: Bélidor, professor of mathematics at the Ecole Royale d'Artillerie, at La Fère, since 1720, initiator of scientific experiments of different kinds in different sectors, from the strength of materials to the statics of vaults (he was expelled from the school in 1741 by Louis Auguste, Prince of Dombes). All his main treatises – that on fortifications, and those on "hydraulic architecture" – go on to become works of reference for 18th-century construction, not just in France but also in other nations, thanks to their translation into several languages. The treatise of Gautier, though it becomes indispensable for the static design of bridges, is far from containing the inestimable quantity of technical and practical information gathered and offered by Bélidor.

At the end of the 1720s, Bélidor has already written *La Science des Ingénieurs dans la conduite des travaux de fortification et d'architecture civile,*

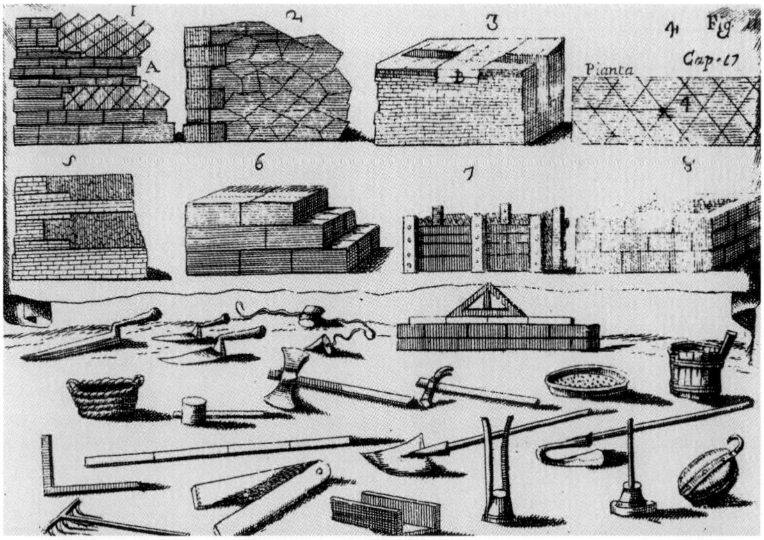

which is published in 1729,[249] and he has partially completed the essay on *Architecture Hydraulique*, which is published in two volumes only in 1737 and 1739, after experiences in the construction of hydraulic works and machines (other volumes of the *Architecture Hydraulique*, fundamental for the study of concrete, are published at the start of the 1750s).[250] Both books, *La Science des Ingénieurs* and *Architecture Hydraulique*, contain precious technical information on masonry made with hydraulic mortar, offering plates of drawings prepared to help explain the technical functioning of the various works. We cannot rule out the idea that some of the most famous inventions of cement, at the start of the 1770s, can be seen as practical applications of certain rules described by Bélidor. In the various subsequent printings, edited by Bélidor himself, explanations and technical drawings are supplied for the most important international hydraulic works.

The 1729 edition of *La Science des Ingénieurs* opens the new chapter in the evolution of the constructive principles of cement masonry. Several considerations offer a glimpse of a certain influence of the constructive outlook of Gautier, when Bélidor describes as the ideal pier of a bridge one made entirely with cut stone, without a nucleus "filled with poor quality rubble"[251] – Bélidor becomes a great admirer of Amédée-François Frézier (1682-1773) as well, and of his treatise on stereotomy published in three volumes from 1737 to 1739. But the pages of *Livre III* devoted to the theory of masonry and entitled *De la Théorie de la Maçonnerie*, leave no doubts about the crucial role of *La Science des Ingénieurs* in the rise of cement masonry. To prepare excellent lime and to correctly select the stones to be calcinated and the way of slaking them in tubs, covering them with sand and soaking them – "lime can be regarded as the soul of masonry," is one of his statements[252] –, Bélidor updates the advice contained in the treatise of De l'Orme and certain techniques prescribed by Vitruvius; he also describes several products possessing particular characteristics, like the lime of Metz which would become, in the course of the 1700s, a goal for inventors and builders engaged in making concrete works.

"There is however a lime of excellent quality that does not run at all," Bélidor writes, "such as that of Metz and its region, where it happens that people with no knowledge about the quality, spread some in holes well covered with sand, and the following year it is found as hard as rock, it was necessary to break it with iron spikes and to use it as rubble, in order to slake this lime, it is covered with all the sand to be taken into the mortar; and one throws some water on it by hand thus watering it, and this is done a number of times, this lime slakes without producing smoke, it makes such good mortar that at Metz almost all the cellars are made of it without any mix other than large river stones, there are no pebbles or brick in it, and it makes a mastic so hard, that the most hardened pikes cannot bite once this mortar has hardened."[253]

Bélidor emphasizes the need to use recently slaked lime, in keeping with the method described by De l'Orme, using a vat with a layer of dampened sand, to produce a strong mortar, following the example of the ancient Romans.

[249] Bernard Forest de Bélidor, *La Science des Ingénieurs dans la conduite des travaux de fortification et d'architecture civile*, Paris, C. Jombert, 1729.
[250] Bernard Forest de Bélidor, *Architecture Hydraulique, ou L'Art de conduire, d'élever et de ménager les eaux, pour les différens besoins de la vie. Premiere Partie, Tome Premier*, Paris, Ch.-A. Jombert, 1737; *Id., Architecture Hydraulique, ou l'Art de Conduire, d'Elever, et de menager, les Eaux pour les differens besoins de la vie. Tome second*, Paris, Ch.-A. Jombert, 1739. In the two volumes of the *Architecture Hydraulique* the question of cement is set aside with a few generic considerations - Bélidor often writes of "cement mortar" and describes a "machine for crushing cement" powered by a horse (Bélidor, *Architecture Hydraulique [...]. Premiere Partie, Tome Premier* cit., 1737,

This recipe indicated by Bélidor later becomes the basis for subsequent discoveries regarding cement.

"In all our observations on lime," Bélidor writes, "we know that the more it is quick, the more it teems when it is slaked, takes more sand, results in a good thick mortar, that having been kept for a long time after slaking, as long as it is kept in sand-covered pits, the better it is. This is why the Romans did not want that it be used for their buildings if it was slaked two or three years previous, and one also notices that lime as powder is worthless, because its salt, having changed it nature and virtue, is no longer right for incorporation in masonry."[254]

Also in *La Science des Ingénieurs*, builders find descriptions of the qualities of pozzolana and *trass*, or *terrasse de Hollande*, which Bélidor probably sees used in hydraulic works. On the etymology of the word pozzolana, subject to many dubious interpretations, Bélidor asserts that "it seems that we call this powder pozzolana because it is found near the city of Pozzol."[255] On the subject of *terrasse de Hollande* he specifies that it is a type of gray stone found along the Rhine, near Cologne, and then ground up (he erroneously states that it is also calcinated). *Terrasse de Hollande* is used not only in Holland but also in France, "in the construction of aquatic works, because of the difficulty of obtaining pozzolana at a fair price," as he points out.[256] Bélidor's explanation of pozzolana – a term he also uses to cover *trass* – sheds light on the properties of the mortar made with such compounds: "A mortar is made from two kinds of powder, the first is Pozzolane, whose color is reddish and which is found in Italy and Holland, this powder is very good for construction, and nothing in the world bonds together stones better than the mortar that is made of it, not just for building masonry carried out in dry places, but in particular for construction at the bottom of the sea and in water, it solidifies shortly after application, because it hardens in water [...], I believe that this powder is nothing other than earth and tuff that are burned by subterranean fire that comes out of mountains, around which we excavate it [...]."[257]

Bélidor reports the existence on worksites of other products with properties similar to those of pozzolana and *terrasse de Hollande*, obtained by means of different techniques, from crushing to pulverizing of "pots and other earthenware vases and pieces of slag from coal burnt in forges," to the firing and grinding of pebbles gathered in fields or along rivers.[258] The practice of firing pebbles to obtain a powder capable of generating a compound with hydraulic properties can be seen as a forerunner of what was to become one of the "artificial" pozzolanas in the second half of the 1700s. In the area of artificial products with characteristics like those of pozzolana, Bélidor also includes *cendrée de Tournai*. "[...] when this stone is in the oven," he explains, "little patches break off and fall through the oven grill where they mix with coal ash, and as this ash is nothing other than small bits of calcinated coal, this is the mixture as it is made that constitutes the cendrée de Tournay that is delivered by the merchants as it is removed from the ovens."[259]

plate IV). It is worth noting that in 1722 the members of the Académie Royale d'Architecture of Paris had been able to read and discuss the handwritten treatise of Bruand, "Architecture hidrolique ou l'art d'élever et construire les écluses et autres ouvrages qui se fabriquent dans les eaux" (*cf.* minutes of the session of 26 August 1722, in, Lemonnier, *op. cit.*, Paris, E. Champion, vol. IV, 1915, pp. 250-51). See also the minutes of the session of 31 August 1722 (*ibid.*, p. 251).

[251] Bélidor, *La Science des Ingénieurs* cit., 1729, vol. I, *Livre II*, p. 50.

[252] *Ibid.*, p. 7.

[253] *Ibid.*, *Livre III*, p. 9.

[254] *Ibid.*

[255] *Ibid.*, *Livre III*, p. 12.

[256] *Ibid.*

[257] *Ibid.*, *Livre III*, p. 11.

[258] *Ibid.*, *Livre III*, p. 12.

[259] *Ibid.*, *Livre III*, p. 16.

**105**. Bernard Forest de Bélidor, various systems of foundations, including those with caissons (fig. 5), and coffer and concrete (fig. 12) (B.F. de Bélidor, *La Science des Ingénieurs dans la conduite des travaux de fortification et d'architecture civile*, Paris, C. Jombert, 1729).

Readers of *La Science des Ingénieurs* also find various recipes used by masons to obtain a quick-setting mortar, from mixing urine into the compound to adding sal ammoniac to the water, a technique tested by Bélidor himself.[260]

Ancient buildings are acknowledged as evidence of construction techniques employing strong mortars, as can be seen in passages where Bélidor reports on the "the difficulty encountered in separating stones that the mortar holds together" during their demolition.[261] His goal is similar to that of the "stone cutters" who want to construct walls and vaults with only perfectly aligned stones, but he wants to achieve it precisely thanks to the mortar the stonemasons tend to reject. Bélidor even envisions a homogeneous masonry, a veritable monolith if the same stone is used to obtain the lime and to cut the blocks for the courses. It is significant that he takes this solution from the man known as the greatest theorist of stone cutting, almost as if to suggest that concrete and stone cutting are competing to achieve a monolithic construction method.

"I believe that even with Philbert [sic] de Lorme," he writes, "it is possible to render the union between stone and mortar almost inseparable if the lime were made from stone with the same quality that one uses in the construction, because the volatile salts that would escape the stones being of the appropriate nature to fill the pores that remain in the lime caused by the loss it undergoes from its own salts, the mortar and the stone would become one single body."[262]

When Bélidor begins the chapter on the foundations of forts on rocky, steep ground, he discusses construction using materials poured into molds made with wooden boards in such technical detail that he even offers recommendations on finishings, making the pages on this theme into the first fundamental writing of the rebirth of concrete in the 1700s. Even in the treatises and pamphlets published decades after *La Science des Ingénieurs*, in France, England or Germany, it is rare to find phrases that bear witness to practical experience and sensitivity to the form of concrete like those written by Bélidor. It is no coincidence that it is precisely Bélidor, describing the pouring operations and the effects of the striking of the mold on the finish, who applies the term *brut* to concrete, an adjective that would return throughout the 20th-century debate on the formal quality of the material.

The masonry made with poured conglomerate discussed in Bélidor's section on the foundations of fortresses is known as "stone mortar." The description starts with that of the molds, which Bélidor calls "*cloisons de charpente*," or wood-frame partitions, to assemble in such a way that "constitutes a chest."[263] Once the "chests" have been made, the workmen pour the "heap of rubble" into them, to "forge it with mortar" that will be "beaten" with special mallets known as *dames*; then, when the stone mortar has set, "the partitions are detached for use elsewhere."[264] This type of foundation is also recommended by Bélidor in the case of partially soft ground, because it can create a continuous system that maintains its strength even if extended over a void caused by erosion of the terrain, "which is what brought Vitruvius to say that

[260] *Ibid.*, *Livre III*, p. 19. "One takes for the composition of ordinary cement three equal parts of crushed roof tile of the hardest sort, chips of millstone, and slag [or pecadin] taken from the furnaces where one casts iron, with each being pulverized separately. Each is then sieved, taking care to wash this powder in tanks in order to remove the carbon that can be found there. After it is well cleaned and dried, these three materials are mixed together forming the edges of a sort of basin in which one slakes about a half equivalent of quick lime, that one leaves to the side for several hours, after which one mashes the whole together, which is then placed on stone platforms or solidly positioned planks, so that it can be beaten once a day with hobnailed bats for seven or eight days, until the

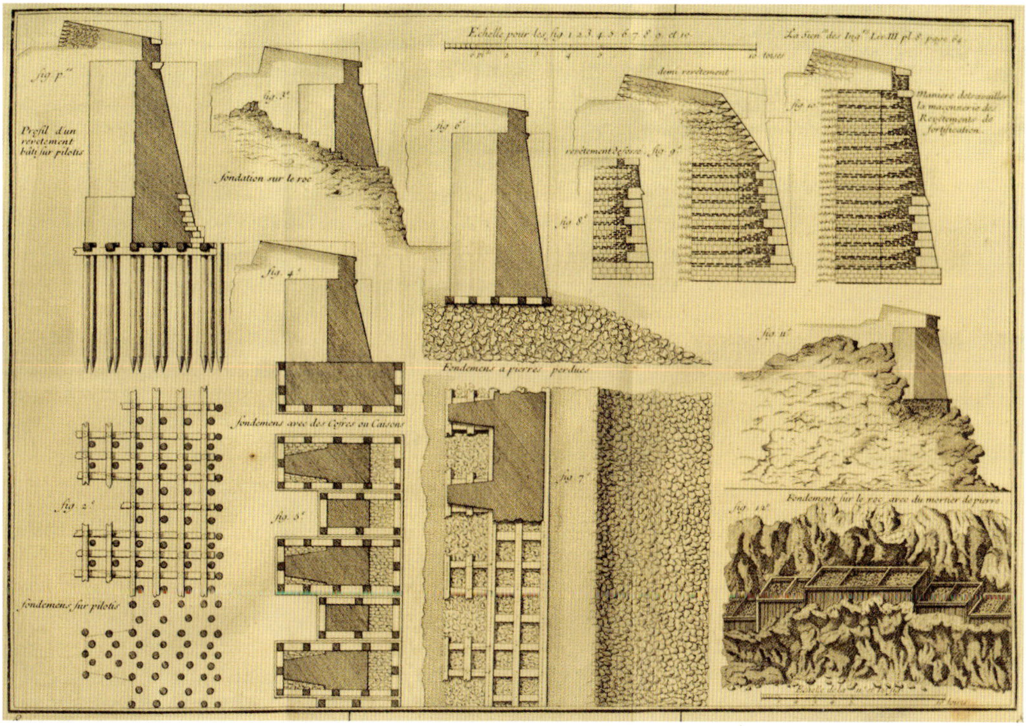


masonry made with small stones is more indissoluble than the others," as Bélidor reports. On the subject of this masonry in "*pierrées*," or small stones, he also refers readers to Perrault's comments on the treatise of Vitruvius.[265]

Bélidor's phrase about the *brut* quality of the finish refers to foundations and bases that, due to the slope of rugged, rocky terrain, remain exposed and therefore visible. To eliminate the *brut* finish of protruding stones caused by the "stone mortar," he advises preparation of a compound of fine aggregates to place directly on the molds, so that their striking will produce a refined finish, possibly to be enhanced after the striking in order to obtain an imitation of a natural stone wall.

"[...] small-stone masonry will never cost as much as cut stone," Bélidor observes, "the only thing that could be said is that, wanting to use it for exposed sub-basements or foundations, it would not be satisfying to the eye to see the brut facing, and a rather ugly surface; but this is easy to avoid by making two types of mortar before the construction, one mixed with rubble like that we have just discussed, the other from course gravel, and if we were in a country where there are two sorts of lime, the better sort should be used for this latter mortar, and the lesser with the other and then used as follows. When working on bedrock, one fills the bottom of the chest with a bed of fine mortar,

cement becomes a paste soft to the touch. If it is not used immediately, it must be reworked from time to time to prevent it from hardening, making sure never to throw water on it during the crushing; but, if necessary, some milk of slaked quick lime can be added when required. In those countries where pozzolana, Holland terrasse or the cendrée de Tournay are available, this can be put to very good use for the construction of works in water [...]" (Bélidor, *Architecture Hydraulique [...]. Premiere partie, Tome Premier* cit., 1737, p. 194).

[261] Bélidor, *La Science des Ingénieurs* cit., vol. I, *Livre III*, p. 20.

[262] *Ibid.*

[263] *Ibid.*, p. 50.

[264] *Ibid.*

[265] *Ibid.*, p. 52.

because it attaches better than the other, then among the workers who will fill the chest, a few will carry fine mortar, telling them to throw it against the inside edge of the chest – here I mean the side that supports the facing – and the rest will be filled with small-stone mortar: if this is done well, with the fine mortar binding to the other, it will form, against the partition, a solid facing that once hardened, will have the same visual effect as stone; if desired, after some time one can decorate it with joints to improve the imitation."[266]

Another type of chest, also made with wooden boards, is described by Bélidor in reference to foundations in ground with problems of water infiltration.[267] After the digging of a trench to prevent landslides at the sides, a "chest" is made, composed of two parallel flanks of wooden boards, two inches thick and known as "madriers," or planks, separated to the necessary distance to make the foundation, and held in position against the ground by means of straps attached to the flanks – "braces that cross the foundation from space to space." Then the conglomerate is poured into the chest, and "when the planks are supported by the masonery, the braces are removed accordingly." Once the operation of filling the mold has been completed, another segment of trench is dug, continuous with the first. "Having made three or four chests in series," Bélidor explains, "and the masonry of the first have well firmed, the braces are removed for use elsewhere, and if the ones at the bottom cannot be had without running the risk of opening a source of water that has been overcome, we decide to leave them."[268] Masonry poured inside stone or brick containers is indicated by Bélidor as being ideal for the construction of military works.[269]

Other recipes for the preparation of hydraulic cements have to do with the creation of the protective surface on vaults of basements and terraces, to guard against rainwater seepage. In this case Bélidor's practical advice is particularly important, if we consider the fact that among the first applications of the cement discovered at the start of the 1700s, there is precisely a covering placed on the vaults of the Orangerie of Versailles to prevent seepage from the panoramic terrace. In *La Science des Ingénieurs* techniques are described that were applied at that time by Jules Hardouin-Mansard to make the protective covering of those very vaults, creating an initial layer of dry gravel and other layers alternating lime and pebbles. "[...] this fabrication worked out so well, that even though the upper part of this Orangerie is only a terrace, no damage was done to the vault," Bélidor points out.[270]

Bélidor also recommends other materials for the protective coating of vaults, to make with a "cement screed" for which he supplies the recipe: "Cement for this use is normally made with the cendrée de Tournay, beaten and prepared once every four or five days for six weeks, being careful not to add any water except for the first time, or else one takes one third good quick lime and two thirds Holland terrasse, which one beats and prepares in the same way; and instead of Holland terrasse, if desired the two thirds can be filled with pozzolana or crushed, well-fired roofing tile, reduced to flour and passed through a baker's sieve; but whether using one or another of these cements, it

[266] *Ibid.*, pp. 52-53.
[267] *Ibid.*, p. 60.
[268] *Ibid.* In the case of foundations, a way of consolidating the ground using poles is in use, with an outer row equipped with grooves for the fastening of the sheet piles – "*palplanches*" (*ibid.*, pp. 57-58).
[269] *Ibid.*, pp. 67-68.
[270] *Ibid.*, p. 81.
[271] *Ibid.*, p. 79. The description of the protective cement coating of the vaults is included, for example, in John Muller, *A Treatise containing The Practical Part of Fortification*, London, A. Millar, 1755, pp. 117-79.
[272] Bélidor, *La Science des Ingénieurs* cit., 1729, vol. I, *Livre III*, p. 79.
[273] *Ibid.*, vol. II, *Livre IV*, pp. 85-86.
[274] *Ibid.*, vol. I, *Livre III*, p. 62.
[275] *Ibid.*, p. 60.

is important to pulverize it with a hand mill and then to beat together the two materials, and then to mix them for a long time in small boarded tubs made for this; the mixing must be repeated several times, without adding water except for the first time."[271]

Bélidor adds a detailed description of the modes of deployment of cement to create the protective coating, from the cleaning of the masonry joints of the vaults to allow the cement to penetrate, to the beating of the layer of cement with "small two-inch-wide bats," to the polishing of the cement layer using an iron of the type used to iron cloth, all the way to the smoothing, repeated for several days, using a cloth soaked in "thinned cement."[272] He has personally made a hydraulic coating of this type, in the construction of a large cistern at Charlemont in 1722, by order of De Breval – in the compound of the "cement mortar" he uses *terrasse de Hollande* due to its superior performance with respect to the powder of "crushed roofing tiles."[273]

Bélidor's recommendations on the composition of cements with pozzolana, *trass* and hydraulic limes form a framework of technical information whose goal is the construction of foundations in the sea or in rivers. Though the subject is part of the books of *Architecture Hydraulique*, certain fundamental notions are already illustrated in *La Science des Ingénieurs*, from the practice of loose stone foundations (Bélidor refers to the sections of Vitruvius on this type of masonry in water),[274] to the explanation of how to prepare a foundation in the sea using layers formed by rough stones or pebbles alternating with layers of lime mixed with pozzolana or *terrasse de Hollande*. "[…] a cement is made on the spot," he writes, "that renders the masonry hard and solid, like that which would be made with more precaution taking advantage of the admirable properties of pozzolana and terrasse."[275] Once the structure has

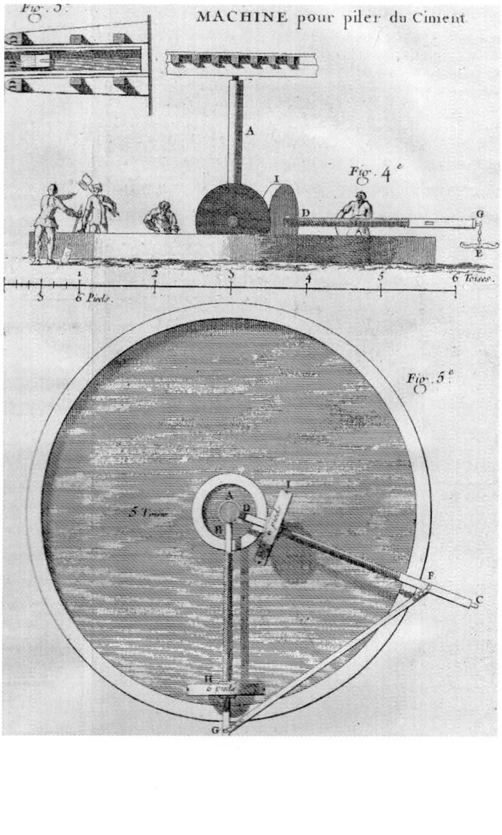


**106**. Bernard Forest de Bélidor, "machine for pulverizing cement" (B.F. de Bélidor, *Architecture Hydraulique, ou L'Art de conduire, d'élever et de ménager les eaux, pour les différens besoins de la vie. Premiere Partie, Tome Premier*, Paris, Ch.-A. Jombert, 1737).

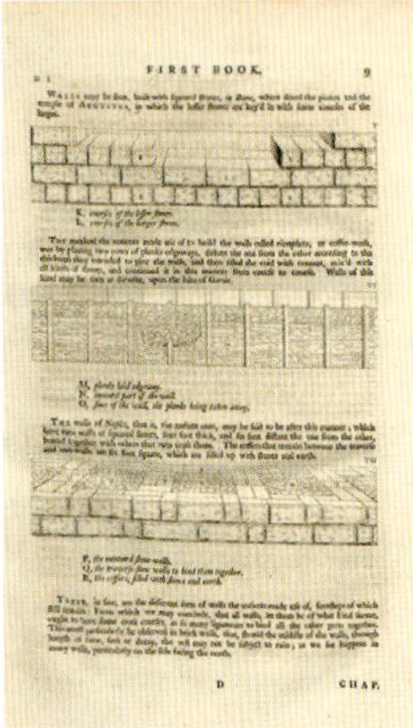
FIRST BOOK. 9

D CHAP.

**107**. Page from the English translation, edited by Isaac Ware, of the Palladio's treatise (*The four books of Andrea Palladio's Architecture*, London, I. Ware, 1738).

emerged above water, the foundation is left loaded with material for years, to test its strength prior to proceeding with the construction of the edifice. Again, it is in the *La Science des Ingénieurs* of Bélidor, and not in the *Traité des ponts* of Gautier, that we find the most accurate description of the method of foundations in deep waters, using the caisson with fixed sides.

The "caisson" is defined by Bélidor as "a well-caulked wood-frame structure";[276] it is equipped with iron rings so it can be anchored by ropes in the sea, at the position where it is to be sunk with its masonry load. While in the system of the loose stones it is necessary to use masonry with hydraulic mortar, thanks to the caisson it is possible to construct the foundation in dry masonry, with "lime and sand."[277] Gradually, as the masons set the layers of the masonry of the foundation in the caisson, it slowly lowers, until it is resting on the bed. To keep the sides of the caisson from being too high, in the case of deep waters, Bélidor recommends the technique of "increasing the height with extensions as it gradually reaches the bottom."[278] Though this method is documented in various ways in early treatises, it is *La Science des Ingénieurs* that becomes the practical guide for the construction of the various caissons used by engineers starting in the 1730s, until the end of the century and beyond.

The information provided by Bélidor on cement masonry is completed by the description of a machine. To produce the powders to mix in order to make hydraulic mortars, Bélidor describes a "machine for "pulverizing cement," composed of two millstones driven by a horse, rotating in opposite directions around a pole placed at the center of a basin – "thus the cement, or any other material, will be crushed along the entire surface of the basin without the usual constraints."[279]

### Local construction practices of walls made in molds

Treatises like those of Viviani and Bélidor form a fundamental compendium on cement and concrete construction, based on knowledge of the working techniques of compounds and their implementation observed in different regions. Important contributions to awareness of the possibilities of poured masonry, besides the recommendations of Viviani and Bélidor, come from the technical observations of travelers like Shaw or Gilij, who tell of building practices for the construction of walls using molds in cities and villages in different parts of the world, from North Africa to South America. The drawing published by Palladio on the question of molds for poured walls remains the central reference point for the culture of construction in the 1700s. The publication, in 1738, by Isaac Ware (1704-66), of the English translation of the treatise of Palladio, contributes decisively to the broadcast of the construction drawing of a concrete wall to be cast in formwork. "The method the antients," we read in the sentence of Palladio translated by Ware, "made use of to build walls called riempiuta, or coffer-work, was by placing two rows of planks edgeways, distant the one from the other according to the thickness they intended to give the walls, and then filled the void with cement, mix'd with all kinds of stones, and continued it in this manner from course to course. Walls of this kind may be seen at *Sirmion*, upon the lake of *Garda*."[280]

[276] *Ibid.*, p. 62.
[277] *Ibid.*
[278] *Ibid.*, p. 63.
[279] Bélidor, *Architecture Hydraulique [...] Première partie. Tome Premier* cit., 1737, pp. 348-59.
[280] *The four books of Andrea Palladio's Architecture* cit., p. 9.
[281] Entry *Boulder walls*, in, Ephraim Chambers, ed., *Cyclopædia, or, An Universal Dictionary of Arts and Sciences*, London, J. and J. Knapton, 1728. *Cf.* also the entry *Boulder walls*, in, Nathan Bailey, *The Universal*

The archaeological discoveries of Fontana at the port of Anzio also make a decisive contribution to awareness of the various construction criteria of concrete in molds. Other sources offer glimpses of the use of molds to build walls. Interesting reading in this sense is provided by the English dictionaries, starting with the *Cyclopædia, or An Universal Dictionary of Arts and Sciences* published in 1728, under the heading *Boulder walls*: "a kind of wall built of round flints or pebbles, laid on a strong mortar; used where the sea has a beach cast up, or where there is plenty of flint."[281] All observers agree that the possibility of making a "strong mortar" is the necessary condition for construction methods using materials poured into molds. But in spite of the proliferation of information on local building processes, the technique of poured cement will not make its comeback as a habitual practice without going through cycles of experimentation and publication on the part of construction specialist.

Treatises like the work published anonymously in The Hague in 1741, *Architecture militaire, ou l'art de fortifier*, written that same year by the captain and engineer Louis De Cormontaingne (1696 circa - 1752), builder among other things of the Moselle and Bellecroix forts in Metz, confirm that after *La Science des Ingénieurs* certain construction manuals, rather than architectural treatises, continue to spread knowledge on the various cements, the different materials for lime, the amounts to use in their preparation and the ways of working with them, including the technique of pouring.[282] De Cormontaingne turns out to be the man behind a series of fertile experiments regarding Metz lime, considered one of the finest in France by a number of technicians, and also much appreciated by Bélidor. De Cormontaingne praises the qualities of Metz lime, but not before having empirically identified the ideal composition process and modes of implementation for hydraulic constructions. His experiments confirm the hydraulic characteristics of that lime, as a substitute for Italian pozzolana, as Bélidor also indicates. De Cormontaingne also explains how to construct masonry works in rivers, using a compound that mixes Metz lime and large gravel, without water. "We throw it in this state," he writes, "as gently as possible into the water [of the river] behind a wooden barrier to prevent it from being overly stirred and washed out by the flow or the current. In less than a year, it hardens like the strongest rock [...]." To contain the poured compound in the water, he advises the use of molds, or coffer, called "coffres," while to make the implementation process less costly he proposes placing inside "these chests a barrowful of lime-gravel mixture alternating with a barrowful of rubble."[283]

From the observations of Fontana, Amico, Shaw, Gilij, Bélidor, Ware or De Cormontaingne we can see that in the first decades of the 1700s there is an awareness that the question of the construction of concrete walls poured into molds comes in part from study of Roman ruins, should not be considered a mere evolution of the *pisé* technique, and is a practice that is to some extent widespread, not limited only to Italian construction. The fact that the construction of walls in the Roman manner, using coffer-work, practiced by masons where it is possible to obtain strong types of lime, is known to engineers is confirmed by certain discussions among the members of the Assemblée

*Etymological English Dictionary*, London, T. Cox, 1731, vol. I, p.n.n.

[282] [Louis De Cormontaingne], *Architecture militaire, ou l'art de fortifier, qui enseigne d'une manière courte & facile la Construction de toutes sortes de Fortifications Régulières & Irrégulières*, La Haye, J. Neaulme, A. Moltjens, 1741. The treatise describes how to make the cement, how to crush old well-fired roofing tiles, and how to create protective coatings for the vaults in terraces (see the section, *Chapes de Ciment*, pp. 87 *et seq.*).

[283] Cit. in, Jean-Elie Bertrand, *Descriptions des arts et métiers*, Neuchâtel, Société Typographique, 1776, vol. IV, p. 329.

des Inspecteurs Généraux des Ponts et Chaussées – one of the recently formed institutions of this type, urged by Daniel-Charles Trudaine (1703-69), and one of the most effective to spread the constructive and engineering know-how.

At the meeting of the Assemblée on 1 October 1747, Pottard reports on the use, in the Généralité d'Auch where he works, of a mortar of such excellent grip that it makes the construction of walls with a thickness of two feet quite possible, at a height of four stories, and "made of stones and mortar." One of the members of the commission, Germain Boffrand (1667-1754), an architect who becomes an engineer of the corps of Ponts et Chaussées in 1723 and an expert on bridges, speaks on a similar practice applied in Metz and Lyon, though he is critical about the construction of walls with *encaissements*. "Mr. Boffrand cited upon this occasion that at Metz and Lyon, the same thing was done; but with less care, since one was satisfied, ordinarily, by the laying of the same masonry by the use of encasings and pouring between planks."[284] Not until the 1750s and 1760s are theoretical contributions published with more detailed considerations than those outlined by Fontana or Bélidor on the system of walls made with materials poured into molds.

## Dutch and German manuals, and the book of Redelijkheid

Italian and above all French technical literature represent the avant-garde when it comes to the spread of knowledge about types of lime and the techniques of fabrication of cement. The grand feats of hydraulic construction accomplished along the coasts of northern Europe generate a comparable output of technical literature in the Netherlands and nearby provinces, but their orientation is towards other questions. The Nordic manuals illustrate works like the Muyden locks, designed by Isaak Stoel, or the locks on the Weser at Hameln, near Hanover, built from 1730 to 1734, also designed by Stoel, under the supervision of Otto Ludwig Voigt (1685-1736) and Picken, first architect of the king-elector. Authoritative hydraulics experts specify the ideal slope of the embankment constructed to counter the force of the sea, like Albert Brahms (1692-1758), in particular, a judge from Friesland and an eminent dike builder, starting in 1718. His first book *Anfangs-Gründe der Deich- und Wasserbaukunst*, published in Aurich in 1754, contains important indications regarding the construction of dikes.[285] Nevertheless, the Dutch authors of the various treatises do not discuss the fabrication of mortars or the components of cement, as is demonstrated by the *Theatrum machinarum universale* of Johannis van Zyl, published in Amsterdam in 1734,[286] and the famous *Theatrum machinarum universale* of Tileman van der Horst, published in 1736-37, again in Amsterdam; Van der Horst references, among others, the writings of Stevin, D'Aviler, Gautier and Bélidor.[287]

Van der Horst's book is a monumental work that sets out, with plates of perfectly engraved and printed drawings, to glorify Dutch hydraulic construction

284 "Précis des assemblées des Ponts et Chaussées, tenues chez M. Trudaine, pour l'examen de differents projets," 1 October 1747, Ms 233, vol. XIII, Ecole des Ponts et Chaussées, Paris.

285 Albert Brahms, *Anfangs-Gründe der Deich- und Wasserbaukunst*, Aurich, H. Tapper, 2 vol., 1754-57.

286 Johannis van Zyl, *Theatrum machinarum universale. Of groot algemeen Moolen-Boek [...]*, Amsterdam, P. Schenk, 1734.

287 Tileman van der Horst, *Theatrum machinarum universale; of keurige verzameling van versheide grote en zeer fraaie waterwerken, schutsluizen, waterkeringen, ophaalen draaibruggen*, Amsterdam, P. Schenk, vol. I, 1736, and vol. II, 1737. For the treatises on hydraulic machines see the exemplary work of the German scientist Jacob Leupold (1674-1727). *Cf.* Jacob Leupold, *Theatrum Machinarum Generale, oder Schauplatz des Grundes mechanischer Wissenschaften*, Leipzig, Ch. Zunkel, 1724.

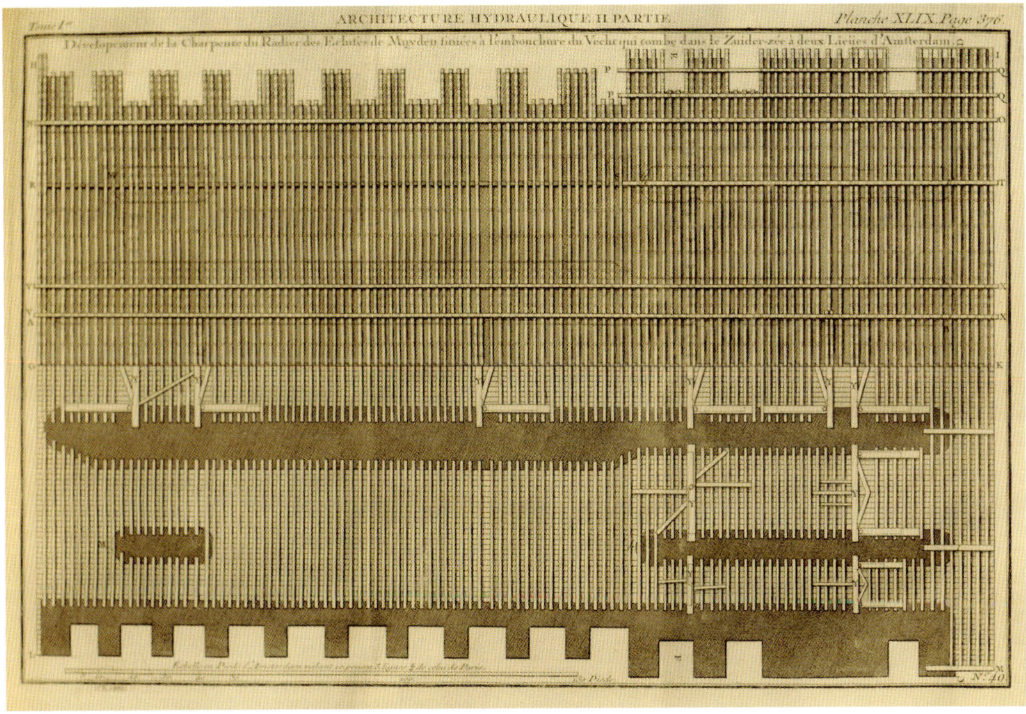


**108**. Isaak Stoel, foundation grille of the lock of Muyden (B.F. de Bélidor, *Architecture Hydraulique. Seconde Partie, qui comprend l'art de diriger les eaux de la mer & des rivieres à l'avantage de la défense des Places, du Commerce & de l'Agricolture*, Paris, Ch.-A. Jombert, vol. I, 1750).

with its systems of locks, its impressive brick masonry on pile underpinnings to protect the earthen levees, and its machinery for the drainage of canals or the sinking of poles. The *Theatrum machinarum universale* is indubitably one of the books that continue that apologia of construction that began with works like *Il Tempio Vaticano* of Fontana, and would proceed with other monumental 18th-century publications.[288] Nevertheless, Van der Horst's book contains only generic indications about masonry.

The German translation of the treatises of Bélidor at the end of the 1740s leads to an evolution of northern European technical culture.[289] The first important discussion on masonry among specialists of the Dutch technical literature is contained in the essay on fortifications published in Rotterdam in 1755, *Verhandeling over de Metselary in Vestingwerken*, written by Cornelis Redelijkheid (1728-87), son of the mason Ary, and a specialist in hydraulic and military construction. The treatise proposes a new way of building walls, partially based on one of the fundamental sources which is repeatedly cited in the text, *La Science des Ingénieurs* of Bélidor.[290] Redelijkheid is the first Dutch

[288] It is significant that a copy of the French version of the book by Van der Horst belonged to Perronet (Tileman van der Horst, *Theatre universel des machines ou recueil choisi de divers grands & beaux ouvrages construits dans l'eau*, Amsterdam, J. Schenk, vol. I, 1737, Ecole des Ponts et Chaussées, Paris).

[289] Bernard Forest de Bélidor, *Architectura Hydraulica, oder, die Kunst, das Gewässer zu denen verschiedentlichen Nothwendigkeiten des menschlichen Lebens zu leiten, in die Höhe zu bringen, und vortheilhafftig anzuwenden*, Augsburg, J.G. Mertz, 1748-50; *Id.*, *Ingenieur-Wissenschaft bey auszuführenden Festungs-Werken und bürgerlichen Gebäuden*, Nürnberg, Ch. Weigel, vol. I, 1757, vol. II, 1758.

[290] Cornelis Redelijkheid, *Verhandeling over de Metselary in Vestingwerken, behelzende een nieuw project met derzelver verdediging en het gene een ingenieur noodig te weten en waar te nemen heeft in het doen bouwen van de Muragie der Vestingwerken, uit beschouwingen en ondervindinge te samengesteld*, Rotterdam, J. Burgvliet, 1755.

**109-110**. Tileman van der Horst, earthen levee with lock in brick masonry, and floating machine to insert the piles (T. van der Horst, *Theatre universel des machines ou recueil choisi de divers grands & beaux ouvrages construits dans l'eau*, Amsterdam, P. Schenk, vol. I, 1737).

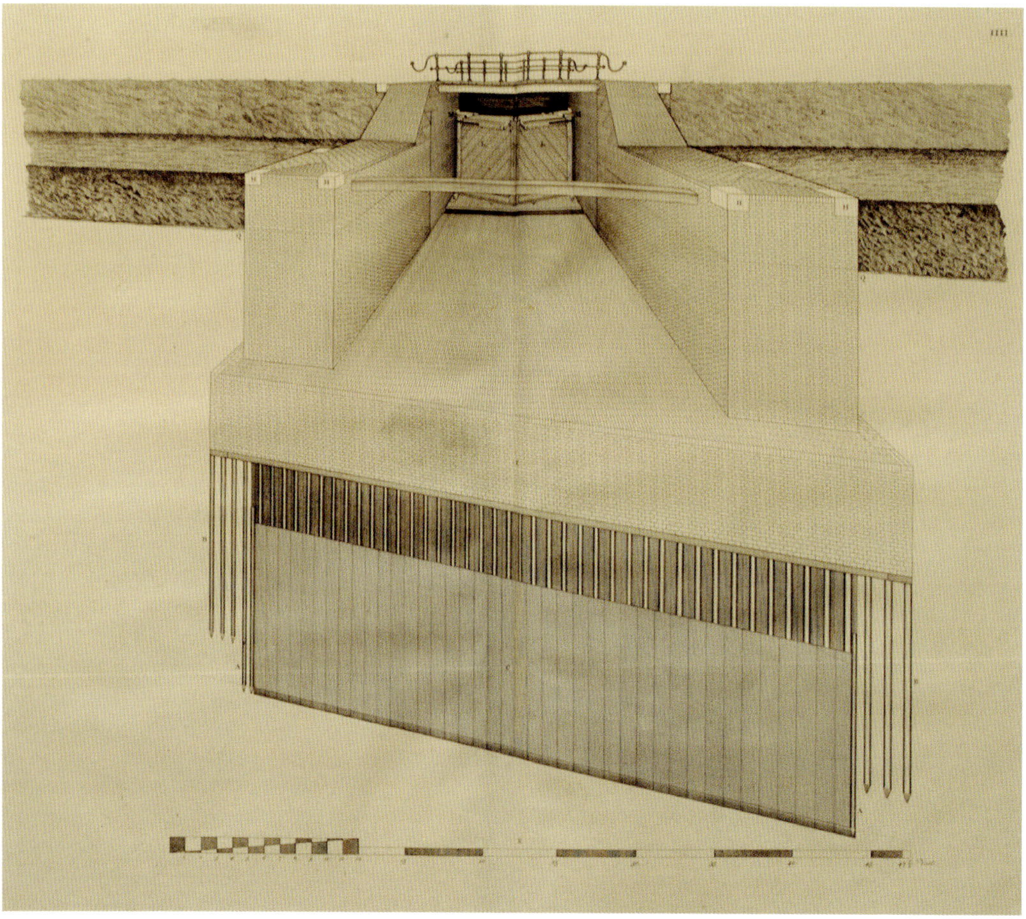

author to provide a detailed discussion of all the aspects of masonry, from mortars to the various forms of brickwork. His reasoning focuses on brick walls, of the type currently in use in Dutch construction, and on their various uses in hydraulic and military construction. The reader finds useful information on foundations made with poles and grids, on ways of connecting the stones of the surface of embankment walls, with metal clamps or particular dovetail or bayonet joints, to prevent slippage.

The most detailed technical recommendations on *trass* and Holland cement are found precisely in Redelijkheid's manual, in the chapters *Handelende over het Zand en de Tras* and *Waar in gehandelt word over het toestellen der Materialen van Kalk en Tras*. Redelijkheid explains the practical rules for fabrication of limes and mortars, and also covers the subject of "cement," both the kind made with various compounds, and the kind made with *trass*. Where *trass* cement is concerned, he indicates the correct composition of Dutch cement, specifying the proportions of lime and *trass*, which vary according to the use to be made of the compound; he explains the fabrication processes and components of the various types of cement, from the type used for hydraulic works, called strong *trass* cement, and the cement for works outside the water, known as "bastard" and characterized by the addition of sand to the compound with *trass*.

"The trass comes from Cologne," he writes, "from trass quarries there, and is milled here [in Holland]. There are blue, dark brown and yellow types of trass. Some think that the blue trass is best; I find the yellow best because, based on some exact tests, I have made the observation that it dries faster and

[291] *Ibid.*, pp. 76-77.

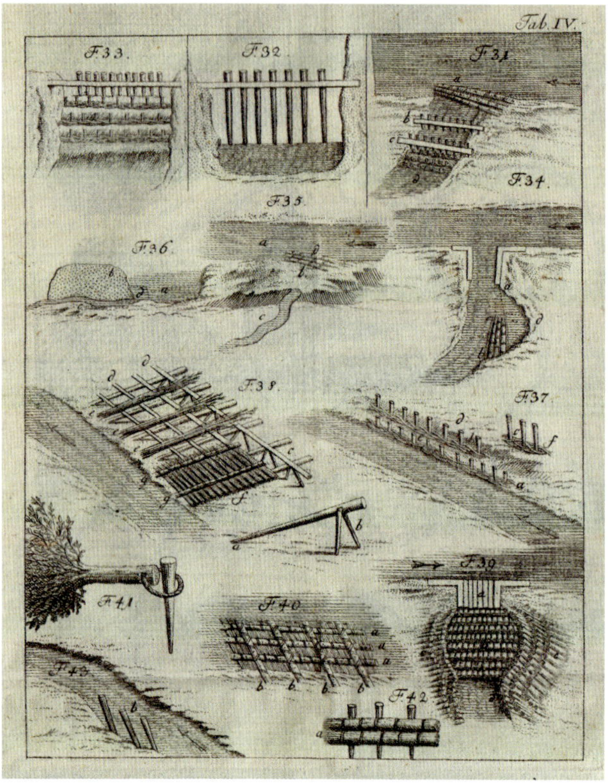


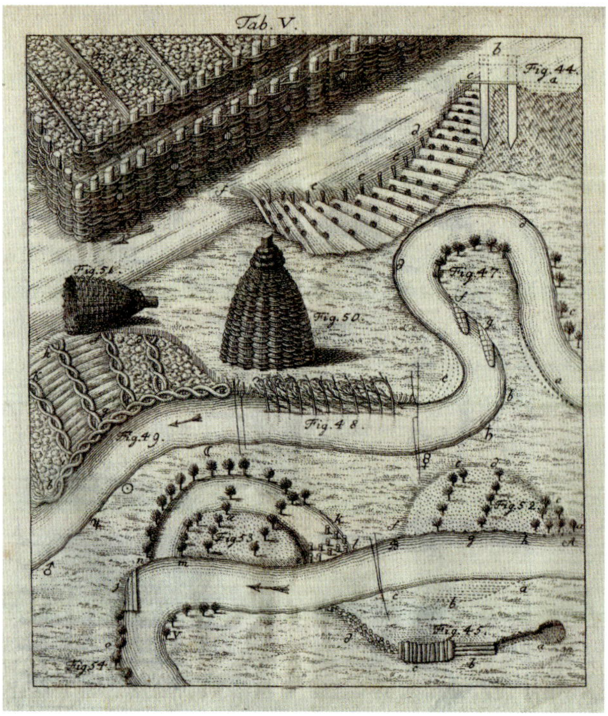


**111-112**. Defense of the banks of rivers with poles and bundles
(J. Bösen, *Anleitung zum Wasser-Bau oder kurzer Unterricht für Landleute*, Göttingen, W. Vandenhoeck, 1769).

**113**. Christian Ludwig Ziegler, project for a hydraulic machine, Oldenstadt, 1764
(B. Adam, T. Albrecht, *Christian Ludwig Ziegler, 1748-1818*, Petersberg, M. Imhof, 2009).

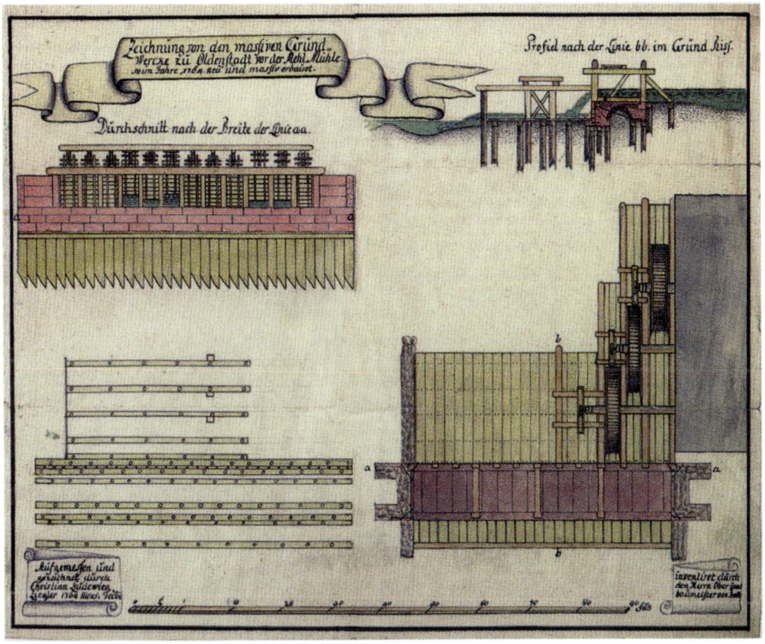

better hardens the work. With the trass that comes milled [from Cologne], there is often cheating, for example, there is a sort of Lamendaalsher earth among them, that often has the color of the yellow trass, but when one rubs it a bit with the hand, one sees immediately, that its quality as a clay is worth nothing. In order to know whether the trass has been cut with sand, I take a clear thin glass with fresh well water and throw one or two handfuls of trass into it, mix it vigorously, and then let it sit until all has settled and the water has become clear again. If there is a bit of sand, it will be at the bottom of the glass – because it is heavier than the well-milled trass it will sink first."[291]

To get an idea of the technical recommendations offered by Redelijkheid, we can read certain passages on the fabrication of compounds for mortars, and the various cases and techniques involved:

"1. Because it makes a significant difference, to which ends the material will serve, whether as containers for oil, lye, rain or foam, each will require a different proportion of the the mixture. 2. Because the cases are never the same, with respect both to the location and the context. 3. And finally because a lime can be richer or thinner, and this makes a noticeable difference; when the lime and the trass cannot be brought together to the one and same work, then the use and purpose cannot be attained, because it is not good enough that one adds too much lime with too little trass, or too much trass and too little lime. The petrifaction, which is created by the compounding of the two, will be very small compared to that obtained from corresponding proportions. Now when the petrifaction by compounding is less strong, so it would be normal

[292] *Ibid.*, p. 122.

[293] *Cf.* Cornelius Redelnkheid, *Abhandlung über die Maurerarbeit an Festungswerken, welche ein Ingenieur nothwendig verstehen muss, nebst einer mit Gründen unterstützten neuen Methode zu mauren*, Breslau, W.G. Korn, 1788. On the success of the manual of Redelykheid in the German technical literature on hydraulic construction see, for example, Reinhard Woltman, *Beiträge zur hydraulichen architectur*, Göttingen, J.C. Dieterich, vol. IV, 1799, pp. 399, 410.

to expect more disadvantages than advantages: because with all types of wall works the petrifaction – which arises through the unification of the parts – is the main point."[292]

Redelijkheid's book meets with great success in the technical culture of northern Europe. In Germany it becomes a manual of reference for comprehension of the working processes of *tuffstein*, for its transformation into *trass* and the fabrication of cements, though translated into German only in 1788.[293] Even prior to the translation, leading German experts on cement in the 1700s, like Christian Ludwig Ziegler (1748-1818), active in Göttingen, make reference to Redelijkheid's book.[294]

Until the translation of the manual of Redelijkheid, the German technical literature on hydraulic construction, represented by manuals like those of Johann Esaias Silberschlag (1721-91), Jobst Bösen (1700-?), Johann Baptist Eberenz (1723-88) and Lukas Voch (1728-83), documents the spread of all the criteria of construction and maintenance of levees, seawalls or dikes with structures in wood, botanical bundles, earth and loose stones, described and depicted in keeping with the models of the Italian technical literature of the late 1600s and early 1700s. In spite of the vast range of constructive solutions covered by the German authors, masonry works remain at the margins, and there are no significant references to concrete construction.[295] A certain influence of masonry construction in the manner of the *murazzi*, though still with stones held in place with metal clamps instead of just mortar, can be observed in the hydraulic architectural manuals starting at the end of the 1750s. It is emblematic that one of the authors of reference, Silberschlag, provides only rare information on the cutting of stones to create barriers capable of standing up to the force of moving water, and on cement, either "hot" or "cold," the latter composed, he explains, "of lime, of stucco and of ground brick"; he also mentions the fabrication of mortar to fill joints, composed of "mastic, mixed with iron filings."[296] Silberschlag's manual, *Abhandlung vom Wasserbau an Strömen*, is translated into French with the warning that the work serves "as the continuation of the second part of the *Architecture hydraulique* by Bélidor."[297] It is not until the early 1770s that Silberschlag takes concrete construction into consideration, also with pozzolana or *trass*, starting precisely with Bélidor's reports on the experiences of Milet de Monville.[298]

## Chemical research on lime

While the construction practices applied on the various worksites and documented by engineering treatises play a decisive role for international knowledge of the criteria of fabrication with the Roman method, during the course

[294] *Cf.* Christian Ludwig Ziegler, *Von dem Terras, der Puzzolana und deren Substitutionsarten*, in "Hannoverisches Magazin," XI, 18 January 1773, no. 6, coll. 81-94; *Id.*, *Fortsetzung des Aufsatzes von dem Tarras, der Puzzolane und deren Substitutionsarten*, in, "Hannoverisches Magazin," XI, 1 March 1773, no. 18, coll. 279-88; *Id.*, *Schluss des Aufsatzes von dem Tarras, der Puzzolane und deren Substitutionsarten*, in, "Hannoverisches Magazin," XI, 5 March 1773, no. 19, coll. 289-300. *Cf.* also, Rudolf August Vogel, *De pulvere cæmentariis nonnullis usitato, tras vulgo dicto, commentatio chemical*, in, "Novi Commentarii Societatis Regiae Scientiarum Gottingensis," vol. III, 1773, pp. 50-59; and *Von dem Cemente aus Trasse*, in, Friedrich Ludwig Aster, ed., *Gesammelte Nachrichten von dem Cemente aus Trasse und Wasserdichten Mauerwerke der Holländer*, Dresden, Leipzig, E. Gerlach, 1791$^3$, pp. 7-30.

[295] *Cf.* Jobst Bösen, *Anleitung zum Wasser-Bau oder kurzer Unterricht für Landleute [...]*, Göttingen, W. Vandenhoeck, 1757; Johann Esaias Silberschlag, *Abhandlung vom Wasserbau an Strömen*, Leipzig, C. Fritsch, 1756 (French transl., *Théorie des fleuves, avec l'art de bâtir dans leurs eaux et de prévenir leurs ravages*, Paris, Ch.-A. Jombert, 1769); Johann Baptist Eberenz, *Erster Gründen der Wasserbaukunst an reissenden*

of the 18th century the evolution of cement is also influenced by scientific research conducted during the first years of the century by chemists, in experiments to explore the nature of the processes of manufacturing of lime. It is precisely in the field of chemistry that the theoretical premises are created for the invention of artificial materials that make it possible to avoid importation of natural products, from *trass* to pozzolana.

Research on limes, on the combustion of minerals and the variation of their weight is conducted in the last decades of the 1600s, in the context of the theory of "inflammable earth," by renowned chemists, mineralogists and pharmacists like Jean-Joachim Becker, Moyse Charas,[299] and Georges-Ernst Stahl (1660-1734), who develops the principle of inflammability of Becker, arriving at the phlogiston theory – a fantastic element thought to be responsible for combustion and calcination. With the publication in 1721 of the previously unreleased 17th-century essay on the hardening of lime by Perrault, *De l'endurcissement de la chaux*, in the volumes devoted to his works on physics and mechanics, the question of a scientific study of the chemical aspects pertaining to construction practices – previously left to the empirical know-how of mason – can be said to be officially posed in scientific circles.

Perrault's essay reveals the tests connected with building practice that had been conducted on lime, and the observations following additions of various products to the compound to find combinations that would accelerate its solidification, such as "slag" – Perrault notes that this product "hardens the mortar," given the fact that it is "a sort of cement," "a fired stone and half burnt," or the "powdered alum, vitriol, or ammoniac salt."[300] Full of general observations and theoretical deductions, Perrault's essay is nevertheless soon surpassed by contributions based on the results of other experiences and direct trials with the materials, all conducted with the aim of achieving useful results for construction practice.

In the context of the analyses of minerals fired in furnaces conducted in the early 1720s by Johann Friedrich Henckel (1678-1744), a German physician, mineralogist and chemist, a pupil of Stahl and director of mines in Freiburg, and continued by his disciple Neumann, the action of volcanoes on rocks begins to be viewed from another angle. Still in the midst of the discussion on the phlogiston theory, Henckel subjects certain stones – the ones usually used to obtain lime – to calcination in order to classify those that are easily melted. His observations on the origin of certain rocks, though generic, can be seen as an introduction to the experimentation on the artificial fabrication of products similar to those produced by volcanic action. "Sand-stone, lime-stone, marle-stone, and shiver," he writes, "which generally form *squats*,

*Flüssen*, Freiburg, J.A. Satron, 1767; Lukas Voch, *Anleitung zum Wasserbau an Flüssen und Strömen*, Augsburg, 1767.

296 Johann Esaias Silberschlag, *Théorie des fleuves, avec l'art de bâtir dans leurs eaux et de prévenir leurs ravages*, Paris, Ch.-A. Jombert, 1769, p. 102.

297 *Ibid.*, p.n.n.

298 Johann Esaias Silberschlag, *Ausürlichere Abhandlung der Hydrotechnik oder des Wasserbaues, zweiter theil*, Leipzig, C. Fritsch, 1773, pp. 413 *et seq.*

299 See in particular, on lime, Moyse Charas, *Pharmacopée royale galenique et chymique*, Lyon, Anisson & Posuel, 1704, pp. 631 *et seq.*

300 *De l'endurcissement de la chaux*, in, Claude and Pierre Perrault, *Œuvres diverses de physique et de mechanique*, Leide, P. Vandera, 1721, vol. II, (pp. 624-32), p. 629.

301 Johann Friedrich Henckel, *Pyritologia, or, a history of the pyrites, the principal body in the mineral kingdom*, London, A. Millar, A. Linde, 1757, p. 70. For considerations on volcanoes see, *ibid.*, p. 67. For the original edition see, *Pyritologia, oder, Kieß-Historie, als des vornehmsten Minerals, nach dessen Nahmen, Arten, Lagerstätten, Ursprung*, Leipzig, J. Chr. Martin, 1725. *Cf.* also, Johann Friedrich Henckel, *Idea generalis de lapidum origine per observationes experimenta & consectaria, succincte adumbrata*, Dresden, Leipzig, Officina Libraria Hekeliana, 1734, pp. 80 *et seq.*

plainly manifest, upon a somewhat more accurate inspection, that they are a concreted sand, or hardened earth."[301] The nascent European volcanism that is an increasing focus of naturalists starting in the 1750s and has a decisive influence on the evolution of construction techniques involving hydraulic mortars originates, in part, in the experiences of mineralogists such as Henckel on the effects of fire on minerals.

In the 1730s true experimental research begins on the nature of lime, from its fabrication in furnaces to the chemical processes of its slaking in water on worksites. This research is conducted above all by French scientists, who combine the practical outlook of Perrault, at times with reference to the contemporary German theory of the phlogiston. It is no coincidence that for the French scientists, the practical experiences and observations of De l'Orme on lime continue to be a point of reference throughout the 1700s.

Starting in 1733, Henri-Louis Duhamel du Monceau (1700-82), an expert on botany and agronomy, performs tests on lime to study its reactions. His results are presented to the Académie Royale des Sciences of Paris only in 1747, and published even later, in 1752. "[...] I wanted to study it during its calcination," he writes about lime, "during the time when quick lime becomes charged with the humidity of the air, fuses together and is reduced to powder; and to observe all the circumstances that accompany its efforvescence with water when it is slaked in this fluid, and finally to study what results from its combination with different acids, with alkaline salts, and with other diverse materials."[302]

The investigations of Duhamel du Monceau range from calcined stones to the slaking of lime. From 1744 to 1746 he fires marbles and stones to determine which are the best products from which to obtain lime. At first, in 1744, he uses the furnace at Courcelles, which is the one that produces lime for construction in Paris – Duhamel du Monceau calcines a block of marble there; then, in 1745, he involves the engineer Charles-François Ribart and Mssrs. Détrée, owners of the furnaces at Senlis, "who have the reputation for making the best lime."[303] Duhamel du Monceau and Ribart record the measurements of the stones, their weight before and after calcination, and observe factors of color, odor and type of firing, preparing a table of data only for the

[302] Henri-Louis Duhamel du Monceau, *Diverses expériences sur la chaux*, in, "Histoire de l'Académie Royale des Sciences. Année MDCCXLVII. Avec les Mémoires de Mathématiques & de Physique, pour la même Année," Paris, Imprimerie Royale, 1752, (pp. 59-81), p. 59.
[303] *Ibid.*, p. 61.
[304] *Ibid.*, p. 62.
[305] *Ibid.*, p. 68.
[306] *Ibid.*, p. 70.
[307] "The shells that partially make up limestone," writes Duhamel du Monceau, "are often sea shells, & so therefore there must still be some sea salt in these stones. Fire will not have much action on this salt, so some must remain in certain limes, while there is perhaps none in others. Vitriolic acid shows up in all fossils, calcination does not lead to its complete dissipation: there must be some therefore in all lime, but in greater quantity in that which is made from yellowish stone, or reddish, which we can consider containing ochre or some other mineral" (*ibid.*). For the experiences of Duhamel du Monceau on lime made with oyster shells see also, Henri-Louis Duhamel du Monceau, Jean Grosse, *Sur les différentes maniéres de rendre le tartare soluble. Seconde parte*, in, "Histoire de l'Académie Royale des Sciences. Mémoires de mathématique et de Physique, tirés des registres de l'Académie Royale des Sciences, de l'année MDCXXXIII," Paris, Imprimerie Royale, 1735, pp. 260-72.
[308] *Sur la chaux et le plastre*, in, "Histoire de l'Académie Royale des Sciences. Année MDCCXLVII. Avec les Mémoires de Mathématiques & de Physique, pour la même Année," Paris, Imprimerie Royale, 1752, pp. 65-72. The reports of Macquer are published in book form and translated into Italian (Pierre-Joseph Macquer, *Osservazioni sulla calce e sul gesso*, Livorno, P.A. Santini e Compagni, 1755). Apart from these contributions, the nevertheless fertile publishing output of Macquer in the 1750s and 1760s makes no further reference to the questions of limes and mortars in

weights.[304] The observations on lime reduced to powder continue for about two years, with the aim of also studying variations of weight over time. Duhamel du Monceau also calcinates ox bones in the furnace, and wonders why they harden (the violent fire or the vicinity, in the furnace, of stones). Then he observes the phenomena produced by water on lime (variation of reactions based on the quantity of water and the immersion times) and on mortars – for this latter investigation, he prepares a mortar with "one pound of well-dryed cement."[305] Probably aware of the production of lime using materials like seashells, he also studies the lime derived from oysters, and starting with the experiments conducted in 1734 he establishes that the "stones that contain shells provide excellent lime."[306] We cannot exclude the possibility that his tests may have contributed to support, with scientific analysis, the Dutch production of the compound for hydraulic lime obtained by calcination of oyster shells.[307]

The experiences of Duhamel du Monceau for the definition of the criteria of fabrication for a perfect lime are carried forward in 1747 by the chemists Paul-Jacques Malouin (1701-78) and Pierre-Joseph Macquer (1718-84), who fire the various types of stones used in Parisian construction.[308] Precisely the results of the work of Malouin and Macquer prompt Duhamel du Monceau to make his findings public. Malouin, like Duhamel du Monceau, studies the lime produced by oyster shells, recording the presence of Glauber's salt.

Towards the 1740s certain German chemists study the nature of rocks and limes, such as Johann Andreas Cramer (1710-77), who starting in 1743 is the Kammerrath of Mines and Smelters at Blankenburg, and conducts experiments on the calcination of spar,[309] or Christlieb Ehregott Gellert (1713-95), who classifies stones according to their reaction to acids, fire and air.[310] Certain scientific institutions sponsor specific studies, like the Akademie der Wissenschaften of Göttingen, which supports research to find the best, most water-resistant lime. In the Göttingen competition, Daniel Gottfried Schreber (1708-77) proposes lime obtained from the gypsum stone.[311]

The criteria of analysis of natural products used for the fabrication of mortars undergo important scientific refinement thanks to the chemist and pharmacist Johann Heinrich Pott (1692-1777), active in Berlin and the author, in 1744, of a series of studies on earths and stones, based on the combustion of

construction (see, for example, Pierre-Joseph Macquer *Eléments de chymie-pratique*, Paris, J.-Th. Herissant, 1751, vol. I).

309 *Cf.* Johann Andreas Cramer, *Anfangsgründe der Probierkunst*, Stockholm, G. Kiesewetter, 1746, vol. I, pp. 118-19. The translation from Latin to German is by Christlieb Ehregott Gellert (1713-95). For the Latin text see, Johann Andreas Cramer, *Elementa Artis Docimasticae, duobus tomis comprehensa*, Lugduni Batavorum, Apud Conradum Wishoff, Georg Jac. Wishoff, Fiz. Cone, 1739. "Large quantities of nitre are extracted at Paris, out of the rubbish of old buildings; which are there all made of stone and stone-mortar. The greatest quantities of nitre, whereof the consumption is now become so great for the making gunpowder, are brought from the East-Indies [...]" (Johann Andreas Cramer, *Elements of the art of assaying metals*, London, T. Woodward, C. Davis, 1741, p. 418, note*).

310 *Cf.* Christlieb Ehregott Gellert, *Metallurgic chymistry, being a system of minerology in general, and of all the arts arising from this science*, London, T. Becket, 1776, pp. 6-17. "By chymical experiments it appears," Gelleret writes, "that those lime-stones contain not only a kind of the spirit of common salt, but at the same time a volatile alcaly" (*ibid.*, p. 7). For the original edition see, Christlieb Ehregott Gellert, *Anfangsgründe zur metallurgischen Chimie, in einem theoretischen und practischen theile nach einer in der natur gegründeten ordnung abgefasset*, Leipzig, J. Wendler, 1750.

311 *Cf.* Daniel Gottfried Schreber, *Sammlung verschiedener Schriften, welche in die öconomischen-, policey- und cameral- auch andere verwandte Wissenschaften einschlagen*, Halle, J.J. Curts, vol. III, 1765, p. 111.

specimens. Here the particularity of the analysis comes from the type of fire used by Pott, of a kind that brings the materials to a state of fusion – not even Henckel had been able to use a comparable fire.

The goal of Pott's experiments is to identify the components and technical processes for the fabrication of fine porcelain, through the use of common German earths. The trials lead Pott to a new classification of earths, of which four are considered the fundamental provisional types: alkaline earth, the only one that can be calcinated; vitrifiable earth, which is rapidly transformed into glass; clay-rich soil, which coagulates when fired; and gypsiferous soil, which becomes plaster when fired. The other soils are said to be compounds of the four fundamental ones, and of other metallic or mineral substances.[312]

During the course of his analysis, Pott combines substances, from spar to marble, calcined oyster shells to clay-rich soil and alabaster, to test the effects of smelting on the various compounds. The analysis leads him to study the nature of limes and of plasters, their various degrees of hardening when mixed with water and with sand – Pott often makes reference to the practical know-how of masons.[313] In some cases, his considerations take construction practices and the making of construction parts into account. "The roof tiles and bricks made from this clay," he writes, regarding the yellowy clay of the terrain of Rüderstorf, near Berlin, "are poor, brittle and breakable, at least when calcareous soil is present in a certain quantity; because this earth, once calcinated, swells with the water it takes up avidly, which makes the bricks burst."[314]

Mineralogists, naturalists and chemists from Sweden play a crucial role in the discovery of processes of artificial fabrication of products with characteristics similar to those of natural volcanic substances like pozzolana and *trass*.

Among the first important Swedish contributions on minerals, their properties and processes of petrifaction, orienting future research on volcanic products, after those of Urbain Hiaerne (1641-1724), Magnus von Bromell (1679-1731), Emanuel Swedenborg (1688-1772), Georg Brandt (1694-1768), Carl Nilsson Linnaeus or Von Linné (1707-78), Daniel Tilas (1712-72) and Adam Leyel (1669-1744), those of the chemist and mineralogist Johann Gottschalk Wallerius (1709-85) stand out, the author of the treatise *Mineralogia, eller Mineral-Riket* printed in Stockholm in 1747, reprinted in an expanded edition in German in 1750, and made internationally famous by the French edition in 1753.

Wallerius discusses the various types of earth, stones and "concretions," and publishes an extraordinary collection of chemical analyses on the main

[312] *Cf. Sur l'analyse de diverses sortes de terres ou de pierres par le moyen du feu*, in, "Mémoires de l'Académie Royale de Prusse pour l'année MDCCXLV," vol. I, 1745, pp. 36-42; Johann Heinrich Pott, *Lithogéognosie ou examen chymique des pierres et des terres en général [...]*, Paris, J.-Th. Herissant, 1753.
[313] Pott, *Lithogéognosie* cit., pp. 53 *et seq.*
[314] *Ibid.*, p. 105.
[315] Johann Gottschalk Wallerius, *Mineralogia, eller Mineral-Riket, Indelt och beskrifvit*, Stockholm, L. Salvius, 1747, French transl., *Mineralogie, ou Description générale des substances du regne Mineral, ouvrage traduit de l'Allemand*, Paris, Durand & Pissot, 1753, 2 vols.
[316] Wallerius, *Mineralogie* cit., vol. II, pp. 164 *et seq.* Wallerius mentions the *marbre artificiel* in a sentence: "It is made with plaster; it is hard, has the same color as marble and is polishable to a certain degree, to the extent to which it has been more or less well prepared" (*ibid.*, p. 172).
[317] *Ibid.*, p. 168.
[318] *Cf.* Actes de l'Académie Royale des Sciences de Suede, 1740, vol. I, p. 209 (cit. in, Wallerius, *Mineralogie* cit., p. 168).
[319] Johann Gottschalk Wallerius, *Expériences sur les chaux*, in, *Collection académique, composée de l'histoire & des mémoires des plus célebres académies & société littéraires de l'Europe, [...] Mémoires de l'Académie des Sciences de Stockholm*, Paris, Panckoucke, vol. XI, 1772, p. 242.

minerals and their compositions. The principle criterion of analysis lies in subjecting the various natural products to firing, observing the variations after calcination, from metallic soils to limestones, spar, slate and seashells.[315] Wallerius also studies the "artificial minerals."[316] In this section of his treatise, he discusses the cement of the ancients, considered by all to be better than contemporary cement due to the quality of their lime.

"[...] but when we look at the matter a bit closer," he points out, "we have reason to believe that our lime is as good and as strong, maybe even better than that used by the ancients, and that the difference is that the lime of the ancients is completely dried, and by consequence more solid and harder than ours which is not at all yet dry; a wall or a building needs about a hundred years in order to be very well dried out, to reach all its consistency; one just needs to observe in the calcination of lime, firstly, to use only pure stones; secondly, to slake with small quantities of water at a time, because the more water one adds, the more the lime is destroyed; and thirdly, when the lime is slaked, to let it rest for a certain period of time before usage, so that the internal parts are also completely slaked."[317]

Certain Swedish experiments on lime are performed under the supervision of the baron Carl Wilhelm Cederhielm (1705-69), and publicized by the Royal Swedish Academy of Sciences, in 1740.[318] Wallerius also works on the calcination of different fossil, animal and vegetable substances, including seashells, eggshells and earth, to observe the reactions of the materials and to identify useful principles for the fabrication of mortars. "The lime of the hardest materials," we read in the conclusions of Wallerius, "is that for which the slaking requires the most water, and for which its use requires the least. That from stone is that which breaks up best during slaking, which requires the least amount of water, and which dries the fastest; it is thus preferable to shell lime for use in buildings. It is important to note that an overly strong calcination removes from lime, as from gypsum, its property of hardening with water: the closer one approaches the vitrification of lime, the more it deviates from it calcareous properties."[319]

Other experiences useful for the comprehension of chemistry connected with lime are those of Joseph Black (1728-99), professor of chemistry at Edinburgh, who discovers the existence in calcareous stone of a considerable quantity of an "*acriform*" substance known as carbonic acid gas, which dissolves at the moment of calcination.[320]

Engineers take part in the research on lime, often transforming their

[320] Joseph Black, *Experiments upon magnesia alba, quicklime, and other alcaline substances*, Edinburgh, W. Creech, 1777 (*cf.* Riche De Prony, Joseph-Louis Gay-Lussac, Pierre-Simon Girard, *Rapport fait à l'Academie royale des sciences sur un ouvrage de M. Vicat, ingénieur des ponts-et-chaussées, intitulé: Recherches expérimentales sur les chaux de construction, etc.*, in, "Mémoires de l'Académie Royale des Sciences de l'Institut de France. Année 1818," Paris, Firmin Didot, 1820, pp. CXXXIV-CLXXVIII).

[321] See, for example, the reports: "Experience faite le 6 avril 1739 à Digny sur la pesanteur de deux pied cube de caillou, sable, ravine etc," Ms 2185, Ecole des Ponts et Chaussées, Paris; "Remarques de M. Le Clerc faites dans le devis pour la construction du pont de Chateauthierry sur l'evaluation de la maçonnerie," Ms 2608, Ecole des Ponts et Chaussées, Paris (narrating the experiments done in 1748 on slaked lime, to establish the exact quantity of lime, sand and water); "Experience sur la chaux d'Alençon," *ibid.*; "Experiences sur la pierre a chaux de Senlis faites au mois de mars 1751 par M. Boulanger," *ibid.* For the construction of the furnaces see the drawings of the "four a platre de Montmartre," with a note of 16 October 1747, and "Memoire sur un four à chaux de Senlis, par le Sr De Brie, le 4 novembre 1753," *ibid.*

[322] Minutes of the meetings of 31 December 1747 and 20 April 1749, in, "Précis des assemblées des Ponts et Chaussées, tenues chez M. Trudaine, pour l'examen de differents projets," Ms 233, vol. XIII, Ecole des Ponts et Chaussées, Paris.

**114**. Jean-Rodolphe Perronet, "Machine for grinding roof tiles for making cement," fig. 3; "Machine to make mortar," fig. 4. The second machine is based on the one invented by Germain Boffrand between 1746 and 1751 (J.-R. Perronet, *Description des projets et de la construction des ponts de Neuilly, de Mantes, d'Orléans & autres [...]*, Paris, Imprimerie Royale, vol. II, 1783).

worksites into laboratories for experimentation on the amounts of components and the criteria for the preparation of mortars.[321] Pitrou has trials conducted on the worksite for the bridge of Orléans, to establish the quantity of quick lime for a cubic fathom of stones, while Mathieu Bayeux (1692-1777) reports on his observations regarding lime – in his view, that made with the "soft stone hardens quickly."[322] Other experiences such as those at Vaucouleurs in 1748 on the "lime mortar with cement and sand" establish quantities and proportions of the various mortar components (sand from the Meuse, slaked lime, water), starting with a specially concocted unit of measure – a cubic box of 12 inches per side – in order to determine the measures starting "the absolute and specific weights" taking what was done in the case of cast iron as a model.[323]

Starting at the end of the 1750s, Jean-Rodolphe Perronet (1708-94), the most highly acclaimed French engineer of the 1700s, focused on the study of the strength of mortars through a series of experiments (Macquer visits his collection of stones).[324] Among the stones weighed by Perronet we find the "mortar of an aqueduct antique of the Romans at Lyons."[325]

With respect to the experiments of the chemists, the innovations of architects and engineers related to lime often take place in the context of practical improvements. The recommendations of Boffrand for the preparation of mortar are significant, in this regard. In the construction of the premises of the Hôpital des Enfants-Trouvés, on the Ile de la Cité in Paris, from 1746 to 1751, to mix the "freshly slaked" lime with sand he has a circular basin built, nine feet in diameter, with a central rod on which to attach a horizontal wooden bar with teeth arranged in an irregular way to mix the compound, moved by a horse.[326] At the time of the work on the Hôpital des Enfants-Trouvés, Boffrand reports on his method of mixing lime to the Assemblée des Inspecteurs Généraux des Ponts et Chaussées, as well as describing a new type of furnace for calcination.[327]

[323] *Cf.* "Remarques sur le mortier de chaux avec ciment et sable, faites à Vaucouleurs en 1748," Ms 161, Ecole des Ponts et Chaussées, Paris. See also the report, "Expérience sur la chaux d'Alençon," Ms 161, Ecole des Ponts et Chaussées, Paris.
[324] Macquer visits the collection of stones of Perronet and remarks on "a kind of efflorescence or crystallization" ("Concernant deux morceaux de pierre presentés à l'Academie par le sr. Perronet le mercredy 9 juillet 1766," Ms 2608, Ecole des Ponts et Chaussées, Paris).
[325] "Tables des experiences faites sur les differentes nature de pierres, granits, marbres et briques, pesée dans l'air et dans l'eau; leur poids reduit au pied cube, et leur degré de dureté, faites par M. Perronet 1.er Ing.r des Ponts et Chaussées, dans l'années 1758, 1759, 1760, et 1762, 1765," Ms 84, Ecole des Ponts et Chaussées, Paris. See also, "Expériences sur la force du mortier composé de cendrée et de chaux de Tournay comparée avec celle de plusieurs autres mortiers ûsités sans la maçonnerie," Ms 137 fol., Ecole des Ponts et Chaussées, Paris.
[326] Pierre Patte, *Mémoires sur les objets les plus importans de l'architecture*, Paris, B. Rozet, 1769, p. 157.
[327] "He also proposed to make mortar in a circular basin in which is mixed by means of a horse, with several planes fixed and vertically attached to a radius or lever which, fixed at the center of the basin, is able to freely turn. He also proposed the construction of a lime kiln" (Minutes of the meeting of 7 February 1751, in, "Extrait du Journal des assemblées des Ponts et Chaussées, tenues chez Mr Trudaine, depuis 1748 jusqu'en 1774," Ms 233, vol. XIII, Ecole des Ponts et Chaussées, Paris). Perronet will use the machine of Boffrand in the worksite of the bridge in Neuilly (*cf.*, Jean-Rodolphe Perronet, *Description des projets et de la construction des ponts de Neuilly, de Mantes, d'Orléans & autres [...]*, Paris, Imprimerie Royale, vol. II, 1783).

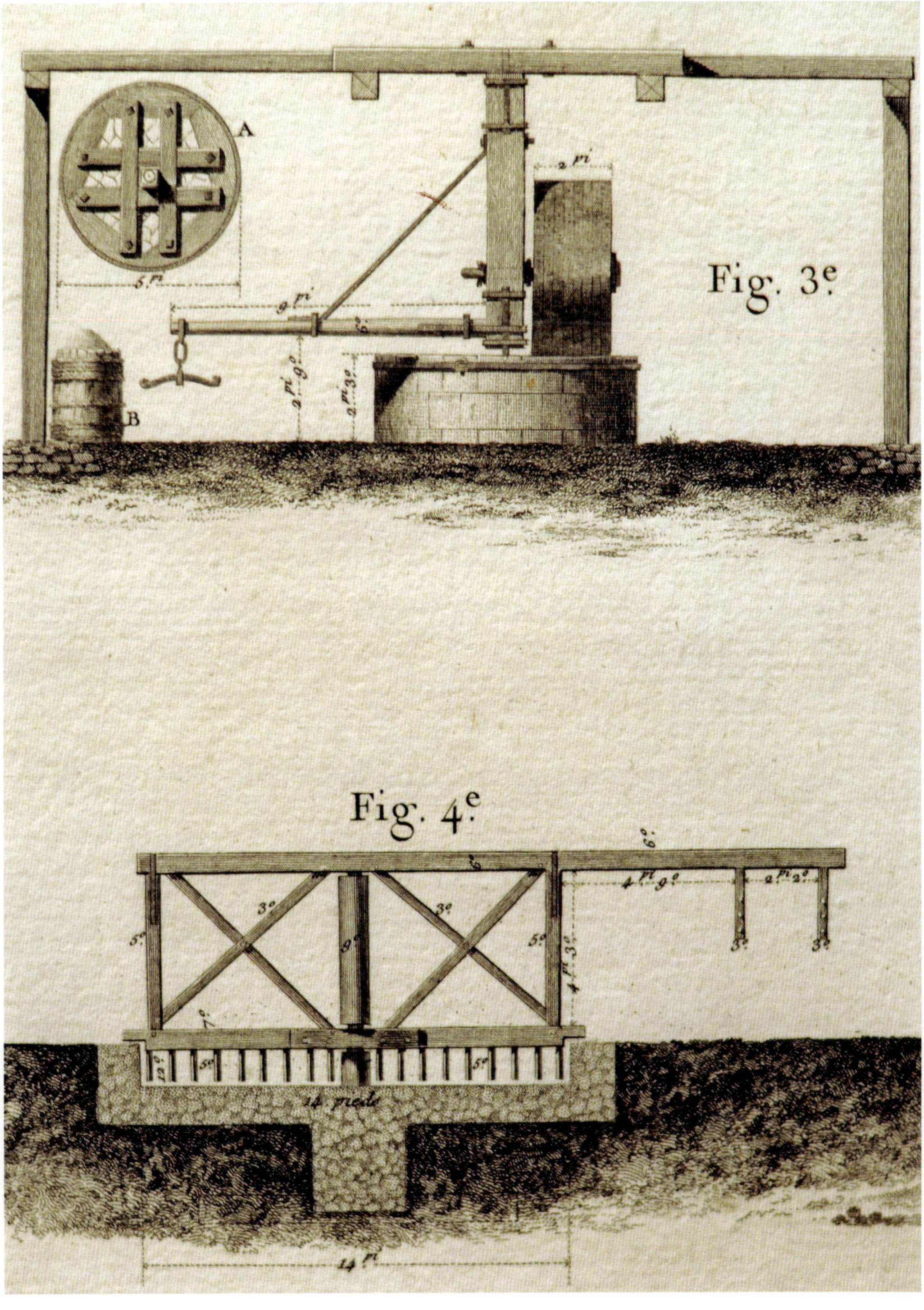
A
B
Fig. 3e.
Fig. 4e.
14 piede
14 pi.

64
28
37
40

# chapter three
# major hydraulic works of the 1730s and 1740s

### The ports of Leghorn and Viareggio and the report of Zendrini

The leading role played by Tuscany in research on cement during previous centuries is confirmed in the early decades of the 1700s, as engineers work on the maintenance and expansion of the ports of Leghorn and Viareggio. A number of these operations become outstanding examples of avant-garde hydraulic works in the first part of the century, serving as models for the projects on the Venice Lagoon.

Leghorn is the destination of ships from Civitavecchia carrying pozzolana for hydraulic works in Tuscany. This port is also a distribution center for pozzolana headed outside Tuscany. As late as in the 1780s important quantities of pozzolana reach Leghorn, as Fea reports – "it is brought to Tuscany by sea as far as Leghorn, and then it also goes to other places."[1] With the supply of pozzolana, Roman cement can be made for use in the hydraulic structures in the ports of the Grand Duchy. A synonym for concrete, the *smalto* is used along the coasts and is not produced with hydraulic lime like that of the inland area, of the students of Galileo. Sometimes, from Leghorn, the pozzolana is transported inland on barges on the Arno, reaching Empoli, as was already the case in the 1600s.[2] It is significant that in the vicinity of Lucca a "strong lime mortar" is used for the walls

[1] Carlo Fea, *La calce, ed in ispecie la pozzolana* (1 December 1760, with additions of 1762), in, Johann Joachim Winckelmann, *Osservazioni sull'architettura degli antichi*, in, *Storia delle Arti del disegno presso gli antichi [...] Tradotta dal tedesco e in questa edizione corretta e aumentata dall'Abate Carlo Fea*, Roma, Stamperia Pagliarini, 1784, vol. III, p. 25.

[2] Pozzolana was transported from Rome to Leghorn and then to Florence on the Arno, at the time of the project of Pietro da Cortona for the church of San Firenze, probably to make the foundations of the building (*cf.* Caroline Coffey, *Pietro da Cortona's project for the church of San Firenze in Florence*, in, "Mitteilungen des Kunsthistorischen Institutes in Florenz," vol. XXII, 1978, (pp. 85-117), p. 86 and p. 110, note 19).

of buildings, which in the second half of the 1720s a German painter of Italian origin, Georg Christoph Martini (1685-1745), reports as being made with "freshly slaked" lime and pozzolana[3] (during his stay in Leghorn he comes into contact with an expert naturalist with knowledge of volcanic materials, Giovanni Targioni Tozzetti). Besides the pozzolana imported from Civitavecchia, in Tuscany there is also trade, later in the century, in pozzolana quarried in Maremma, at Campiglia, in the quarries of the old ferric sulfate mines.[4]

The construction of hydraulic works with caissons sunk in the sea after having been filled with stones and pozzolana mortar is used for the works done at the port of Viareggio. It is precisely when a caisson blocks the mouth of the Serchio River at the sea, that the man who would become the protagonist of the new phase of construction of the levees of Venice, Zendrini, was called on for advice.[5] He is recommended to the Ufficio della Foce of Viareggio by a certain Antonetti, a merchant from Lucca residing in Venice.[6] This is probably Giovanni Domenico Antonetti, who is to become the main importer of pozzolana to Venice, used precisely for the works supervised there by Zendrini.

In the spring of 1735 the port of Viareggio is inspected by Zendrini, requested by the Ufficio della Foce to provide a technical proposal for the improvement of the bed at the mouth of the river, and the circulation of air. During his inspection Zendrini has a chance to swap views with Giuseppe Natalini, the engineer who possesses all the surveys and documents of the port, who had consulted with the experts called in by the Ufficio della Foce prior to Zendrini. Natalini completes some of the main points of the project drawn up by Zendrini for Viareggio. Thanks to the visit, the knowledge of the documents and the discussions with Natalini, Zendrini gains firsthand knowledge of Tuscan construction methods for hydraulic works not mentioned in the writings of Viviani on the *cantoni di smalto*. Precisely because in his discussions with Tirali, starting in 1721, Zendrini has understood that the technical solution to improve the defenses of the Venetian Lagoon might depend on a particular type of pozzolana-based cement, his stay in Tuscany takes on particular value, not just for the improvements to the port of Viareggio, but also for the works he would soon be proposing to the Magistrato delle Acque in Venice. Yet the fundamental element discovered by Zendrini in the hydraulic works at Viareggio is not pozzolana at all; it is the caisson, filled with pozzolana concrete. It is precisely this type of caisson that influences Zendrini's first proposals for Venice.

The report prepared by Zendrini in May 1735 and published the following year,[7] represents, after the treatise of Viviani, another fundamental chapter in Italian hydraulic construction. The models discussed by Zendrini are those of Roman antiquity – the ports of Miseno, Gaeta, Terracina, Anzio, Ostia and Civitavecchia – along with suggestions on ways of making foundations in water contained in *De Architectura* by Vitruvius (among the works of ancient literature mentioned on the subject of the construction of ports, there are writings by Suetonius, Rutilius, Pliny the Younger, Cassius Dio and Virgil). To reconstruct the possible sources of the idea of Zendrini to use a solid cement mortar, we should not overlook the treatises of Bélidor, like *La Science des Ingénieurs*.[8]

[3] Georg Christoph Martini, *Viaggio in Toscana (1725-1745)*, Massa, Modena, Palazzo di S. Elisabetta, Aedes Muratoriana, 1969. For other uses of pozzolana in Tuscany see, *ibid*, p. 70.

[4] Pietro Maestri, *Nuova statistica dell'industria italiana*, in, "Annali Universali di Statistica, economia pubblica, legilsazione, storia, viaggi e commercio," vol. I, series IV, January 1860, (pp. 240-82), p. 251.

[5] Bernardino Zendrini, *All'illustrissimo ufficio della Foce di Viareggio. Relazione, che concerne il miglioramento dell'Aria e la riforma del Porto, con una appendice intorno agli effetti delle Macchie, per rapporto all'alterazione dell'aria*, Lucca, S. & G.

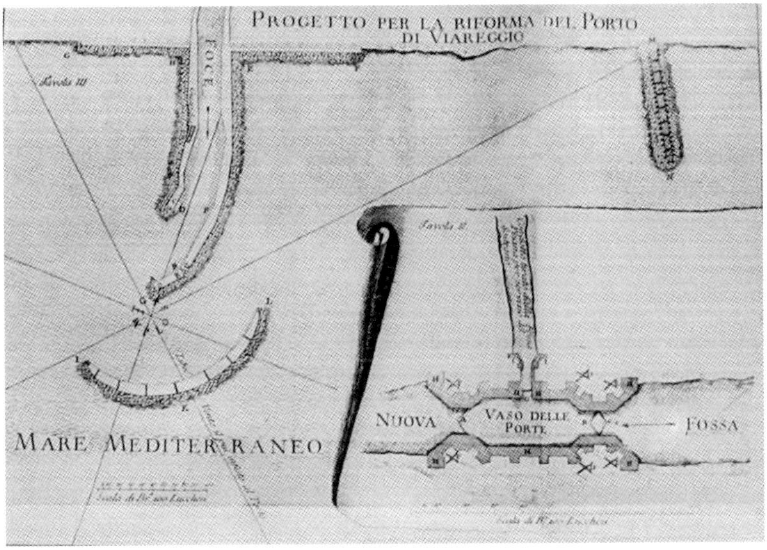


**115**. Bernardino Zendrini, "Project for the reform of the port of Viareggio," 1735 (C. Benzio, *Viareggio. Storia di un territorio. Le Marine lucchesi tra il XV e il XIX secolo*, Pisa, Pacini Editore, 1986).

Rereading the quotations from the Roman authors, what interests Zendrini is the system of caissons used by the Tuscan engineers at the port of Viareggio. Here one significant contribution is the description provided by Suetonius of the port of Ostia during the life of Claudius. Zendrini does not fail to underline the similarities with contemporary construction:

"He had two moles installed, one to the right, the other to the left, and at the entrance, though at a great depth, he set a mass whose construction, he adds, was done by sinking that ship that from Egypt had brought the great Obelisk, then putting Caissons with stones, and lime in the way and form that is still in use today, taking these Caissons as being described by the expression *congestisque pilis*. Cassius Dio also speaks extensively of this during the life of Claudius. The pier, then, could not be bigger than twenty-nine to thirty geometric paces, in keeping with the length of the sunken ship."[9]

Zendrini cites other passages, again to demonstrate the use of caissons on the part of the ancient Romans – "they were none other than the modern Caissons,"[10] the ones he sees made in the ports along the Roman and Tuscan shorelines. His proposal to prevent the silting up of the port of Viareggio consists precisely in the placement of an antemural, or arched jetty, "a mole of stones, and of Caissons,"[11] about 200 fathoms in length, to be made with seven and a half caissons – each caisson with a length of about 25 fathoms.[12] Zendrini explains that the caissons should be "left straight, or with a few stones against them," in the part facing the port, "and left towards the Sea with the ordinary reef."[13] Though in the report no references are made to the use of pozzolana, precisely the properties of this earth described in texts cited by Zendrini and known to the Tuscan technicians working on the hydraulic projects of the ports are what make the system of caissons Zendrini develops for Viareggio effective.

Marescandoli, 1736, p. 44. For 17th-century works at the port, and for the contribution of a Venetian technician, Sebastiano Roccatagliata, in 1639, see, Benzio, *op. cit.*, p. 83.

6 *Cf.* Benzio, *op. cit.*, p. 121.

7 *Cf.* Zendrini, *All'illustrissimo ufficio della Foce di Viareggio* cit.

8 On the publications of Zendrini that mention Bélidor, see, Angelo Zendrini, *Elogio di Bernardo Zendrini, matematico della Repubblica di Venezia*, Venezia, Stamperia Palese, 1807.

9 Zendrini, *All'illustrissimo ufficio della Foce di Viareggio* cit., 1736, pp. 40-41.

10 *Ibid.*, p. 41.

11 *Ibid.*, p. 42.

12 *Ibid.*, p. 57.

13 *Ibid.*, p. 44.

## Caisson and pozzolana masonry: works by Vanvitelli for the port of Ancona

The construction system involving the sinking of caissons in water is widespread in Mediterranean ports, as confirmed by the works at Viareggio, Zendrini's projects for its port, and the writings of Bélidor in *La Science des Ingénieurs*. But the caisson will now be sufficiently updated in its constructive details to become the typical, renowned foundation of the most demanding hydraulic works of the 1700s, in the wake of the experiences of two great builders engaged in two of the most outstanding works of marine and fluvial foundations during the 1730s: Vanvitelli, destined to become one of Italy's most famous architects, at the service of the Holy See and the King of Naples; and Charles Labelye (1705-81), the Swiss engineer and mathematician behind the construction of the first grand 18th-century bridge over the Thames, that of Westminster. Both works must rely on the quality of mortars and their composition, the decisive factor of the different construction system of Vanvitelli and Labelye for the caisson.

As in the ports of the Grand Duchy of Tuscany or the Republic of Genoa, the main hydraulic works of the Papal State and the Kingdom of Naples continue to be made with pozzolana concrete, from the Fountain of Trevi designed by Salvi, begun in 1732 and completed in 1762 (its rock formation is made with pozzolana-base mortar),[14] to the treatment of segments of the Tiber waterfront. Concrete made with pozzolana is also used for the quay and road along the riverbank at the port of Ripetta,[15] built in 1703-06 and designed by Alessandro Specchi (1688-1729).

Hydraulic works of the Papal State are also indicated by certain themes of the Concorso Clementino at the Accademia di San Luca, in Rome. "Plaza located on the heights, overlooking a Seaport" is the theme chosen in 1728, for the competition won by the Ancona-based architect Carlo Marchionni (1702-86), who would later become a specialist on hydraulic architecture, working with Vanvitelli, and was the father of Filippo (1732-1805), another architect specializing in hydraulic works in the employ of Vanvitelli. It seems to be no coincidence that the competition of 1732 for the first architecture class, won by Bernardo Antonio Vittone (1704/05-70), is on the subject "a city in the middle of the sea," and that the guidelines specify that on one part of the artificial island there should be a leper hospital, with the competition participants required to explain the "how this city would be constructed,"[16] if we consider the fact that a few months later work began in Ancona on the most important hydraulic construction undertaken by the Papal State during the course of the 1700s: the leper hospital, or Lazzaretto, to be built in the sea, which would also serve to protect the port. The technical question of the creation of an artificial island in the sea involved specifying the methods of foundation in water, from the provisional structures to the nature of the binder. Some of the participants in the Concorso Clementino also go into detail on methods for hydraulic foundations, like Carlo Sala, who wins the second prize *ex-aequo*. Sala explains the four ways of making foundations in the sea, among

[14] *Cf.* Maria Grazia Chilosi, *L'uso delle malte e dei metalli nella costruzione e nella decorazione*, in, Luisa Cardilli, ed., *Fontana di Trevi, la storia, il restauro*, Roma, Carte Segrete, 1991, (pp. 149-55), p. 149.

[15] Nicoletta Marconi, *I legni e le pietra. Gli approdi per i materiali edili tra XVI e XVIII secolo sul Tevere a Roma*, in, Carlo M. Travaglini, *La città e il fiume (secoli XIII-XIX)*, Roma, Ecole Française de Roma, 2008, (pp. 181-95), p. 190.

[16] *Concorso Clementino 1732*, in, Paolo Marconi, Angela Cipriani, Enrico Valeriani, *I Disegni di Architettura dell'Archivio Storico dell'Accademia di San Luca*, Roma, De Luca, 1974, vol. I, pp. 15-16.

**116-120**. Carlo Sala, scholastic project "a city in the middle of the sea," Concorso Clementino, 1732, Accademia di San Luca, Rome (P. Marconi, A. Cipriani, E. Valeriani, *I Disegni di Architettura dell'Archivio Storico dell'Accademia di San Luca*, Roma, De Luca, 1974, vol. I).

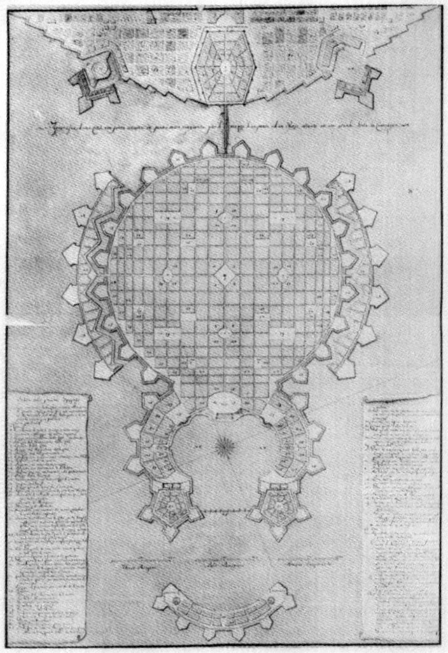

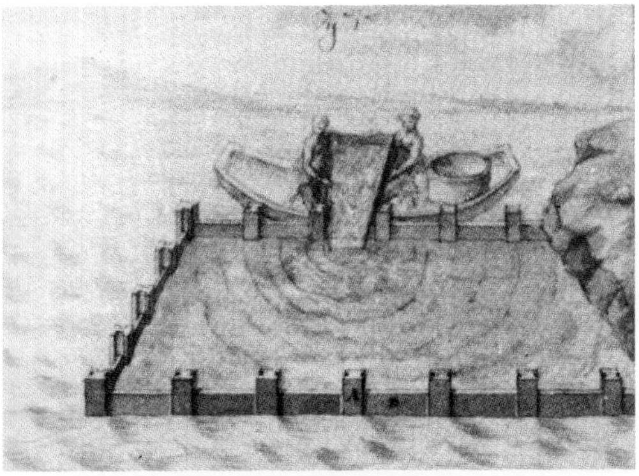

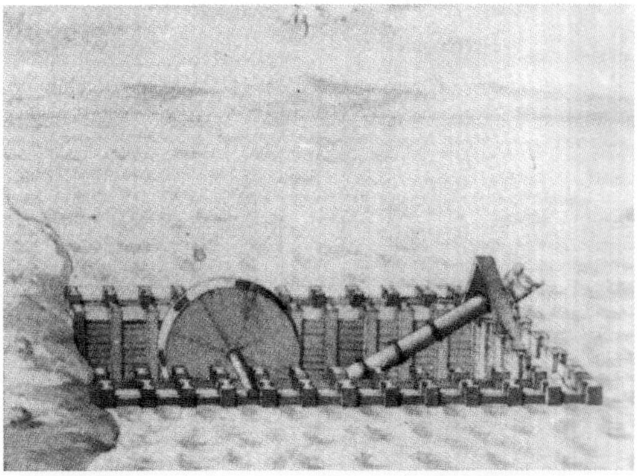

Veduta della Fabrica de nuovi Lazzaretti d' Ancona adi 4 Giugno 1734.

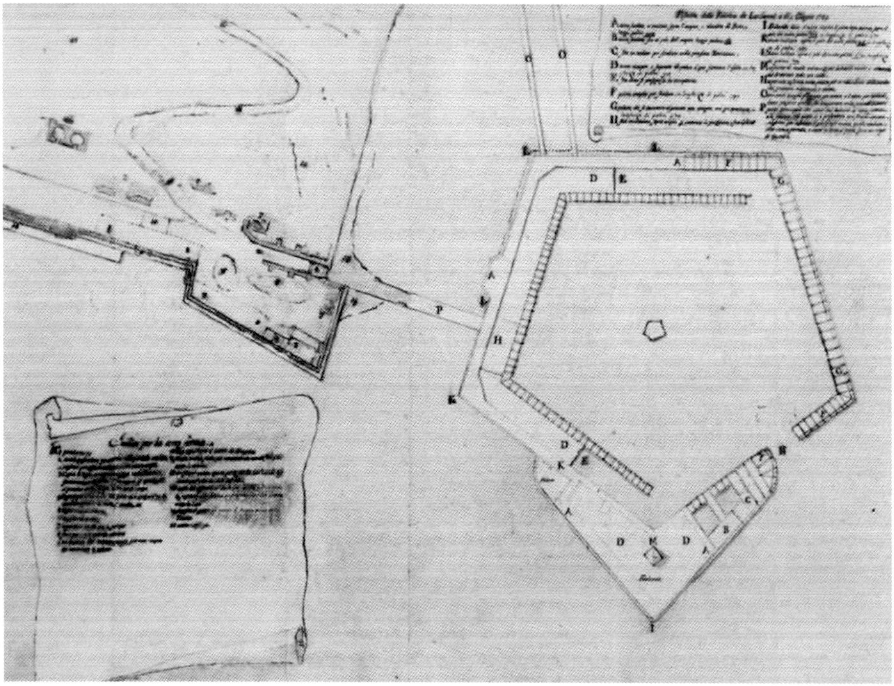

**121-122**. Luigi Vanvitelli, view and plan of the works in progress in June 1734 for the construction of the Lazzaretto at the port of Ancona (vol. 44 H 32, no. 70434, Gabinetto Nazionale delle Stampe, Roma; "Palladio," IV, 1953).

which the first is that of enclosures, rows of piles or coffers, formed by poles and boards, to fill with stones and lime mortar without removing the water, thanks to the use of a "material that can harden like Pozzolana."[17]

Starting in April 1733 the architect and engineer of Pope Clement XII, Vanvitelli, is in Ancona to supervise the work on the construction of the Lazzaretto – the commission probably dates back to October 1732; the ceremony of the laying of the first stone takes place on 15 August.

Vanvitelli's expertise in the field of hydraulic construction is not unrelated to his Dutch background. His family's interest in this type of construction is demonstrated by a codex of drawings by Luigi's father, Gaspar Adriaensz van Wittel, known as Gaspare Vanvitelli (1653-1736), on hydraulic works and based on solutions put into practice by another famous Dutch engineer active in Rome at the end of the 1600s, Cornelio Meyer,[18] supervisor of important works done along the bed of the Tiber and made with wooden piles. Meyer is also the author of an essay published in 1685, containing a chapter entitled *Dell'inalzamento de i fondi, che seguono nel Porto l'Ancona e del rimedio per sgravarlo delli riempimenti.*[19]

The creation of the artificial island for the Lazzaretto of Ancona is done by erecting, thanks to the shallow seabed, a *passonata* or *palata*,[20] of fir poles, beaten with a pile-driver and reinforced with iron. The arrangement of piles is organized in large compartments, to be filled with pozzolana masonry,[21] and borders an enclosure of water that is filled with stones and earth, starting from the center. It is not out of the question that Vanvitelli's system for the erecting of the enclosure in the sea has been influenced by the examination of Fontana of the construction of the piers of Anzio at the time of the Romans by means of "dammed up voids." The filling material for the foundation of the Lazzaretto is brought from near Ancona, from a place on the coast near Falconara – "filler of earth, which composes the Island," as is written in a document from 1736.[22] Other "earth and chips of stone" to "form the land filling" is taken "from around the Lazzaretto building," from the "bottom of the sea."[23]

For the masonry works inside the enclosure of piles a large quantity of pozzolana from Baia is used, supplied by Balì Papino Bussi[24] and transported on barges from the other side of the Italian peninsula. A load of pozzolana also arrives from Fiumicino, by way of Civitavecchia, against the wishes of

[17] See the drawing with legend by Carlo Sala, on systems of construction in the sea (*ibid.*, fig. 396; Giorgio Bucciarelli, *Storia e architettura della mole vanvitelliana*, in, Carlo Mezzetti, Giorgio Bucciarelli, Fausto Pugnaloni, *Il Lazzaretto di Ancona, un'opera dimenticata*, Ancona, Cassa di Risparmio di Ancona, 1978, p. 164).

[18] Eugenio Battisti, *Il Vanvitelli e i lavori portuali di Ancona*, in, *Atti del XI congresso di storia dell'architettura, Marche, 6-13 settembre 1959*, Roma, Centro Studi per la Storia dell'Architettura, 1965, pp. 429-52.

[19] *L'arte di restituire a Roma la tralasciata navigazione del suo Tevere, Cornelio Meyer Olandese*, Roma, L. Varese, 1685.

[20] Legend in, Luigi Vanvitelli, "Veduta della Fabbrica de nuovi Lazzaretti d'Ancona adi 4 Giugno 1734," vol. 44 H 32, n. 70434, Gabinetto Nazionale delle Stampe, Roma (*cf.* Armando Schiavo, *Disegni inediti di G.L. Barberini e di L. Vanvitelli*, in, "Palladio," IV, 1953, (pp. 153-64), p. 161, and Elisabeth Kieven, *Ferdinando Fuga e l'architettura romana del Settecento*, Roma, Multigrafica Editrice, 1988, p. 110).

[21] Pietro Carreras, *Gli interventi vanvitelliani nel porto di Ancona. I. Il lazzaretto*, in, "Storia dell'arte," 1972, no. 14, (pp. 163-91), p. 167, note 22. The term "*passonata*" is used in a document of 1740, from which it appears that the structure is formed of fir poles with the point and the "head rounded off, to make it possible to drive them with a sledgehammer, in the passonate to make the foundations of the Lazzaretto." On the use of the term *passonata* in 1600s see, Pietro Frosini, *La 'passonata' sul Tevere davanti alla Vigna di Papa Giulio dell'ing.re olandese Cornelio Meyer*, in, *Strenna dei Romanisti*, XVII, 1956, pp. 61-65.

[22] Carreras, *op. cit.*, pp. 186-87.

[23] Cit. in, Giorgio Bucciarelli, *Storia e architettura della mole vanvitelliana*, in, Carlo Mezzetti, Giorgio Bucciarelli, Fausto Pugnaloni, *Il Lazzaretto di Ancona, un'opera dimenticata*, Ancona, Cassa di Risparmio di Ancona, 1978, p. 169. *Cf.* Pietro Carreras, *Studi su Luigi Vanvitelli*, Firenze, La Nuova Italia, 1977.

[24] Luigi Vanvitelli, report on the works at the port of Ancona, 4 September 1754, in, Franco Strazzullo, *Le lettere di Luigi Vanvitelli della Biblioteca Palatina di Caserta*, Galatina, Congedo Editore, vol. I, 1976, (pp. 361-65), p. 364.

**123**. Giuseppe Vasi, view of the Lazzaretto of Ancona (*Il secondo libro del nuovo teatro delle fabbriche et edifici fatti fare In Roma e fuori Roma dalla Santità di N.S. papa Clemente XII*, Roma, Calcografia della R.C.A, 1739).

Vanvitelli, because he believes that pozzolana to be inferior to the pozzolana of Baia. It is interesting to see what Vanvitelli writes about the nature of the two types of pozzolana, relying on the opinion of Vitruvius, also because the pozzolana traded in Europe comes from Civitavecchia: "because it can remain 10 years underwater without setting; while there is the excellent pozzolana of Baia, which is sand thrown up centuries ago from the volcanoes that were there. This sets with an iron grip under water, and it is what Vitruvius speaks of, and it gave pozzolana its reputation [...]."[25]

The pozzolana shipped for the Lazzaretto is stocked on dry land, near the wooden bridge leading to the pile structure, in the "pozzolana store" next to the "wooden shack to conserve worksite provisions and irons for the piles," where "wooden poles and planks for the pile structure" are also stored.[26] The pozzolana is processed on a special wooden platform mounted on the "pile structure" itself – "a wooden deck above the water to unload the pozzolana and to make lime mortar."[27] A compound of pozzolana, lime and gravel is poured into the gaps of the pile structure, to create a solid barrier against the sea – in the worksite jargon, when the "*palate*" have been filled with the pouring they are said to be "*compite*."[28] At the end of 1736 the embankment, the sub-foundations and the two wings of the Lazzaretto have almost been finished (the entire work is completed in 1738).

Following damage caused by a storm on 15 September 1733, in 1734 Vanvitelli designs a pier to protect the Lazzaretto and the port. Dutch, English, Swedish, Danish, French and Papal technicians and sea captains take part in the development of his two proposals: a mole that follows the layout of the old one begun by the emperor Trajan, the selected solution, or an "antimural," or "curved mole," based on the model of Civitavecchia, and which as Vanvitelli

[25] *Ibid. Cf.* also, Emilio Rufini, *Ricerche sull'attività del Vanvitelli nelle Marche*, in, *Atti del XI congresso di storia dell'architettura, Marche, 6-13 settembre 1959*, Roma, Centro Studi per la Storia dell'Architettura, 1965, (pp. 453-67), p. 459. Milizia's opinion differs: "the first chest, which was poured under his supervision, was ruined during the first storm, as he himself had foreseen, who had uselessly protested against that pozzolana, or Baia ash, that had been used" (Francesco Milizia, *Memorie degli architetti antichi e moderni. Quarta edizione*, Bassano, Remondini, 1785, vol. II, p. 264).

**124**. Giuseppe Vasi, plan diagram of the port with the indication of the "Nuovo Braccio," Ancona (C. Mezzetti, G. Bucciarelli, F. Pugnaloni, *Il Lazzaretto di Ancona, un'opera dimenticata*, Ancona, Cassa di Risparmio di Ancona, 1978).

himself recalls in 1754, "had to fill a large bed, so the work began with the sinking of many large ships filled with stones, and then continued with the pouring of stones, and caissons as is done in Genoa and elsewhere, where the ports are situated above the stones, and finally with caissons to join the work together, making it so sturdy that it can stand up to any impact of storms [...]."[29]

To make the foundations of the mole in deeper water than the zone where the pile structure of the Lazzaretto stands, Vanvitelli makes use of the caisson system familiar to him at least in the case of the foundation of the moles of the port of Anzio. Vanvitelli's caisson is 100 palms long (about 20 meters) and formed by a "skeleton" of "*bardonali grossi*," or girders,[30] made with fir wood "for lack of other types"; it has four large openings to facilitate sinking. To limit the risk of major water infiltration during the operation of filling with masonry, certain experts advise installing four "spouts" at the corners – Vanvitelli does not share their opinion.[31]

Constructed in 1735, the caisson remains exposed to the weather for about two years, due to a suspension of the worksite ordered by the Vatican. When work resumes, in 1737, the operation of caulking of the lower part of the caisson is done quickly, to guarantee that it will float. The caisson is placed in the water on 18 August, reinforced by "chains and fasteners of wood held in place at the ends with iron bands" ("*Risposta e relazione veridica*"), and then filled with stones, "lime, and Puzzolana as far as the height of 10 palms of wall" ("*Relazione apocrifa*"). On September 1st the caisson is moved to the selected place for its sinking, where it is moored using ropes and lines anchored in the sea or attached to the walls of the Lazzaretto. In this position the workmen continue to "make it caulk" and to fill it with the same masonry (600 workmen are on the job by day, 300 by night). But on the night of 4-5 September the caisson lists dangerously on one side, because it is "poorly caulked" according to an anonymous critic, while Vanvitelli believes it is because the workmen have not noticed that the masonry is not distributed in a uniform way. Unable to remove the water and to spread the masonry properly to restore the horizontal position of the caisson, and given the onslaught of a storm that could ram the caisson up against the walls of the Lazzaretto or the port, Vanvitelli orders that it immediately be sunk. When it strikes the bottom, some

[26] Legend in, Luigi Vanvitelli, "Veduta della Fabbrica de nuovi Lazzaretti d'Ancona adi 4 Giugno 1734," Gabinetto Nazionale delle Stampe, Roma, vol. 44 H 32, no. 70434 (*cf.* Armando Schiavo, *Disegni inediti di G.L. Barberini e di L. Vanvitelli*, in, "Palladio," IV, 1953, (pp. 153-64), p. 161).
[27] *Ibid.*
[28] *Ibid.*
[29] "Trascrizione della relazione su i lavori del Porto di Ancona," 4 September 1754, in, Rufini, *op. cit.*, p. 458.
[30] "Relazione apocrifa," in, Battisti, *op. cit.*, p. 442.
[31] *Ibid.*

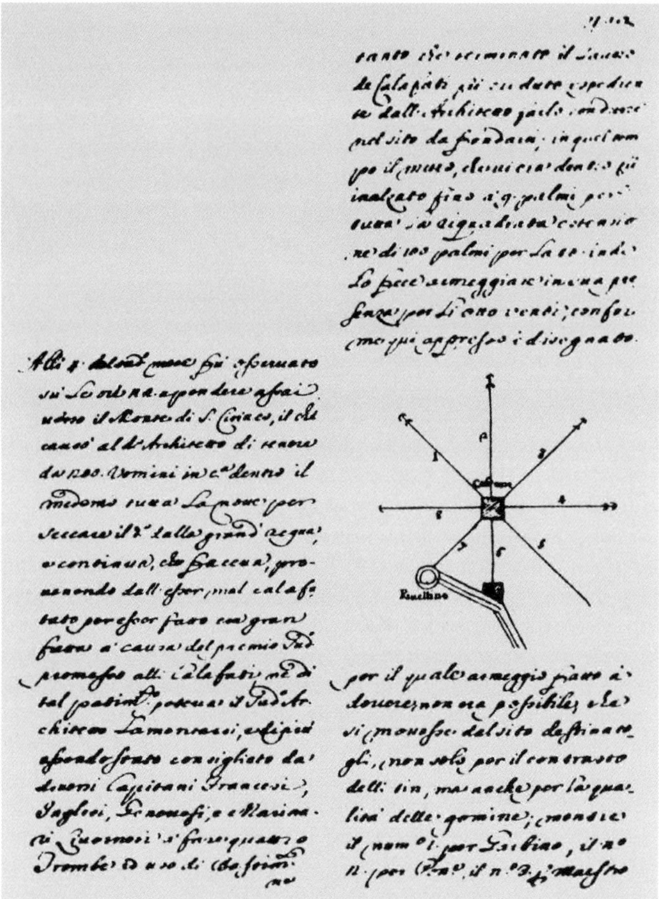

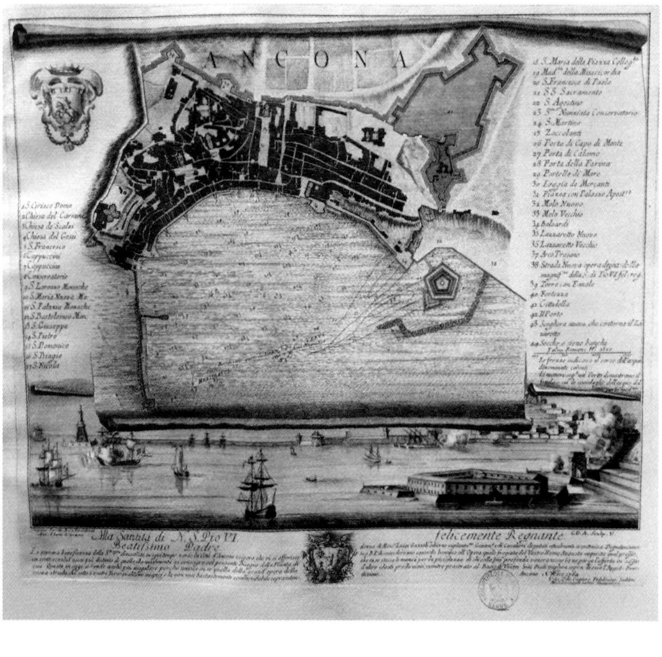

**125**. Luigi Vanvitelli, sketch regarding the positioning in the sea of the caisson for the construction of the "Nuovo Braccio," following the accident in 1737, Ancona (P. Carreras, *Studi su Luigi Vanvitelli*, Firenze, La Nuova Italia, 1977).

**126**. Scipione Paretti, view and plan of the port and city of Ancona, 1784 (Cartes et Plans, Ge CC-1376 (54), Bibliothèque Nationale de France, Paris).

of the girders and boards of the inclined side break; others snap during the next few hours, due to the power of the storm. In the days to follow Vanvitelli studies a plan to strengthen the caisson with small satellite caissons anchored to it, in order to complete the foundations of the pier. First he has two smaller caissons built, with a length of 40 palms, to attach to the sunken caisson, and he sets up a "pile structure" in order to be able to "finish filling the wall up to the surface of the sea, in the way he had already put into practice in the piece of the pier already made, and for all the foundations of the Lazaretto."[32] Underwater inspection of the condition of the caisson suggests construction of four small coffers, 15 by 15 palms, to fill with masonry, and with a side 30 palms high, or "small caissons made to resemble chairs with high backs," to form, also thanks to "two decks," the destroyed side of the caisson. To complete the masonry inside the caisson, Vanvitelli invents "suitable devices" that allow him to lower the pozzolana cement into the water "in such a way that it does not lose even the fine grain of the lime in the water."[33] Nevertheless, he believes that part of the failure of the first caisson can be blamed precisely on the pozzolana, the material shipped from Fiumicino – "it was the reason of the loss of the Caisson that would have advanced the pier about 130 palms more [...]."[34]

In the mid-1750s resumption of the work on the unfinished pier of Ancona is discussed. Vanvitelli is again consulted regarding the orientation of the mole – "do not deviate from the direction that was begun," is his view.[35] The question of Baia pozzolana is again the focus of his concerns regarding how to make the pier foundation. "[...] then, at the time of the German government," he writes, referring to the years of his first works in Ancona, "it could be easily obtained, but now I cannot say, because it is needed for the royal public buildings."[36] By "royal buildings" he means those he is constructing himself in Caserta for Charles of Bourbon, including the aqueduct. In 1755, as discussion continues on new pozzolana masonry works to be done at the port of Ancona, with Carlo Murena (1713-64), Vanvitelli refers to what he had done to repair the caisson in 1737: "Carlo recalls the remedy I made to repair the first Caisson. And this method is new to all even now, even to those who saw it in Ancona, but have never understood the effect in order to be able to put it into practice."[37]

**127**. Giuseppe Vasi, view of the "Nuovo Braccio," Ancona (C. Mezzetti, G. Bucciarelli, F. Pugnaloni, *Il Lazzaretto di Ancona, un'opera dimenticata*, Ancona, Cassa di Risparmio di Ancona, 1978).

[32] *Ibid.*, p. 450.
[33] *Ibid.*, p. 451.
[34] "Trascrizione della relazione su i lavori del Porto di Ancona," 4 September 1754, in, Rufini, *op. cit.*, p. 459.
[35] *Ibid.*
[36] *Ibid.*
[37] Luigi Vanvitelli, 18 November 1755 (cit. in, Rufini, *op. cit.*, pp. 464-65).

Work on the mole continues under the supervision of Filippo Marchionni, with the pozzolana from Civitavecchia and the system of piles where the seabed permits it (the work is interrupted in 1770 precisely due to a lack of pozzolana, and of piles). The type of workers involved demonstrates the different potential of the various construction systems. About five hundred convicts continue to pour stones and pozzolana lime mortar, up to water level, inside the mole structure formed by two rows of long poles as "large as the biggest mast of a ship";[38] then masons intervene, laying the squared stones of the parapet. The worksite demonstrates that construction with pozzolana cement does not require the use of specialized workers.

"[...] the pouring of stones and lime," an observer reports, "done [...] by those many prisoners does not cease by day or by night, once a section has been started, until all the filling between the two rows of piles has been completed [...]. And the reason for this is that all the lime is prepared with a mixture of course sand, known as pozzolana, transported here in many ships from the vicinity of Rome [...] all that great volume of stones and lime, thrown down into the water at a furious pace by those prisoners, very quickly forms a surpassingly strong boulder, to which another boulder made with the same method would never join itself perfectly if added later [...]."[39]

Vanvitelli also makes use of pozzolana for the construction of the Palace of Caserta. At the start of 1752, experiments are even conducted on pozzolana dug from the garden of the Palace, and considered "excellent."[40]

The Carolino aqueduct, built from 1753 to 1762 by Vanvitelli as part of the Palace complex, to bring water from the springs of San Martino Valle Caudia, is another fundamental example of historical construction using pozzolana cement. In the spectacular segment, about half a kilometer in length, of the so-called Ponti della Valle, it forms a viaduct with three orders of arches, based on the model of that of Pont du Gard.[41] The aqueduct, like the Lazzaretto, the mole of Ancona and the Palace, can be seen as one of the fundamental manifestations of the continuity between the ancient technique of conglomerates and its reutilization and spread in the form of modern cement. Nevertheless, for there to be a true international spread of the ancient constructive forms being put into practice in Italy, foreign travelers have to report on their characteristics, emphasizing the possibilities for contemporary European construction. It is interesting that Jean-Baptiste Rondelet (1743-1829), following a visit to Vanvitelli's aqueduct, makes a detailed description of the masonry pattern of the cladding, pointing out the similarities with ancient construction in Rome and Pompeii, and the quality of the cement – a "very hard cement."[42]

Works in Naples and its vicinity are also done by following the constructive criteria of the Roman tradition. The writing of Pierre-Louis Moreau in 1757, when he sees the grand worksite of the Albergo dei Poveri in Naples,

[38] Goffredo Franzini, letter to Jacopo Durandi, in, Pietro Custodi, ed., *Lettere e scritti varj di Giuseppe Baretti*, Milano, Società Tipografica de' Classici Italiani, 1839, vol. IV, (pp. 371-79), p. 374.

[39] *Ibid.*, pp. 375-76. In April 1768, Antonio Lecchi takes soundings, with the collaboration of Marchionne (or Marchionni), at the port of Ancona by order of the cardinal Perelli to establish "if the Port was truly raised at the bottom by the sand, which the Sea at times may bring there [...]" (Antonio Lecchi, *Memorie idrostatico-storiche delle operazioni eseguite nell'inalveazione del Reno in Bologna, e degli altri Torrenti per la linea di Primaro al Mare dall'anno 1765 fino al 1772*, Modena, Società Tipografica, 1773, vol. I, pp. 186 *et seq.*). With the soundings, Lecchi tests the nature of the seabed: "Only near a large boat was the sounder raised once, with attached a few small pieces of Puzzolana, which a short time earlier had been unloaded from the boat into small vessels to transport it to the warehouse used by the pier worksite" (*ibid.*, p. 188).

[40] Luigi Vanvitelli, letter to Urbano Vanvitelli, 5 February

by Ferdinando Fuga (1699-1782), confirms the pragmatic nature of construction with pozzolana cement, and the fact that it can be done without specialized laborers. "[...] the construction is quite like that practiced in Rome. Stone is used sparingly, the rubble used is soft and of mediocre quality, the pozzolana is white and maybe of even better quality than that used in Rome. They use soldiers, convicts and other people who can be gathered together for this work, who are able to carry out all the manual labor."[43]

## The Westminster Bridge and the caisson of Labelye

Technical information on the floating caisson and the pozzolana masonry used in Ancona is not spread by publication of pamphlets or reports, and European engineers in the 1700s, though carefully considering all the main techniques of construction in the sea, make no significant references to what had been accomplished by Vanvitelli.[44]

Instead, the Westminster Bridge in London, designed and built in 1734-50, becomes the fundamental reference in the evolution of the caisson for underwater foundations, also thanks to the coverage in the London press and then in international publications, during all phases, from the design to the construction. On the worksite for the bridge in London the caisson is of the type with removable sides, perfected in such detail that it becomes a model for other underwater foundations. Thanks to the Westminster experience, the caisson gains remarkable success during the course of the 18th century, until it becomes a system of foundation for cement masonry. In the case of the Westminster Bridge, the particular conditions posed by the Thames, the geological nature of its riverbed and the phenomenon of the high tide, which reaches 17 feet, create the premises for research on a foundation technique that would be an alternative to the cofferdam.[45]

The first projects for the Westminster Bridge trigger intense debate on the various methods of construction on riverbeds; the question of the type of masonry is also discussed, though without going beyond generic assertions of the need to use strong mortars. Study of the components of hydraulic mortars does not begin until the construction of the piers of bridges in the 1750s.

Different pamphlets are written during the discussion of the design of Westminster Bridge, starting in 1734. In 1735 John Price, who proposes his own bridge design, describes the two possible ways of found the piers of the new bridge: "by digging up the gravel [...] into lighters, as they do ballast, at every ebb tide and low water," and then laying on the bed "the largest sort of Portland blocks, with iron pins, by machines," to cramp together using metal anchors; or to erect in the water, where the piers are to be placed, "two frames of timber-work, secured by piles drove into the river," following the model of the carpentry of Caesar for the bridge on the Rhine, equipped with frame-

1752, in, Strazzullo, *op. cit.*, (pp. 106-109), p. 109.
[41] *Cf.* Luigi Vanvitelli, letter to Urbano Vanvitelli, 2 May 1752, *ibid.*, pp. 48-49.
[42] Jean-Baptiste Rondelet, *Traité théorique et pratique de l'art de bâtir*, Paris, Chez L'Auteur, 1803, vol. I, p. 345.
[43] Cit. in, Sophie Descat, *Le Voyage d'Italie de Pierre-Louis Moreau: Journal intime d'un architecte des Lumières (1754-1757)*, Pessac, Presses Universitaires de Bordeaux, 2004, p. 119.
[44] Mention of the works of Vanvitelli at Ancona is made by Louis-Alexandre de Cessart (*cf.* Louis-Victor Dubois d'Arneuville, ed., *Description des travaux hydrauliques de Louis-Alexandre de Cessart*, Paris, Baudouin, 1806, vol. I, p. 42).
[45] Jean-Rodolphe Perronet, *Mémoire sur les différentes méthodes qui ont été employées pour fonder les ouvrages de Mâçonnerie dans l'eau, & principalement sur celles qui tendent à supprimer les batardeaux & épuisemens dans la construction des Ponts*, in, "Histoire de l'Académie Royale des Sciences. Année MDCCLXVI. Avec les Mémoires de Mathématiques & de Physique, pour la même Année," Paris, Imprimerie Royale, 1769, (pp. 139-48), p. 141. The report is dated 13 November 1765.

work, to then construct, once the water has been eliminated using a special machine, the masonry of the pier with "work well cemented and join'd together with proper ligaments."[46] The fact that he specifies, regarding the masonry, that "all joints" should be "set in Tarris," indicates a use of hydraulic mortar to be made with trass imported from Holland.[47]

The foundation methods of bridge piers contained in the treatises of Palladio, Scamozzi, Serlio and Gautier are discussed in the pamphlet published in 1736 and written by the authoritative Nicholas Hawksmoor (1661 circa - 1736), disciple and collaborator of Wren – the pamphlet also contains his design for the bridge.[48] Thanks to Hawksmoor, English readers find explanations of the various types of wooden frameworks to construct on riverbeds, as well as solutions such as that of the deviation of the river's course: the "coffer-dams, made of timber and boards," named "batterdeaux" [*sic*] by the French, that make it possible to lay masonry "well cramped and tied with Iron";[49] the "grillage of oak," made solid by means of iron bars, made to float at the position where the pier is to be built thanks to "large hulks or lighters, with cables and other machines," equipped with a layer of stones "being well cramp'd together" and then lowered "gently to bed of the river, by proper machines and cables."[50] Hawksmoor's indications on the various possibilities of composition of the cement to place between the stones together with metal anchors are also important, as they point to knowledge of the various compounds gleaned from the treatises, but certainly not to their use in London and vicinity: "the joints filled with strong *Padua* lime, mortar, *Puzzoli* terrace, or cement."[51] Probably the hydraulic properties of the lime obtained from Padua stones are known to Hawksmoor thanks to the treatise of Palladio.

Hawksmoor's focus on masonry construction with the possible use of pozzolana to make a Roman-type mortar is not surprising if we consider the observations of Wren on the concrete masonry of the foundations discovered during excavation in London for the reconstruction of the church of Saint Bartholomew, the church of Saint Mary-le-Bow,[52] and Saint Paul's cathedral (Hawksmoor was also involved in the latter).[53]

Also in 1736, another protagonist of the discussions on the Westminster Bridge is Langley, also working on the fabrication of artificial stone, who addresses the technical details of the construction of "coffer-dams" and "grillage of oak," not explicitly included in the explanations of Price and Hawksmoor (Langley proposes different bridge designs, in masonry and wood).

If the riverbed on which to place the piers of the Westminster Bridge is deep, Langley suggests using a floating carpentry structure to be sunk after have begun the masonry work above water level. His system is linked not so much to the "grillage of oak" proposed by Hawksmoor, as to the English experiences in the 1600s conducted by Henry Shere for the construction of the port of Tangiers, starting with the coffers used by De Mari at the port of Genoa. To describe his system, Langley uses the definition "concave parallelopipedon" or

[46] John Price, *Some Considerations Humbly offered to the Honourable Members of the House of Commons, For Building a Stone-Bridge over the River Thames, from Westminster to Lambeth [...]*, London, 1735, pp. 5-7.
[47] *Ibid.*, p. 5.
[48] Nicholas Hawksmoor, *A Short Historical Account of London-Bridge; with A Proposition for a New Stone-Bridge at Westminster [...]*, London, J. Wilcox, 1736.
[49] *Ibid.*, pp. 26, 30-31.
[50] *Ibid.*, p. 31.
[51] *Ibid.*
[52] *Cf.* George Godwin, *Prize Essay upon the Nature and Properties of Concrete, and its application to Construction up to the present period*, in, "Transactions of the Institute of British Architects of London. Sessions 1835-36," vol. I, part I, London, 1836, (pp. 1-35), p. 8.
[53] *Cf.* Charles Robert Cockerell, *On the Foundations*

"parallelopipedon vessel."[54] His floating machine would be equipped "with double joist and plank" and high flanks, and once anchored to four piles sunk at the site where the pier is to be built, the first layer of masonry would be laid, with large stones and cramp irons, after which the whole thing would be slowly lowered into the water. After the construction of the pier, the sides of the "parallelopipedon vessel" would be removed.[55] Instead, in the case of "strong clay" found on a shallower riverbed, Langley advises using "coffer-dams," which he explains as being formed by a wooden structure with spaces "filled with cases of stiff clay" – the water in the "coffer-dam" would be pumped away.[56]

Also for the masonry, Langley recommends solutions that do not correspond to the ideal method indicated by Gautier, such as that of possibly making use of the pouring of materials inside stone sides. Certain indications reveal English construction practices, like the use of Dutch trass also suggested by Price.

"It will be best to build the piers with Moor-stone," Langley explains, "as being of greater strength than Portland, which is best for the upper part. If this bridge be built with piers of Moor-stone unto the low water level, the sides and vaults of the arches faced with Portland-stone, about 3 feet in thickness, and the internal part filled with found brick-work laid in terrace, etc., the expense will be greatly abated, and the whole fabrick equally (at least, if not more) durable, as it would be were it entirely of Stone."[57]

An architect specializing in carpentry, collaborator of Hawksmoor and Wren also active on the Saint Paul's project, John James (1673 circa -1746), is the author of another pamphlet in which he describes his own bridge design and criticizes the assertions of Price and Langley.[58] James goes into foundation methods – translating the definition of Langley of the "concave parallelopipedon" into that of "chest or coffer," also used in the 1600s for the coffers of Shere.

When Langley publishes his bridge design, discussion is already in progress on the proposal of Labelye, which contains a solution for the pier foundation similar to the "parallelopipedon vessel" – Langley lays claim to the invention, but we cannot rule out the possibility that he based his work precisely on the ideas of Labelye.

After a series of studies, deliberations and proposals, and after demonstrations with models, in 1738 the works are entrusted to the supervision of Labelye; he is joined by the mathematician John Theophilus Desaguliers (1683-1744) for the calculations of the loads and flow and the subsequent dimensioning of the piers, by Henry Herbert, Ninth Earl of Pembroke (1693-1750), an architect with engineering skills, and by Richard Graham (? - 1749), a technician and expert on the financial management of construction sites.

One of the crucial questions of the project has to do with the foundations and construction of the piers of the bridge. In September 1737 and May 1738, Labelye is asked to explain, for the commission charged with assessment of the

*of the late Church of St. Bartholomew*, cit. in, *Architecture. Institute of British Architects*, in, "The Gentleman's Magazine," vol. XXI, January 1844, (pp. 75-79), p. 76.

[54] Batty Langley, *Design for the Bridge at New Palace-yard, Westminster*, London, J. Milan, 1736, pp. 26-27.

[55] *Ibid.*

[56] *Ibid.*, p. 25.

[57] *Ibid.*, p. 23. See also, Batty Langley, *A Survey of Westminster Bridge, as 'this now sinking into ruin' with remarks on the pratichal method used for building the piers*, London, M. Cooper, 1748.

[58] *Cf.* John James, *A short review of the several pamphlets and schemes, that have been offered to the publick*, London, H. Woodfall, 1736.

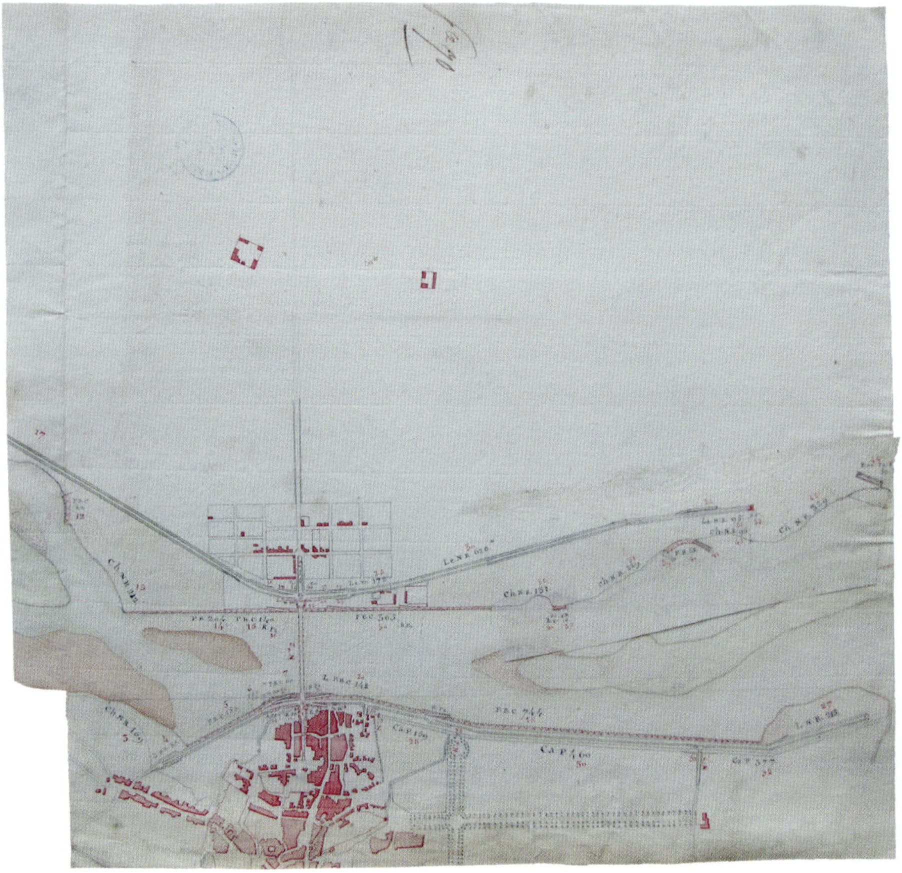

**128**. [Charles Labelye], plan of the sector of London with the project for the Westminster Bridge and the treatment of the riverbank (F.14.10208.80, Archives Nationales, Paris).

**129**. Charles Labelye, project for the balustrade of the Westminster Bridge, London, 12 March 1739 (F.14.10208.3, Archives Nationales, Paris).

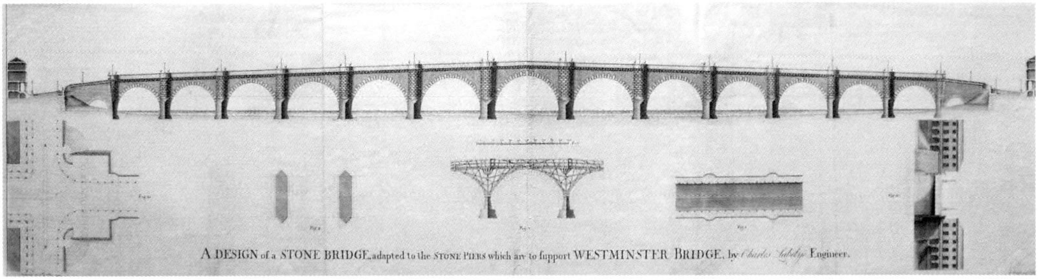


**130**. Charles Labelye, project for Westminster Bridge, London (F.14.10208.1, Archives Nationales, Paris).

project, the main points of his method based on the caisson system, for which he consults with the carpenter James King and the mason Andrews Jelfe.[59]

Labelye's method for pier foundations made with caissons differs from the usual caisson procedure, and involves carpentry refinements, a considerable increase in the caisson size and a worksite logic based on precision of execution and deployment. While the traditional caisson is sunk without controlling its position on the bed, Labelye invents a method that can be seen as an evolution of the cofferdam, to be built according to the logic of the floating caisson.

Precisely because the pier has to be built inside the compartment, the size of Labelye's caisson must be exceptional if compared to earlier specimens (80 by 30 feet). The structure also requires more accurate carpentry than the conventional versions, because it must simultaneously float, form the foundation grille of the pier – the "grating or bottom" – and stand up to the pressure of the water of the river during the construction of the masonry. This is why the structure of Labelye's caisson, perhaps based on Swiss building solutions of wooden carpentry, is composed of a crosswise vertical and horizontal arrangement of the fir beams and the boards of the sides, which form the solid mobile flanks anchored to the lower grille, which is also wrapped in boards – the connections between the pieces are made with oak. Reinforcements mounted on the upper edge prevent the water pressure from bending the sides when the caisson is lowered into the river. Another particular feature of Labelye's invention is the device for lowering and raising, in order to sink the caisson at the desired position, given the fact that the grille of its base will constitute the foundation left ready for the pier, and will therefore determine its slope.

"When *chest-work* is used," Labelye explains, "the most common method is to fill the chests with stone-work as high as they can without sinking, and then suffer them to sink to the bottom, and bed there as chance will let them: whereas this *Caisson* was contrived to be let down and raised up again without any difficulty, until it should be found to bed itself sufficiently level upon the surface of the hard gravel on the bottom of the pit, dug in the river for that purpose. In the construction of works laid in water by the help of *chest-work*, the chests are sunk one after another, as near as they can be, but the whole work has no bond to keep it together except its own weight, whereas each of the piers of this bridge are built, course after course, with the same bonding, the same cementing and cramping, as if built upon dry ground, and every pier is, as it were, like one solid stone, laid none less than 5 feet, and none more than 14 feet under the surface of the bed of the river."[60]

[59] *Cf.* Charles Labelye, *A short account of the methods made use of in laying the foundation of the piers of Westminster-Bridge*, London, A. Parker, for the author, 1739.

[60] Charles Labelye, *A Description of Westminster Bridge*, London, W. Strahan, 1751, pp. 44-45.

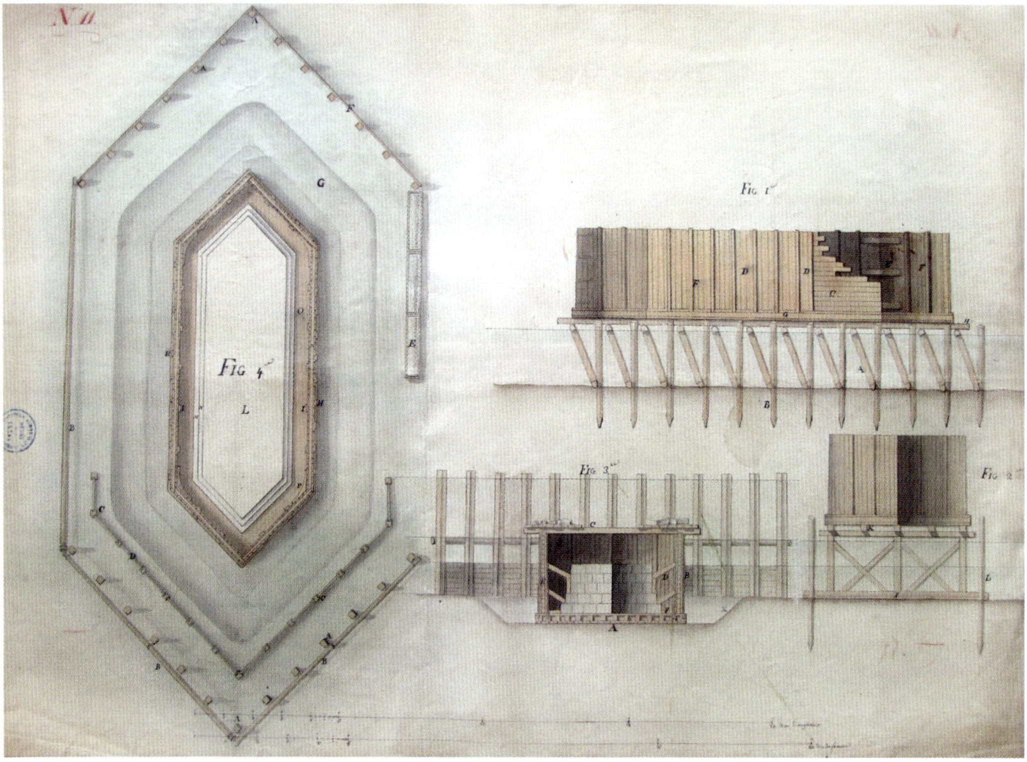

**131**. Charles Labelye, project for Westminster Bridge, London. Drawing of the caisson and its position for the construction of a pier (F.14.10208.80, Archives Nationales, Paris).

**132**. Charles Labelye, project for Westminster Bridge, London. Drawing of pier with wooden arches (F.14.10208.65, Archives Nationales, Paris).

**133**. Charles Labelye, project for Westminster Bridge, London. Drawing of the caisson (Ms 2516, Ecole des Ponts et Chaussées, Paris).

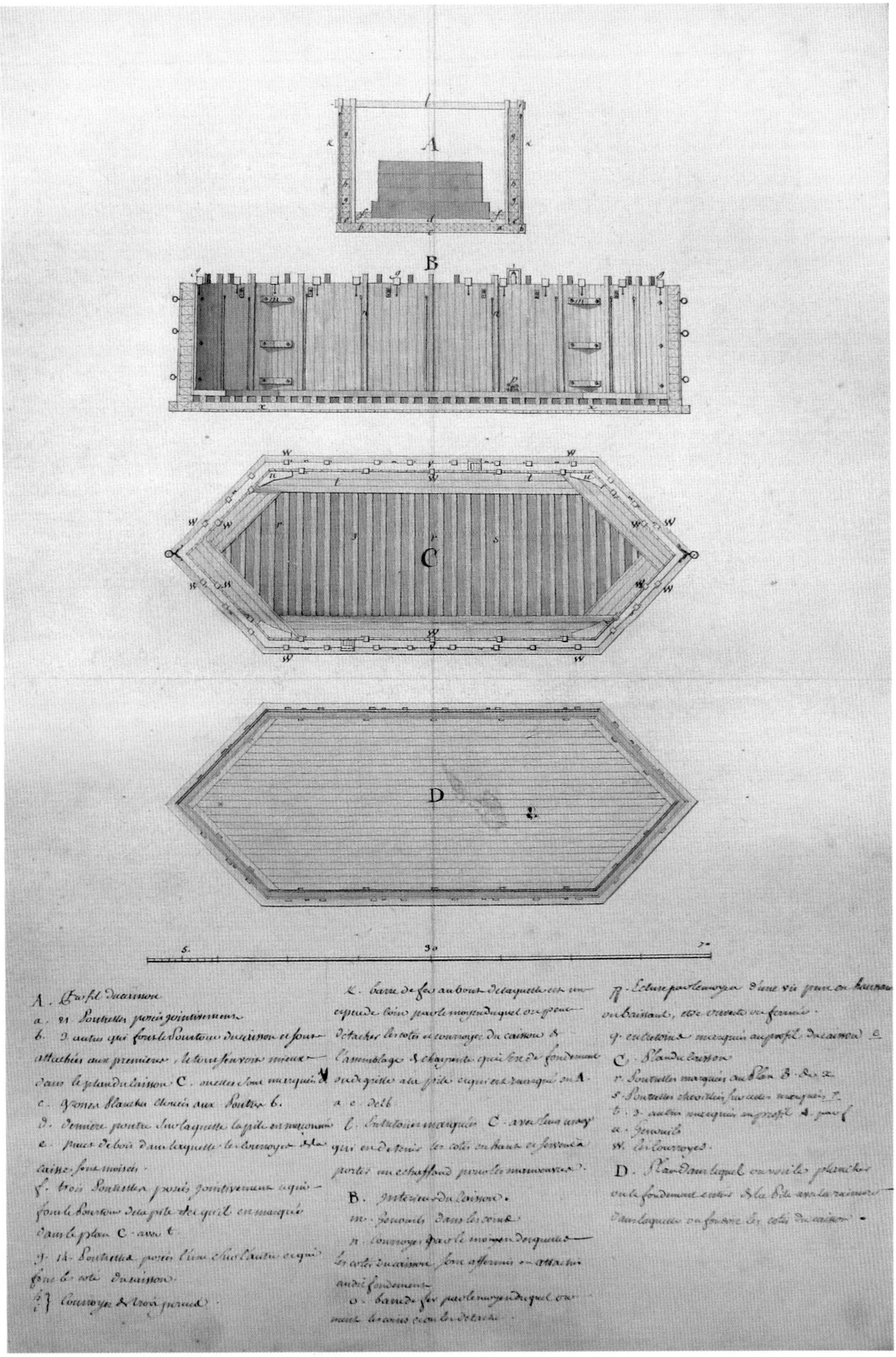
A
B
C
D

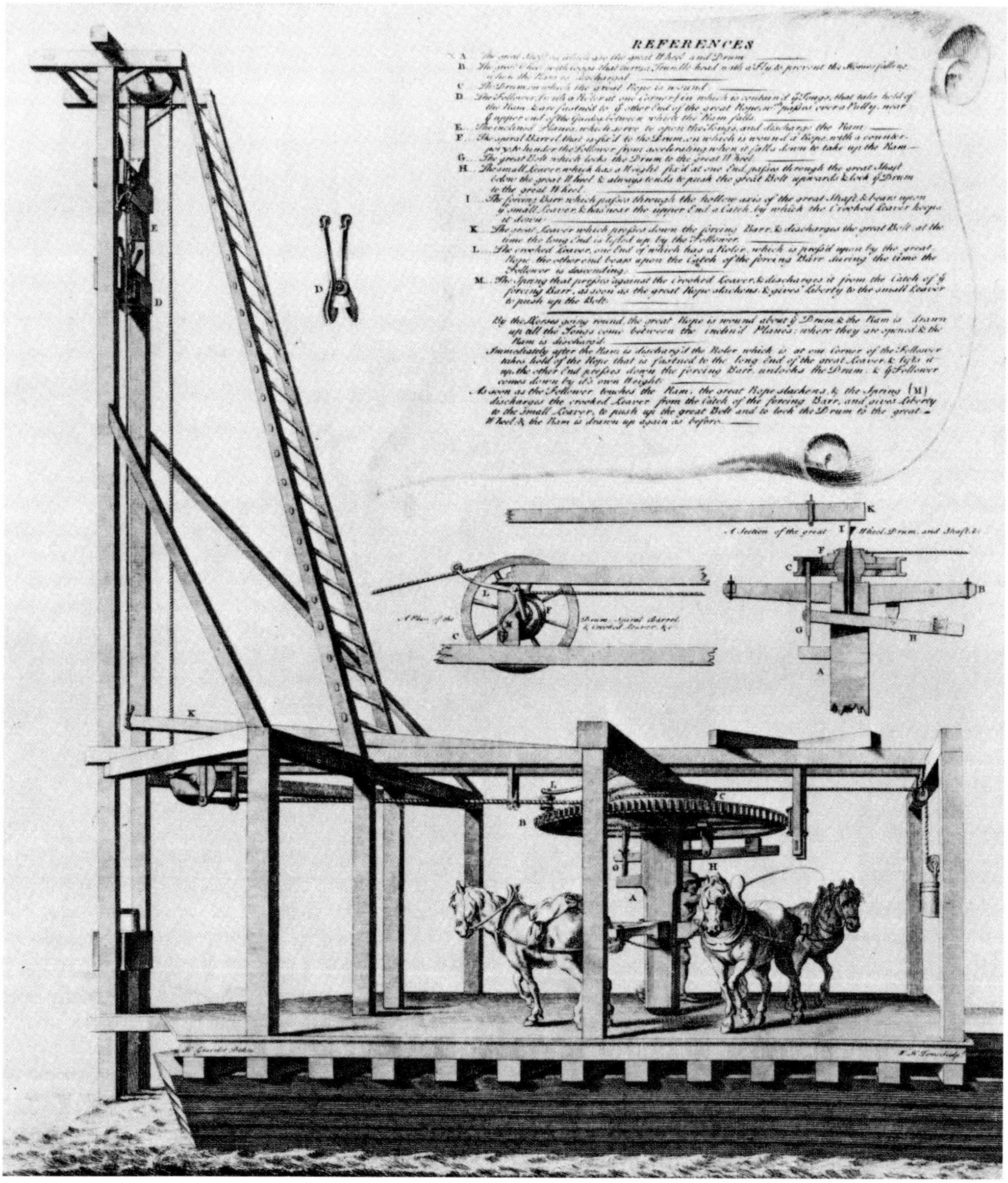


**134**. James Valoué, pile driver for the construction of the Westminster Bridge, London, 1737-38 (T. Ruddock, *Arch Bridges and Their Builders, 1735-1835*, Cambridge, Cambridge University Press, 1979).

The statement of the technical origin of his method coming from solutions practiced in France is contained in the explanations presented to the commission and published in several essays by Labelye himself: "That the foundation [...] should be laid on a strong grating of timber, planked underneath; that this grating of timber should be made the bottom of a vessel, such as is called *Caisson*, by the *French*; (which term I shall make use of); that the sides of this *Caisson*, should be so contrived, as to be taken away, after the pier should be finished: that the bed of the river should be dug to a sufficient depth, and made level, in order to lay thereon the bottom of the *Caisson* [...]."[61]

Labelye indicates the forerunners of his invention: "The method I have made use of is not absolutely new: something like it is often practised in the construction of *moles*, *dikes*, *ramparts*, and other works erected in the sea, or in lakes, or in deep fens or morasses, and I have assisted in making use of it, where all other methods had proved ineffectual [...]."[62]

As of May 1738, when the site for the bridge has been definitively established, two solutions exist: one with masonry piers supporting a grand arched structure in wood, designed by King – a "wooden superstructure"[63] – and another in which the arches too are in masonry (to implement if the construction of the piers turns out to be sufficiently solid), proposed by Labelye and by a member of the commission, perhaps Herbert.

In June Labelye goes with his masons to Portland to select the stones for the piers. From June 1738 to January 1739 the carpenters construct the two main worksite machines for the pier foundations: one to ram the piles bordering the area of the pier, invented in 1737 by the clockmaker James Valoué (powered by three horses),[64] and the "new sort of *Caisson*" invented by Labelye.[65] The worksite for the construction of the caissons is set up on the banks of the Thames, near the site of the new bridge, and is formed by a platform on stilts, equipped with rotating cylindrical trunks for the lowering of the caisson into the water. The pile driving machine of Valoué is at work in September 1738, attached on two boats – the piles border the area of the pier, preventing boats passing on the Thames from running into the worksite.

"About this time," Labelye writes, referring to August 1738, "the carpenters began to make and erect, under the *surry* shore, 12 frames of timber, supported in a vertical situation, parallel one to another, and kept in their places by short stakes or piles, driven into the ground. These frames reached about 2 feet above the common high-water mark, and were braced together, so as to be kept upright and steady, till the *Caisson* should be built and finished, on the top of them. The ground-cills, or bottom pieces of these frames were made cylindrical, that by taking the braces away, they might serve as so many axes, or centerpieces, round which the frame could move, or revolve altogether, for the lowering or launching of the finished *Caisson*, at any hight-water without any danger of racking or straining. The grating for the first pier, was also begun this month [...]. About the beginning of *october*, the grating belonging to the first pier was finished, and the sides of the *Caisson* 16 feet height, began to be raised upon it. These sides were made of fir timbers, laid horizontally and close one over another, pinned with oaken trunnels, and framed together at all corners

61 *Ibid.*, p. 10.
62 *Ibid.*, p. 44.
63 [Charles Labelye], *The Present State of Westminster Bridge, Containing a Description of the said Bridge [...]*, London, J. Millan, 1743, p. 13.
64 Labelye, *A Description of Westminster Bridge* cit., p. 31. In 1738 Valoué's machine receives the Copley Medal of the Royal Society of London.
65 [Labelye], *The Present State of Westminster Bridge* cit., p. 14.

of the *Caisson* except the two points, or saliant angles, where they were secured by proper iron work; which being unskrewed, would permit the sides of the *Caisson* (in case it should be found necessary) to part asunder into two halves. These sides were planked across the timbers, inside and outside, with three inch planks in a vertical position. The thickness of those sides was 18 inches at bottom, and 15 inches at the top: and, in order to be further strengthen'd, every angle but the two points had three oaken *knee timbers*, properly bolted and secured. These sides when finished, were fastenend to the bottom or grating, by 28 pieces of timber on the outside, and 18 withinside, called straps, about 8 inches broad, and about 3 inches thick, reaching and lapping over the top of the sides; the lower part of these straps had one side cut *dove-tail fashion*, in order to fit the *mortoises* that were cut to receive them in the outermost kerb of the grating, and kept in their respective places by iron wedges. The use of these straps and wedges is as follows. When the pier is built up so much above low-water, as no longer to want the *Caisson*, for the masons to work in, the wedges being drawn up, give liberty to clear straps from the mortoises; the consequence of which is, that the sides must rise by their own levitation or buoyany, leaving the grating under the foundation of the pier, which grating extends several feet all round, by way of footing or basis, as is, or should be always served in such building. The sides of the *Caisson* were hindered from being pressed in by the ambient water, towards the stone work that was to be placed in it, by means of ground timber or ribbon 14 Inches wide, and 7 inches thick, pinn'd upon the upper row of timbers of the grating, which exactly fill'd the space contain'd between the sides of the *Caisson*, and the firs course of lower plinth of the stone pier; and the top of the sides was secured by a sufficient number of beams laid across, which also served to support a floor, on which the masons and labourers stood to hoist the stones out of the lighters, and to lower them into the *Caisson*: several other frames of timber had been contrived to keep the side from being crushed inwards; but those sides alone proved strong enough, and all the rest were omitted, excepting some few struts or props, placed occasionally between the sides and the stone work, which saved both time and expence."[66]

The operations completed for the construction of the first pier are emblematic of the constructive precision of Labelye's method. The riverbed on which to place the caisson is smoothed and packed using a special tool formed by a rectangular stone attached to a long graduated post that makes it possible to gauge the level of the riverbed from the surface of the water (enough of these tools are made to occupy the entire surface of the pier; this work of flattening of the riverbed begins in October 1738). The usual piles are to be driven only if the riverbed turns out to be unsuitable for direct positioning of the caisson. On 15 January 1739 the caisson is placed in the water and moored at the position of the pier; on the 29th of that same month the masons begin to lay the stones of the first row, which is finished in February. At this point the valve is opened to sink the caisson in order to check its position on the bed, and it is observed that silt deposits are preventing it from resting horizontally; therefore, after closing the valve, pumping out the water and bringing the caisson back to the surface, the workmen level the riverbed

66 Labelye, *A Description of Westminster Bridge* cit., pp. 28-33.
67 Charles Labelye, *A Description of Westminster Bridge*, London, W. Strahan, 1751.
68 [Labelye], *The Present State of Westminster Bridge* cit., p. 12.
69 *Ibid.*, p. 7. *Cf.* also Labelye, *A Description* cit., pp. 19-20. For Labelye's references to Dutch constructive culture see, Charles Labelye, *The Result*

in two days, and then proceed successfully with a new trial of the lowering of the caisson. After having finished, on 16 February, laying the third course of the stones, the caisson is definitively sunk and the construction of the masonry continues only during low tide – including night shifts for the masons. When the high tide comes in, the valve is opened to let the water enter slowly; at low tide, it is closed and the water is removed with four pumps, to resume work on the masonry. The first pier is finished on 24 March; on the 30th, the sides of the caisson are removed, to be positioned on the bottom grille of the second pier (this caisson is placed in the water on 11 April and moored between the piles at the position of the pier on 13 April). On 30 April the scaffolding of the first pier is dismantled. With this system, in May the first three piers of the bridge are already complete.[67]

The construction principle of large stones adopted by Labelye for this type of masonry fits into the line indicated by Gautier: "That instead of chalk, small stones, or rubbish, with which the inside of most buildings are filled, the piers are entirely built with solid blocks of *Portland* [...]."[68] The stones are also fastened with stone chucks (Labelye also calculates the arrangement of the stones in the courses at the base of the pier). The masonry work is entrusted to Jelfe and Samuel Tuffnell. The use of trass from Holland is documented by Labelye himself in the description of the masonry of the bridge piers: "All the piers are built the same in the inside as on the outside, of solid *Portland* block stones, none less (by the contract) than one ton, or twenty hundred weight, unless here and there a smaller called a *Closer*, placed between four other larger stones, but most of them are two or three tons weight, and several of four and five tons. All the stones are set in (and their joints filled with) a *cement* called *Dutch Tarris*, and they are fastened together besides with iron *cramps* run in with lead, and so placed that none of those *cramps* can be seen, or ever be affected by the water, as soon as the arches are begun."[69]

The fact that the whole construction machine of the Westminster Bridge is conceived like clockwork is also confirmed by the initial proposal of Labelye to make the masonry of the possible stone arches with wedges all laid in keeping with the same arch center, avoiding masonry filler, based on the model seen in the Rialto Bridge in Venice, or in bridges in Paris. Nothing could be further from the Roman constructive logic of concrete than the system invented by Labelye. Nevertheless, his caisson is destined to become the source of inspiration for similar floating machines to be filled with pozzolana masonry.

Only on 12 March 1740, following the results obtained with the construction of the first pylons, is it decided to dispense with the "wooden superstructure" and to build the bridge with arches that are also in masonry, of the same type as that of the pylons. The first three arches are made between the summer of 1741 and that of 1742, with centerings designed by King and resting on special piles inserted up against the piers. The system for striking is discussed by the members of the commission, based on proposals advanced by the technicians. This is the context of the invention, in 1745, by a collaborator of King, the carpenter William Etheridge (1709-76), of a machine capable of sawing piles with a horizontal movement done under the water.[70] Etheridge's saw is composed of two floating rectangular wooden platforms,

*of a Particular View of the North Level of the Fens, Taken in August, 1745*, London, C. and J. Ackers, 1748, p. 22.

70 For the drawing see, William Etheridge, "A perspective view of the Engine made use of for sawing off, under water, the piles which help'd to support the centers for turning the arches of Westminster Bridge [...]," 1 May 1745, Ms 2170, Ecole des Ponts et Chaussées, Paris.

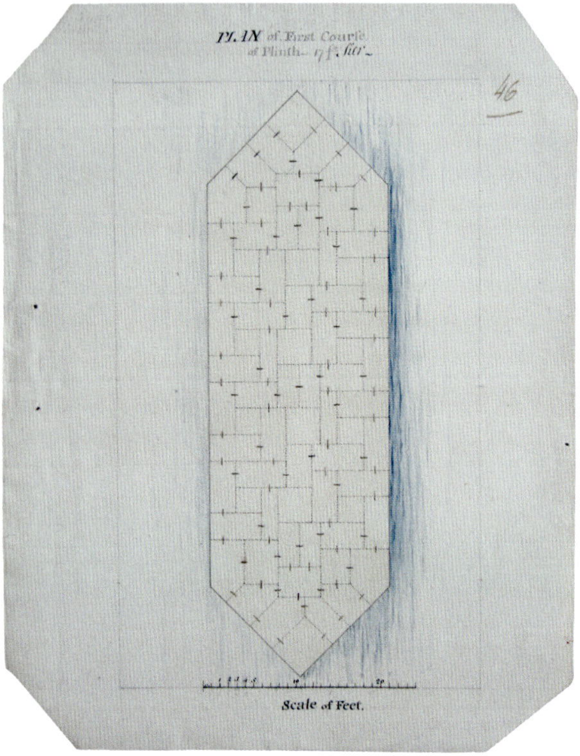


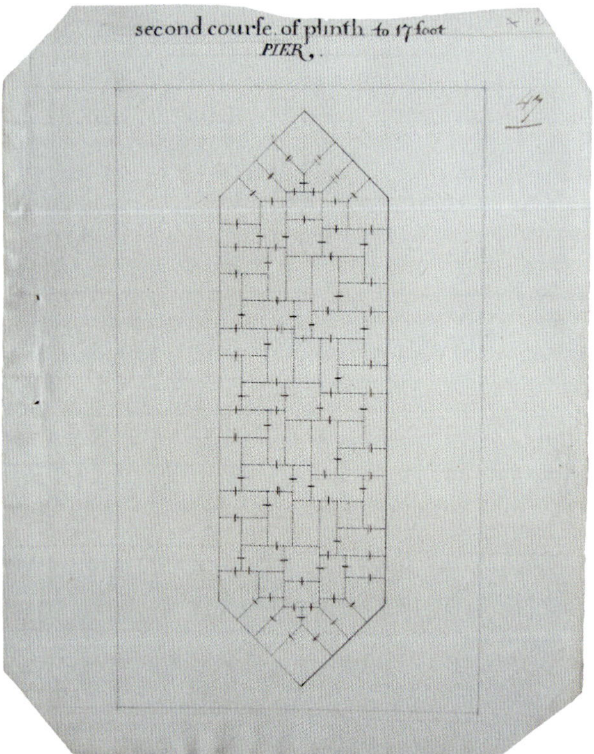


**135-137**. Charles Labelye, project for Westminster Bridge, London. Plan of the first plinth course of a pier (F.14.10208.57, Archives Nationales, Paris); plan of the second plinth course of a pier (F.14.10208.58, Archives Nationales, Paris); drawing of the anchoring of the caisson and the centering (Ms 2516, Ecole des Ponts et Chaussées, Paris).

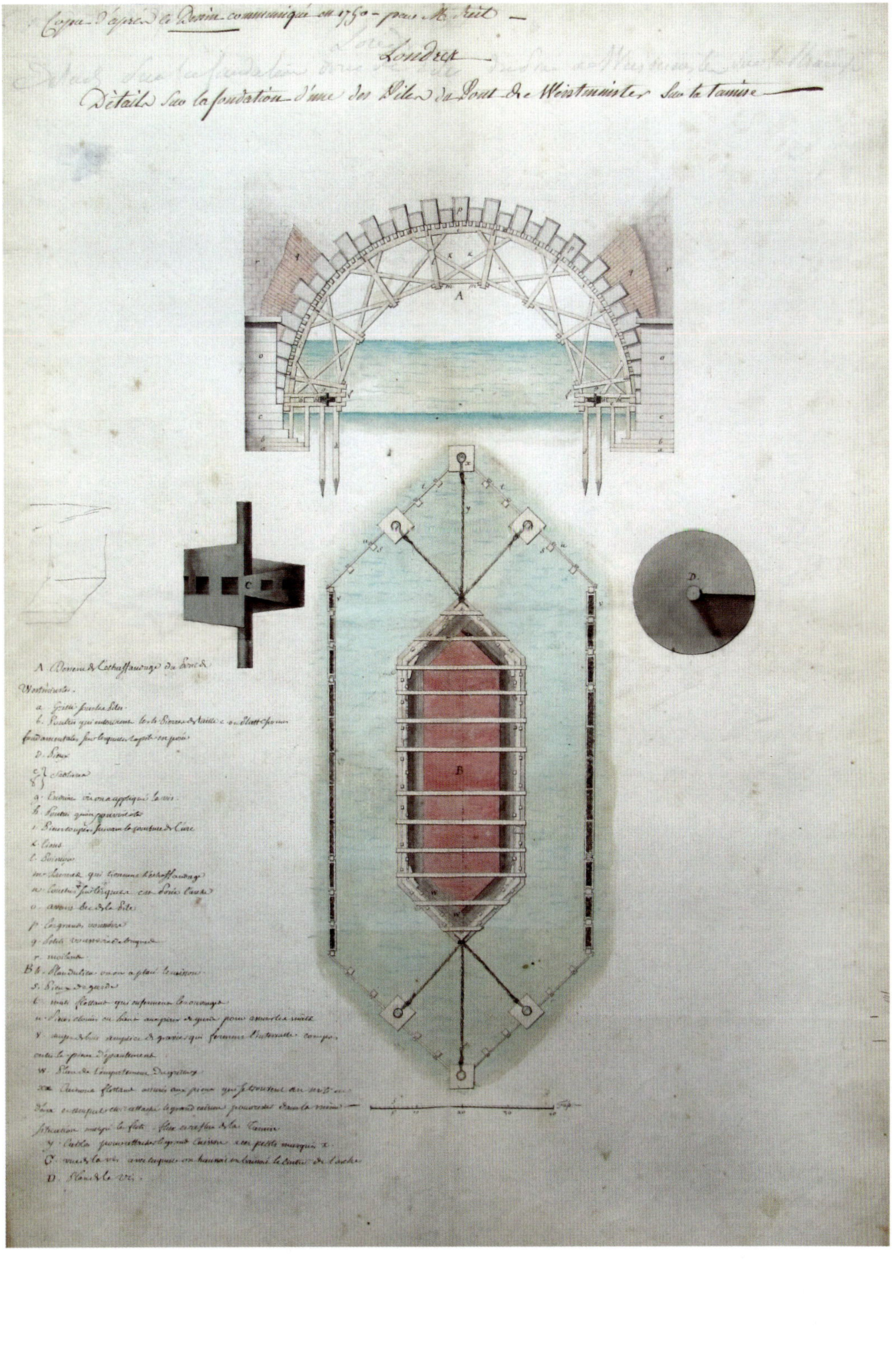
Londres
Détails sur la fondation d'une des Piles du Pont de Westminster sur la Tamise
A
B
C
D

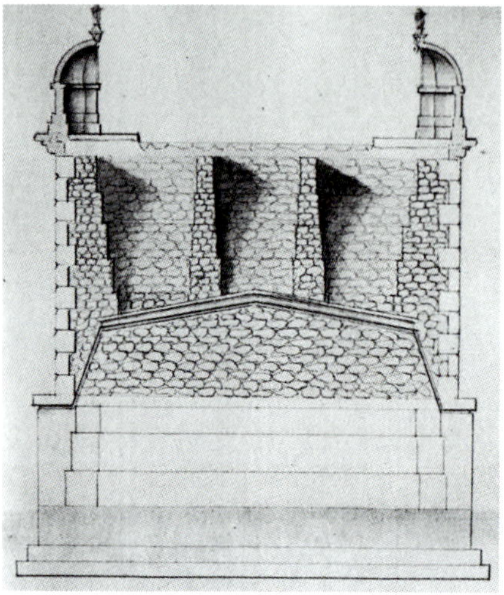

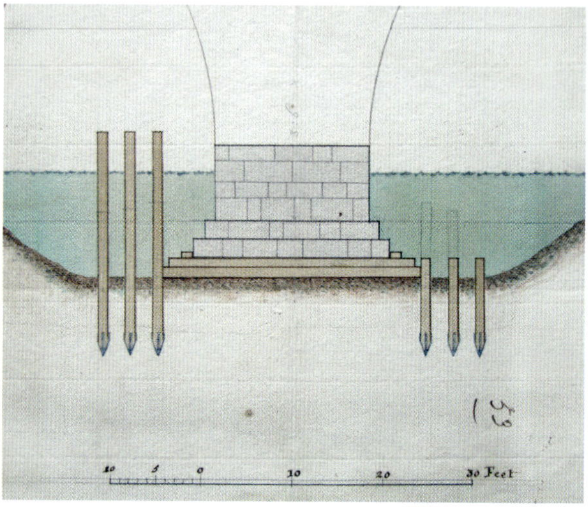


**138**. Charles Labelye, project for Westminster Bridge, London. Cross-section of an arch (T. Ruddock, *Arch Bridges and Their Builders, 1735-1835*, Cambridge, Cambridge University Press, 1979).

**139**. Charles Labelye, Westminster Bridge, London. Indication of the level at which to cut the piles of the enclosure protecting the caisson (F.14.10208.66, Archives Nationales, Paris).

**140**. William Etheridge, "A perspective view of the Engine made use of for sawing off, under water, the piles which help'd to support the centers for turning the arches of Westminster Bridge [...]," 1 May 1745 (Ms 2170, Ecole des Ponts et Chaussées, Paris).

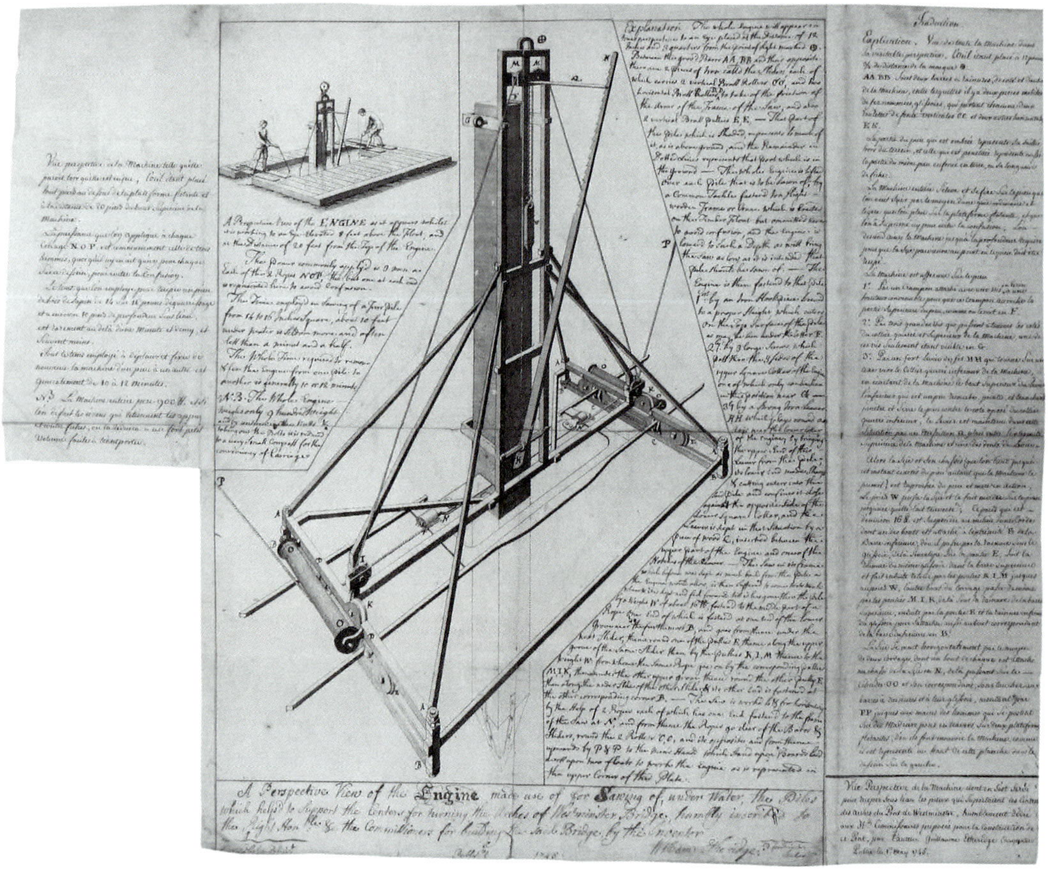

**141**. Charles Labelye, Westminster Bridge, London. Section with plan of half of the bridge (F.14.10208.78, Archives Nationales, Paris).

attached at the sides of the pile to be cut, with a metal device equipped with a horizontal saw that is lowered into the water as far as the height of the desired cut (mechanisms make it possible to operate the saw by means of ropes pulled by two men standing on boards placed between the two platforms). This invention is used to cut the temporary piles around the place of construction of the pylon to protect the structure from possible impact by boats. Etheridge's saw goes on to become a fundamental machine for the rise of Labelye's caisson method in the constructive culture of French engineers.[71]

The masonry over the arches of the Westminster Bridge, up to the level of the deck, is an unusual feature for English construction. In the space between the arches and the ends, Labelye creates a series of compartments filled with gravel, poured in without any mortar, using dry masonry walls. The aim of this dry filler is to create a drainage system that makes it possible to expel the water that might enter from the ends and the deck. The masonry of the abutments is closely studied by Labelye to prevent what he sees as one of the causes of the ruin of bridges: "between every two arches a drain is managed to carry off the water and filth, which, in time, might penetrate and accumulate in those places, to the great detriment of the arches (some bridges having been ruined for want of this precaution) which should be observed in all considerable stone or brick bridges: and yet (as far as I have been able to learn) it has been always omitted."[72]

During the course of 1747 an accident happens that leads to the need to improve Labelye's method of the caisson resting directly on the leveled riverbed. Due to a yielding zone at one point along the bed of the Thames, the foundation of one of the piers sinks, threatening the stability of two arches.

During the discussion that continues even in December 1748 about the measures to be taken (the repairs are not completed until 1750), in 1749 a short pamphlet is published by Charles Marquand, on the foundations of the piers of bridges and the unexpected threats to them caused by the geological nature of riverbeds. The pamphlet is written with the aim of providing expertise for the repair of the pier of Labelye's bridge, and to convince people, "for the future," to "set aside the use of coffer-dams and Cassons."[73] Marquand has discussed things with French engineers and architects who are experts in the construction of bridges – Gautier, Gabriel and René Lambotte (1696-1766), a building contractor working for Gabriel.[74] For the foundations of bridges at

[71] The drawings of the saw of the Westminster Bridge circulate above all among French engineers. See the report with drawings, "Scie employée au Pont de Westminster pour récépage des pieux d'enceinte des Piles. Déscription," DG 1968, Ecole des Ponts et Chaussées, Paris.

[72] [Charles Labelye], *The Present State of Westminster Bridge, Containing a Description of the said Bridge [...]*, London, J. Millan, 1743, p. 9.

[73] Charles Marquand, *Remarks on the Different Constructions of Bridges and Improvements To secure their Foundations [...]*, London, The Author, 1749, p. 8.

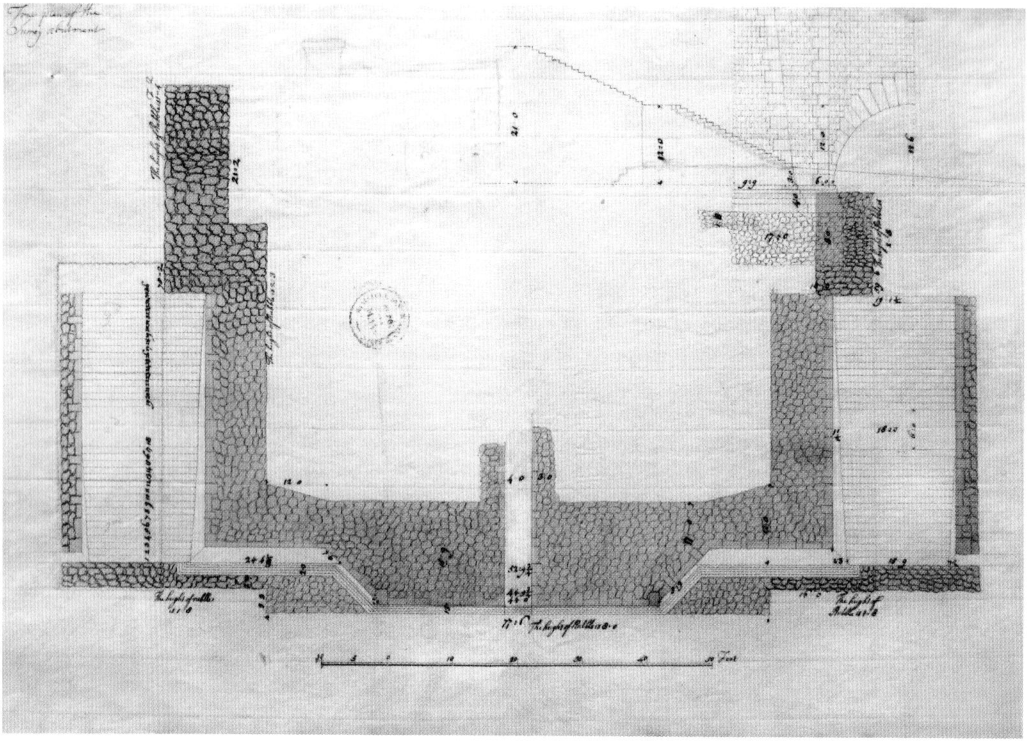

**142**. Charles Labelye, Westminster Bridge, London. Drawing for the reinforcements of the abutment of the bridge with indication of the masonry made with pebbles embedded in mortar (F.14.10208.55, Archives Nationales, Paris).

any depth, Marquand proposes the insertion around the perimeter where the pier is to be built of a series of square-section piles with grooves on two sides, in which boards can slide to create a large compartment from which to remove the water with the usual pumps (a similar solution can already be found in the treatises of previous centuries). To repair the pier of Westminster, he proposes a sort of raft to attach to the pier itself – a "frame of timber."

"By the method I propose," Marquand explains, "the piers of any bridge may be prepared a mile or two from the place of erecting such an edifice, especially where the tide ebbs and flows: for the piers may be floated thither whole [...] and so set down on a sound foundation, well secured [...]. In which case, the piles may be cut off under water, smooth and level, by a particular machine, the model of which may be viewed at the place of sale of this pamphlet, or the author's house. He hath likewise an instrument, invented for driving piles under an arch, where there is no room for a gin [a sort of barge] to come to work."[75]

After the repairing of the pier, with the dismantling of the two arches also documented in drawings by Canaletto, the bridge is finally opened to traffic on 18 November 1750.[76] The phases of the construction of Westminster Bridge, the floating machines and the masonry are also observed and studied by technicians not involved in the construction, as demonstrated by the fact that the

[74] *Ibid.*, p. 7.
[75] *Ibid.*, pp. 8-9.
[76] Following the construction of Westminster Bridge, a book is published by the master mason who worked on Labelye's site, Thomas Gayfere the elder, criticizing the use of the caisson for the foundation of the piers (*cf.* Thomas Gayfere, *Gephryalogia. An historical account of bridges, antient and modern, from the most early mention of them by authors, down to the present time. Including a more particular history and description of the new bridge at Westminster. And an abstract of the rules of bridge-building, by the most eminent architects*, London, C. Corbett, 1751).

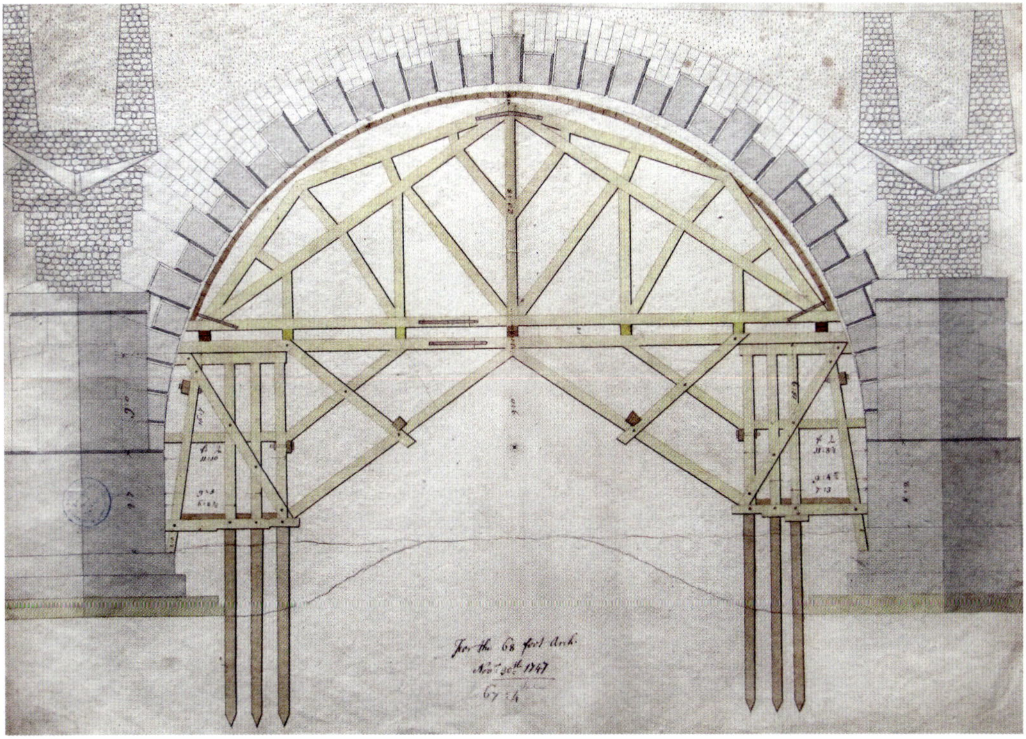


**143**. Charles Labelye, Westminster Bridge, London. Design of centering for the defective arch, 30 November 1747 (F.14.10208.47, Archives Nationales, Paris).

**144**. Canaletto, view of the repair work on the Westminster Bridge, London, 1749 (Royal Collection, Windsor Castle, London).

young Smeaton is among those who monitor the progress of the worksite – he also has copies of the technical drawings for the centerings.[77]

While the piers of Westminster Bridge are constructed in continuous rows, the technical efficacy of another solution is discovered by Labelye during a visit to a medieval bridge to prepare an expert's report. In 1746, together with George Dance the Elder (1700-68) and the carpenter Bartholomew Sparruck, Labelye has to prepare a report on the state of conservation of the London Bridge, built from 1176 to 1209, and propose alterations to reinforce the original piers and create two large central archways to facilitate shipping traffic. On this occasion, Dance presumes that the piers were built with stone exteriors and a nucleus of filler, adding with regard to the latter: "the rubble appears to me to be laid with good mortar."[78] The discovery of the solidity of masonry core walls, made according to criteria taken from the Romans, leads English builders over the years to question the efficacy of coreless masonry using large stones, as theorized by Gautier. Labelye is also aware of the core masonry nature of the piers, as he demonstrates in his drawings of the archways. He intervenes with underwater reinforcements using "rubble" in the abutments of the Westminster Bridge. Dance's hypothesis about the London Bridge is confirmed in 1826 during the demolition of the piers, as part of the construction of the new bridge designed by John Rennie the Elder (1761-1821). It is on this occasion that William Knight (? - 1832), a collaborator of Rennie, makes careful surveys and observations on the masonry during the demolition of a pier: "The inside of the pier was filled with a quantity of loose stones, chiefly Kentish rag and flint, below the level of high water, but above it were large chalk-stones, evidently used in a liquid state, and so compact was this mass of material, that it was with much difficulty separated."[79] The filler between the arches also turns out to be formed by a similar cement compound "of so hard a nature that it was taken out with great difficulty."[80]

In April 1752 Labelye, suffering from asthma, leaves London to spend some months in the south of France, at Béziers. During this stay, he comes into contact with some French engineers, and they discuss the caisson foundation system with him in person.

[77] *Cf.* Ted Ruddock, *Arch Bridges and Their Builders, 1735-1835*, Cambridge, Cambridge University Press, 1979, ed. 2008, p. 18.

[78] George Dance, *Queries concerning London Bridge: with the answers by George Dance, Esq.*, cit. in, Charles Hutton, *Tracts on mathematical and philosophical subjects; comprising among numerous important articles, The Theory of Bridges [...]*, London, F.C. and J. Rivington, 1812, vol. I, (pp. 115-20), p. 120. *Cf.* Ruddock, *op. cit.*, p. 55.

[79] *Old London Bridge*, in, "The Gentleman's Magazine," vol. CXLVII, April 1830, pp. 294-95.

[80] *Cf.* Richard Thomson, *Chronicles of London Bridge*, London, Smith, Elder and Co., 1827, p. 539.

**145-148**. Charles Labelye, survey drawings of the London Bridge with proposal for the expansion of its central arches, 1746 (F.14.10208.28, 29, 9, 15, Archives Nationales, Paris).

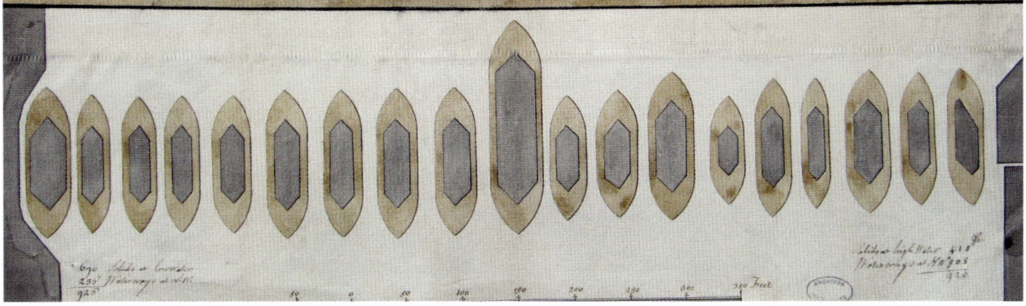

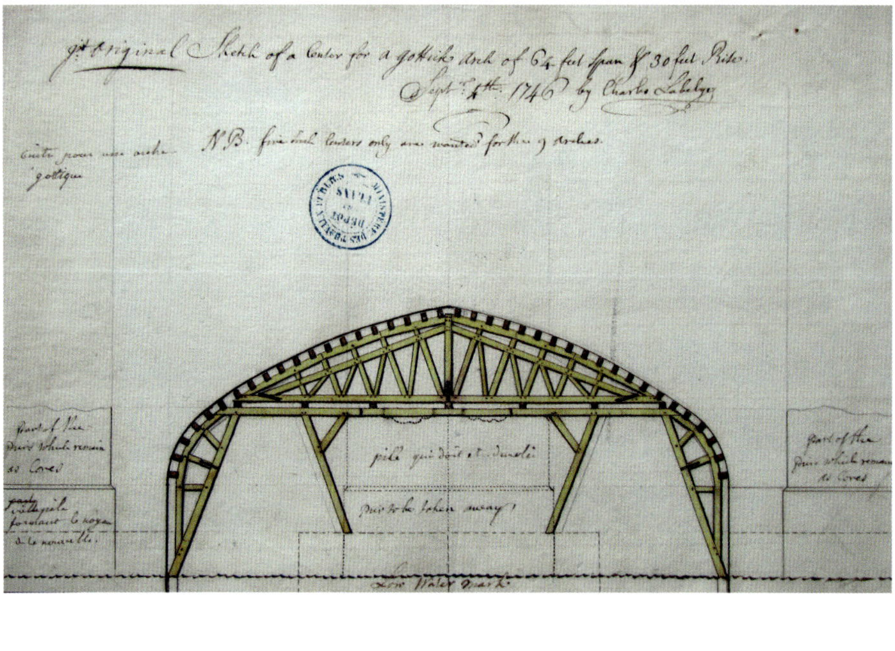


**149-150**. Charles Labelye, survey drawings of the London Bridge with proposal for the expansion of its central arches, 1746 (F.14.10208.13, 24, Archives Nationales, Paris).

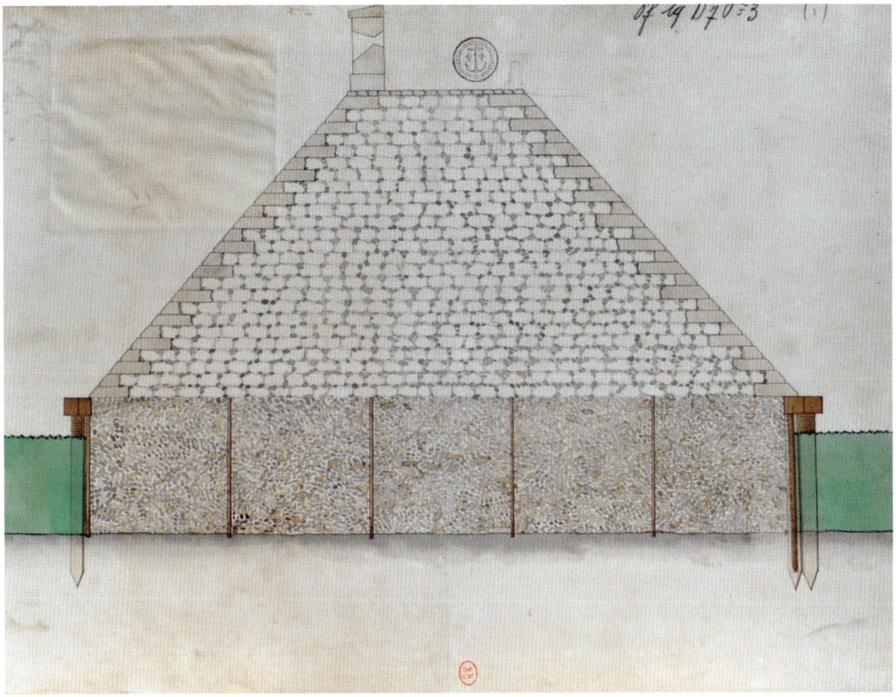

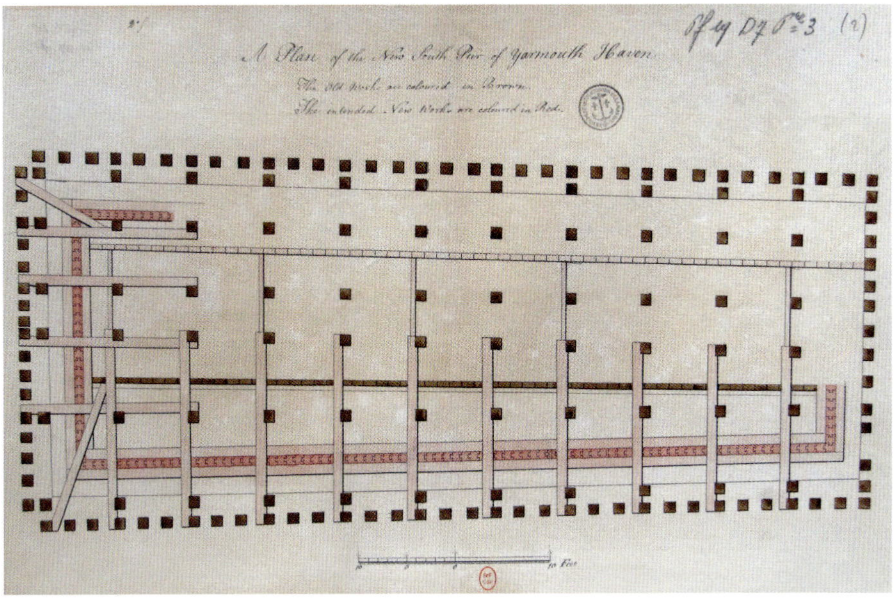


**151-152**. Charles Labelye, project for the mole of the port of Yarmouth, near Sandown Castle, Kent, 1744 (Cartes et Plans, GE SH 18 PF 19 DIV 7 P 3 D, Bibliothèque Nationale de France, Paris).

## From the caisson to the "murazzo a pozzolana": Zendrini's experiments for Venice

A new and fundamental chapter in the history of hydraulic masonry made with pozzolana concrete begins in the field of construction in Italy towards the end of the 1730s and the start of the 1740s, alongside the experiences of Vanvitelli in Ancona, and in the wider context of the habitual use of this kind of masonry in the various ports that are part of the pozzolana trade network. The protagonist of this further spread of pozzolana-base cement masonry is Zendrini, while the place of the experimentation is the sea of the Venetian Lagoon, with the levees along the line of defense that includes Malamocco, Pellestrina and Chioggia.

The theoretical backdrop of the construction of the Venetian breakwaters does not only have to do with mathematical calculations of hydraulics, but also with the theory of the origins of architecture. Just consider the extent to which the gradual metamorphosis of the construction system of the levees, from the original structure of only earth and vegetation to the stone and wood of the Guberni system, and then to the pure masonry barriers of Tirali and Piccoli, is reflected in the theoretical outlook on the three originating systems of architecture elucidated in 1731 by Tommaso Temanza and Matteo Lucchese (1705-76), both pupils of Poleni, collaborators of Giovanni Antonio Scalfarotto (1700 circa -1764), and employed by the Magistrato alle Acque. According to Temanza and Lucchese, the three systems depend on construction materials and their combinations, that significantly coincide with those being studied for the levees until the 1730s: stone and wood used separately, and their combination.[81]

Already, towards the end of the 1720s, when Tirali is still active, joined by Zendrini, Piccoli, Cornello, Boschetti and Filippini, many "boats of Istria stones" reached the Lagoon, with which to carry on the work of consolidation of the sea defenses that began with the prototypes at Caroman and the Magre.[82] Tirali's report of 20 July 1733 contains an initial, significant assessment of the experiments with the new sizes of Istria stones, conducted at Caroman. Thanks to the weight of the "large Istria stones," Tirali says he has been able to oppose "force with force," forming "a solid base" against which the sea can break without "destroying such a solid, real work." "Even at the median center of this shoreline, which at present is the most vulnerable part, such work has been initiated, though of different bulk," he adds.[83]

Works like the planned multiple groyne with pile structures arranged to create a series of large compartments to fill with loose Lispida stones, envisioned by Tirali, Piccoli, Filippini, Boschetti and Lucchese in 1734 to defend the port of Malamocco, can be seen as the extreme point of evolution of the technique of construction of sea barriers using wooden structures, prior to the use of the new hydraulic binder for the stones proposed by Zendrini the following year.[84] Zendrini himself, in April 1735, a few days after reaching Viareggio,

[81] *Cf. Lettera del signor Tommaso Temanza architetto veneziano indirizzata al signor M. Lucchese architetto pure veneziano*, in, Angelo Calogerà, ed., *Raccolta d'opuscoli scientifici e filologici*, Venezia, vol. V, 1731, pp. 177-97.

[82] Andrea Tirali, report, 16 May 1730, Savi ed Esecutori alle Acque, 75, Archivio di Stato di Venezia. Zendrini's opinion is critical of the works at Caroman, as shown by what he writes in 1727, after having observed the "disorder" produced in the "mass of ordinary stones from Lispida, and from Istria," with which it was attempted to divide the Lagoon from the sea "with a slender strip of land" – "but the fact did not bear out the good intention"; therefore he advises the construction of groynes (Bernardino Zendrini, report, 31 May 1727, Savi ed Esecutori alle Acque, 75, Archivio di Stato di Venezia).

[83] Andrea Tirali, report, 20 July 1733, Savi ed Esecutori alle Acque, 75, Archivio di Stato di Venezia. From the report, it can be seen that in some segments Tirali also uses "large Lispida stones."

[84] Andrea Tirali, Domenico Piccoli, Giovanni Filippini, Lorenzo Boschetti, Matteo Lucchese, report, 2 September 1734, Savi ed Esecutori alle Acque, 75, Archivio di Stato di Venezia.

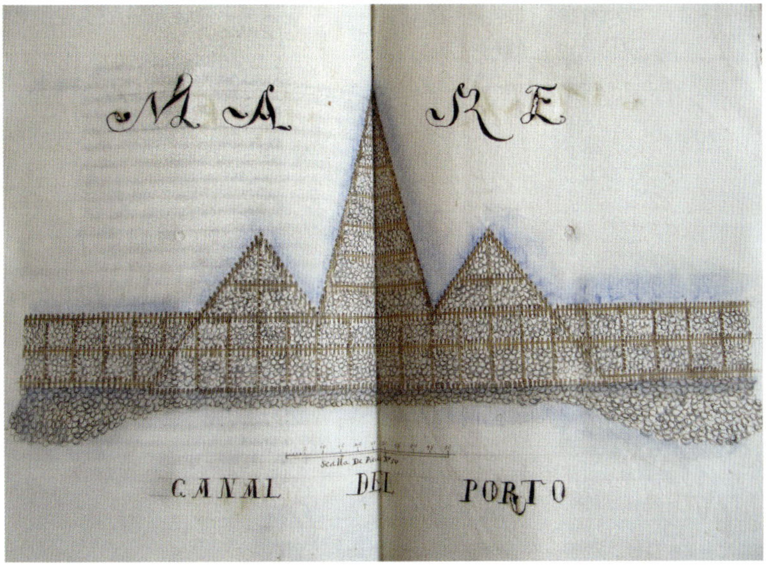


**153**. Andrea Tirali, Domenico Piccoli, Giovanni Filippini, Lorenzo Boschetti, Matteo Lucchese, project for groynes at the Malamocco port, 1734 (Savi ed Esecutori alle Acque, 75, Archivio di Stato di Venezia).

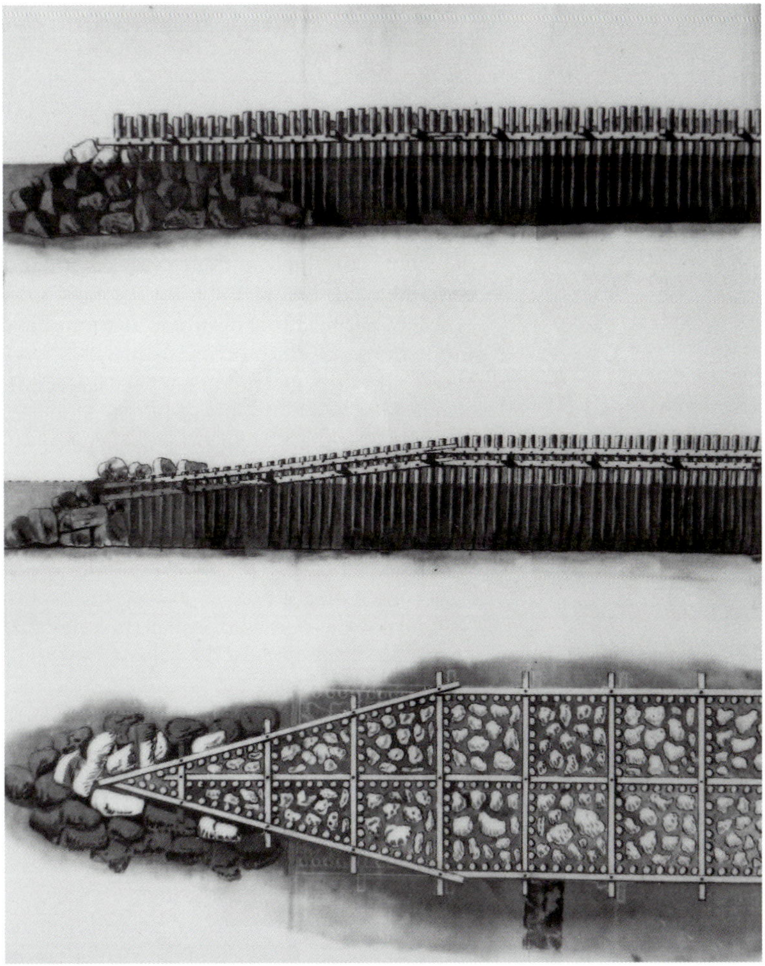

**154**. Project for a groyne with piles and loose stones, Venice (Savi ed Esecutori alle Acque, 76, Archivio di Stato di Venezia).

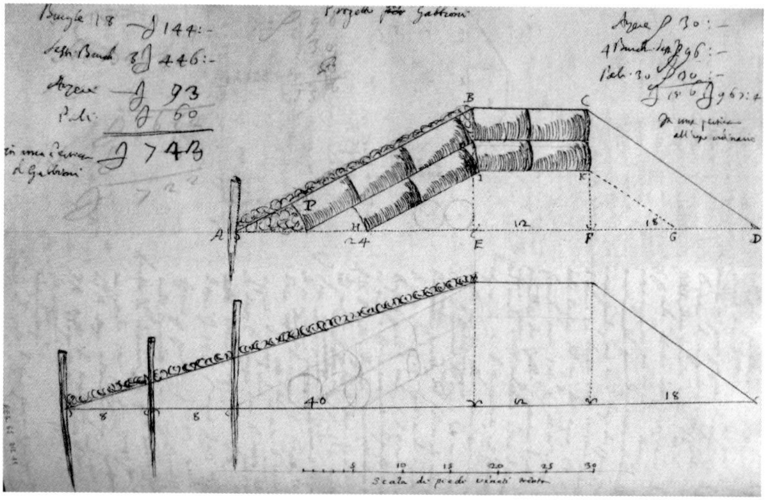


**155**. Bernardino Zendrini, project for "gabbioni," 1735 (Archivio Proprio Zendrini, 9, Archivio di Stato di Venezia).

suggests protecting the shores with cylindrical "large cages" filled with stones, citing ancient examples – a description from Cassiodorus.[85]

Information on what is happening at the port of Ancona, brought in July 1735 to the Savi ed Esecutori alle Acque by the captain of a ship, ushers in the latest phase of the evolution of the levees of the Lagoon, bringing together the questions of Istria stones and of the most important worksite for the use of pozzolana concrete. The captain reports that many ships of the fleet of the Papal State are being used to transport stones from Dalmatia to Ancona, where they are used "to build the dock, and Lazzaretti."[86] For the moment, the Venetian authorities are primarily concerned about possible competition from the Papal State for a supply of those stones that, for the Republic of Venice, have been indicated as indispensable for the construction of the "public shores."[87]

After returning from his trip to Tuscany, Zendrini immediately submits a report to the Savi e Esecutori alle Acque, on 4 December 1735, containing the features of the "new design" for a type of levee.[88] The proposal goes against the principle advanced by Tirali and Piccoli, and also by Zendrini in the past, of resistance to the power of the sea accomplished solely through the use of large, heavy blocks of Istria stone. Now Zendrini attributes the strength of levees to the grip of binders, with which to bond stones of any kind, indirectly linking back to the cement conglomerate solution outlined by Minorelli. After discussions with Tirali, leading to the hypothesis of using pozzolana from Vicenza, and relying on his experiences in Tuscany, Zendrini identifies the ideal product for the cement to be fabricated at the Lagoon: the pozzolana of the ancient Romans. But his project for the Lagoon now calls for the construction of levees by means of caissons and pozzolana concrete, no longer using large Istria stones. The idea, as he writes when he specifies the source of the pozzolana, is to "abandon the use of piles on the public shores and to replace them with barriers formed by pebble of any nature, lime and pozzolana taken from the quarries of Rome or Baia."[89] The fact that he writes "pebble of any nature" demonstrates the diversity of this proposal from what is done on the Lagoon. Regarding the efficacy of pozzolana, Zendrini cites ancient literary sources – first of all the books of Vitruvius, in the passages in which the author is "speaking of the Pozzolana powder"[90] –, ancient hydraulic works, and the works recently completed in the ports of the Mediterranean. Zendrini himself lists the models: the ports of Baia, Anzio and Cuma, for the ancient examples, "and for the modern those of Naples, Messina, Viareggio, Genoa and Leghorn" (the name Viareggio is crossed out).[91]

"[...] but a long time has passed in implementation, on an experimental basis, of those most humble suggestions I made to His Excellency the Magistrate starting in 1735, after my return from Tuscany, where I had the chance to observe the way in which with great industriousness those people defend beaches and ports against the force of the sea," Zendrini writes at the start of 1739.[92] In the time between the proposal and its realization, Zendrini has to convince Venetian *proti*, masons and technicians of the superiority, with respect to breakwaters made with large Istria stones, of his "new method," consisting in the use of caissons in which to pour a conglomerate of pebbles, lime and pozzolana, given the fact that the Venetians have no experience of this type of masonry.[93] It is precisely with the outlook of constructing levees with caissons containing a pozzolana cement mixture Zendrini conducts his first experiment at Malamocco.

As usual, before deciding whether to conduct experiments, the Savi e Esecutori alle Acque call for expert technical opinions. One assessment that leans toward acceptance of Zendrini's recommendations is written by Gabriel Badoer, on 22 February 1736. After having outlined the construction systems for the Lagoon and their shortcomings, including those caused by the use of "large Istria stones" weighing 500 pounds, Badoer approves of Zendrini's proposal of intervention at the Magre and Caroman with the "mode applied by the people of Leghorn and of Genoa to repair their shores," "a defense method consisting in the use of certain caissons filled with pebbles and Pozzolana powder [...]."[94]

Only in 1737 do the conditions arise through which the Savi e Esecutori alle Acque permit Zendrini to conduct an experiment to test the efficacy of the method of caissons with pozzolana-based cement conglomerate. The required material, that pozzolana used by the ancient Romans and contemporary builders, is purchased at the port of Ancona, where in the meantime Vanvitelli has resumed work on the construction of the pier with the caisson and pozzolana cement.

The caisson experiment with cement takes place on the beach of Malamocco, on the site known as the Capitello, corresponding to segment number 63, and is conducted under the supervision of Zendrini, with the technical assistance of Temanza. The caisson has a length of two "pertiche"; it is "broad and tall in proportion" and constructed with oak boards. Once completed, the caisson is filled with stones and bricks, mixed with "lime and pozzolana."[95] Zendrini specifies that the whole operation should happen in keeping with the

[85] Bernardino Zendrini, report, 9 April 1735, Proprio Zendrini, registro 9, Archivio Storico di Venezia (Grillo, *op. cit.*, p. 66).
[86] Report, 16 July 1735, Savi ed Esecutori alle Acque, 302, Archivio di Stato di Venezia.
[87] The Venetians "prohibit the transport from their province of Istria of those great boulders that serve to border the mole" of Ancona (Pietro Custodi, ed., *Lettere e scritti varj di Giuseppe Baretti*, Milano, Società Tipografica de' Classici Italiani, 1839, vol. IV, p. 378).
[88] *Cf.* Bernardino Zendrini, report, 8 January 1739, Tomo XI, 961, Archivio Storico di Venezia.
[89] *Ibid.*
[90] *Ibid.*
[91] *Ibid.*
[92] *Ibid.*
[93] *Ibid.* "[...] many difficulties in overcoming them," writes Zendrini, "first for the shortcomings of the workers here, and for the distance of the pozzolana, which is the material in which the novelty of the operation consists, and secondly due to the various remarks of those that cannot be convinced that such a sand can form, with lime, a cement so strong as to stand up to any impact of the stormy sea, encouraged by indication of those practical experts [...]."
[94] Gabriel Badoer, report, 22 February 1736, Savi ed Esecutori alle Acque, 3, Archivio di Stato di Venezia. In the title of the report, it is stated that Badoer supports "the suggestion of the Public Professor Zendrini, in his writing of 4 December 1735, on the project of using stones and pozzolana for the coastal protection, used by the people of Leghorn and of Genoa."
[95] Bernardino Zendrini, report, 8 January 1739, Tomo XI, 961, Archivio Storico di Venezia.

procedures he has directly observed at the port of Viareggio: "filled [...] in that way that I explained in my report on practices at the ports of Tuscany [...]."[96] The caisson of concrete is left on the shore without any protection; exposed to the eroding action of the waves, it is "stripped of the oak covering," leaving the "nude masonry" visible.[97] Following the usual coastal inspection, in August 1737 Filippini writes a description of what remains of Zendrini's test: "Moving on to number 63, above this shore, to observe the results of the experiment done with the assistance of the Vice-Proto Temanza of the Caisson formed with pebbles, Stones and Pozzolana, to my great disappointment I was unable to find it, as it had been broken up and dispersed by the untimely stormy seas in the past days, and nothing could still be recognized of the material; as the base of the Caisson was found thrown up by the waves to the earth, and was recovered there by [Francesco] Nordio, with a few scattered pebbles."[98]

The failure of the caisson is documented in the report of 31 August, and blamed on "various accidents." The author of the report informs the Savi ed Esecutori alle Acque that Zendrini is "contacting persons of practical expertise in this work, employed in foreign places, who can provide some advises," and expresses the opinion that it is necessary to proceed with "the proper experiments," still under the guidance of Zendrini, to see if it is possible to put into practice, on the Venetian shores, construction systems that have been used elsewhere, "if it is useful to follow their practices, and if it is possible to obtain, as in foreign places, success with this method here in these parts as well."[99] A few years after that failure, Zendrini explains that with the first caisson he was only trying to test the quality of the binder of the pozzolana. "[...] but whereas the aforesaid experiment was rapidly conducted to see if the binder I had proposed of the lime with the pozzolana would stand up to the force of the waters driven by the wind, and not to form a real breakwater," he explains.[100]

Still in 1737, Francesco Rossi, a contractor praised for work performed on other worksites of the Republic of Venice, builds another work "with stones in pozzolana," at the segment of the shore indicated with the numbers 44-45, at the "Magre."[101] At the same time, in 1737-38, *murazzi* are planned along the rivers in the Treviso area, while one is built by Rossi along the Brenta at Stra (though perhaps they were all made without pozzolana cement).[102]

The importing of large stones for the works of Zendrini for the Lagoon is done by Antonio Girardi, under a contract that sets the quality and the weight – no less than 500 pounds each.[103]

96 *Ibid.*

97 *Ibid.*

98 Giovanni Filippini, report, 19 August 1737, Savi ed Esecutori alle Acque, 75, Archivio di Stato di Venezia.

99 Report, 31 August 1737, ASV7, Savi ed Esecutori alle Acque, 200, Archivio di Stato di Venezia.

100 Bernardino Zendrini, report, 8 January 1739, Tomo XI, 961, Archivio Storico di Venezia.

101 *Cf.* report, December 1737, Savi ed Esecutori alle Acque, 4, Archivio di Stato di Venezia.

102 Francesco Moscenigo, report, 13 March 1738, Savi ed Esecutori alle Acque, 3, Archivio di Stato di Venezia.

103 *Cf.* report, 13 August 1737, Savi ed Esecutori alle Acque, 408, Archivio di Stato di Venezia.

104 Giovanni Filippini, report, 26 November 1737, Savi ed Esecutori alle Acque, 75, Archivio di Stato di Venezia. "It remains to seriously reflect," Filippini writes, "on the defense of the shores, in particular those of Chioggia, where in a short time the depths are corroded, making the piles useless. An excellent remedy would be that suggested by Professor Zendrini of forming the levees with Caissons, composed of stones, lime and pozzolana of the kind brought from Naples, but since the bulk of the levees would be very large, and the expense far from moderate, it will be sufficient to reduce the part of the lower embankment between the first and the last row in Caissons, since if the foot is formed as the fundamental principle of the embankments of the earthen levees, coverings of pebbles will be able to survive; and these caissons, taking the place of the pile structures, would not be subject to rotting. [...]. But reflecting thoroughly on the pozzolana, which must be brought from afar, and carefully guarded

The transformation of the experimentation on the construction system in 1737 is not prompted only by the yielding of the caisson with the pozzolana cement. After the negative results of the first trial, Filippini raises the issue of the cost of obtaining pozzolana. The "forming of levees of caissons, composed of stones, lime and pozzolana from Naples," as envisioned by Zendrini, becomes the subject of Filippini's report of November 1737, in which he proposes either reducing the bulk of the pozzolana work, to be concentrated in "caissons" at the levee toe, or using local products with characteristics comparable to those of pozzolana imported from foreign states. Filippini is referring to the lime of Monselice – a type of the Padua lime previously suggested by Minorelli – and he proposes the creation of a caisson with that material, on an experimental basis.[104] Filippini's suggestions are accepted, as can be seen in the initiatives taken by the Magistrato alle Acque in the first half of 1738, ordering the making of three caissons with different kinds of masonry, two of them with "Vicenza porcelain" (pozzolana probably from two different locations) and one with "porcelain of Rome." The fact that pozzolana from Vicenza is used, and not Monselice lime, can be attributed to the personal intervention of Zendrini, who had already suggested this type of local pozzolana for the masonry of the levees of Venice. The deliberations of 4 June assign the superintendent of Malamocco, Nordio, the task of "lending all assistance" to these "caissons."[105] Reports of the testing of different types of pozzolana are extremely important, as they prove that already in the preliminary phase of the construction of the *murazzi* according to Zendrini's method, there was awareness of the Vicenza pozzolana that the naturalists of the Academy of Sciences, Letters and Arts of Padua would later claim as their discovery at the start of the 1780s, also subjecting it, once again, to comparative trials with Roman pozzolana, done at the Lagoon under the supervision of technicians of the Savi ed Esecutori alle Acque who were unaware of what had happened at the time of Zendrini.

In the early days of August 1738, following the arrival of the load of pozzolana from Civitavecchia ordered on 7 May from the company Giovanni Domenico Antonetti e Figli,[106] precisely to "conduct an experiment,"[107] Zendrini undertakes the second trial ordered by the Savi e Esecutori alle Acque, to be done at the place known as the Magre, on the shore at Pellestrina, "in the most vulnerable and battered place of the public coast," and not at Malamocco, as he had requested, in order to have easier access to the worksite and pay closer attention to its progress.[108] The proto in charge of the shorelines,

against the rain prior to use, and in the quantity that would be required, I humbly advance my very humble suggestion, that is to form an experiment with caissons formed with lime of Monselice, and sand taken from these hills near the Madonna di Manna, and with the materials of Istria stones, and Lipsida stones, to form a specimen suitable for these sites of the coast of Chioggia; thus I have the courage to advance this humble suggestion to Your Excellencies, a project based on the experience of the embankment of stone formed at the Magre of material thus composed, and that nevertheless survives after 25 years. The size of that one is very big; and of great expense, but since the matter is but one of 25 feet of embankment interrupted by banks formed with the same caisson, the rest of the embankment up to 40 feet can be made with earth, with its surface covered with two layers of stones. That earthen levee will always survive, when the foot of the embankment is formed by such material it will be permanent. As the beginning of the ruin of levees is precisely the failure of the foot. Which in my view would be of assistance for the security of the levees and their duration, and for savings in terms of piles and other materials that are consumed in a short time [...]."

105 "Terminazione," 4 June 1738, Savi ed Esecutori alle Acque, 408, Archivio di Stato di Venezia.

106 *Cf.* "Terminazione," 9 September 1738, Savi ed Esecutori alle Acque, 408, Archivio di Stato di Venezia. The company Andrea Betti & Giovan Domenico Antonetti operates by order of the Conti family of Lucca, which at the start of the 1700s founds, in Venice and Verona, with the Colognola and Betti families, a trading store under the name Andrea Betti & Giovan Domenico Antonetti. *Cf.* Anna Vittoria Migliorini, *I Conti, una famiglia di collezionisti del Settecento lucchese*, in, *Le Dimore di Lucca. L'arte di abitare i palazzi di una capitale dal Medioevo allo Stato Unitario*, Firenze, Alinea, 2007, (pp. 257-61), p. 257.

107 "Terminazione," 10 September 1738, Savi ed Esecutori alle Acque, 408, Archivio di Stato di Venezia.

108 Bernardino Zendrini, report, 8 January 1739, Tomo XI, 961, Archivio Storico di Venezia.

Piccoli, is made responsible for the work. The experiment consists in the construction of a "groyne or pier," significantly indicated as a "*molo a pozzolana*," whose purpose is to "make a last trial of the strength of our cement," as Zendrini writes.[109] The seabed on which to place the structure is of "pure sand" and the site is covered with water. Zendrini's narrative of the initial phases of the worksite should be understood against the backdrop of the exceptional power of pozzolana to set in water:

"It was a curious thing to see the way with which this construction set under water, not when it was made dry, where the foundations should have yielded, which would have been impossible on that site, but with the pouring in of the lime mixed with pozzolana, placing marble blocks cut sufficiently on the outside, but inside pouring all sorts of stones. [...] such is the bond, that resists any force that attempts to separate it, and what is more remarkable, that mixture hardens in water, and in a short time becomes not unlike a large stone. The operation was terminated not prior to the end of September, and as soon as ten days had passed a strong sirocco tempest put it to its first furious test."[110]

The results of the experiments with local "soils" seem not to have achieved results that would guarantee the success of Zendrini's undertaking, so in September 1738 the Savi e Procuratori alle Acque set conditions for the importing of sufficient quantities of pozzolana, also thanks to negotiations with the company Antonetti e Figli.[111] Confirming the exceptional character, for Venice, of the financial and commercial system organized to make the trials for the "new project" of Zendrini possible, on 4 October a Senate decree is issued stating that if funds are lacking in the accounts of the shoreline authorities – the "Cassa dei Lidi" – the money for the purchase of the pozzolana can be withdrawn from the "Cassa della Laguna." This dispensation is put into practice not a few times during the course of the 1700s for payment of pozzolana shipments, right from its enactment. In 1738 and 1739 shipments of pozzolana arrive in Venice, measured differently depending on their origin in the Papal State or the Kingdom of Naples, in cartloads from Civitavecchia and barrels from Baia.[112]

The renewed confidence in Zendrini prompts the Venetian Senate to issue a decree, on 29 September 1738, to authorize a second experiment at Capitello, still with a caisson to insert "on the seashore" and "formed by pebble, lime and pozzolana";[113] but by now the construction technique has changed, as compared to the initial test. Zendrini's definitive proposal for the masonry levee has to take the established method of experimentation followed

[109] *Ibid.*
[110] *Ibid.*
[111] *Cf.* report, 10 September 1738, Savi ed Esecutori alle Acque, 302, Archivio di Stato di Venezia. The document includes a passage about "a wall, formed with pozzolana mixed with lime, and that the experience of many centuries brings credit to this operation used at the beaches and ports of the Mediterranean, the Adriatic as well, and particularly at Genoa, Leghorn, Naples, and Civitavecchia, and for the defense of shores threatened by the sea"; therefore it is ordered to "have such an ingredient, to conduct an experiment [...]."
[112] See, report, 22 December 1739, Savi ed Esecutori alle Acque, 670, Archivio di Stato di Venezia.
[113] For the decree see, Tiepolo, *op. cit.*, p. 130.
[114] See, for example, "Terminazione," 29 July 1738, Savi ed Esecutori alle Acque, 408, Archivio di Stato di Venezia. This is a contract with Rossi for the construction of the *murazzi*.
[115] See, "Terminazione," 20 May 1757, Savi ed Esecutori alle Acque, 413, Archivio di Stato di Venezia.
[116] Bernardino Zendrini, report, 8 January 1739, Tomo XI, 961, Archivio Storico di Venezia.
[117] Report, [September 1738], Savi ed Esecutori alle Acque, 302, Archivio di Stato di Venezia.
[118] Bernardino Zendrini, 17 January 1739, Savi ed Esecutori alle Acque, 75, Archivio di Stato di Venezia.

by the Savi ed Esecutori alle Acque into account, namely the gradual development of a practical solution by means of prototypes made on site. Starting in the mid-1710s this method had generated a type of levee made with large stones as the ideal solution to block the fury of the waves, and the size and weight of these stones had even been theorized by outstanding hydraulic construction experts like Guglielmini. So following the two *murazzi* made under the supervision of Tirali and the considerable results achieved over the years in terms of strength and durability in time, Zendrini has difficulty inverting the progress of the tests back toward small pebbles, even if they are embedded in pozzolana cement. The Roman concrete technique, using pozzolana cement, and that of the caissons seen in the work of Tuscan engineers or Vanvitelli, which Zendrini wants to introduce in the construction of the Venetian levees, run up against the increasingly well honed machine developed by Tirali, the failure of the experiment in 1737 and the cost of importing pozzolana. The clash between these different techniques and their respective economic logic leads to the hybrid solution of the sea wall implemented after the first trial of the caisson with pozzolana cement: a bulky construction of large stones in the manner of Tirali and Piccoli, but held together using the method of Zendrini, and set on a foundation of pozzolana concrete, again in keeping with Zendrini's views. The masonry work is assigned to Rossi,[114] who from this point on becomes the contractor hired for the construction of all the sea walls. In 1757 the pozzolana work for the levees is assigned to Rossi's son, Domenico, confirming the confidence of the Savi ed Esecutori alle Acque in the contractor who had done such work under Zendrini.[115]

At the start of January 1739 Zendrini describes the masonry mole erected at Capitello in these terms: "until today it has resisted so well that not even the smallest shard has been detached [...] it seems like a small cliff on that beach [...]."[116] The Magistrato alle Acque sums up Zendrini's operation: "he has placed, as the first experiment, the Caisson at Malamocco at the site of the Capitello on the shore of the sea, formed by stones, lime and pozzolana."[117]

Also in January, the Savi ed Esecutori alle Acque send Zendrini, accompanied by Filippini, Piccoli and Temanza, to inspect the groynes and pile structures of the shores of Malamocco, Pellestrina and Chioggia, with the aim of drawing up a plan for possible intervention to reinforce those structures.[118] In the report of 17 January, Zendrini outlines certain principles of orientation for the construction of the levees of the Lagoon, and to achieve the goal of what he calls "impenetrable shores": he establishes that the slope of the embank-

**156**. Remains of the "murazzo," at Capitello, Malamocco shoreline, Venice (photograph by the author, 2010).

**157**. Masonry with pozzolana concrete, "murazzo," at the Pellestrina shoreline, Venice (photograph by the author, 2010).

ment at the sea should be "gentle" ("a surface less inclined is preferable to a steep one"), based on observations of the movement of the waves against the levees and in keeping with the "principles of the true statics, and of the mechanics of inclined planes" (Guglielmini had already proposed a similar slope); he criticizes, after having described their virtues and drawbacks, the traditional practices of earthen levees, with the use of "loose stones" and of "poles subject to decay," and urges instead the spread of masonry construction with pozzolana, as tested at Malamocco and Pellestrina; finally, he outlines a strategy of progressive extension on the shores of his own "new project" or "new method," to apply first in the most vulnerable sectors, in order to limit costs. When he describes his experiments, he evokes "the ancient practice taught by Vitruvius with pozzolana and pebbles." Then he proceeds to indicate the interventions considered most urgent, for which he recommends the construction of the "*muro a pozzolana*," or pozzolana masonry: at the Capitello site, the location of the "first experiment"; at Pellestrina, between numbers 44 and 46, near the "new stone pier"; and at numbers 49 and 50, near the church of Sant'Antonio, and number 58, "as the test has been undone, I believe it would be very opportune to remake it with only stones in pozzolana." Also in the case of the large levees built as an experiment at Caroman, "with big squared Istria stones, though without any mortar to held them together," Zendrini emphasizes the need for urgent intervention to secure the stones, which are in danger of being shifted by the sea: "these too, one day, should be taken in hand to secure them with pozzolona cement, to prevent the sea from entering their gaps, they should be fortified with a base sunken 4 feet below the ordinary level of the sea, and then the work will be eternal."[119]

It is against the backdrop of these reports that the Savi ed Esecutori alle Acque decide to undertake another experiment, between July and August 1739, following the indications supplied by Zendrini, regarding the composition of the mortar, to be made with lime mortar from the Monti d'Este and pozzolana. The site for the experiment is again the Pellestrina shoreline, at the

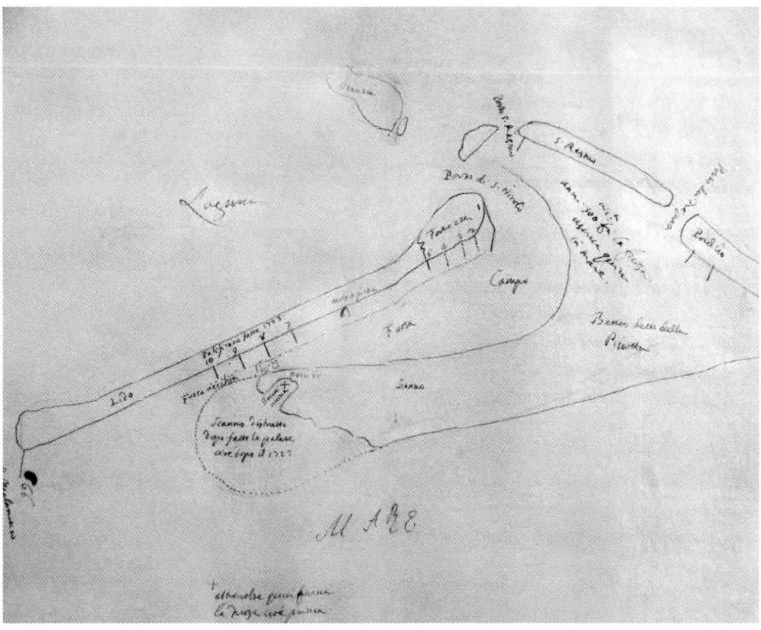


**158**. Bernardino Zendrini, drawing of the Malamocco shoreline with the masonry segment, or "molo di pietra," indicated by the number 6, next to sectors 7-10 which are "pile structures made [in] 1723," Venice (Archivio Proprio Zendrini, 65, Archivio di Stato di Venezia).

Magre, between numbers 44 and 45, at one of the points indicated by Zendrini as critical. The worksite supervision is assigned to Piccoli, the masonry work to Rossi, the cutting of the stones to Andrea Carrara and the making of the lime mortar to Pierolamo Monferà. Near the worksite, Piccoli has a shed built with oak poles, an iron door and a roof of wooden boards and mats, to protect the materials, especially the pozzolana and the lime. Then he orders Antonio Busetto to transport to the site the pozzolana conserved at Punta di Malamocco, the caisson used at number 58 of the Malamocco waterfront, after dismantling it, and the Istria stones. The cooperation of the various types of workmen can be seen in the fact that Rossi and Carrara, together with Piccoli, decide "how to begin the work on the said massif."[120]

The characteristics of pozzolana are precisely the basis of the capacity to "form with lime a cement so strong as to stand up to the impact of the stormy sea," Zendrini writes in 1739.[121] After the deliberations of December 1739, Zendrini is officially appointed to be in charge of the defense of the public shoreline and the lagoon, assisted by the proto Lucchese.[122] During the course of 1740 Zendrini again expresses his opposition to defensive structures based on the use of wooden piles,[123] insisting that breaches should be closed up, wherever possible, with stones and pozzolana masonry ("*muro di pozzolana*"[124] or "*argine di pietra a pozzolana*"[125]) and he requests an enclosure to protect the pozzolana from theft, where it is stored at Malamocco.[126]

Following the decision of 4 March 1740 to build a pozzolana sea wall, or "*murazzo a pozzolana*," on the site "near Capitello," still according to Zendrini's indications, on 14 March a contract is signed with Rossi. The formula, which would be repeated many times in the years to come for other sea walls, indicates the supply of the materials and labor, and the general measurements of the work: "the Murazzo to be built should have a base four feet below the normal water level, ten feet wide, to join which it will be necessary to dig as far as necessary into the sea bed, making use of all things that may be necessary for such a work"; "once the said base has been made, over it should be erected a Murazzo ten feet above the normal water level, nine feet wide at the bottom over the base, and five feet wide above so that the embankment of four feet is all on the side of the sea"; "this entire Murazzo with base should be made for the length of about thirty pertiche"; "the lime should be mixed with pozzolana powder, excluding all others, and this pozzolana will be supplied to Rossi by the most Excellent Magistrato alle Acque, in the required quantity"; the Magistrato will also supply "all the stones needed for the work," and to facilitate Rossi "they will be brought to the Capitello towards the Lagoon, so he can more easily go to get them and take them to the work"; Rossi will also be supplied with "all the wood needed to make coffers and reinforcements," while all the lime, "which will have to be of excellent quality, that is needed for this work," must be obtained at Rossi's expense, just as he is responsible for paying the workers; the price of "ninety-three lire per cubic pace, as measured at completion"; the work has to be

119 *Ibid.*
120 Domenico Piccoli, report, 20 July 1739, Savi ed Esecutori alle Acque, 75, Archivio di Stato di Venezia.
121 Bernardino Zendrini, report, 4 September 1739, Proprio Zendrini, 11, Archivio di Stato di Venezia (cit. in, Grillo, *op. cit.*, p. 68).
122 Bernardino Zendrini, report, 18 September 1740, Savi ed Esecutori alle Acque, 75, Archivio di Stato di Venezia.
123 Bernardino Zendrini, report, 6 May 1740, *ibid.*
124 Bernardino Zendrini, report, 22 August 1740, *ibid.*
125 Bernardino Zendrini, report, 18 September 1740, *ibid.*
126 *Ibid.*

"terminated in the coming month of June" and Rossi "will see to its maintenance during the course of one year from completion."[127]

In June 1740, for the construction of the "stone embankment" and the "wall base with Istria stones and pozzolana" in various sectors of the levee, at Malamocco, Pellestrina and Chioggia, it is stipulated that Rossi will have to obtain the pozzolana at his own expense from the "public stores" and "mix it according to practice with excellent lime."[128] In certain documents it is specified that the "excellent lime" is that of Monselice.[129] On 20 September it is decided to continue the "pozzolana sea wall near the Capitello," for about 39 *pertiche*.[130] To build the masonry foundation of the base, Rossi uses enclosures made with wooden boards to protect against the water, or "coffers" and "reinforcements."[131]

On 28 June 1740 the new *murazzo a pozzolana* at the Capitello is about to be completed. On that date, the Savi ed Esecutori alle Acque send a report to the Senate of the Venetian Republic. Following the initial trials done at the Capitello, the building of the pier at the Magre and the work in progress on the Capitello sea wall, the success of Zendrini's method has been certified, and a new era of construction of the waterfront to protect the Venetian Lagoon begins. What emerges from the observations contained in the report of the Savi ed Esecutori alle Acque regarding the pier at the Magre is precisely the awareness that by now all the strength to counter the force of the waves, previously sought in the arrangement of pile structures or in the connections and sizing of the stones, now shifts to the nature of the cement compound introduced by Zendrini.

"It would be of little import to reassure [...] the most Excellent Senate that the said Caisson at Malamocco has already demonstrated valid resistance in two winter storms, undergoing no damage of any kind, but we are also very pleased to be able to report to Your Excellencies the success of another much more extensive experiment. Along the whole coast the most vulnerable place, which undergoes the greatest impact of the sea, is that of the Magre on the coast of Pellestrina. This was the site chosen for the enterprise of the new trial of last August. Here a mole of 16 geometric paces was formed, with pebbles, lime and pozzolana, set in six and seven feet of water at the normal level near the shore, and ten feet of water in the sea where it has its front. The way of erecting it was particular. As it was impossible to dry the place where the foundations must reach, the lime mixed with pozzolana was poured, with suitably hewn blocks of marble on the outside, submerged in the inner part. The base was kept at twelve feet of width, and the summit at six feet, the embankment increasing in proportion, kept at such a height as to rise above the normal water level by four feet. The grip of this cement is so strong that no force can separate it, and what is more admirable is that this mixture sets in water in such a way that in a very short time it becomes not unlike the very stone. This work was completed at the end of September, so that in the space of about eight months it can truthfully be stated that the greatest experiment has been done, while the season brought a series of storms from the Greek Sirocco, and from the Levant, all winds that more than any others flay this coastline and have severely damaged the pile structures, and groynes, uprooting poles and carrying away stones. But the pozzolana pier, though without a stone barrier

[127] "Terminazione," 14 March 1740, Savi ed Esecutori alle Acque, 670, Archivio di Stato di Venezia.
[128] "Terminazione," 3 June 1740, Savi ed Esecutori alle Acque, 409, Archivio di Stato di Venezia.
[129] "Terminazione," 23 July 1740, *ibid.*
[130] Report, 27 September 1740, *ibid.*
[131] *Ibid.* and 10 July 1741, *ibid.*

by its side, has not only stood up to any sea storm without undergoing any damage whatsoever, even on the outside, but has also, it can be observed, where inserted at a depth of 10 feet from normal water level, gathered so much sand, both upwind and downwind, that the sea bed at its end, and around it, are now four feet less deep. This is even greater evidence that it is not only sufficient to stand up to the most robust and repeated blasts of a stormy sea, but can also reduce the force of the waves, making them deposit the sand they carry with them, and this is one of the most beneficial things that can be said regarding such barriers. This mole has also attracted a quantity of stones, of those that in past storms had been removed from nearby beaches, so that a sort of rocky coastline has taken form. We can assure the most Excellent Senate of all this thanks to firsthand experience, having gone to view the results on a visit, accompanied by the most worthy Marchese Poleni, who expressed his great pleasure, and we are confident that a type of work has been found, proven by experience to be the best and the most secure. While such a work may lead to greater than usual expenditures, it is nevertheless a better choice, as its durability knows no limit in proportion, while the operations of pile structures built with poles, irons and stones, last at most for five years, and in many more exposed sites for but one or two years, and at times as soon as they are made they are lost, so that over the years very large expenditures are needed, whose effects are often not even to be seen. [...]. Where expenditures are concerned, the said pier made of stone, lime and pozzolana has a cost of one hundred sixty-eight ducats per geometric pace, and an equal sum would have had to be used to make it with piles, nails and stones, as can be seen in the exact accounting existing at our Magistrato. The defenses of the shorelines, as Your Excellencies are well aware, consist of two types of works, that is of levees

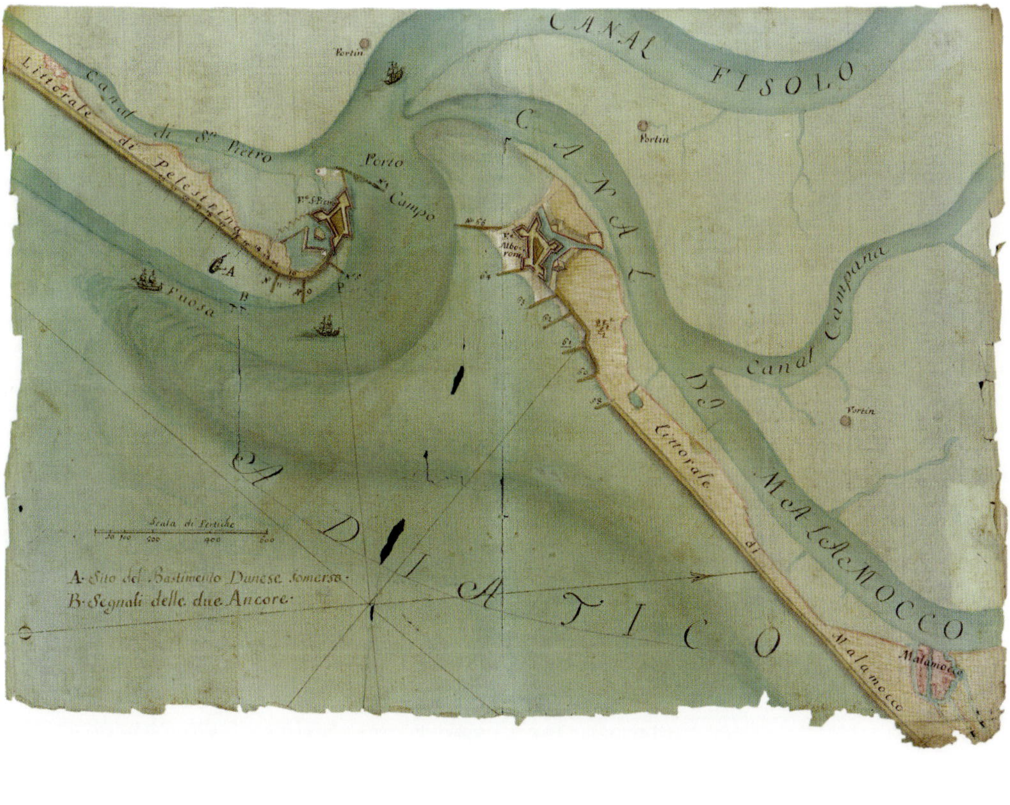


**159**. Plan of the Venice Lagoon with the Pellestrina and Malamocco shores (Archivio di Stato di Venezia).

and piers, which until today have been made with piles and stones. For the piers made with pozzolana the most Excellent Senate now has the information, and knows the strength and the cost. Where the levees are concerned, which due to their bulk require more materials, we have ordered a real experiment for the closing of a breach at Malamocco, where work is presently in progress, and when it is completed it will be reported to the most Excellent Senate, concerning both the expenses and the outcome. In this way, we hope to initiate a work that in the future will form on the coasts an impenetrable defense, not subject to so many episodes, which one day will put an end to excess expenditures of public funds, making the temporary become perpetual."[132]

The report of the Savi ed Esecutori alle Acque concludes with praise of Zendrini, granting him "all the credit for this idea." "It was he who in the year 1735 in Tuscany, there observed the defense of those shores, and those ports with works of pozzolana as practiced in Baia, Civitavecchia, Messina, Leghorn and Genoa, and with such examples, and with the observations made of them, and brought with good will the project, also availing it of the experience of the most remote ages, in which it is found that this type of work was done with Pozzolana of Rome and of Baia, not only in use in the ports of the Mediterranean, but taught from the time of Vitruvius, Master of architects at the time of Augustus, and all the ports of the ancient Romans, as marvelous as any of their buildings, were defended by such works. Not satisfied, the said Professor, with having made the project, but led by the zeal of public service, he granted his work and assistance to carry out the design, also training persons of the State to make such works, who were previously not expert in this area."[133]

On the basis of the report of the Magistrato alle Acque, the Senate issues a decree on 4 August 1740, approving the progressive and definitive transformation of the levees along the coasts of Pellestrina, Malamocco and Chioggia, with pozzolana sea walls.[134] In 1742 Lucchese observes that no other operations "excepting those of pozzolana" are done, and the material is used to repair other breaches at Chioggia, while at Pellestrina "grand pozzolana works" are made, as well as the "pozzolana embankment at Magre."[135] That same year, at the moment of the renewal for the next five years of Zendrini's appointment as director of the works of pozzolana sea walls, it is noted that the undertaking has already attracted "great interest and admiration from abroad."[136] The Roman origin of the works of Zendrini is commemorated by a plaque with a Latin inscription in praise of pozzolana, ordered by the Venetian Senate in October 1742, to apply "above the murazzo of Malamocco," where the

[132] Report, 28 June 1740, Savi ed Esecutori alle Acque, 302, Archivio di Stato di Venezia.

[133] *Ibid.*

[134] See, "Terminazione," 4 August 1740, Savi ed Esecutori alle Acque, 360, Archivio di Stato di Venezia. The document reports on the "good effects of the caisson inserted in 1738 on the shore of the sea at Malamocco in the Capitello site, formed with lime, stones and pozzolana, which stood up well to the impact of all the two recent winter storms, without any damage whatsoever." "After such a useful experience," we read, "it was followed by a second at the Magre site, the most vulnerable and beaten by the sea, at the shore of Pellestrina, also made with lime, stones and poured pozzolana, constructing in the manner explained a pier of over sixteen geometric paces, which even in strength at the Cape of Malamocco mentioned above has produced the described good effects of reducing the force of the wave, which can thus deposit, upwind and downwind, and around its front, such quantities of sand and of stones attracted by the storms from the nearby shores, that where the seabed was at ten feet of water below the normal level, now it has become just six [...]." The Magistrato, in any case, orders another experiment "to close a breach in the levee of Malamocco." "And since around it work is presently proceeding, we await the completion of the work by those experts and builders, to have new information on the effects of the works with pozzolana."

[135] Matteo Lucchese, report, 10 July 1742, Savi ed Esecutori alle Acque, 75, Archivio di Stato di Venezia.

Magistrato had "since the year 1739 commanded the formation of the sea wall with stones and pozzolana at Capitello."[137] "Quod Urbis Aeterne Aestuaria/ Libertatis Sacrosancta Sedes/ Aeternum Conservaentur/ Muri contra Maris Impetum/ Puteolano Pulvere consolidati/ Hic primum ex S.[enatus] C.[onsultu]/ Presidibus Aquarum Curantibus/ a Fundamentis excitantur/ Anno R.[eipublicae] S.[aulits]/ MDCCXXXIX."[138] In 1751 Temanza complains that the plaque has not yet been put into place – "I know not due to what accident the order has not yet been carried out."[139]

### Poleni and pozzolana

The University and Academy of Sciences, Letters and Arts of Padua had offered services and scientific consulting for decades for the improvement of the construction systems of the shores of the Lagoon and other hydraulic works done in the territories of the Venetian Republic. Guglielmini was but one of the professors involved in providing expertise on this type of work. Zendrini himself, at that University, had learned about the ideas of Guglielmini on hydrology.[140] One of the outstanding personalities of the University of Padua and the local Academy is the Marchese Giovanni Poleni (1685-1761), professor of astronomy, then of physics with hydraulic applications, and later of mathematics. Poleni is also a great connoisseur of Latin technical literature. The series of essays he publishes in Padua from 1717 and 1722 made him the leading expert on theoretical questions of hydraulics of the Venetian Republic: *De motu Aquæ mixto libro duo*, the result of observations of the Venetian Lagoon, become the theoretical-mathematical basis for his later contributions on the movement of waters, the banks of rivers and the levees of the Lagoon;[141] *De castellis per quæ derivantur fluviorum aquæ habentibus latera convergentia*, on the calculation of structures to deviate the waters of rivers;[142] and, finally, an annotated edition of the Latin treatise of Frontinus, *De aquæductibus Urbis Roma*.[143] The research conducted by Poleni for the publication in 1739 of the first volume of the *Exercitationes Vitruvianæ primæ* confirm his wealth of technical knowledge on the construction practices of the Romans.[144]

So it should come as no surprise that in the initial phases of the construction of the Venetian shorelines according to the "new method," Poleni has a role to play, alongside Zendrini. Even the fact that Zendrini is assisted by two disciples of Poleni, Temanza and Lucchese, in the first experiments on pozzolana masonry is important as an indication of the circulation of ideas and solutions in Venetian circles. Besides the passage in the report of 28 June 1740 of the

[136] Report, 5 July 1742, Savi ed Esecutori alle Acque, 360, Archivio di Stato di Venezia.
[137] Report, 6 October 1742, Savi ed Esecutori alle Acque, 410, Archivio di Stato di Venezia.
[138] *Ibid.* Zendrini is flattered that the Senate has decided to honor the "muraglioni" and the "lavorari a pozzolana" (Bernardino Zendrini, report, 15 September 1742, Savi ed Esecutori alle Acque, 75, Archivio di Stato di Venezia).
[139] "Relazione Temanza circa il murazzo à pozzolana sopra il lido di Malamocco del giorno primo aprile 1751," 31 March 1751, Savi ed Esecutori alle Acque, 76, Archivio di Stato di Venezia.
[140] Having graduated from the University of Padua in 1701, Zendrini was personally acquainted with Guglielmini (*cf.* Angelo Zendrini, *Elogio di Bernardo Zendrini matematico della Repubblica di Venezia*, Venezia, Stamperia Palese, 1807, pp. 9-10).
[141] Giovanni Poleni, *De motu Aquæ mixto libro duo*, Padova, G. Comino, 1717.
[142] Giovanni Poleni, *De castellis per quæ derivantur fluviorum aquæ habentibus latera convergentia*, Padova, G. Comino, 1718.
[143] Giovanni Poleni, *Sexti Julii Frontini. De aquæductibus Urbis Roma Commentarius antiquæ fidei restitutus, atque explicatus, opera et studio Joannis Poleni*, Padova, G. Manfrè, 1722.
[144] Giovanni Poleni, *Exercitationes Vitruvianæ primæ*, Padova, G. Manfrè, 1739.

Savi ed Esecutori alle Acque that bears out the inspection made by Poleni of the works of Zendrini – another visit by Poleni to the pozzolana works is documented in 1741[145] – the decree issued by the Venetian Senate on 4 August of that same year associates Poleni with Zendrini's undertaking, with a rather surprising indication: "the solid masonry work in pozzolana," "as recalled by the mathematician Zendrini," seems to have been "suggested by Professor Poleni."[146]

Pozzolana returns to the field of interest of Poleni when, in 1743, he is commissioned by Benedict XIV to prepare a report on the dome of St. Peter's at the Vatican. For the restoration of the dome, Poleni is in contact with the most famous Italian builder of pozzolana works, Vanvitelli. In the analysis of the dome and his stays in Rome, Poleni studies the Roman pozzolana masonry. He describes the strength of the Roman pozzolana cements, which "in buildings [...] acts as a very valid binder."[147] "I recall having observed, among other places, near Castel Gandolfo [...] at the Barberini garden, a very beautiful underground structure, in the form of a large hall, in which the stones had been eaten away by humidity and other damages, while the pozzolana was still conserved intact."[148] Nevertheless, Poleni warns that "for the production of such hardness a very long time is required," and that the drying is a slow process, delayed by the fact that the compound with the stones is placed "enclosed inside walls."[149] The notes on the yielding of masonry works caused by the process of setting are written by Poleni against the backdrop of the research on the causes that led to the cracks in the dome of St. Peter's, given that part of this structure is made with "bricks, and with cements."[150]

## Importation and trade of pozzolana in the Venetian Republic

Zendrini's "new method" leads to radical change in the system of supply of construction materials in Venice. Besides increasing the incoming shipping of large Istria stones, the Venetian Republic sets in motion administrative machinery to permit the importing of pozzolana for the cement of the *murazzi*.

For the supply of pozzolana, at first the Venetian consulates at the ports of Genoa, Leghorn and Civitavecchia are consulted. After the first decree of the Senate on 4 October 1738 regarding payment for pozzolana, and when the project of the levees is already in its implementation phase, as established on 4 August 1740, the Senate takes further measures regarding pozzolana. With the decree of 4 August 1741 the Magistrato alle Acque is put in control of the purchasing procedures of pozzolana, its price and its quality, since at this point pozzolana is considered an asset of public benefit for the Republic.

Though some shipments are imported from Baia and Naples, the major part of the pozzolana unloaded in Venice comes from the quarry of San Paolo in Rome, transported on ships from Civitavecchia. During the period of the papacy of Benedict XIV, from 1740 to 1758, precisely when the first major

145 *Cf.* report, 28 December 1741, Savi ed Esecutori alle Acque, 360, Archivio di Stato di Venezia.

146 In the decree of the Venetian Senate issued on 4 August 1740, we read: "It is praised, in the Magistrato, the thinking that begins to cultivate the making of every study, for the long line of the shores, which absorbs, etc. etc., there is introduced the solid work of a wall of pozzolana recalled by the mathematician Zendrini and suggested by Professor Poleni, which by now in three sites has been tested and found to correspond perfectly to the needs etc. It is to be encouraged, therefore, that the Magistrato cultivate the idea, even more so since also from calculations made under the supervision of Professor Poleni, it is found that the construction of one geometric pace in pozzolana will have a cost of 168 ducats, equal to the cost of that same geometric pace if it were defended by the usual pile structure made with wood, nails and stones. The Magistrato is urged to promote shipments of the materials needed for this end, and praise is due to the said professor, author of the suggestion,

**160**. Anonymous, "Golf van Venetien" (1049B12.083, Atlas Van der Hagen, Koninklijke Bibliotheek, Den Haag).

Venetian works are being done with the new masonry, the port of Civitavecchia sees a sizeable increase in shipments to Venice, especially those of pozzolana.[151] The *proti* in the employ of the Savi ed Esecutori alle Acque have to monitor the operations of receiving of the pozzolana, measuring the quantities of every shipment – the measurements are by the cartload, according to the custom of Civitavecchia. The unloading of the pozzolana is done by the contractor for the masonry work, Rossi. The port of arrival is Malamocco; but in order to be able to transport the material to the various worksites along the entire coast, the pozzolana is stored at various places, in the public warehouses of Malamocco and Pellestrina, where it is "sealed and stamped,"[152] and in the warehouses of the Franciscan fathers at Chioggia.[153] To reduce the supply costs of pozzolana, it is decided that the unloading and transport in the public warehouses should be done at the expense of the ship captains, monitored by Rossi.[154]

The company Antonetti e Figli, the supplier of pozzolana for Zendrini's first experiments, is not the only firm involved in the transport of pozzolana from Civitavecchia to Venice. In the early 1740s there are three other suppliers bringing pozzolana to the Republic of Venice: Girardi, already involved in

meritorious in the supervision of these works and of the workmen in engaged in this construction" (cit. in, *Gli antichi murazzi e le nuove dighe della laguna veneziana, descritti dalla gazzetta privilegiata di Milano*, in, "L'Appendice della Gazzetta di Venezia," vol. II, 1837, pp. 224-38, pp. 231-32)

147 Giovanni Poleni, *Memorie Istoriche della Gran Cupola del tempio Vaticano e de' danni di essa, e de' ristoramenti loro*, Padova, Stamperia del Seminario, 1748, p. 67r.

148 *Ibid.*

149 *Ibid.*, p. 68r.

150 *Ibid.*, p. 73v.

151 Odoardo Toti, *Storia di Civitavecchia, da Sisto IV a Pio VI*, Ronciglione, A. Spada, 1996, p. 211.

152 "Terminazione," 18 April 1743, Savi ed Esecutori alle Acque, 410, Archivio di Stato di Venezia.

153 Report, 15 December 1740, Savi ed Esecutori alle Acque, 409, Archivio di Stato di Venezia.

154 "That every ship's captain must provide that number of men sufficient for the unloading, taking the Pozzolana to the sites where it is to be used, keeping distinct note of the cartloads that will be unloaded together with the public superintendent, ship by ship" (Report, 22 August 1740, Savi ed Esecutori alle Acque, 409, Archivio di Stato di Venezia).

trade of Istria stones for the works of 1737; the company of Giovanni Martini and that of Angelo Revedin with Bartolamio and Giacomo Gianfrancesco Zanchi (Bartolamio is the captain of the ships *Madonna del Rosario* and *Anime del Purgatorio*).[155] But the system of several suppliers chosen to ensure supply of pozzolana for the waterfront worksites immediately proves to be ineffective. The pozzolana trade is not adequately regulated, in this way, and the quantities required by the Magistrato alle Acque are not accounted for. A Senate decree on 23 February 1743 urges the Magistrato alle Acque to make use of all possible means to bring to Venice the "largest possible quantity of pozzolana," both with "public ships" and private operators, without allowing a monopoly to be established, as long as the price is fixed at 22.5 lire per cartload.[156] The Magistrato must also set the quantities of pozzolana required for one year (using "all circumspection and every precaution to prevent deliveries from multiplying to the point of exceeding demand"), and must see to it that only pozzolana from the San Paolo quarry is imported to Venice.[157] In the context of these measures the figure of the Venetian consul at Civitavecchia, Giovan Battista Buccella, or Bucelli, enters the picture. In May 1743 he decides to get involved in the pozzolana trade, and signs a contract with the Venetian Republic in which he commits himself to provide an ongoing supply of pozzolana from San Paolo until 1749, for a final total of 9000 cartloads, split into 1500 per year. The entire responsibility for the transport of the pozzolana, "not adulterated, mixed nor changed" during the voyage, is placed on Buccella.[158] But the activation of the contract is delayed until April 1744, with its expiration shifting to April 1750.[159] The various organisms of the Republic, when consulted following Buccella's proposal, point to the use of "public frigates" as the best and most economical means of transport from Civitavecchia of the "pozzolana required for the public works," but they also petition the Magistrato alle Acque to take urgent steps to bring the necessary quantities of pozzolana during the course of the year.[160] So the importing of pozzolana from Civitavecchia continues to be done by the company Antonetti e Figli (590 cartloads arrive in September 1744),[161] and by Girardi, who brings 300 cartloads in 1745, while the initiatives of the Venetian consul at Civitavecchia come to naught.[162]

For the years of the construction of the *murazzi*, the largest quantities of pozzolana are imported by Antonetti e Figli. The procedures are those established in the early 1740s. The pozzolana is stored in public warehouses, depending on necessities, and on the condition of the warehouses themselves. Control of the quality of the pozzolana being delivered happens, at times, with rather vague criteria, based on comparison with the color of the pozzolana already in stock.[163]

Sometimes the ships bringing pozzolana from Civitavecchia to Venice, as in September 1766, in the case of a Polish vessel under the Venetian captain Tommaso Rosconichi, arriving from Genoa, call at Ancona to deliver pozzolana for the pier of Vanvitelli.[164]

[155] Report, 16 May 1743, Savi ed Esecutori alle Acque, 360, Archivio di Stato di Venezia.
[156] Report, 18 April 1743, *ibid.*
[157] Report, 16 May 1743, *ibid.*
[158] "[...] the voyage to Civitavecchia, the loading of the ships, the entire risk of the sea passage, shall all be his responsibility, as it will be his task to make the captains fulfill their duty, that they maintain the quantities they have received, and that will be stated on the Bill of Lading" ("Terminazione" of the Magistrato, 23 November 1743, Savi ed Esecutori alle Acque, 410, Archivio di Stato di Venezia). *Cf.* Maria Francesca Tiepolo, *Difesa a mare, in Mostra storica della Laguna Veneta, Venezia, Palazzo Grassi, 11 luglio - 27 settembre 1970*, Venezia, Stamperia di Venezia, 1970, (pp. 133-38), p. 137.
[159] "Terminazione," 21 March 1744, Savi ed Esecutori alle

**161**. Giovanni Battista Piranesi, "Veduta del Porto di Ripa Grande," Rome (G.B. Piranesi, *Vedute di Roma*, Roma, 1757).

Traffic on the Tiber does not always make it possible to deliver shipments of pozzolana on schedule. The Venetian consul at Civitavecchia, Domenico Natale, sends information to Venice about the delay of a Venetian ship, forced to remain in port for two weeks in June 1767, "because to take on the pozzolana different loads will have to be put together to achieve the intention."[165] The traffic is intense at the port of Ripa Grande in Rome, where the "grain boats" are found, that prevent "the same from bringing the pozzolana down to Fiumicino." "To take on cargo last week, the two French vessels of said pozzolana [...] a note [...] was issued, to the effect that one or more two-masted ships be allowed, as they go to Fiumicino, to bring with them the cargo of said pozzolana, to be transferred to the tartanas, which return empty here after having been loaded with grain." Thus the captain of the Venetian ship should be supplied, according to Natale, with a similar pass to put into the hands of Antonio Deri, the contractor of the pozzolana, so that the latter can negotiate with the owners of the tartanas to the effect that once the grain has been unloaded at Ripa Grande, they can return to Fiumicino with the pozzolana to be shipped "for the canals of Venice."[166]

The pozzolana stocks in Venice are not used only for the shoreline worksites, though most of the product is earmarked for the *murazzi*. Zendrini's "new method" of fabrication of pozzolana cement spreads widely, starting in the early 1740s, throughout worksites in the city and those of bridges, rivers and canals in the inland zones of the Republic. Thus the pozzolana imported from Civitavecchia by order of the Magistrato alle Acque is also sold, though in small quantities, to other Magistrates of the Republic for public works. The

Acque, 410, Archivio di Stato di Venezia.
160 Report, 18 April 1744, Savi ed Esecutori alle Acque, 360, Archivio di Stato di Venezia.
161 Report, 17 September 1744, Savi ed Esecutori alle Acque, 410, Archivio di Stato di Venezia.
162 Report, 18 July 1745, Savi ed Esecutori alle Acque, 360, Archivio di Stato di Venezia.
163 Stefano Foin, report, 13 March 1766, Savi ed Esecutori alle Acque, 76, Archivio di Stato di Venezia.
164 Domenico Natale, letter, 22 September 1766, Archivio Proprio Ambasciata Roma, 155, Archivio di Stato di Venezia. Natale reveals the fraudulent practices of the captains, who usually do not pay the taxes owed to the consulate; he reports on the case of Captain Tommaso Rosconichi, who sailed from Civitavecchia on 21 September with a cargo of 320 cartloads of pozzolana for the construction of the pier of Ancona and for Venice; finally, he underlines the fact that there are very few "Venetian ships that come to this Port."
165 Domenico Natale, 8 June 1767, Archivio Proprio Ambasciata Roma, 155, Archivio di Stato di Venezia.
166 *Ibid*.

**162**. Canaletto, "Le Porte del Dolo," near Venice, 1742 circa (Museo Correr, Venezia).

fact that the construction of the levees with pozzolana marks a turning point in the fabrication of hydraulic lime mortar in Venice and the territories of the Serenissima in the direction of cement made in the Roman way is confirmed by a range of documents. Zendrini himself, while still working on the tests on the shoreline, intervenes in 1740 at Dolo, in the so-called "Porte del Dolo" – the sluice gates located at the confluence of two branches of the Brenta – to protect the masonry from water infiltration, applying Roman-type cement. "The pozzolana with which the cements are mixed," he writes, "and the many blocks of marble placed anew in large bulk to form the support for the spindles of the sluice gates, permit the most solid strength of said Royal structure in the future."[167] About 25 cartloads of pozzolana are used for the restoration of the steeple of St. Mark's, in Venice, in June-July 1745.[168] That same year, for the works in progress at Ponte Canale della Battaglia, near the settlement known as Ritratto di Monselice, in the Euganean Hills, Temanza is ordered to deliver six cartloads of pozzolana.[169] Temanza also supplies the 200 cartloads ordered by the Magistrato alle Fortezze in 1745.[170] In August 1750 the Magistrato alle Acque informs Filippini and Lucchese that the pozzolana stocked at Malamocco should be used to restore the piers of the wooden bridge at Bassano, destroyed by a flood in August 1748, making a "wall of pozzolana, to establish a firm, sturdy base capable of standing up to any impact and force."[171]

The use of pozzolana in the constructions of the Venetian Republic leads to a sort of emulation for the construction of hydraulic works in nearby provinces. The mathematician, astronomer and hydraulics expert, Father Paolo Frisi (1728-84), recalls that at the time of the breach, in the 1760s, of a bank of the canal in Bologna, at the Casalecchio lock, the repairs are done "at great expense, and with a double wall of stones, and pozzolana."[172]

[167] Zendrini, *Leggi e fenomeni* cit., p. 348.
[168] See, report, 1 June 1745, Savi ed Esecutori alle Acque, 410, Archivio di Stato di Venezia.
[169] See, report, 27 September 1745, *ibid.*
[170] See, report, 9 March 1745, and report, 18 March 1745, *ibid.*
[171] See the reports of 12 August 1750 and 27 August 1750, Archivio Proprio di Giovanni Poleni, cit. in, Giovanni Mognato, Stefano Tosato, *Dalla scienza all'architettura: 'Il ponte di nuova invenzion' del fisico-matematico Giovanni Rizzetti (1749)*, in, Lionello Puppi, Ruggero Maschio, *Giovanni Rizzetti, scienziato e architetto (Castelfranco Veneto 1675-1751)*, Stocco di Vedelago, Banca Popolare di Castelfranco Veneto, 1996, (pp. 139-95), p. 152. *Cf.* also, Bruna

## Variations to save pozzolana in the Venetian "murazzi"

The great effort to transform the levees of Venice into defenses made with pozzolana cement continues throughout the 1700s, segment by segment, responding to breaches caused by sea storms, with some modifications that never reach the point of challenging the substance of the "new method," namely its reliance on large stones and pozzolana cement. "Over the public shorelines of Venice," Zendrini can write, with pride, "I have introduced piers and levees of Istria marble sealed by lime and pozzolana, so that where the stone, in spite of its bulk, could not withstand the impact of the see, now it is connected to remain solid in the face of any storm, and in time the entire coast will be made utterly impenetrable and eternal [...]."[173]

Zendrini's statements are contained in his treatise, *Leggi e fenomeni, regolazione ed usi delle acque correnti*, one of the fundamental texts of Italian hydraulic construction, notably published in Venice in 1741, when the success of his "new method" was at its height. Though it contains few remarks on pozzolana, *Leggi e fenomeni* is one of the first books in the technical literature of the first half of the 1700s, after the fundamental contribution of Bélidor, to point to the potential of pozzolana in the preparation of cement for hydraulic constructions. It is precisely pozzolana, Zendrini explains, that makes it possible to improve the qualities of limes. A "good lime," he points out, "if mixed with pozzolana, would be even better, and this is because it resists the flow of water and its impact, which it often exerts as it passes, with greater force."[174]

The fact that the *murazzi* made with pozzolana, in Zendrini's vision, can be seen also as an evolution of the *cantoni di smalto* can be gleaned from the sequence of reasoning outlined in *Leggi e fenomeni*. After having referred to the pages of Vitruvius, Bélidor, Barattieri and the experiences of Guglielmini in Piacenza, Zendrini recalls the recommendations on *cantoni di smalto* written by Viviani in *Intorno al difendersi da' Riempimenti, e dalle Corrosioni de' Fiumi*. "Two types of these protections, then, are proposed, the first with said Viviani, with the above-mentioned cantoni di smalto, and the second with the large containers, when stones and lime are lacking to make the former," is the way Zendrini sums it up.[175] As in the case of the pozzolana sea walls, the works using *cantoni di smalto* also prevent the constant maintenance expenses required for levees made with wooden pile structures. "[...] it may be that one day," Zendrini hopes, "all the piles will be left behind, and engineers will replace them with this other [system], that can be said to be a perpetual defense, which besides offering assurance of good results, does not claim, we may say, any other expenditure for its conservation, where on the other hand the piles call for serious expense for their construction and not mediocre expense for their maintenance, even during those few years in which they survive."[176]

In spite of the success of the first implementations of the "new method," from the early 1740s Zendrini and the technicians of the Savi ed Esecutori alle Acque have to consider various possibilities to make the construction more economical, to save pozzolana and to speed up the masonry work. The first radical modification in this sense had already been made in 1737, after the failure

Tessarolo, *La ricostruzione del ponte di Bassano dopo la piena del 1748. Polemiche e diatribe*, in, "Antichità viva," XXIII, 1984, no. 6, (pp. 16-24), p. 18.
172 Paolo Frisi, *Dei canali navigabili*, in, *Raccolta d'autori italiani che trattano del moto dell'acque*, Bologna, J. Marsigli, 1833, p. 279.
173 Zendrini, *Leggi e fenomeni* cit., p. 355. Also for other hydraulic works, Zendrini trusts in the power of pozzolana, for example to form solid foundation slabs, or to make the lime mortar with which to lay the bricks in the foundations of buildings. The experience of the construction of levees with large stones and pozzolana binder has effects on the principles of fluid mechanics (*cf.* Zendrini, *Leggi e fenomeni* cit., p. 288).
174 *Ibid.*, p. 355.
175 *Ibid.*, p. 329.
176 *Ibid.*, pp. 329-30.

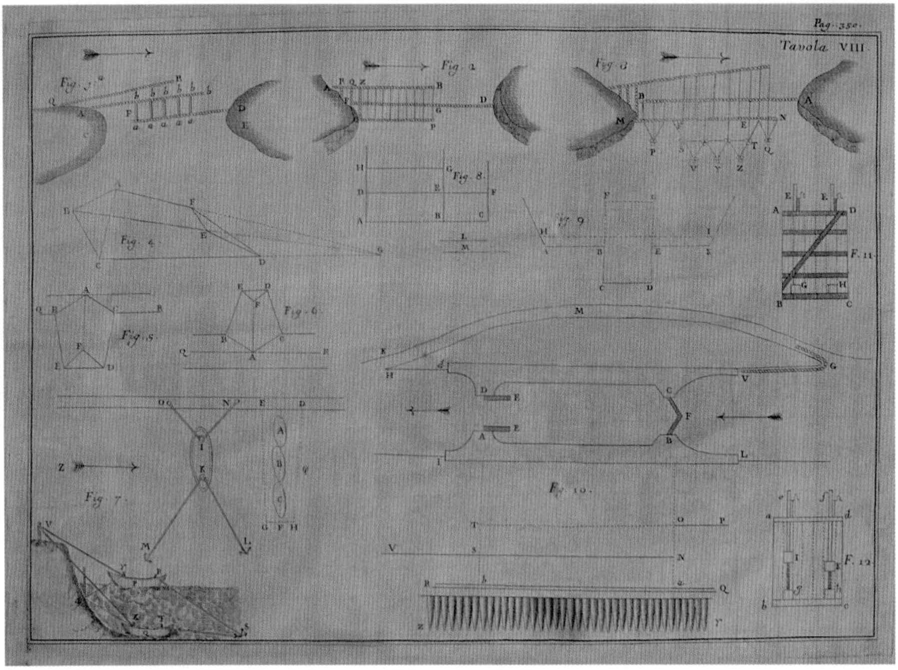


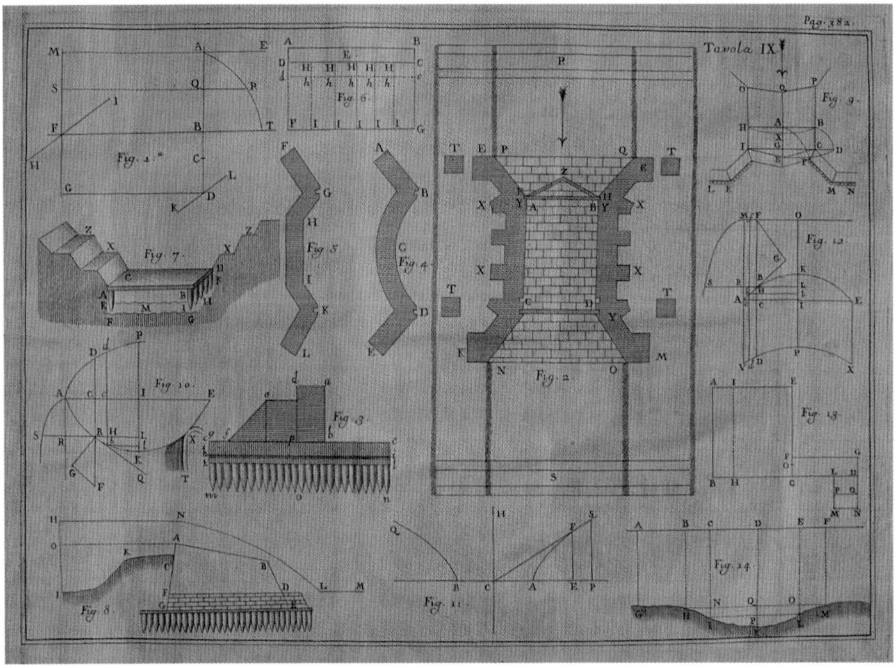


**163-164**. Bernardino Zendrini, *Leggi e fenomeni, regolazione ed usi delle acque correnti*, Venezia, G. Pasquali, 1741.

of the caisson with pozzolana cement. In 1741 Zendrini alters certain measurements of the new levee structure,[177] while the high costs involved in importing pozzolana prompt Lucchese to revive, in July 1742, the discussion on the use of different products from the pozzolana imported from Civitavecchia, an issue already raised by Filippini in 1738.[178] "[...] it would be worth considering whether it is advantageous, in the works using pozzolana, in place of it to use to some extent also quarried sand of good quality, and it seems to me that this could be done inside the volume of the construction," is Lucchese's suggestion, offering a glimpse of a type of breakwater built with different mortars, depending on which part is most exposed to the force of the sea.[179] A notation in the margin of the proposal indicates the authority to whom any decision on the cement for the sea walls must be submitted: "Works of pozzolana, to be submitted for the examination of the Public Mathematician Zendrini."[180]

In the report to the Savi ed Esecutori alle Acque on 15 September, Zendrini takes a position on Lucchese's proposal. On this occasion, he can state that the layer of pozzolana cement that forms the foundation of the levee, a "pure mixture of pozzolana," as he explains, has proven to be so strong in the various experiments already completed as to make the barrier of loose stones placed at the base, or "*zoccolo*," no longer necessary – a solution, that of the rocks, used not only in the traditional levees with the poles, but also in similar pozzolana structures Zendrini reports as having been made in Genoa and Leghorn.[181] This technical specification regarding the base of the *murazzo* made with pozzolana is pertinent to the discussion of the merits of Lucchese's proposal. In fact, Zendrini accepts the suggestion of concentrating the pozzolana work precisely at the points of the levee in direct contact with the sea, in that *zoccolo* made stable by the layer of "pure mixture of pozzolana" and on the embankment, leaving the *proto* in charge with the task, case by case, of selecting the points where the pozzolana should be used, though without ever straying from the principle of the "necessary solidity of these sea walls."

"[...] therefore," Zendrini continues with his opinion, "I would consider it sufficient to make in pozzolana that part that remains exposed to the sea and to the air, with greater need of protection, but the inner part of the structure can be formed with ordinary sand and lime, as that bond is enough, and much more the enormous weight of the marble blocks, to stand up to any external force, when the resistance on the outer part is of greater activity, because formed with pozzolana; operating in this way, great will be the saving on that material, reducing the expense it brings [...]."[182]

177 "[...] having to reduce the base in width at the bottom to eight feet, in place of ten, and above to six feet in place of four, and at a depth below normal water level of four feet, already established, and above it of two feet, and of three at the middle, like the crown above the whole, and with everything set at the already established measurements, with the technique, however, of forming it in such a way that it will stay straight, and at the level of the old one" (Bernardino Zendrini, report, 18 May 1741, Savi ed Esecutori alle Acque, 409, Archivio di Stato di Venezia).

178 Matteo Lucchese, report, 10 July 1742, Savi ed Esecutori alle Acque, 75, Archivio di Stato di Venezia.

179 *Ibid.*

180 *Ibid.*

181 Bernardino Zendrini, report, 15 September 1742, *ibid.* "As for the works I have introduced anew in pozzolana [...] their strength is such," Zendrini writes, "that each of them is recognized as the only means for keeping the levees permanent, also without the barriers of loose stones that are nevertheless in use in Genoa and Leghorn, and that I myself felt should still be used on these shores to shield the foot of the levees, with all this to sink them for four feet below normal water level, with a certain pure mixture of pozzolana, that is being made along the foot of those protections, has then proven that it is possible to dispense with those stone barriers, and as a result to eliminate great expense, and all that which might still be removed, since the sea can very easily overturn any stone that is not bound, breaking it in a short time and rendering it useless."

182 *Ibid.*

At the end of the report Zendrini returns to the need to "bond" with "pozzolana cement" the large Istria stones of the experimental levees made under the supervision of Tirali, and to flank those two levees with new "*muraglioni*," also made with pozzolana.[183]

The works in progress following Zendrini's method meet the test of a storm in November 1742, which causes serious damage to the levees of Tirali, produces breaches in the levees made with the traditional systems of earth, piles and stones, and clearly demonstrates that only the works done with pozzolana can survive, even in the state of incompletion with which they were overtaken by the storm. The *murazzi in pozzolana* "over the shore at Malamocco have resisted, and no damage can be discovered in them"; "the pozzolana groyne over the shore at Pellestrina, between 44 and 45, for 87 pertiche, though its covering and short gap were not yet perfectly completed, has remained fully intact after the storms"; "the sea walls over the shore at Chioggia, that is the one already completed, between 17 and 18, for 400 pertiche, remained undamaged by the forceful blows of the sea": these are passages from the report prepared by Filippini after an inspection of the levees.[184] Instead, regarding Caroman and the structures of the epoch of Tirali, Filippini states that everything "has become a precipice."[185]

Observation of the effects of the storm creates a state of urgency in the progress of the work using the Zendrini method, against the backdrop of awareness of the costs of pozzolana and the difficulties involved in its importation. Charged with indicating the segments that require immediate attention, and with establishing "if it is best to proceed with pozzolana, or with piles," Filippini and Temanza repeat their urging, in March 1743, for intervention along all the breaches, using pozzolana, and they warn the Savi ed Esecutori alle Acque against continuing to use the traditional levees with wooden poles, earth and loose stones.[186] After the suggestions of Filippini on the use of local soils, and above all following the proposal of Lucchese, backed by the positive view of Zendrini, to concentrate the pozzolana works in the construction of certain sectors of the levees most exposed to the force of the sea, the search for technical solutions to save pozzolana again becomes a timely issue among the technicians working for the Savi ed Esecutori alle Acque. Evidence of these developments can be seen in what Filippini and Temanza write regarding the works to be done at Caroman: they advise using pozzolana only to construct the base of the levee – "*scarpon in pozzolana*" – to achieve "savings of pozzolana in the internal parts, using it only for the external parts."[187]

The masonry in bricks and lime mortar made by order of the Magistrato alle Fortezze and completed in 1726, of the "tooth" at Forte di San Pietro della Volta, built to defend the port of Malamocco, becomes the example for possible transformation of the masonry works of Zendrini's method into a structure in Istria stone and pozzolana with a nucleus similar to that of the Fort.[188] We cannot rule out the idea that Filippini may be the main person behind the transformation of Zendrini's *zoccolo* into masonry with Istria stone with a core of bricks and lime mortar, which the Savi ed Esecutori alle Acque submit, on 30

[183] *Ibid.*
[184] Giovanni Filippini, report, 8 December 1742, *ibid.*
[185] *Ibid.*
[186] Giovanni Filippini, Tommaso Temanza, report, 28 March 1743, *ibid.*
[187] *Ibid.*
[188] See, report, 30 April 1740, and, Bernardino Zendrini, Giovanni Filippini, Matteo Lucchese, Tomaso Temanza, report, 1 June 1743, *ibid.*
[189] Report, 30 April 1740, *ibid.*
[190] Bernardino Zendrini, Giovanni Filippini, Matteo Lucchese, Tommaso Temanza, 1 June 1743, *ibid.*

April 1743, to the examination of a commission formed by Zendrini, Filippini, Lucchese and Temanza, and set up for "varied research regarding the works in pozzolana": "For the structure of the Zoccolo, it should be determined if it needs to be improved in some part, and it if can be sufficient to make with pozzolana its covering towards the sea, of the necessary size; and to thus make also the top, filling the inner part with stones, or baked stone, using only lime binder without pozzolana, which would cost less [...]."[189]

The questions to be examined by the commission also have to do with the proposal to construct pozzolana works in two phases, in order to be able to consolidate, with Zendrini's structure corrected through the lime mortar nucleus, the largest length of levee, at least in the most vulnerable zones: in an initial phase the pozzolana work would be limited to the crust of the base, with the core in brick and lime mortar (the core can be entirely wrapped in Istria stones, including the base, or protected only on the two sides exposed to the sea, excluding the base); while the embankment would receive the traditional covering of Lispida stones, in keeping with the loose stone technique, with three layers and, possibly, "the addition of some barriers of old piles, to hold the stones."[190] Only in a second phase, when much of the shoreline would be protected by the *zoccolo*, would the whole embankment be strengthened with the pozzolana work.

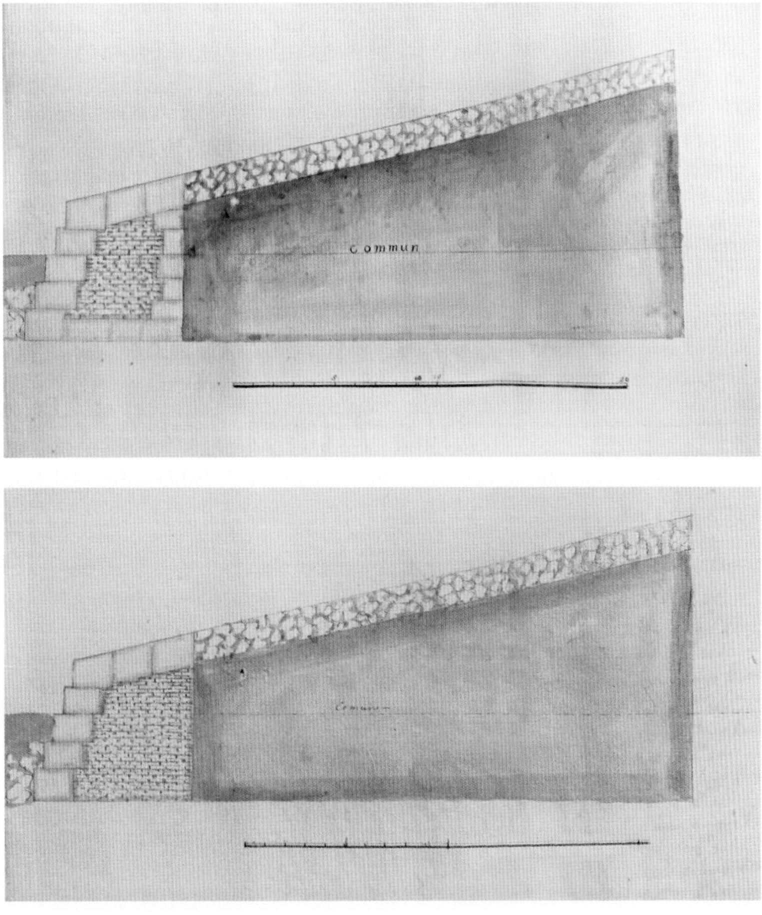


**165-166**. Giovanni Filippini, attributed to, proposals for the construction of the base of the "murazzo," or "zoccolo," with a core of bricks, Venice, 1743 (Savi ed Esecutori alle Acque, 75, Archivio di Stato di Venezia).

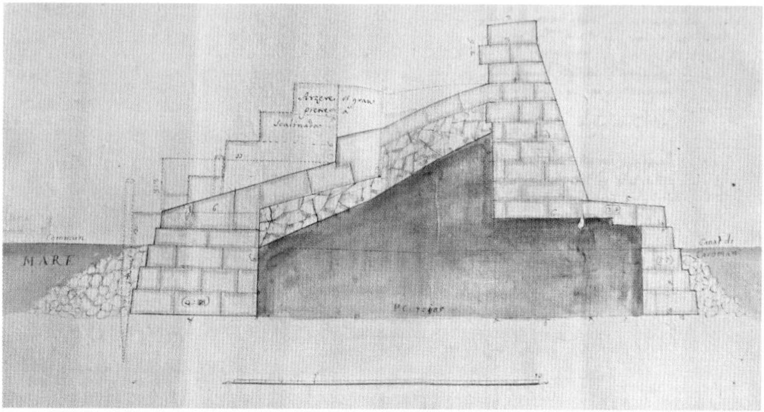


**167**. Bernardino Zendrini, Tommaso Temanza, "murazzo," Venice, 1743 (Savi ed Esecutori alle Acque, 75, Archivio di Stato di Venezia).

After the inspection of the shorelines and examination of the various proposals, during the course of May the commission headed by Zendrini issues an unfavorable opinion on the suggested transformations. It is even stated that "when the entire protection cannot be made from top to bottom in pozzolana, we think it would be much preferable to work with ordinary material, without making any zoccolo." The opinion of the commission expresses no doubts regarding a critical evaluation of the solution of the *zoccolo* made with a brick core: "for the perpetual conservation of the shorelines, the strength of these works lies both in the very strong bond given them by the mixture of pozzolana with lime mortar, and by the weight of the materials [...] the protection will always be stronger when the specific gravity of the material that forms it is greater than that of another, so that for example if the specific gravity of brick with respect to that of Istria stone is 14, to 21, it is clear that this will be the reason for the strength of the wall of stone as opposed to that of bricks, if the latter is not of much greater bulk. If the brick is placed in pozzolana to give it a stronger bond, and greater consistency, using a much larger quantity of that material than the quantity required of blocks of marble, the expense would also be significantly greater than that incurred by making the zoccolo with true stone. With our proposal to be able, inside the structure of the shoreline protections, to save large amounts of pozzolana by using only lime mortar with sand, for some savings of costs, we have never meant that this should be done in the immediate sea defenses, but only in those further from the sea, as in the inner bulk of the embankments, and in the solidity of the walls from the part of the Lagoon, and always above water; never in the zoccolo, which is the first part to be struck by the sea, and never in brick, but always with true stone, and above the shore."[191]

After the discussions on the various levee projects using composite materials, precisely for a sea wall to be built at Caroman, contiguous with that of Tirali with the steps, Zendrini, in collaboration with several technicians, including Temanza, proposes a structure in Istria stone and pozzolana, as illustrated on 1 June 1743 to the Savi ed Esecutori alle Acque – the "*murazzi a pozzolana*" with "steps."[192] In this version the barrier becomes a powerful struc-

[191] Bernardino Zendrini, Giovanni Filippini, Matteo Lucchese, Tommaso Temanza, 1 June 1743, *ibid.*

[192] *Ibid.*

[193] Tommaso Temanza, report, 29 March 1745, *ibid.*

[194] Bernardino Zendrini, Tommaso Temanza, report, 5 June 1745, *ibid.*

ture formed by a crust of large Istria stones that wraps the entire levee, both on the side of the sea and on that of the Lagoon. The foot of the wall in the water is composed of a *zoccolo* in only Istria stone – this *zoccolo* is wider on the side of the sea than on the side of the Lagoon. The squared blocks of stone that form the embankment toward the sea are laid over a ballast of Lispida stones, and form an intermediate wavebreak step, harking back to the solution introduced by Tirali. The embankment is topped by a wall that forms the end of the powerful inclined buttress set up on the side of the Lagoon – between the buttress wall and the *zoccolo* in the Lagoon, a walkway runs the length of the whole structure. This is the most highly evolved system of Zendrini, the one of greatest strength, invented by observation of the impact of the waves, the principles of fluid mechanics and the inclined plane, the constructive potential of works in pozzolana and the strength of the specific weight of stones. This version of the *murazzo* is put into practice at Caroman, and 37 *pertiche* have been constructed when, in March 1745, Temanza makes an inspection of the levees to assess the damage at the end of the winter season.

Temanza lists the portions requiring construction with "stones and pozzolana," and which groynes to extend "with the pozzolana trial," while he indicates the damages at Caroman to the "first mass of stones upwind made by Tirali" and concludes with a warning: "for some time the established very valid defense of an embankment in pozzolana similar to the other has been delayed."[193] Therefore Zendrini and Temanza are ordered to examine the works of Tirali and to suggest repairs – the two "masses" at Caroman with the large "stones without any lime binding.[194] Zendrini and Temanza believe it is necessary to construct a "*zoccolo a pozzolana*" in continuity with that of the "neighboring pozzolana embankment that is presently being built," and to consolidate the stones of the "upper parts, and perhaps also the internal parts, with the use of pozzolana."[195] In their phrases, there is a clear respect for those "moles," to be "conserved" as the forerunners of the new period of construction of the levees of the Lagoon.[196]

The drawing attached to the report of Zendrini and Temanza, which can be attributed to Temanza, is the fundamental document of the state of the levee built at Caroman starting in the 1710s. It shows the two masses of stones made without pozzolana, indicated as work of Istria stones ("lavoro di pietre d'Istria"), the new pozzolana sea wall of 37 *pertiche*, indicated as pozzolana work ("lavoro a pozzolana"), and the earthen levee, indicated as corroded ("argine corroso").

For the construction of the sea walls, until the end of the 1740s the methods vary, from segment to segment and from year to year, depending on the different conditions of exposure of the levees to the force of the waves, and on the availability of materials in a given moment – the supply of pozzolana remains the critical factor on the books of the waterfront worksites. In certain stretches the procedure follows the recommendations of the various experts, making only the base in pozzolana masonry, to protect an earthen levee; only at a later date are the earthen levees transformed into sea walls made with pozzolana. "The murazzo that follows, made with pozzolana, holds the earthen embankment from behind," Lucchese explains in 1746, regarding a *zoccolo*.[197] In certain

[195] *Ibid.*
[196] Bernardino Zendrini, Tommaso Temanza, report, 5 June 1745, Archivio Proprio Zendrini, 65, Archivio di Stato di Venezia.
[197] Matteo Lucchese, report, 14 November 1746, Savi ed Esecutori alle Acque, 75, Archivio di Stato di Venezia.

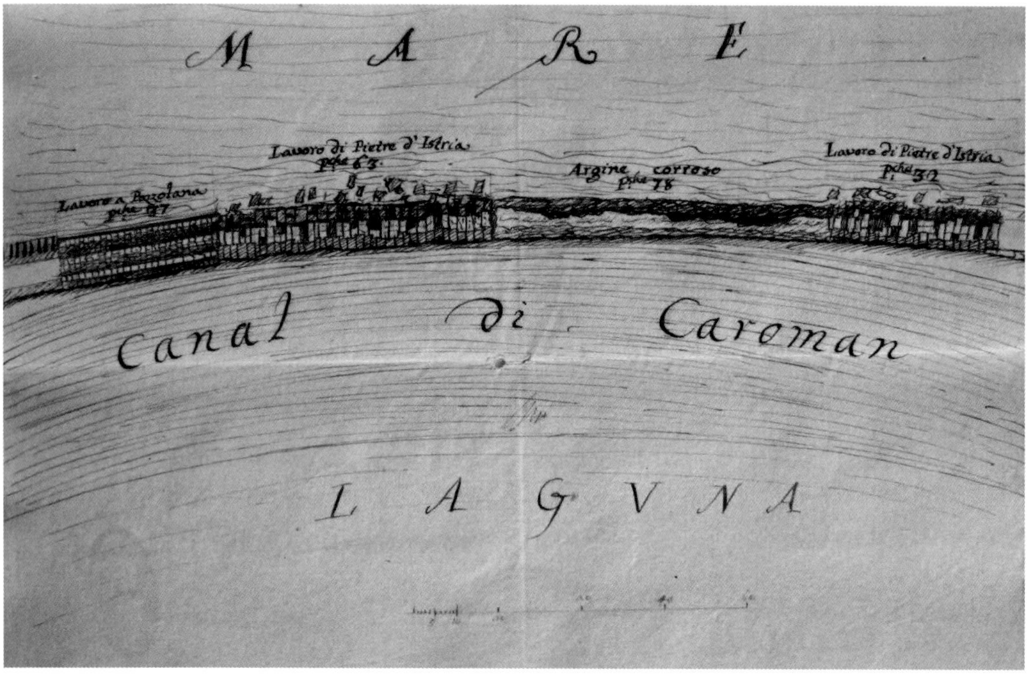


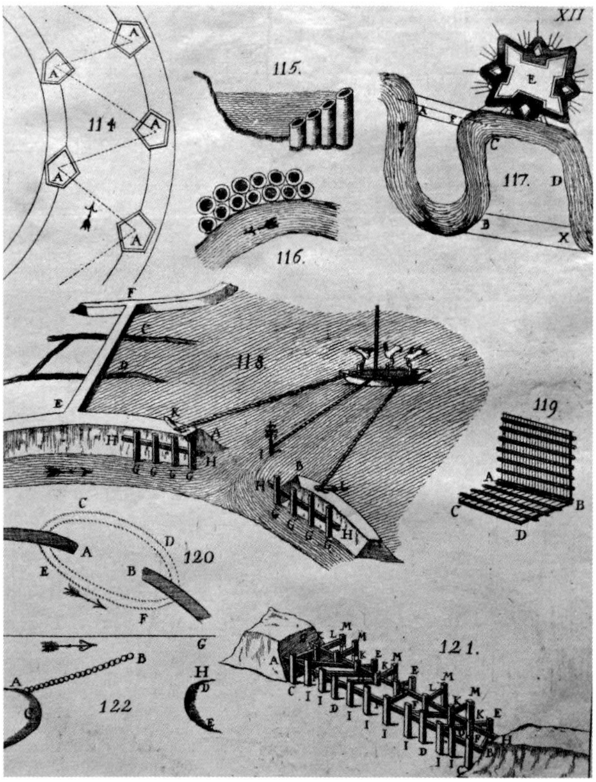


**168**. Tommaso Temanza, drawing of the levee at Caroman, Venice, attached to Bernardino Zendrini, Tommaso Temanza, report, 5 June 1745 (Archivio Proprio Zendrini, 65, Archivio di Stato di Venezia).

**169**. Various levee repair systems (G.A. Alberti, *Istruzioni pratiche per l'ingegnero civile, o sia Perito Agrimensore, e Perito d'Acque*, Venezia, G.B. Recurati, 1748).

periods the rebuilding of the sea defenses is not done in masonry, due to shortages of pozzolana.[198] At times Zendrini himself recommends intervention with the old systems, followed by application of his method "as soon as the pozzolana arrives."[199] Also due to the lack of pozzolana, in 1747 Zendrini advises the construction, along the Chioggia shore, of a "crown of stone in pozzolana" for 40 *pertiche*, using just 100 cartloads of pozzolana.[200]

After the first works of Zendrini, in the documents on the constructions at the Lagoon, throughout the 1700s, the term pozzolana is joined by the term *murazzo* – *murazzo a pozzolana* – and the term *zoccolo* – *zoccolo a pozzolana* – and plaques are made with epigraphs, like the one in 1749, in praise of the walls cemented with pozzolana ("muri cementati con pozzolana").[201] Those works are even defined in singular terms – "machines, as they are called."[202] It is no coincidence that during the course of the 1750s it is precisely a Venetian, trained in the context of the *proti* working on the construction of the *murazzi a pozzolana*, who discovers the magnificence of the foundations of the monuments of ancient Rome.

The era ushered in with the construction of the levees in Istria stone and pozzolana cement still does not definitively put an end to the constructive practice of reinforcement of the levees by means of piles and wicker caissons. All the constructive knowledge gained prior to the pozzolana sea walls is combined in manuals like the one published in 1748 and written by Giuseppe Antonio Alberti (1712-68), an expert land surveying and hydrostatics, and an architect. An important section of this manual, however, is devoted to methods of consolidation using concrete, for which Alberti cites those in use since the 1600s and described by the engineer Giovanni Battista Baratteri, or Barattieri, as "cylinders filled with concrete with pebbles,"[203] as well as the more recent works by Zendrini – the "moles with a three-sided truncate pyramid form; but with the section at the top oblique to the base, to terminate in a gently sloped embankment,"[204] though Alberti does not make reference to *murazzi* or to pozzolana, but only to "good lime."[205]

## The concrete of Milet de Monville at Toulon and the pozzolana for the ports of Brest and Cartagena

The destinations of exports of pozzolana from Civitavecchia continue to indicate the main centers of experimentation with concrete. After Ancona and Venice, the trading accounts for pozzolana indicate shipments from Civitavecchia, in the 1740s, of large quantities to Toulon, Brest and, in subsequent years, also to Cartagena. These imports of pozzolana are for major modernization projects under way at the two most important naval dockyards in France, that of Toulon, for the Mediterranean fleet of warships, and that of

[198] Bernardino Zendrini, report, 1 April 1746, *ibid.*
[199] *Ibid.*
[200] Bernardino Zendrini, report, 8 May 1747, *ibid.*
[201] Cit. in, Grillo, *op. cit.*, p. 87.
[202] Matteo Lucchese, report, 14 November 1746, Savi ed Esecutori alle Acque, 75, Archivio di Stato di Venezia.
[203] Giuseppe Antonio Alberti, *Istruzioni pratiche per l'ingegnero civile, o sia Perito Agrimensore, e Perito d'Acque*, Venezia, G.B. Recurati, 1748, p. 152. For the use of concrete structures during the 1600s to defend shorelines see, Giovanni Battista Baratteri, *Architettura d'acque*, Piacenza, G. Bazachi, 1656, p. 61. It is in the context of research on solid masonry levees that the essay is republished, in 1763, by Barattieri (Giovanni Battista Baratteri, *Architettura d'acque*, Piacenza, Stampa Ducale di G. Bazachi, 1763). See also, *Trattato teorico-pratico circa la divisione degli incrementi fluviali ossia il terzo, e quarto libro della prima parte dell'opera di Gio. Battista Barattieri, preceduti da un discorso preliminare e corredati da annotazioni da Francesco Sartorio*, Piacenza, G. Tedeschi, 1783.
[204] Alberti, *Istruzioni pratiche* cit., p. 147.
[205] *Ibid.*, p. 198.

**170**. Plan of the port of Toulon (Ministerio de Educación, Cultura y Deporte, Archivio General de Simancas, Mapas Planos y Dibujos, 20,005).

**171**. François Milet de Monville, machine for pouring concrete used in 1748 for the work at the port of Toulon (B.F. de Bélidor, *Architecture Hydraulique. Seconde Partie, qui comprend l'Art de diriger les eaux de la Mer & des Rivieres à l'avantage de la défense des Places, du Commerce, & de l'Agriculture*, Paris, Ch.-A. Jombert, vol. II, 1753).

Brest, for ocean-going vessels, and for the works in Spain at the port of Cartagena on the Mediterranean coast.

The fact that the rebirth of concrete during the course of the 1700s can be attributed precisely to the construction of hydraulic works in seas and rivers is confirmed by the experiences of a military man who was destined to play a crucial role, like other colleagues and engineers: the officer of military engineering François Milet de Monville, employed from the 1740s in the making of defenses along the Mediterranean coast, as director of the fortifications of Provence (he is responsible for the design of Fort Lamalgue, in Toulon, on which work begins in 1764). The experiences of Milet de Monville and other hydraulic projects of significance for the application of "*beton*" (or "*béton*") take place around the naval shipyard of Toulon. To prepare for the construction of a new dockside in 1748, Milet de Monville makes tests that become milestones in the rediscovery of concrete in the 1700s, also thanks to their documentation by Bélidor. Milet de Monville's objective is to identify a process of fabrication capable of producing a concrete for foundations built in the sea. It is probable that for the percentages of the ingredients of the compound he is guided by the indications of Vitruvius, who, "when speaking of concrete wants that it be composed of two parts pozzolana and one part lime."[206]

The hypothesis utilized by Milet de Monville during the trials is that the lime should be used quick, and slaked during the operations of preparation of the concrete, "so that through the effervescence that will take place it might liberate the largest fraction possible of the salts contained in the rubble and the pozzolana that must be incorporated."[207] For the composition of the concrete, Milet de Monville has an area prepared with a solid surface, where he forms a ring of about 5 or 6 feet in diameter, with twelve parts of hydraulic powder (pozzolana, *terrasse de Hollande* or *cendrée of Tournai*), and six parts of sand; in the disk bordered by the ring he spreads nine parts of quick lime, to be pulverized with iron mallets, moistened with sea water and mixed every so often with poles until a paste has been obtained, into which to blend the hydraulic powder and the sand.

"Once all is well mixed," we read on the subject of the experiments of Milet de Monville, "we throw in thirteen parts of chopped-up stone and three of crushed slag, when it can be had, otherwise one must be satisfied with the use of sixteen parts, instead of thirteen, of chopped-up stone & rubble, or small stones whose size should not exceed that of a hen's egg. One then mixes by hand the entire composition for an hour by dragging it out here and there with shovels to better incorporate the parts, after which one forms masses which one lets harden for 24 hours in summer in warm countries; but in winter it must be left sometimes three or four days, making sure to keep it covered from the rain, and to use it only once it is firm enough so that it can't be removed except with a shovel."[208]

To test the efficacy of the compound, Milet de Monville has the concrete placed in a wooden chest measuring 27 cubic feet, which is kept immersed in the sea for two months. The state of solidity of the masonry is such "it would be more difficult to separate it into parts than a block of the best stone."[209] This

[206] This is what Bélidor writes in the explanation of the experiences of Milet de Monville (Bernard Forest de Bélidor, *Architecture Hydraulique. Seconde Partie, qui comprend l'Art de diriger les eaux de la Mer & des Rivieres à l'avantage de la défense des Places, du Commerce, & de l'Agriculture*, Paris, Ch.-A. Jombert, vol. II, 1753, p. 185).

[207] *Ibid.*, p. 186.

[208] *Ibid.*

[209] *Ibid.*, p. 188.

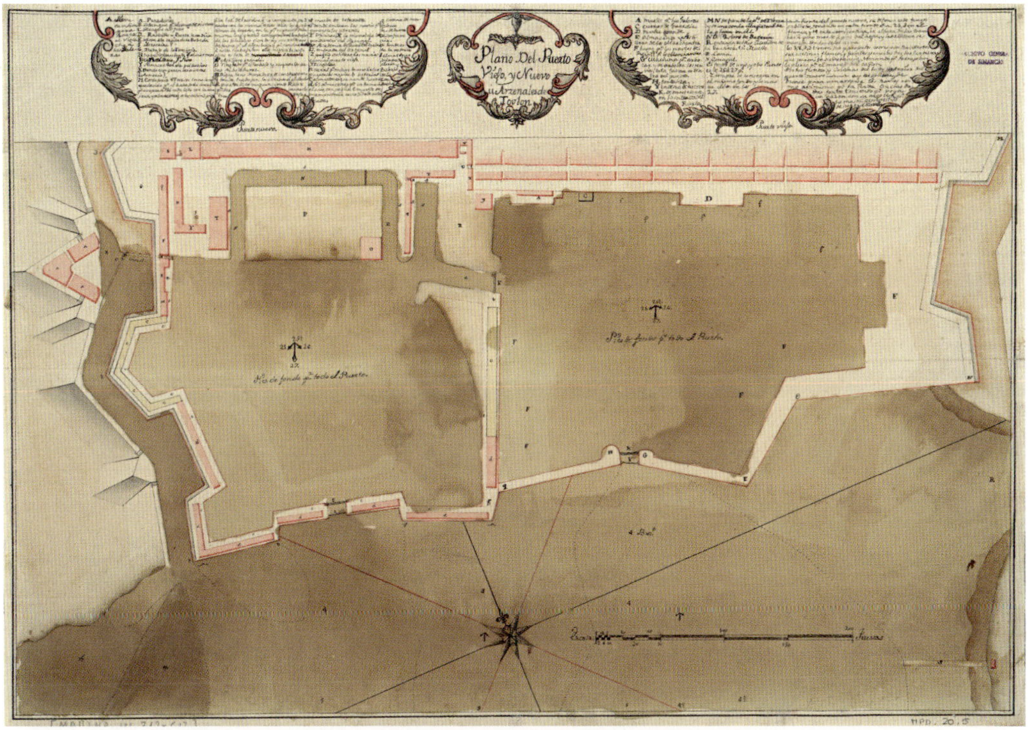
Plano Del Puerto Viejo y Nuevo y Arzenal de Tolon

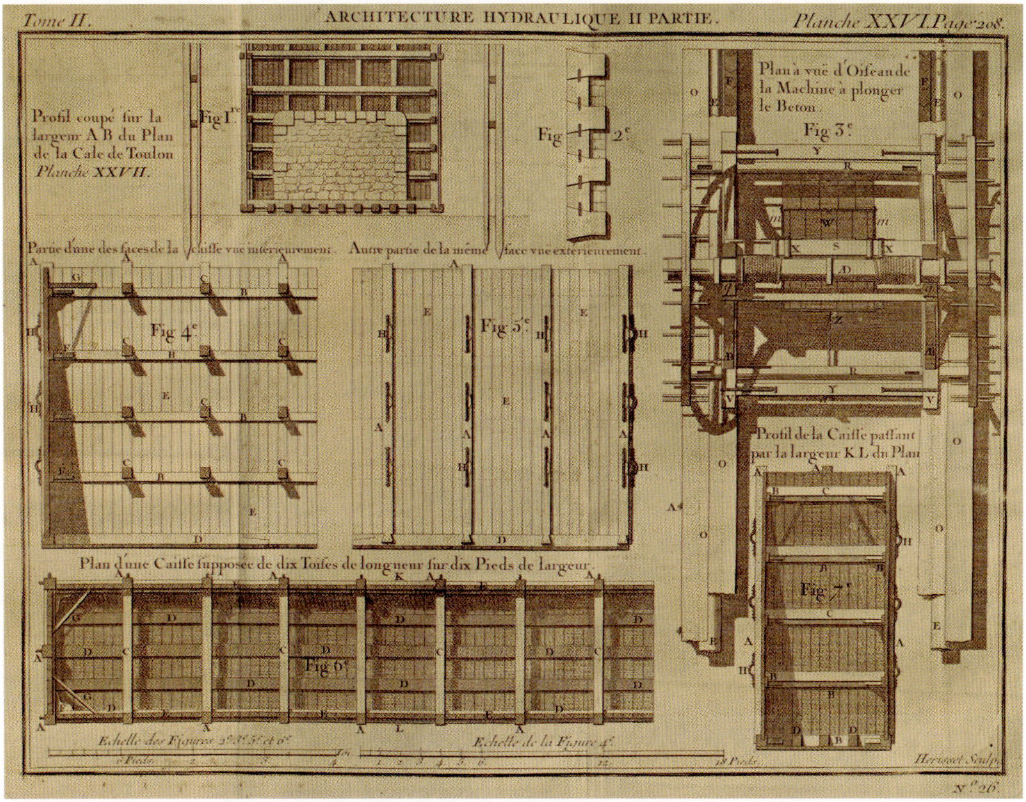
Tome II.
ARCHITECTURE HYDRAULIQUE II PARTIE.
Planche XXVI. Page 208.
Profil coupé ſur la largeur A B du Plan de la Cale de Toulon Planche XXVII.
Fig 1re
Fig 2e
Plan à vuë d'Oiſeau de la Machine à plonger le Beton.
Fig 3e
Partie d'une des faces de la caiſſe vuë interieurement.
Autre partie de la même face vuë exterieurement.
Fig 4e
Fig 5e
Profil de la Caiſſe paſſant par la largeur K L du Plan
Fig 7e
Plan d'une Caiſſe ſuppoſée de dix Toiſes de longueur ſur dix Pieds de largeur.
Fig 6e
Echelle des Figures 2e 3e 5e et 6e.
Echelle de la Figure 4e.
Heriſſet Sculp.
N.o 26.

**172-173**. François Milet de Monville, machine for pouring concrete used in 1748 for the work at the port of Toulon (B.F. de Bélidor, *Architecture Hydraulique. Seconde Partie, qui comprend l'Art de diriger les eaux de la Mer & des Rivieres à l'avantage de la défense des Places, du Commerce, & de l'Agriculture*, Paris, Ch.-A. Jombert, vol. II, 1753).

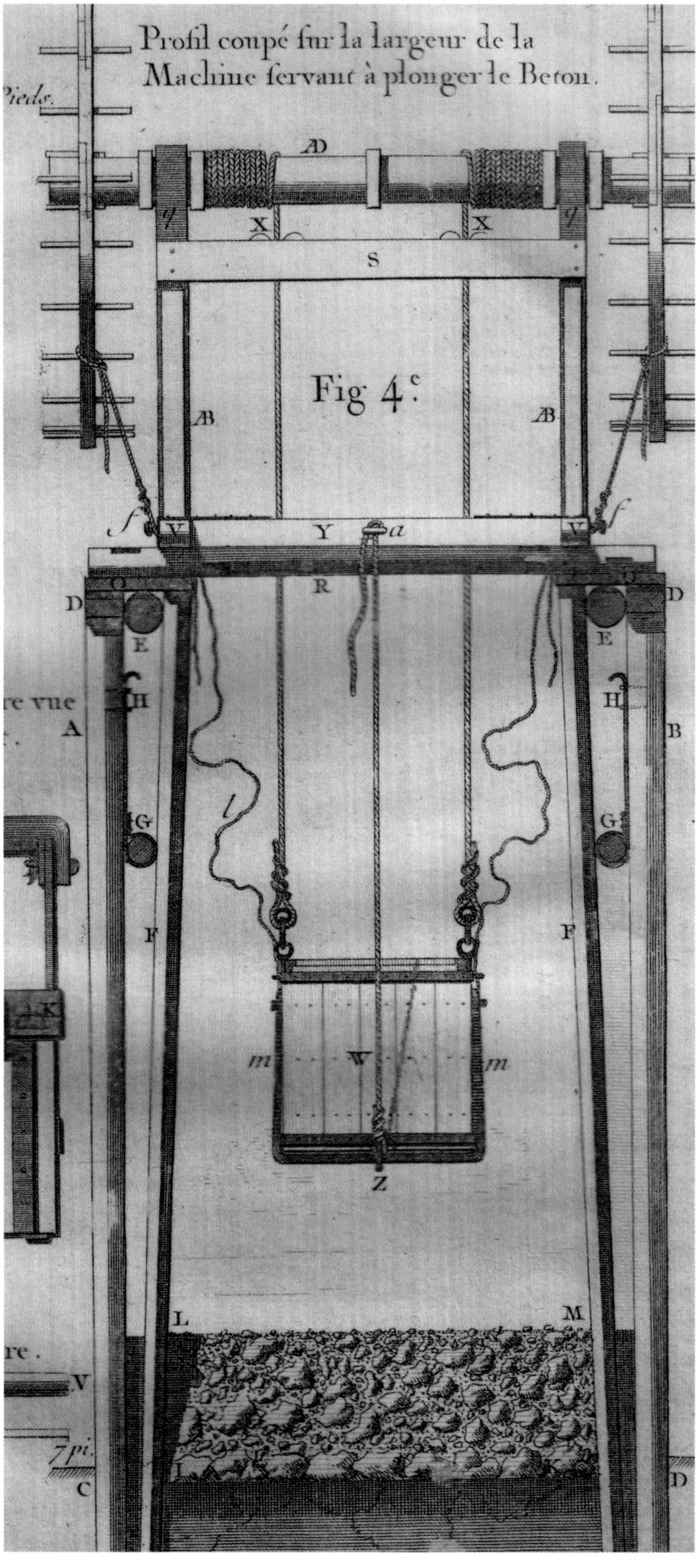

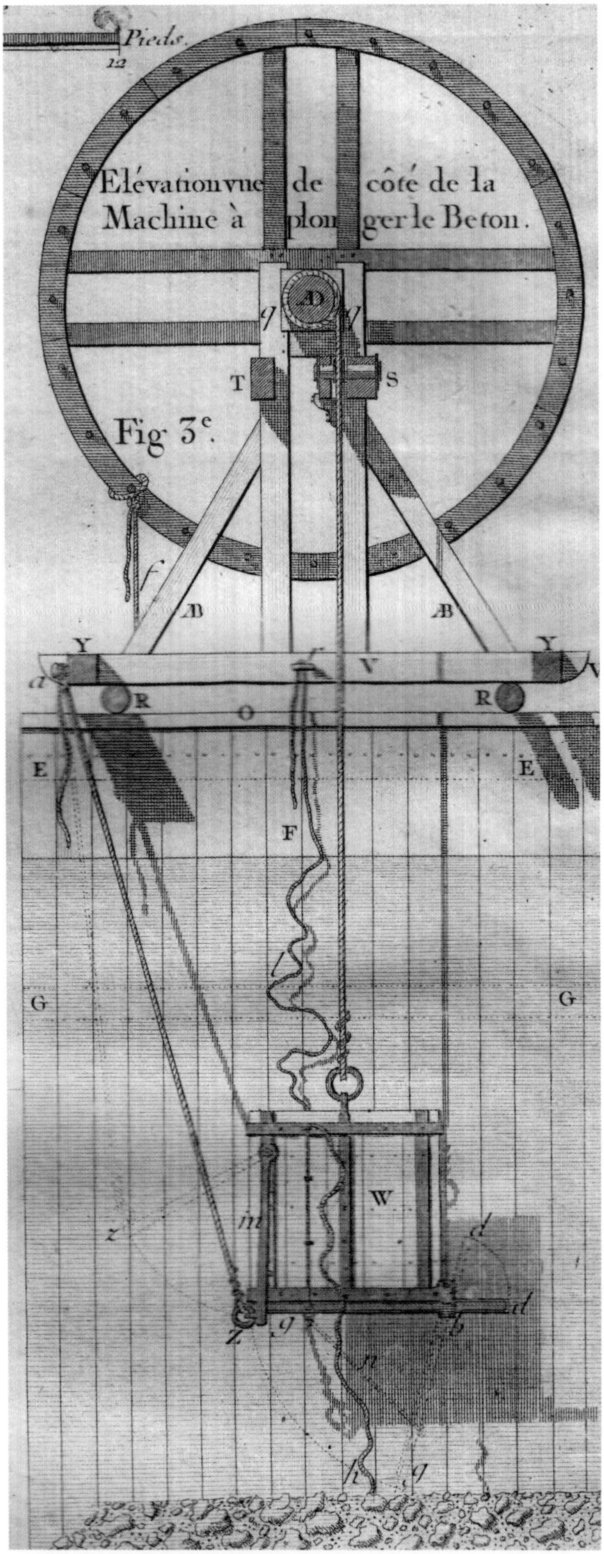
Pieds
12
Elévation vue de côté de la
Machine à plonger le Beton.
Fig 3e.
T
S
Y
Y
V
R
R
O
E
E
F
G
G
W
Z

test with the wooden box becomes the experiment of reference for all subsequent tests of hydraulic mortars conducted by French engineers. Even in the 1780s, mention is still made of what was done in Toulon in 1748.[210]

With a concrete based on red pozzolana from Civitavecchia, Milet de Monville has 16 cubic feet of masonry made at a depth in the sea of 10-18 feet, with the aim of calculating the quantity of the materials and the time required for construction (the fact that the workers are prisoners, chained in pairs and thus hampered in the movements, increases the time needed for the procedure).[211]

Bélidor describes the operations of foundation in the sea with the concrete of Milet de Monville, which he sees in Toulon in 1748. The construction of the foundations happens in keeping with a system similar to the "dammed up voids" described by Fontana in the case of the Roman piers of the port of Anzio. Once the concrete has been prepared and allowed to dry, Milet de Monville has the compound spread inside the void of the special formwork with sides made with boards held in place with poles, bordering the thickness of the undersea foundation. The operation of pouring of the concrete is done with a special machine equipped with a chest whose bottom can be opened; in contact with the water the concrete "spreads out and sags";[212] then a layer of stones is poured, which penetrate into the liquefied concrete; alternating layers of concrete and stones are then poured, obtaining a masonry that in the span of three or four months "becomes inseparable, and even more so as it ages."[213] It is worth recalling that precisely at Toulon, at the time of Colbert, experiments had been conducted to make the foundations of defense bastions in the water using, coffers, including advice to the technicians to "travel to Italy" to learn about fabrication procedures.[214]

Bélidor grasps the radical transformation of the worksite and the expertise of the workmen with the technique of concrete – there is no longer any need of the stone cutters, but of carpenters to assemble the formwork and of laborers to prepare the compound and pour it; furthermore, he envisions a sort of permanent formwork that can contribute to the solidity of the foundation, probably based on the model of the wooden grilles of different geometric orders used for stone moles without mortar.

"One can see by this way of building," he writes, "that the masonry can be considered as having been executed according to the best rules of the art, because, being composed of very little material with a large exposed surface for the parts in concrete, a larger quantity of the salts specific to the concretion separate away; and it is convenient that it can be done in the absence of a mason, even the least skilled workers are capable of carrying it out; only a

[210] Antoine-Chrysostome Quatremère de Quincy, *Encyclopédie méthodique. Architecture*, vol. I, Paris, Lièges, Panckouke, Plomteux, 1788, p. 278.
[211] Bélidor, *Architecture Hydraulique. Seconde Partie* cit., vol. II, 1753, pp. 189-90.
[212] *Ibid.*, p. 186.
[213] *Ibid.*, p. 187.
[214] Jean-Baptiste Colbert, letter to Arnouls Fils, 5 August 1678, in, Pierre Clement, ed., *Lettres, instructions et mémoires de Colbert*, Paris, Imprimerie Impériale, 1864, vol. III, (pp. 117-20), p. 118. The use of "*encaissements*" for the construction of piers is also documented in a comment to Suetonius on the life of the emperor Claudius, regarding the port of Ostia – later Civitavecchia: "The bulwark being formed by *encaissements* of stone and lime, similar to what we do today [...]" (*Projet du sieur Bernardin Zendrini, Ingénieur de la République de Venise, pour réparer le Port de Viareggio, & en corriger le mauvais air*, in, "Journal Oeconomique, ou Memoires, notes et avis sur l'Agriculture, les Arts, Le Commerce," Paris, A. Boudet, January 1760, (pp. 28-35), p. 33). The pier of Viareggio and the way of correcting its orientation are also described: "It might be difficult to demolish: however, as it is solid and well built in lime and pozzolana, I believe that by mining it we would finish by breaking it up" (*ibid.*, p. 32).
[215] Bélidor, *Architecture*

few carpenters are required to build the encasings. In addition, the encasing in which the masonry has been formed can remain, particularly in place exposed to the waves of the sea, or to very rapidly flowing waters."[215]

The concrete works of Milet de Monville, thanks to the accurate descriptions of Bélidor, become the model for the application of the material in hydraulic constructions. It is significant that Perronet, in turn, describes the work of Milet de Monville as an example of the use of concrete, precisely that of the "mole built in 1748 at Toulon," familiar to him thanks to Bélidor's description, where the concrete is spread thanks to "small cases whose bottoms can be opened."[216] Perronet underlines the existence of two different deployment modes of concrete, which he sums up with two verbs – "descend" and "cast": the mode of Milet de Monville, using a chest with a bottom that opens to deposit the compound on the seabed, and that of the ancient Romans, who instead pour the compound into formworks, as Perronet infers from Book V, Chapter 12 of *De Architectura* by Vitruvius: "they cast [the masonry] into the water, after having surrounded the position of the foundation to be built by a row of stake and sheet piling, when the depth of the ocean would allow."[217] It is worth noting that Perronet, in his description of the method practiced by Milet de Monville, neglects to mention the existence of an *encaissement* in which the concrete is spread; furthermore, he writes about multiple *caisses* for the construction, and not just one – the one that is part of the machine.

Major concrete works are also made, in parallel to those at the dockyard of Toulon, at the naval shipyard of Brest. The trade in pozzolana from Civitavecchia, which has trouble reaching Brest, documents the importance of the worksites and the construction techniques utilized, which are probably very similar to those used by Milet de Monville. The pozzolana is used in the works done in 1738 under the supervision of the engineer Antoine Choquet de Lindu (1712-90), employed by Frézier (the shipbuilder Blaise Ollivier is also involved).[218] The first works of Choquet de Lindu in which pozzolana may have been used are the four shipbuilding docks of Bordenave, built in 1738-44. The great size of the worksites is documented by the arrival in Brest, in 1749, of 961 forced laborers for the projects;[219] among other things, work is begun on the excavation of the hill of Keravel for the construction of a prison, completed in 1751, while during the 1750s works are completed on the drydocks at Pontaniou, already planned at the end of the 1730s but not finished until 1756-57.[220] In 1764 Choquet de Lindu also builds the second pier of Pontaniou – and in that same year he has repairs done on the foundations of the drydock of Troulan.

"It is the late Mr. Ollivier, chief building engineer of the Navy who corrected the defects of the ancient forms of Brest; he gave the plans and had

*Hydraulique. Seconde Partie* cit., vol. II, 1753, p. 187.

[216] Jean-Rodolphe Perronet, *Mémoire sur les différentes méthodes qui ont été employées pour fonder les ouvrages de Mâçonnerie dans l'eau, & principalement sur celle qui tendent à supprimer les batardeaux & épuisemens dans la construction de Ponts*, in, "Mémoires de Mathematiques et de Physique tirés des registres de l'Académie Royale des Sciences. Année MDCCLXI," published with, "Histoire de l'Académie Royale des Sciences, Année MDCCLXVI," Paris, Imprimerie Royale, 1769, (pp. 139-48), p. 144. The source used by Perronet to describe the work of Milet de Monville comes from Bélidor.

[217] *Ibid.*, p. 144. Rondelet also draws on Bélidor text, describing the works of Milet de Monville as being done with the presence of "encaissemens as those which are done for batardeaux," "by a row of stake and sheet piling" (Jean-Baptiste Rondelet, *Traité théorique et pratique de l'art de bâtir*, Paris, Chez L'Auteur, vol. III, 1814, p. 61).

[218] *Les ports militaires de la France. Notices Historiques et descriptives. Cherbourg, Brest, Lorient, Rochefort, Toulon*, Paris, Challamel Ainé, A. Bertrand, 1867.

[219] Prosper Levot, *Histoire de la ville et du port de Brest*, Brest, Paris, L'Auteur et Bachelin-Deflorenne, 1865, vol. II, p. 104.

[220] For the description of the works done in Brest see, Antoine Choquet de Lindu, *La Description des trois formes du port de Brest, bâties, dessinées et gravées en 1757*, Brest, R. Malassis, 1757, and, *Id.*, *La Description du bagne pour loger à terre les galériens ou forçat de l'arsenal de Brest*, Brest, R. Malassis, 1759.

**174**. Amédée-François Frézier, “Plan de la Ville de Brest,” 1754 (Ms-6459(599), Bibliothèque Nationale de France, Paris).

**175-176**. Antoine Choquet de Lindu, drydocks in the Brest Arsenal (A. Choquet de Lindu, *La Description des trois formes du port de Brest, bâties, dessinées et gravées en 1757*, Brest, R. Malassis, 1757).

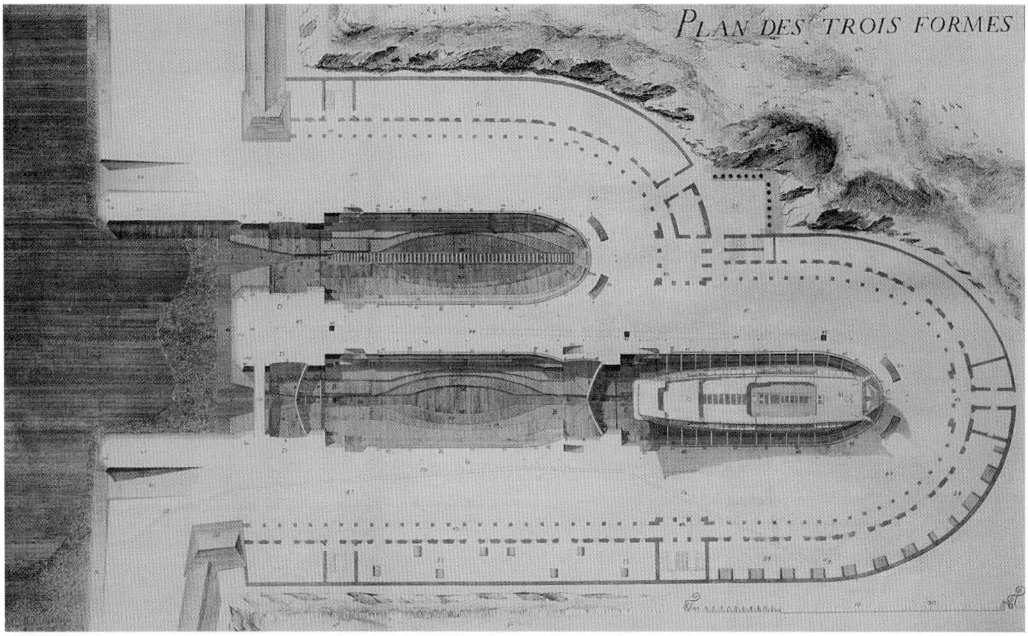
PLAN DES TROIS FORMES

PROFIL DE LA SECONDE FORME
ET
La Vue de la Tête de la Troisieme

Plano de Cartagena enelqual se deseri be el camino del Aqueducto i sus Fuentes
A. Venero ò Manantial delas fuentes.
B. Primer Arca ò Deposito donde se recojen i Juntan las Aguas del Manantial.
C. Segunda Arca ò Deposito.
D. Terzera i ultima Arca.
E. Fuente dela Puente de Sn Roque.
F. Fuente dela Puerta de Murzia
G. Fuente de Sn Sebastian.
H. Fuente de Sn Francisco.
I. Fuente de Sn Agustin
L. Fuente de la Puerta del Mar.
M. Fuente delas Monjas.
N Fuente dela Plaza del Muelle.
O. Fuente dela Cabeza del Muelle.
Capilla de las Galeras
Parte del Puerto
El Castillo
Parte dela Ziudad de Cartagena
Arrabal de Sn Roque
Linea que Demuestra por donde passa la Cañeria
Rambla
Sn Anton.
MARINA-375
MPD, 16, 42

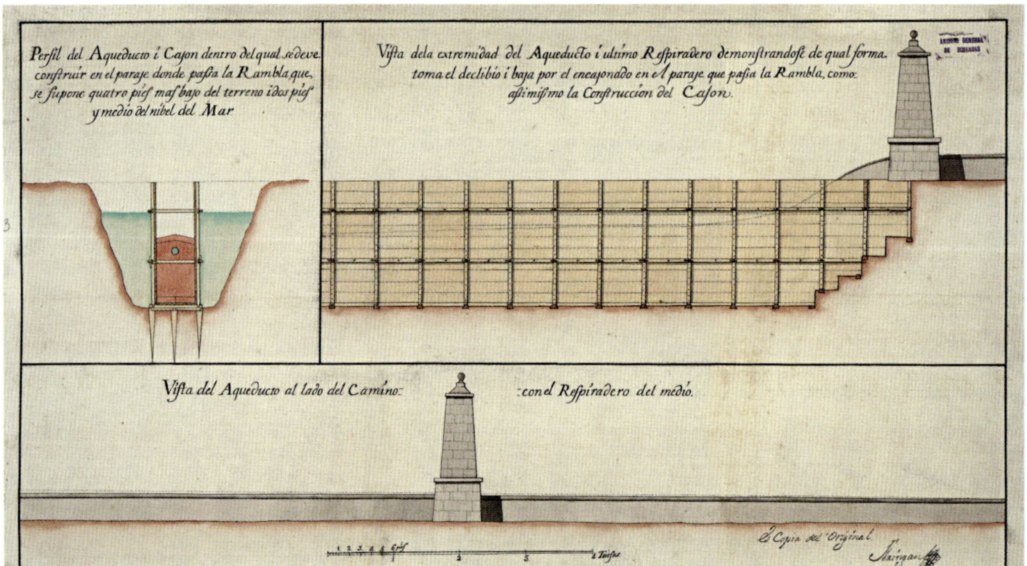
Perfil del Aqueducto i Cajon dentro del qual se deve construir en el paraje donde passa la Rambla que se supone quatro pies mas bajo del terreno i dos pies y medio del nibel del Mar
Vista dela extremidad del Aqueducto i ultimo Respiradero demonstrandose de qual forma toma el declibio i baja por el encajonado en el paraje que passa la Rambla, como assimismo la Construccion del Cajon.
Vista del Aqueducto al lado del Camino con el Respiradero del medio.
Es Copia del Original
SMA- 375
MPD-25, 3

**177-179**. Sebastián Feringan Cortés, project for the aqueduct, Cartagena, 1732 (Ministerio de Educación, Cultura y Deporte, Archivio General de Simancas, Mapas Planos y Dibujos, 16,042).

work started on the three new forms of Brest, by Pontaniou, that Mr. Choquet Lindu had finished after the death of Mr. Ollivier and that one considers, correctly, as one of the best examples of its type." It is significant that these words of praise for the construction at Brest are written by Antoine Groignard (1727-98), the military naval engineer and collaborator of Leonhard Paul Euler (1707-83) involved in the building of ships in Brest, who would become the maker, in the 1770s, of one of the most spectacular floating caissons with pozzolana masonry.[221]

The large quantities shipped from Civitavecchia to Spain, that reached a peak in the 1750s, are intended for the works of the dockyards of the port of Cartagena, beginning in 1732 and continued by the military engineer Sebastian Feringan Cortés (1700-62), and where the drydocks are being built exactly in the 1750s. The pozzolana is also used for the construction of the underground aqueduct, designed in the same 1732 by Feringan Cortés. To build the aqueduct, concrete is poured into wooden formworks.

[221] See, "Etablissement des formes dans le port de Toulon, par Mr. Groignard, Toulon 1774," in, "Antoine Groignard," 41AP1, Archives Nationales, Paris.

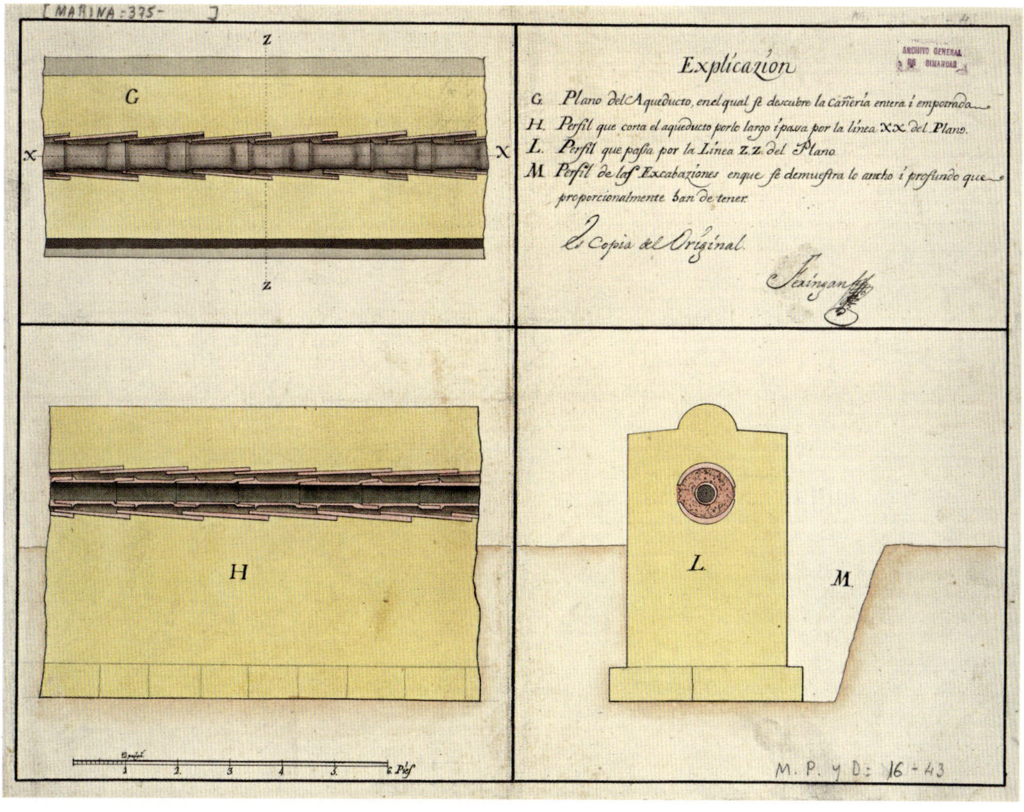

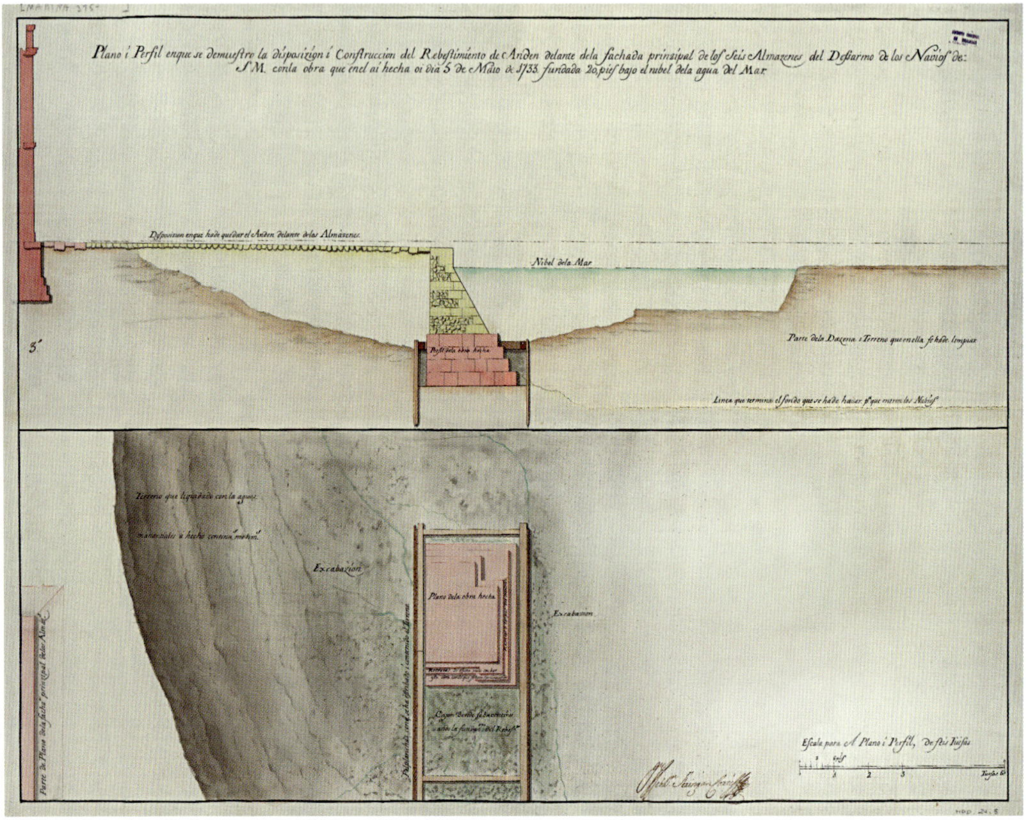


**180-182**. Sebastián Feringan Cortés, project for the arsenal of Cartagena, 1757 (Ministerio de Educación, Cultura y Deporte, Archivio General de Simancas, Mapas Planos y Dibujos, 06,069 and 38,116).

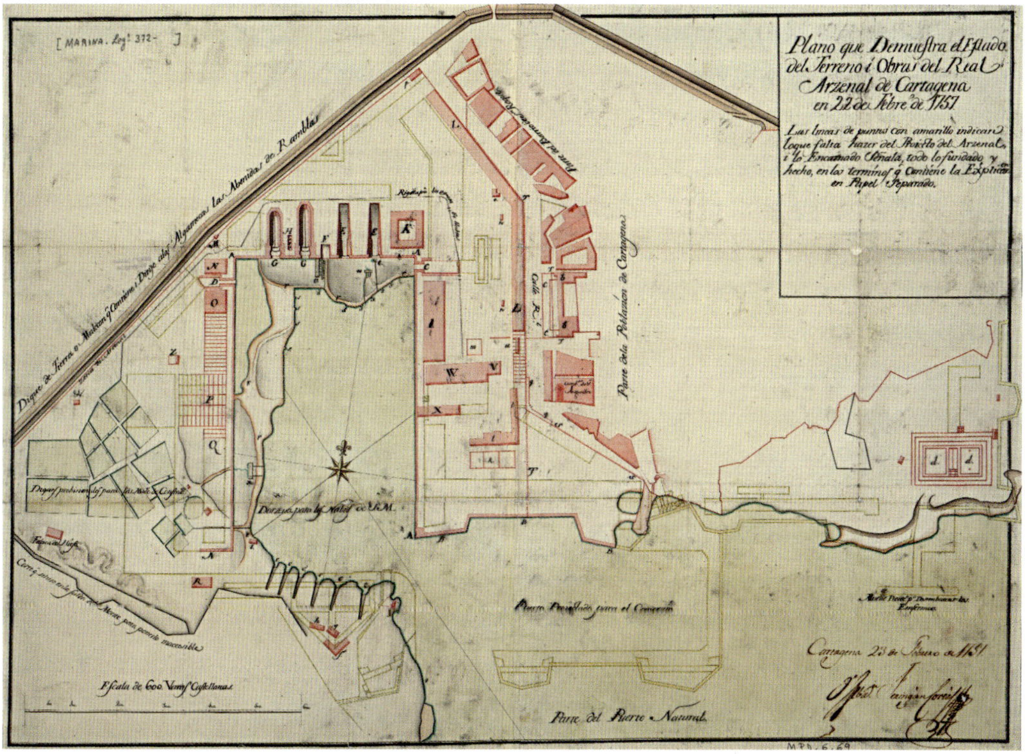
Plano que Demuestra el Estado del Terreno i Obras del Real Arzenal de Cartagena en 22 de Febre de 1751
Escala de 600 Varas Castellanas
Parte del Puerto Natural.
Parte de la Poblazion de Cartagena


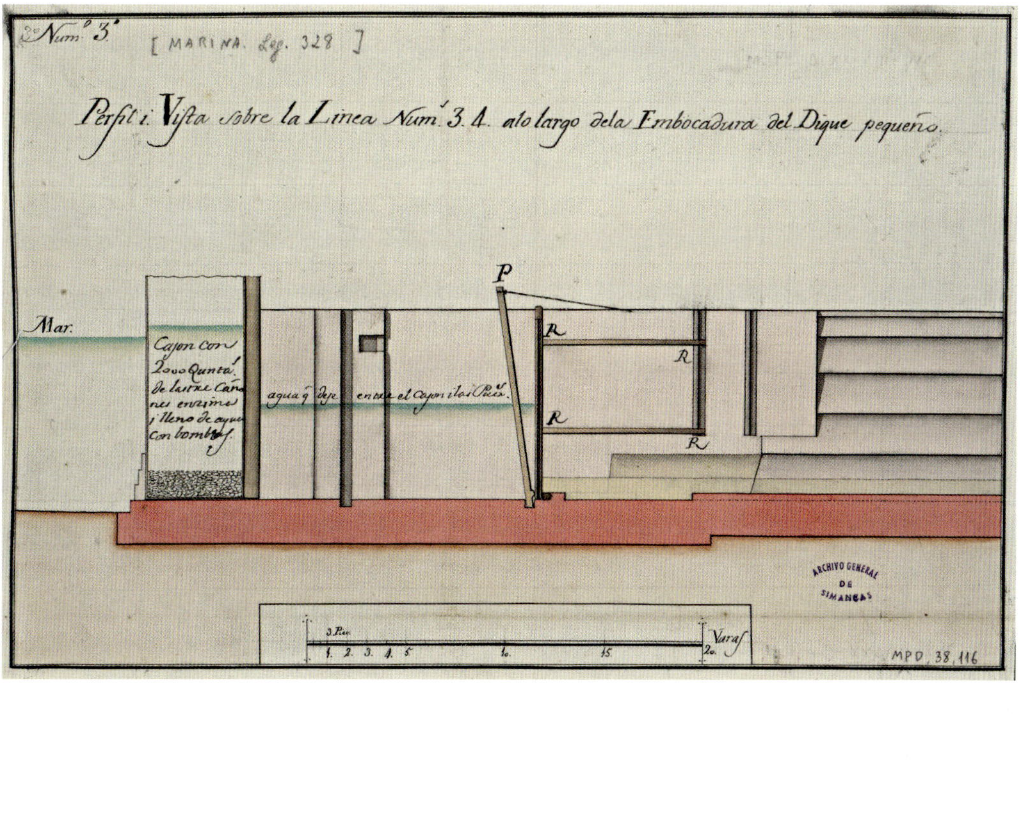
Num.o 3.
[MARINA. Leg. 328]
Perfil i Vista sobre la Linea Num. 3.4. alo largo dela Embocadura del Dique pequeño.
Mar.
Varas
MPD. 38, 116
ARCHIVO GENERAL DE SIMANCAS

### Portable mold for pisé: the contribution of Delorme

The building of conglomerate structures happens by pouring the material into cavities made directly in the ground, in the case of foundations and cellars, or created with a system of boards, or formwork, to erect walls. Starting in the early 1700s travelers report on concrete walls that bear visible signs of the boards of the temporary structure for their pouring, seen in different parts of the world, while treatises lead to the circulation of drawings of formwork, though they are still exceptional items in the literature.

After Fontana's writing about the "dammed-up voids" of the Anzio, Bélidor stands out for having paid particular attention to this aspect of concrete construction, discussing the fabrication of "wooden partitions," or "cases" for the pouring.[222] But in the 1700s formwork use spreads, and not just in Europe, above all for the construction of walls rammed earth, or *pisé*, in rural houses. It is in this context that the system goes through its most important evolution, and the formwork becomes the focus of study of technicians. Memoirs, articles, treatises and, towards the end of the century, patents are devoted to the fabrication of *pisé* walls using wooden formwork. As concrete becomes a technique no longer applied primarily for foundations, it finds an previously evolved system of formwork already at its disposal for the construction of walls.

One of the main contributions to the technical literature on *pisé* is the paper on the construction of earthen walls, *Mémoire pour la construction des murs de terre*, presented to the Académie des Sciences, Belles-Lettres et Arts of Lyon on 17 March 1745 by Guillaume-Marie Delorme (1700-82), an architect, hydraulics expert and archaeology enthusiast.[223] Delorme sums up, in practical recommendations, the current practices for this type of construction that is very widespread in the countryside around Lyon: he explains the empirical methods to recognize the best earth for the *pisé*, namely the soil that conserves its form when packed between the hands (however he nevers uses the term *pisé*); he outlines how to build the *encaissement* of wood, or *moule portatif* (portable mold), a type of "long case without bottom," into which the earth is to be packed, how to build the mould with four iron bolts and the way to raise it to make the various layers of the wall (for the planks, he recommends fir wood because of its lightness); he specifies the best time of year to build a work in *pisé*, the number of workers needed, how to apply stucco, or "*rustiquer*," the earthen walls so they will last for centuries, and the operative modes for the demolition of *pisé* structures. It is no coincidence that Delorme becomes the author, little more than a decade after the *Mémoire pour la construction des murs de terre*, of an essay on the construction of the Roman aqueducts, containing the first and most accurate description of the ancient methods of construction with formwork.

Halfway through the 1700s, treatises on agricultural economics and the construction of houses and other buildings in the country begin to include succinct descriptions of the making of earthen structures. In the reprints after

[222] Bélidor, *La Science des Ingénieurs* cit., 1729, vol. I, *Livre III*, p. 50.

[223] Guillaume-Marie Delorme, *Mémoire pour la construction des murs de terre*, Académie des Sciences, Belles-Lettres et Arts de Lyon, 17 March 1745 (Ms PA 194, Delandine, 959, Mémoires d'architecture, Fol. 61, Académie des Sciences, Belles-Lettres et Arts, Fonds de la Bibliothèque du Palais des Arts, Lyon). *Cf.* Antoine-François Delandine, *Manuscrits de la Bibliothèque de Lyon, ou Notices sur leur ancienneté, leurs auteurs [...]*, Lyon, F. Mistral, 1812, vol. II, p. 446; Jean-Baptiste Dumas, *Histoire de l'Académie royale des sciences, belles-lettres et arts de Lyon*, Lyon, Gilberton et Brun, 1839, vol. I, p. 257. The author thanks Pierre Crépel for allowing him to consult Delorme's memoir.

[224] Louis Liger, *La nouvelle maison rustique*, Paris, Ch. de Sercy, 1700.

[225] Louis Liger, *La nouvelle maison rustique, ou économie rurale, pratique et générale de*

Delorme's contribution of the most well known of these treatises, *La nouvelle maison rustique*, written by the agronomist Louis Liger (1658-1717) and published for the first time in 1700,[224] mention is made of the fabrication of "earthen walls": "To build the wall, the earth is cast little by little into molds or portable cases made of pine boards of two and one-half feet in height, and about twelve feet in length; each layer is rammed and pressed with a sort of club designed for this use. When a mold is filled to the top, it is taken apart to be used again to construct another section of wall in the same fashion."[225] The molds are reutilized and moved each time. In the successive editions of the *Nouvelle maison rustique* also it is explained that the use of pebbles should be limited to the construction of foundations and walls built to border properties or fields – "the small masonry stones are called rubblework."[226] But it is not until the 1770s that the construction of walls with earthen compounds becomes a subject of study in technical manuals, offering significant stimuli for the evolution of the construction methods of concrete.

## Compounds for bitumen sealing, or "spalme-mastic"

The corrosion of wood caused by marine worms has an impact not only on breakwaters made with poles, but also on ships, which have to be protected with products that are spread on the hull. Often the various mastics invented to seal hulls are also used in construction, precisely for their waterproofing properties, as cements to seal the gaps between stones in roofs. It is significant that one of the terms used until the early years of the 1700s to indicate the binder capable of gripping pebbles in a conglomerate is "mastic," and that the various cements invented in the 1770s are also defined as such. Some of those cements are even used in place of mastics to protect the wood of ships.

Towards the end of the 1730s in England, a type of protective seal for wood begins to circulate, both in naval and on-land construction, produced by Alexander Emerton according to a recipe patented in 1737, which includes stone and glass powder, sand and oil.[227] "He prepared the planks or boards with boiling oil in the then old way, and afterwards coated them with compounded poisons, powdered glass, and sand, cemented with painting colors and oils, laid on as paint [...]."[228]

In the area of the research on mastics, variants of the tar, or *goudron*, and the hull coating, or *courroi*, an outstanding substance is the one based on pitch, to apply when hot, invented in 1718 to protect hulls by Jean Maille (? -1754), and tested by him simultaneously in France and Holland – the *spalme*, from the term "espalmer" which in shipbuilding jargon means coat the hull ("enduire de courroi").[229] After various attempts in Dutch ports and at Rouen, Maille is able to found the first large *spalme* factory in 1751, at Carriere-Saint-Denis, near Chatou – in 1753 the factory is run by his son-in-law, Besnard, and then moves to Port du Pecq, near Saint-Germain-en-Laye.[230]

*tous les biens de campagne*, Paris, Knapen, 1762[8], p. 51. The explanation is not included in the edition of 1708 (*cf.* Louis Liger, *La nouvelle maison rustique*, Paris, Cl. Prudhomme, 1708).

[226] Liger, *La nouvelle maison rustique* cit., 1762[8], p. 28.

[227] Alexander Emerton, *Covering and painting timbers, planks, and boards of ships, yachts, barges, lighters, boats, or vessels; and timbers, planks, boards, and plastering in building, or other things suited to this method*, 13 June 1737, in, Bennet Woodcroft, ed., *Subject-matter index of patents of invention, from March 2, 1617 [...], to October 1, 1852*, London, G. E. Eyre and W. Spottiswoode, 1867, p. 562.

[228] *Prevention of decay and oxidation in ships*, in, "Journal of the Franklin Institute," vol. LXXVIII, July 1864, no. 1, (pp. 61-67), p. 61.

[229] *Exposition des propriétés du spalme, considéré comme courroi, pour la conservation des Bâtimens de Mer: comme enduit, pour celle des bois d'œuvre & des corps en général: comme mastic, pour la jonction des Marbres, des Pierres & des Métaux: avec la maniere de l'employer sous ces trois rapports*, Paris, Le Breton, 1763, p. 4.

[230] The *spalme* is sold in barrels at the factory in Pecq, in Rouen at Robert & Louis Durand, and in Le Havre at Louis Dupont.

Though the first and most important applications of *spalme* are for boats, Maille also conducts experiments to demonstrate its value as protection for the beams of houses and bridges, the wainscoting of ceilings, the boards of water basins and even masonry – "the undersides of entablements, of pediments, of plinths, of cornices, and of any structure protruding from buildings made of plaster, stucco, mortar, or any other material less solid than stone."[231] In 1729 in Rouen, *spalme* is successfully used to cover the wood with which toys are made, the chests for orange trees and the planks covering storage facilities.[232] The *spalme* is also used to seal the gaps between the stones of a balcony, and in the 1750s and 1760s to coat water basins and terraces.[233] Its most prestigious applications in the construction are done from 1734 to 1738 by Gabriel in the palace of Versailles, to coat the sheets of water in the Latona fountain and the roof of the chapel (here it is spread over the stones of the vault and the dripstones).[234] Four commissioners of the Académie Royale d'Architecture – Jacques-Hardouin Mansart de Lévy (1703-1788), Louis-Adam Loriot (?- 1767), Claude-Guillot Aubry (1703-1771) and Pierre-François Godot – on 3 May 1751 write down the account of the inspection to examine the condition of the *spalme* seals at Versailles: "they are in a good condition as if just applied, and of better sort and quality than those that have come to their attention until now."[235] In the 1750s and 1760s, *spalme* is publicized as an alternative to "cement" for basins, tanks and cisterns; it is considered so waterproof as to eliminate the construction of double walls to make watertight structures. While in Europe many mastic recipes are in circulation, often publicized as cements to apply to create waterproof coverings for small hydraulic constructions, from cisterns to tanks,[236] *spalme* is even recommended for large hydraulic works – "for moles, dikes, bridge foundation rafts, and buildings of all types" – to use as a replacement for mortar, almost as

[231] *Exposition des propriétés du spalme* cit., p. 39. In the context of bituminous substances used to seal the joints of stones we should also mention a product, or "matière bitumineuse," in use at Gaïac (or Gajac) near Bordeaux, in place of mortar and cement in the construction of flat roofs (minutes of the session of 4 May 1711, in, Lemonnier, *op. cit.*, vol. III, 1913, p. 356). See also what is stated on roof terraces in stone (*cf.*, minutes of the session of 4 April 1707, *ibid.*, p. 269).

[232] *Exposition des propriétés du spalme* cit., pp. 43-45.

[233] *Ibid.*, pp. 54 and 57.

[234] *Ibid.*, p. 59.

[235] *Ibid.*, p. 61.

[236] *Secrets concernant les arts et métiers. Nouvelle edition*, Bruxelles, La Compagnie, 1758, vol. I, p. 66. Various recipes for mastic and cement are published, including the "Cold cement for cisterns and fountains": "Take litharge and bowl powder of two pounds each, resin and yellow earth of each four ounces, five ounces of mutton tallow, two ounces each of putty and turpentine, enough walnut oil to make it manageable: we need then to knead it all together, and then use" (*ibid.*).

[237] *Ibid.*, p. 68.

[238] [Anon.], letter, 11 July 1763, B18, Archives de l'Académie Royale d'Architecture, Institut de France, Paris.

if it were a substitute for cement made with pozzolana or *trass*. "[...] it will unite with stone," it is written about *spalme*," it will give to wood shims the durability of bronze, and thereby render monuments eternal."[237]

The cases of possible application of *spalme* in construction have to do with the large monuments of Paris and Versailles, which at the end of the 1750s are in a severe state of decay due to water seepage from the flat roofs, terraces and vaults. It is significant that these very monuments become the sites of the first experiments on the new cements made in the Roman way. One report indicates that in 1763 a request was made to try using *spalme* to correct water seepage in the Louvre.[238] The brochure that explains the qualities of *spalme* describes the state of conservation of the monuments of Paris and Versailles, and the failures of cement products to protect masonry from infiltration of rain water:

"It was necessary to rebuild the terrace of the Château of the Tuileries, and to recently waterproof the slabs of the Château of Luxembourg, with a compound that has not entirely stopped the infiltration of the waters through the joints of their vaults; and we will soon be confronted with the ruin of the Observatory, and that of the Orangerie of Versailles, where we need find a mastic that, becoming integrated with the mass, can prevent the humidity from penetrating and destroying them. Such a mastic, if it exists, will necessarily have an adhering property that is now missing with asphalt and the other mastics; and its durability will necessarily be the same of the slabs that form the surfaces of the terraces, aqueducts and sewers; and necessarily the same as the stones that constitute the cornices, plinths, steps, balconies, and all the other protruding parts of buildings; the walls of the vaults and of the docks, and in general of underground or immersed walls."[239]

[239] *Exposition des propriétés du spalme* cit., pp. 52-53. *Cf. Exposition des propriétés du splame, considéré comme courroi, pour la conservation des bâtimens de mer: comme enduit pour celle des bois d'œuvre & des corps en général: comme mastic pour la jonction des marbres, des pierres & des métaux*, Paris, Le Breton, 1758; *Traité du spalme, comme courroy et comme mastic, réduit à ses justes propriétés pour la marine & pour les Bâtimens de terre, avec la maniere de l'employer à l'un & à l'autre usage*, Paris, A, Boudet, 1769. On the proposals for the restoration of the surface of the roof terrace of the Paris Observatory with small brick vaults on which to rest the stone slabs, see, Jacques-Germain Soufflot, report, 8 May 1767, O.1.1554, Archives Nationales, Paris. *Cf.* also, "Extrait des devis faits en 1751 pour les reparations de la terrasse de l'Observatoire," O.1.1554, Archives Nationales, Paris. See also the entry *Terrasse de bâtimen*, in, Noël Chomel, Jean Marret, Pierre Roger, *Dictionnaire œconomique, contenant divers moyens d'augmenter son bien, & de conserver sa santé*, Commercy, H. Thomas, 1741.

ttoni quadrati segati a linee diagonali per formare i Mattoni triangolari. Di questi mattone tria
ostruiti i muri di Portici dell'Ustrino; siccome ancora la maggior parte delle Fabbriche antiche. B Muro d
strino, composto de detti mattoni triangolari a riserva degli angoli, i quali sono formati di mattoni
drati. Di dentro egli è riempiuto di calce, pozzolana, scaglie, e pezzi di mattone; posta ogni cosa sempre
in piano. C Superficie formata dai lati maggiori de mattoni triangolari. D Superficie formata da
Tufi tramezzati da altrettanti corsi di mattoni. E Muro della Fabbrica suprema del M

# chapter four
# apologia of roman construction, from soufflot to piranesi to winckelmann

## Research on pozzolana in Italy, from Nollet to Soufflot

The 1750s witness the circulation of important knowledge on the nature of pozzolana-based mortars, formwork and the technical potential of Roman cement construction. Because the supposed discovery, halfway through the 1760s, of the composition of Roman cement follows so soon after the spread of observations and reports of an archaeological character, it would appear that this discovery is the logical consequence of these reports, and that the two developments represent steps in the same renaissance of cement in the 18th century.

The sequence of fundamental events leading to deeper international comprehension of hydraulic cements produced according to historical methods begins with the trip to Italy of Jean-Antoine Nollet, known as Abbé Nollet (1700-70), a physicist and member of the Académie Royale des Sciences of Paris, known for his studies on electricity, and with the journey of Jacques-Germain Soufflot (1713-80), accompanied by Charles-Nicolas Cochin (1715-90) and of Abbé Jérôme-Bernard Le Blanc (1706-81), in the entourage of Abel-François Poisson de Vandières (1727-81), future Marquis de Marigny.

During his voyage in Italy in 1749, Nollet is able to investigate certain aspects of construction, such as the fabrication of limes and mortars, and the method of pouring vaults.[1] He makes the his most detailed observations in Turin and Rome. Nollet continues to Naples, where he climbs the slopes of Mount Vesuvius in the company of Giovanni Maria Della Torre (1712-82), a Neapolitan theologian and amateur naturalist, author of many essays on Vesuvius and very familiar with local construction practices.[2]

During his stay in Piedmont, Nollet gathers information on the making of the so-called “strong lime” by “architects & other craftsmen” – above all thanks to discussions with Isaac-François Matthey, from Switzerland, *mecanico* to the King of Sardinia.[3] With that type of lime, obtained by firing pebble-like stones taken from the mountains around Turin, between Moncalieri and Casale, hydraulic mortars are made (mixing the "strong lime" with silt from the Dora River, which Nollet examines under the microscope to understand its composition). These mortars are used to build bridges or dams, and to make the artificial blocks, or *cantoni*, to position along rivers to strengthen their banks and to control the flow of the water.[4] Also in Piedmont, the same mortars are used to coat brick walls, obtaining a facing that Nollet describes as durable in time and perfectly white, thanks to "a thin layer of slaked lime [*chaux douce*], mixed with sieved white marble powder" – "I've seen it last more than one hundred years under the worst outdoor conditions."[5]

It is worth re-reading Nollet's painstaking explanation of the criteria applied in Piedmont to slake the "strong" lime, because it contains previews of certain fabrication processes in France for a similar hydraulic cement.

"One makes onsite, with sand or else," he reports to the assembly of the Académie Royale des Sciences in 1749, "a flat basin the size of which is proportional to the quantity of lime to be slaked each time: a hole is dug, not far away, that is large enough to hold anything that has to be slaked in several times & to be readily used. Some lime pieces are then thrown into the basin and sprayed with water as evenly as possible, and in such a way that all the water is absorbed and there is no more water apparent in the basin: several minutes later, when the lime pieces start to split open they are sprayed again in the same way and this is repeated as often as possible, all the while making sure that the lime pieces neither turn into dry powder, nor conversely, drown in water while they are still in chunks. Once the spraying has sufficiently opened and divided the lime, then as much water as necessary is poured over the pieces so that it takes on a light mud-like liquid nature, and when it has been stirred and beaten like ordinary lime, to complete its dissolution, the basin is opened and a specially made chute directs the liquid into a ditch below. One must not slake more lime than what can be used in one month, because the longer it is kept in the ditch, the harder it becomes [...]. While watching the lime being slaked, I noticed that it produces a lot of heat, and much more, according to appearances, than slaked lime [...]."[6]

Besides the strong lime of Piedmont, Nollet is interested in the various types of pozzolana, from the ash-gray version of Naples to the reddish material from Rome – both "look like crushed stone, quite similar to our cement."[7]

[1] The rediscovery of concrete construction for the making of large vaulted structures also depends on the possibility of erecting solid wooden centering onto which the concrete can be poured. Though the first large vaulted concrete structures are made only starting in the first decades of the 1800s, it is in the first years of the 1700s that centering becomes a fundamental chapter in treatises, studies and teachings on engineering. Until that moment, in general, the question of framework for the centering of vaults and domes had been neglected, and the rare drawings in circulation in the treatises are those spread by Mathurin Jousse and François Blondel. The calculation of the strength of the centering, including the dynamic and variable weight of the masonry that was progressively added for the construction of the arch of a bridge or the dome of a church, is studied in 1726 by Henri Pitot (1695-1771). *Cf.* Henri Pitot, *De la force qu'il faut donner aux cintres dont on se sert dans la construction des grands voutes, des arches des ponts, &c.*, in, "Histoire de l'Académie Royale des Sciences. Année MDCCXXVI. Avec les Mémoires de Mathématiques & de Physique, pour la même Année," Paris, Durand, 1753, pp. 216-36. Pitot's report is dated 6 July 1726. In the case of poured construction, the premises exist for the construction of framework, both as examples of those for masonry domes and vaults, and in calculations to scientifically establish their strength, starting with Pitot. Pitot also warns that in order to exactly determine the strength of the framework it is necessary to know the

As in Piedmont, in Rome Nollet talks with workmen at construction sites and sees the structures that can be built with mortars based on strong lime and pozzolana. During one of these worksite visits, he observes all the phases of the construction of a vault in concrete, from the preparation of the centering to the pouring of the compound and the laying of the final layer of stones. Nollet's accurate description is one of the precious documents regarding the survival, in building traditions, of ancient methods that had been gradually modified over the course of centuries.

"Pozzolana mixed with strong lime," Nollet writes, "creates a mortar that is much more reliable than our own: the most striking example that I can cite is the way I saw vaults being built in Rome. A convex platform is made using boards supported by wooden pillars: then a six to seven inch thick vault is built on this sort of scaffolding, using the mortar I just mentioned and raw stones placed any which way, without any regard to their shape: five or six days later, the board structure that is underneath is removed, and the vault is found to be solid. The stones that are used for these vaults are from a soft and light rock found everywhere around the city where it is gathered and is known as *mollasse* [soft sandstone]: its color is very close to pozzolana, one would easily believe that these are portions of sand bound together by a sort of stone sap [*suc pierreux*]."[8]

It is no coincidence that Nollet's interests during his trip to Italy also include the so-called Venetian *terrazzo* – "a sort of plaster that is well-polished and almost as hard as marble," which some spread in such a way as to create designs and colors "like the most beautiful of carpets."[9] "These floors are made out of powdered marble, well-sieved cement, strong lime and gypsum soaked with, some say urine, while others say only with water," he explains, adding that the compound must be beaten with great force, and then treated with linseed or walnut oil to make the surface shiny and waterproof.[10]

Looking ahead, we can see that the itinerary and the construction materials observed by Nollet in Turin, Venice, Rome and Naples outline a sort of map of the questions on Italian 18th-century cement mortars that is confirmed, later, by one of the greatest construction theorists, Rondelet.

A fundamental step in the rise of cement during the 1700s is the trip to Italy of Soufflot, from December 1749 to December 1750, in the retinue of Poisson de Vandières considering the roles both would play after their return to France: as the prestigious, authoritative architect to the king, for the former, working on the most important projects, and as superintendent of the Bâtiments du Roi, for

materials of the vault or dome (not with reference to the pouring, but only to the stones). For the centering of bridges during the early 1700s, another important point of reference is offered by the three drawings contained in the treatise of Mathurin Jousse published at the start of the century (Mathurin Jousse, *L'art de charpenterie*, Paris, T. Moette, 1702, pp. 154-56).

2 Jean-Marie Della Torre, *Histoire et phénomène du Vésuve*, Paris, J.-T. Hérissant, 1760, p. 28. "This is done in Naples," Della Torre writes, "with large slag and pumice stones for vaults in houses where the walls are not walls in tuff, but with stones that can support the weight of these materials; and they are so well bound with lime and sand that it would be very difficult to separate them" (*ibid.*, p. 237).

3 Jean-Antoine Nollet, *Maçonnerie & Architecture*, in, *Id.*, *Expériences et observations faites en différens endroits de l'Italie*, in, "Histoire de l'Académie Royale des Sciences, Année MDCCXLIX, avec les Mémoires de Mathématique & de Physique, pour la même Année," Paris, Imprimerie Royale, 1753, (pp. 444-88), pp. 473-83, p. 473.

4 "With this mortar, one can very solidly build anything; its greatest merit resides in the fact that it can be accomplished in wet areas and even under water, in such a way that after a certain amount of time it becomes one with the hardest of quarry stone, and it stands up to even the wrecker's chisel. It is a big help in this country for building bridges, embankments, and masonry assemblages that are used to deter corrosion and river erosion as well as protect against all of the water that comes down the mountains" (Nollet, *Maçonnerie & Architecture* cit., p. 478).

5 *Ibid.*

6 *Ibid.*, pp. 476-77.

7 *Ibid.*, p. 479.

8 *Ibid.*

9 *Ibid.*, p. 481.

10 *Ibid.*, p. 482.

**183-184**. Jacques-Germain Soufflot, Gabriel-Pierre-Martin Dumont, survey of the crater of Mt. Vesuvius, 1750 ([G.-P.-M. Dumont], *Les ruines de Paestum, autrement Posidonia, ville de l'ancienne Grande Grece, au Royaume de Naples [...]*, Londres, Paris, Ch.-A. Jombert, 1769).

the latter, from 1751 to 1773 as "General Director for buildings, gardens, arts, academies, and royal works," also in charge of the work of Soufflot. Soufflot and Poisson de Vandières were to become leading figures, though in different ways and with different forms of expertise, in the early French experiments on cement made in the Roman way, as observed during their trip to Italy. Soufflot's Italian voyage becomes famous for his surveys of the temples of Paestum, drawn by his companion on these excursions, Gabriel-Pierre-Martin Dumont (1720-91), drawings that contribute to the international spread of information of architecture in the Greek manner. Indeed, the journey familiarizes Soufflot and Poisson de Vandières with the materials and techniques for the fabrication of cement in the Roman way, both ancient and contemporary. The question of pozzolana – already raised in the observations of Nollet – lies behind Soufflot's particular interest in all the volcanic products he is shown during his stay in Italy. It is worth emphasizing, to illustrate the interest of the French naturalists, that Soufflot makes an excursion on 9 June 1750 to Mount Vesuvius, to study and survey the structure of the volcano and the formation of the lava, accompanied by Dumont and Jérôme-Charles Bellicard (1726-86), architect-in-residence at Villa Medici (Dumont publishes the section on the volcano in 1764).[11] Such excursions had been fashionable since the first decades of the century, although not always with the scientific intentions of Soufflot.[12] In November Poisson de Vandières completes the observations on Vesuvius, by request of Soufflot himself.[13] The survey of the mouth of the volcano is done by Soufflot with the participation of Alexis-Jean-Eustache Taitbout de Marigny de Fontanelle (1695-1778 circa), consul of France in Naples, and De Lair, his chancellor. Confirmation that Vesuvius is the subject of study by the group of French travelers is found in an etching showing the volcano and the observations of the lava done by Cochin, at the start of his book on Herculaneum and Pompeii.[14] Soufflot's sojourn in Viterbo, where he is forced to remain for reasons of health, also offers him a chance to study volcanic stones used in local construction. Among other things, Soufflot recognizes the volcanic origin of the flows containing leucite at Capo di Bove, the site of the tomb of Cecilia Metella on the Appian Way, and reports his finding to the Roman physician Giovanni Girolamo Lapi.[15]

It is not possible to precisely document Soufflot's role in French research on pozzolana-based mortar, but many clues indicate that he is one of the main

[11] See, Georges Bunel, ed., *Piranèse et les français, colloque tenu à la Villa Médicis, 12-14 mai 1976*, Roma, Edizioni dell'Elefante, pp. 43-66, and, Gilles Chomer, *Le second voyage en Italie (1750)*, in, Michel Gallet, Daniel Ternois, ed., *Soufflot et son temps, 1780-1980*, Paris, Caisse Nationale des Monuments Historiques et des Sites, 1980, p. 44. About the surveys of Vesuvius done by Soufflot, see, [Gabriel-Pierre-Martin Dumont], *Les ruines de Paestum, autrement Posidonia, ville de l'ancienne Grande Grece, au Royaume de Naples [...]*, Londres, Paris, Ch.-A. Jombert, 1769, plates 15, 16.

[12] See, Elisabeth Chevalier, *Les voyageurs du XVIIIe siècle et le Vésuve*, in, Lucette Desvignes, ed., *Travaux comparatistes*, Saint-Etienne, Publications de l'Université de Saint-Etienne, 1978, pp. 55-79.

[13] *Cf.* Gallet, Ternois, *op. cit.*, p. 50.

[14] *Cf.* Charles-Nicolas Cochin, Jérôme-Charles Bellicard, *Observations sur les antiquités d'Hercolanum*, Paris, Ch.-A. Jombert, 1754. For the description of the excursions on Mount Vesuvius and the observations on the lava, see, *ibid.*, pp. 1-7. On the "Agrippa reservoir, locally know as the admirable pool," at Capo Miseno, Cochin describes the cladding of the reservoir, whose composition is discussed by many – as to whether it is "a mastic bound with egg white," or a "deposit made by the water." "[...] it seems to me," is his position, "that it truly was a cement composed of marble powder and sand from the country, with some ingredients that are unknown to us. Whatever is the case, this coating, only two lines thick (4.5 mm), and with distinguishably different layers, is so perfectly hard that even iron and steel can barely penetrate it" (*ibid.*, pp. 80-81)

[15] *Cf.* Charles-Marie de La Condamine, *Extrait d'un journal de Voyage en Italie*, in, "Histoire de l'Académie Royale des Sciences, Année MDCCLVII, avec les Mémoires de Mathématique & de Physique, pour la même Année," Paris, Imprimerie Royale, 1762, (pp. 336-410), p. 382, and, Francesco Rodolico, *Esplorazione naturalistica dell'Appennino*, Firenze, Le Monnier, 1963, p. 344. See, Giovanni Girolamo Lapi, *Discorso intorno l'origine de' due laghi Albano, e Nemorense*, Roma, Accademia dei Quirini, 3 September 1758 (*cf.* Emmanuele Lucidi, *Memorie storiche dell'antichissimo municipio ora terra dell'Ariccia, e delle sue colonie Genzano, e Nemi*, Roma, Lazzarini, 1796, p. 78), and, Giovanni Girolamo Lapi, *Della selce Romano. Ragionamento mineralogico*, Roma, Salomoni, 1784.

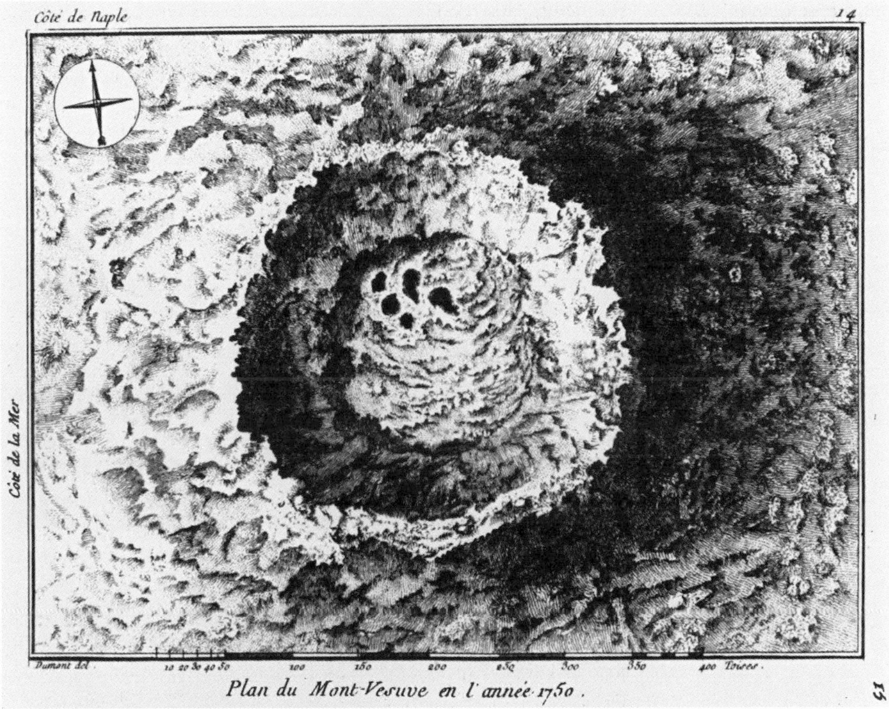
Côté de Naple
14
Côté de la Mer
Dumont del.
10 20 30 40 50 100 150 200 250 300 350 400 Toises.
Plan du Mont-Vesuve en l'année 1750.
15

1700. Pieds
Coupe du Mont Vesuve à deux lieues ½ de Naples, tel qu'il étoit le 9. Juin 1750. Parcouru et mesuré par M.M. Soufflot et Dumont.
Au mois d'Octobre 1751. la Montagne s'entrouvrit et vomit une quantité prodigieuse de Lave.
Germain Sculp.
16

promoters of French experimentation on cement. It is true that the inventors of 18th-century cement, from Antoine-Joseph Loriot de la Rivière (1716-82) to Barthélemy Faujas de Saint-Fond (1741-1819), are often in close contact with Soufflot, and ask him for information on pozzolana and its manufacturing methods as used to make cement in Naples. It is likely that Soufflot is interested in the Neapolitan way of constructing pavement in view of the works to be executed in France, where the fashion of flat Italian-type roofing is on the rise, making the question of techniques for the building of waterproof terraces a crucial one. Also in the context of interest in pozzolana and its applications, we should note that the group of French travelers also visits the Lazzaretto of Vanvitelli in Ancona, "built in the sea."[16]

Poisson de Vandières, following Soufflot's lead, also studies with great interest Italian construction and takes notes on pozzolana, lapilli and their applications in Neapolitan building, for terraces and cisterns. These notes are later transmitted to the first French discoverer of what would be considered the recipe for Roman cement, Loriot. In the context of the discoveries of ancient and contemporary cement construction techniques, it is significant that Poisson de Vandières visits the workshop of the greatest admirer of Roman construction, Piranesi, to purchase etchings.[17]

**185**. Jacques-Germain Soufflot, nymphaeum, at the La Rivette estate of Jean-Baptiste Pitra, Caluire-et-Cuire (J.-M. Pérouse de Montclos, *Jacques-Germain Soufflot*, Paris, Editions du Patrimoine, 2004).

After his return to France Soufflot makes a significant contribution to the knowledge of volcanoes and their products. At the Académie des Sciences, Belles-Lettres et Arts of Lyon he delivers a lecture on Vesuvius, on 12 April 1752 – *Diverses remarques sur l'Italie. Etat du Mont Vésuve dans le mois de juin 1750 et dans le mois de novembre de la même année* (Diverse remarks about Italy. The state of Mount Vesuvius in the month of June in 1750 and also in the month of November of the same year)."[18] The "Mercure de France" of September 1753 publishes the main observations of Soufflot on Vesuvius and advises that the *Diverses remarques* contain "everything one would like to know about materials liquified by fire, what Neapolitans call lava."[19] In December "Mercure de France" also publishes the continuation of the *Diverses remarques* of Soufflot, with his observations on the geological formations of the hot springs of Bollicame, near Viterbo, where the water produces artificial prismatic petrifactions, used for construction and to make lime.[20] The observations published by "Mercure de France" are discussed by the most important volcanologists, and Soufflot becomes an authority for the naturalists specializing in volcanology.[21]

A construction made with the methods of the ancient Romans, in France, seems to indicate an interest in conglomerate walls that may have depended precisely on Soufflot's trip to Italy. It is a nymphaeum with a Borromini-like profile, built on the La Rivette estate of Jean-Baptiste Pitra (1695-1756), at Caluire-et-Cuire, and attributed to Soufflot, where the customary motif of the ruin is staged with faux-decrepit masonry made of layers of bricks, stones similar to *tuff* and *peperino*, and cement with Saône gravel. The layers of cement are done with the Roman technique, demonstrating that the craftsman has direct knowledge of the ruins of Rome. This nymphaeum might even be considered one of the first applications of the nascent taste for Roman concrete construction.

## Studies of Guettard on the pozzolana of Auvergne

The increasingly widespread knowledge of the characteristics of Roman cement leads international research to focus on pozzolana, which is recognized, also thanks to ancient literary sources, as the fundamental component of the solidity and durability of Roman constructions. The idea of Colbert, developed in the wake of the proposals of Bernini for the Louvre, of conducting explorations in France to find reserves of pozzolana, leads to no immediate concrete results. But the reports on pozzolana and various cements written after travels in Italy by Nollet, Poisson de Vandières and Soufflot bring Colbert's question back to the fore.

[16] Charles-Nicolas Cochin, *Voyage d'Italie, ou recueil de notes*, Paris, A. Jombert, 1758, vol. I, p. 96.
[17] For the purchases of De Marigny of etchings by Piranesi in 1750, in Rome, see, Francesca Lui, *Des ruines aux bibliothèques. Piranesi e il mondo francese: fortune, suggestioni, eredità*, in, Mario Bevilacqua, Mario Gori Sassoli, ed., *La Roma di Piranesi. La città del Settecento nelle Grandi Vedute*, Roma, Artemide, (pp. 79-90), pp. 86-87.
[18] *Cf.* Gallet, Ternois, *op. cit.*, p. 50. For the manuscript see, Ms 136, folios 2-12, Bibliothèque de l'Académie des Sciences, Belles Lettres et Arts, Lyon. To the manuscript is attached a "map of Mount Vesuvius's summit in June, 1750."
[19] *Diverses Remarque sur l'Italie. Etat du Mont Vesuve, dans le mois de Juin 1750, & dans le mois de Novembre de la même année*, in, *Assemblée publique, de la Société Royale de Lyon, du 19 Avril 1752*, in, "Mercure de France," vol. II, September 1753, (pp. 25-27), pp. 26-27.
[20] *Assembée publique de la Société Royale de Lyon, du 7 Décembre 1752*, in, "Mercure de France," vol. II, December 1753, pp. 7-29.
[21] *Cf.* Jean-Etienne Guettard, *Neuvième mémoire. Sur la Minéralogie de l'Italie*, in, *Id.*, *Mémoires sur différentes parties des sciences et arts*, Paris, L. Prault, 1768, vol. I, (pp. 347-439), p. 379.

**186**. Jean-Etienne Guettard, volcanic rock ("Histoire de l'Académie Royale des Sciences, Année MDCCLII, avec les Mémoires de Mathématique & de Physique, pour la même Année," Paris, Imprimerie Royale, 1756).

**187-188**. Jean-Etienne Guettard, volcanic rocks ("Histoire de l'Académie Royale des Sciences, Année MDCCLIII, avec les Mémoires de Mathématique & de Physique, pour la même Année," Paris, Imprimerie Royale, 1757).

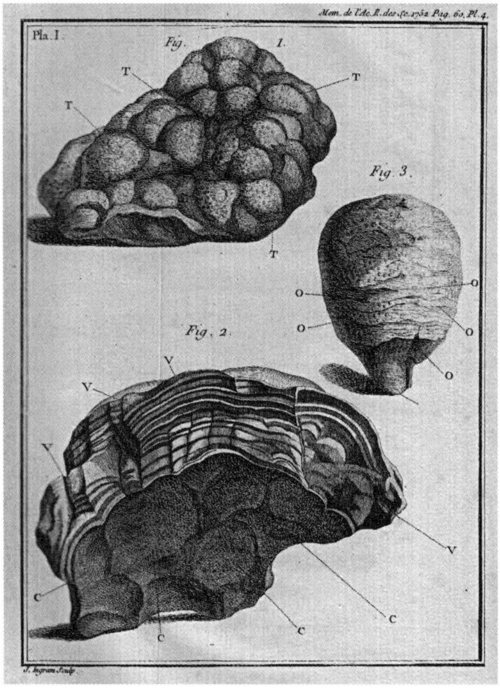


The discovery of the existence, in the central-southern Auvergne region of France, of deposits of volcanic materials dates back to the first explorations done in 1751 along the volcanic chain known as Chaîne des Puys, near Clermont-Ferrand, by the naturalist Jean-Etienne Guettard (1715-86), a member of the Académie Royale des Sciences of Paris, along with Chrétien Guillaume Lamoignon de Malesherbes (1721-94).[22] The discovery was driven by the pressure to find construction materials similar to those of Italy, as well as by the resumption of major volcanic activity in different regions of the planet. At Moulins and then at Riom, Guettard observes for the first time, in the masonry of walls, the use of a volcanic stone similar to the lava of Vesuvius, which he has seen in samples from the mineral collection of the Duke of Orléans, as he states on 10 May 1752 at the Académie Royale des Sciences, while reporting on his journey.[23] Guettard observes that the stones of Mont-Dore are "of the same type of stone from Viterbo, Italy,"[24] probably referring to the observations of Soufflot, and demonstrates that the stones of Auvergne are of volcanic origin by comparison with the samples of stones from active volcanoes made available by various personalities close to him. "Making this comparison is easy for me," Guettard writes, "thanks to having material spewed from Vesuvius, which I received from Mr. de Montigny and from Mr. l'Abbé Nollet, as well as those from volcanos on the Bourbon Islands (La Réunion), sent to me by the late Mr. Lieutaud, Surgeon for the Compagnie des Indes."[25]

[22] *Cf.* François Ellenberger, *Précisions nouvelles sur la découverte des volcans de France: Guettard, ses prédécesseurs, ses émules clermontois*, in, "Histoire et Nature," 1978, nn. 12-13, pp. 3-42.

[23] Jean-Etienne Guettard, *Mémoire sur quelques montagnes de la France qui ont été des Volcans*, in, "Histoire de l'Académie Royale des Sciences, Année MDCCLII, avec les Mémoires de Mathématique & de Physique, pour la même Année," Paris, Imprimerie Royale, 1756, pp. 27-59. *Cf.* also, Jean-Etienne Guettard, *Mémoire sur plusieurs morceaux d'histoire*

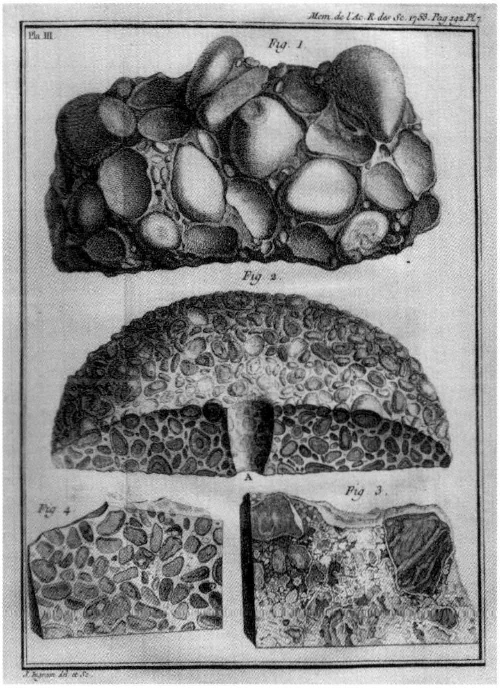


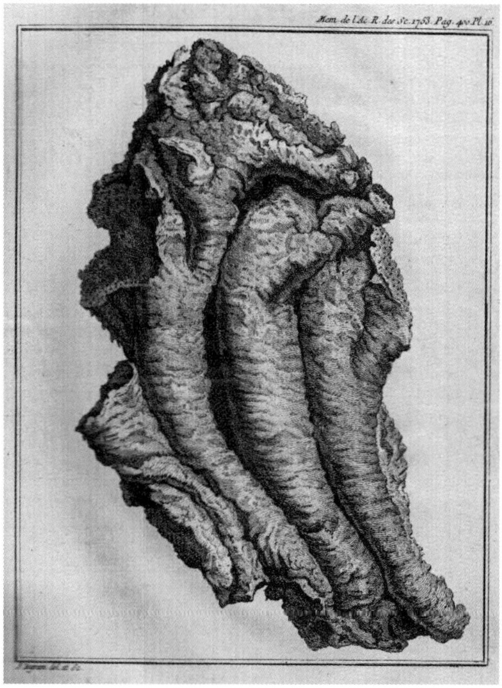

Guettard makes many observations on the use of volcanic stones in the construction of the zones of the Massif Central visited in 1751. The fundamental comparison is still to what has been done in Italy, and in particular near Naples. Guettard even imagines the construction of vaults no longer with the French method of the *coupe des pierres*, but with the Roman method, with pumice and pozzolana-based mortar. Moreover, the formworks and materials in use in Rome for the construction of concrete vaults are described by Nollet, the authoritative friend of Guettard. For the progress of research on pozzolana in France, it is significant that Guettard also identifies deposits of lavic materials similar to Italian pozzolana.

"Those from Riom and other areas surrounding the mountains that burned," he writes on this matter, "would be able to take even greater advantage of pumice stone if they used them in their buildings in the same way as the Italians do: they could be used in vault construction; they are suitable for several purposes; they are light and even more spongy than lava and therefore absorb a greater quantity of mortar; and by consequence, they are more strongly bound to each other, and form a vault that becomes, so to speak, one unique piece and lasts longer than ordinary vaults, especially if the mortar is made with the type of sand called *pozzolana*, which seems to me to only consist of ash and small pumice stones. This pozzolana is readily available in the county, I believe it is found in abundance even in the areas surrounding Riom near Paris; there I observed a dark and ashy terrain, which I believe to be suitable

*naturelle tirés du cabinet de S.A.S.M. le duc d'Orléans*, in, "Histoire de l'Académie Royale des Sciences, Année MDCCLIII, avec les Mémoires de Mathématique & de Physique, pour la même Année," Paris, Imprimerie Royale, 1757, pp. 369-400.

[24] Guettard, *Mémoire sur quelques montagnes de la France qui ont été des Volcans* cit., p. 47.

[25] *Ibid.*, p. 48.

189. Jean-Etienne Guettard, "carte minéralogique de l'Auvergne," 1759 ("Histoire de l'Académie Royale des Sciences, Année MDCCLIX, avec les Mémoires de Mathématique & de Physique, pour la même Année," Paris, Imprimerie Royale, 1765).

for the same uses as pozzolana: it can also be found in territory surrounding the Puy-de-Domme."[26]

In 1759 Guettard updates his observations on the types of stone of Auvergne and prepares the "mineralogical map of Auvergne" with indications of volcanic deposits.[27] The map marks the beginning of a series of explorations fundamental for the search for French pozzolana. Thus, after the discoveries of Guettard, the list of locations for the quarrying of pozzolana is extended, beyond those in Italy and in places under French rule, to include Auvergne.[28]

During that same year, 1759, at the Assemblée des Inspecteurs Généraux des Ponts et Chaussées, Etienne Mignot de Montigny (1714-82), a member of the Académie Royale des Sciences and of the Assemblée – he is the one who shows Guettard the samples of lava from Vesuvius, obtained during his trip to Italy in 1740 – discusses the qualities of pozzolana as compared to other powders used for the fabrication of cements: "It has also been said that the most strongly fired materials are suitable for making the best cement, a reason for which roofing tiles, butter pots, etc. are better than brick in general, which is not heavily fired. Mr. De Montigny has observed, incidentally, that sand from Pozzolane, so renowned in Italy for making good mortar, is apparently composed of stone and pumice debris: it is perhaps for this reason that ashes, in general, and especially those from Tournay, Boulogne and also ashes from coal, are highly reputed for making mortar: their mineral salts could also play a large role."[29]

## Concrete and "architecture hydraulique" of Bélidor

The role of Bélidor in the growth of concrete use does not end with the fundamental indications he provides in *La Science des Ingénieurs* and *Architecture Hydraulique.* Another book in his *Architecture Hydraulique* series, published in 1753, contributes to spread word of the most recent practices of fabrication of the compound and the use of formwork – though only regarding foundations in seas and rivers – in a crucial moment for the evolution of concrete, when the interest of naturalists, architects and engineers is concentrated precisely on the properties of the various pozzolanas of Italy and France.[30] His pages on concrete circulate all over Europe and become a point of reference for authors of manuals on foundations and concrete published until the early decades of the 1800s. In no other treatise published prior to that of Bélidor is it possible to find a detailed description of concrete-masonry fabrication techniques, from the compound to the formwork to the machinery for pouring the conglomerate. His analysis has the precise aim of serving as a manual for engineers and technicians, becoming a milestone of the literature on concrete, referenced in the most significant texts and the various encyclopedias. There is no doubt that the instructions supplied by Bélidor are the result of his observations on the worksite run by Milet de Monville at Toulon. Also in the book of 1753, Bélidor reports in France on the system of caissons used by Labelye for the foundations of the piers of the Westminster Bridge, and the pile drivers of Valoué.[31]

Technicians using the volume of the *Architecture Hydraulique* published in 1753 find precise instructions on the potential of conglomerate masonry. The construction of the "jetties made out of wooden coffers ["*encaissements*"] filled with stones," with dry masonry, according to the construction methods of the era of Vauban, is considered inferior to mortar-based masonry, like that of the piers of Cherbourg and Bayonne, particularly when taking into account the rotting of wood in the sea, resulting in costly replacement after about forty years.[32] It bears emphasizing that this was precisely the crucial question facing Zendrini and the *proti* of Venice in the invention of the pozzolana sea walls (though the grand "hydraulic architecture" of the Venetian Lagoon is not mentioned by Bélidor). For hydraulic masonry, Bélidor advises that it is possible "to successfully use a mortar composed of lime made out of many different types of calcinated seashells, mixed fifty-fifty with Dutch *trass*, and which can be used marvelously

[26] *Ibid.*, p. 34.

[27] Jean-Etienne Guettard, *Mémoire sur la Minéralogie de l'Auvergne*, in, "Histoire de l'Académie Royale des Sciences, Année MDCCLIX, avec les Mémoires de Mathématique & de Physique, pour la même Année," Paris, Imprimerie Royale, 1765, pp. 538-76.

[28] "This name is given," we read at the entry *Pozzolane,* "to a type of sand that is found in the territory of Pouzzol, an Italian city near Naples: it is also found in Guadeloupe, in Martinique, l'île de France and even in Auvergne" (Jacques-Christophe Valmont de Bomare, *Dictionnaire raisonné universel d'histoire naturelle, contenant l'histoire des animaux, des végétaux et des minéraux*, Suisse, Librairie Associés, 1780$^5$, vol. IX, p. 312).

[29] Minutes of the meeting of 27 May 1759, in, "Extrait du Journal des assemblées des Ponts et Chaussées tenues chez mr. Trudaine, depuis 1748 jusqu'en 1774," Ms 233, vol. XIII, Ecole des Ponts et Chaussées, Paris. *Cf.* also, Eugène-Jean-Marie Vignon, *Études historiques sur l'administration des voies publiques en France aux Dix-septième et Dix-huitième siècles*, Paris, Dunod, 1862, vol. II, p. 304.

[30] Bélidor, *Architecture Hydraulique. Seconde Partie* cit., vol. II, 1753.

[31] *Ibid.*, pp. 199 *et seq.*

[32] *Ibid.*, pp. 104 *et seq.*

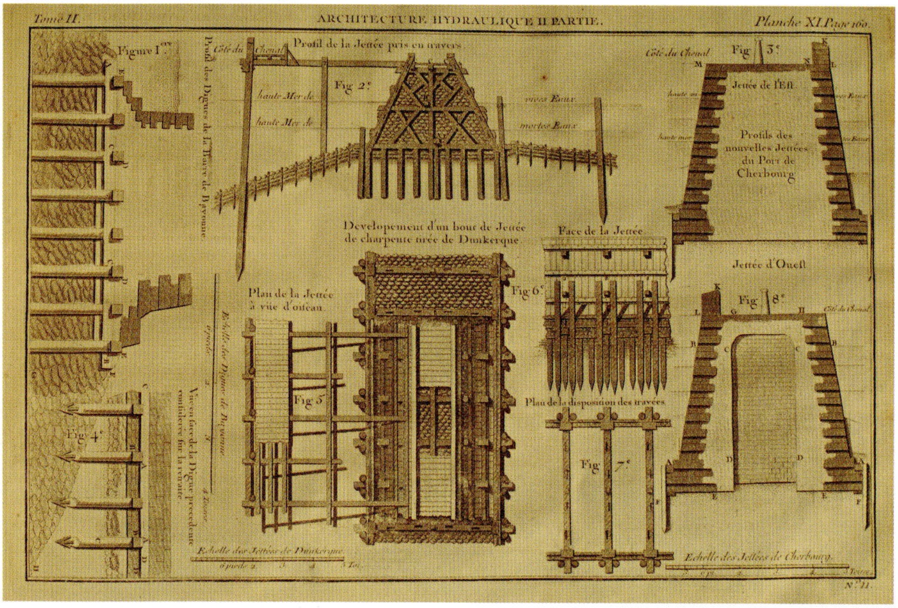
Tome II.
ARCHITECTURE HYDRAULIQUE II PARTIE.
Planche XI. Page 160.
Figure Ire.
Profil des Digues de la Barre de Bayonne
Côté du Chenal
Profil de la Jettée pris en travers.
Fig 2e.
haute Mer de
vives Eaux
haute Mer de
mortes Eaux
Côté du Chenal
Fig 3e.
Jettée de l'Est.
Profils des nouvelles Jettées du Port de Cherbourg
Developement d'un bout de Jettée de charpente tirée de Dunkerque.
Face de la Jettée
Jettée d'Ouest
Plan de la Jettée à vûe d'oiseau.
Fig 6e.
Fig 8e.
Echelle des Digues de Bayonne
Vue en face de la Digue precedente construitte sur la retraite
Fig 5e.
Plan de la disposition des travées.
Fig 4e.
Fig 7e.
Echelle des Jettées de Dunkerque.
Echelle des Jettées de Cherbourg.
N.° 11.

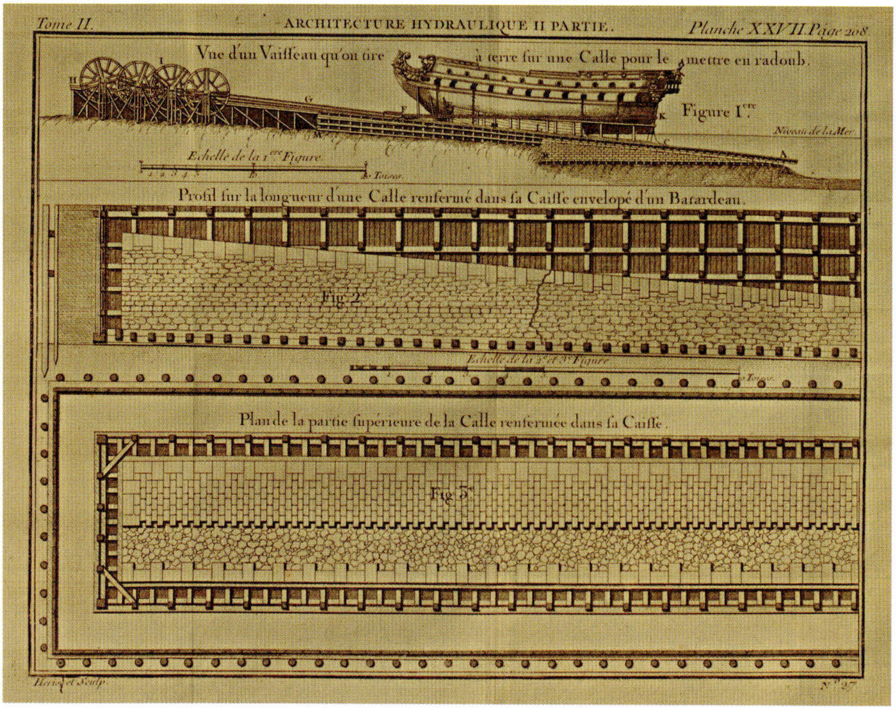
Tome II.
ARCHITECTURE HYDRAULIQUE II PARTIE.
Planche XXVII. Page 208.
Vue d'un Vaiffeau qu'on tire à terre fur une Calle pour le mettre en radoub.
Figure Ire.
Niveau de la Mer
Echelle de la 1ere Figure.
Toises
Profil fur la longueur d'une Calle renfermé dans fa Caiffe envelopé d'un Batardeau.
Fig 2e.
Echelle de la 2e et 3e Figure.
Plan de la partie fupérieure de la Calle renfermée dans fa Caiffe.
Fig 3e.
N.° 27.

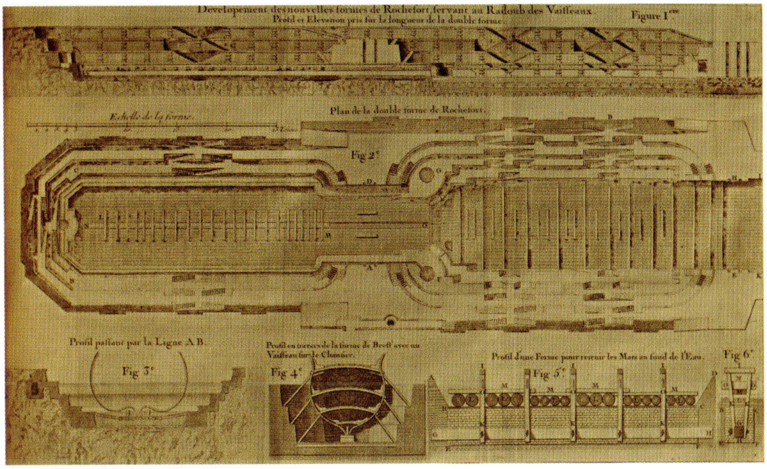


**190-192**. Bernard Forest de Bélidor, *Architecture Hydraulique. Seconde Partie, qui comprend l'Art de diriger les eaux de la Mer & des Rivieres à l'avantage de la défense des Places, du Commerce, & de l'Agriculture*, Paris, Ch.-A. Jombert, vol. II, 1753.

as long as it can be protected from the sea for only several hours, because it then hardens enough to hold up against the violence of the waves."[33] To build foundations on rocky beds, he recommends preparing a mortar "composed of one-half sand and large gravel and as much quicklime and pozzolana," in which to embed large quantities of "small rocks and large pebbles about the size of walnuts"; the compound should be transported on barges to the location of the foundation; following the method of Milet de Monville, Bélidor advises that the compound should be left to "dry for twenty-four hours, and when it is ready to be shoveled, to transfer it to drums with an opening hatch in the bottom to be able to empty them as close as possible to the summit of the riprap in order to avoid the dispersing of the mortar when passing through the water"; layers of concrete should be alternated with layers of pebbles until water level is reached; if an edifice is to be built on this foundation, a wooden mesh should be erected, with compartments to be filled with "pozzolana mortar, Dutch trass, or cinders from Tournay," depending on availability of the materials.[34]

In the chapter on how to use concrete for foundations at the bottom of the water without cofferdams nor dewatering, *De la maniere d'employer le Beton pour les Fondations faites au fond de l'eau sans batardeaux ni épuisement*, Bélidor recommends, for the foundations of a fort or a jetty at depths from 15 to 25 feet, bordering the base with a "framework" of poles and "sheet piling" and filling the interspace with "concrete mortar mixed together with stones,"[35] spreading it with a special machine equipped with a chest whose bottom can be opened, to avoid scattering "the best part of the lime."[36] In spite of the studies of Fontana on Anzio and famous works like the Lazzaretto of Vanvitelli, involving conditions that correspond to the theme discussed in *Architecture Hydraulique*, Bélidor continues to overlook the examples of Italian concrete construction.

Bélidor addresses the technical aspects he sees as "less well-known," like the "caisson for containing masonry,"[37] the method of proportioning the poles and connecting the two rows at intervals, through "keys or braces that cross the

[33] *Ibid.*, p. 122.
[34] *Ibid.*, p. 174.
[35] *Ibid.*, p. 178.
[36] *Ibid.*, p. 179.
[37] *Ibid.*

upper part of the chest in order to prevent separation; they can be removed when the concrete has solidified and is topped off with a leveling layer of ordinary masonry."[38] "These rows of piles will be further linked together with two internally-placed purlins to support the sheet pile walls."[39] Bélidor recommends checking that the boards of the formwork are well sealed, to prevent "some of looser lime and pozzolana from escaping, which would create holes in the masonry."[40] Then he illustrates the machine that "is used to immerse the concrete," with the chest containing 27 cubic feet of compound, "but it can be made bigger and in proportion to the size of the foundation to be filled."[41] The machine consists of two trestles installed above the formwork, with wheels allowing it to to be rolled, that can deposit the concrete in the desired location. The chest with its opening hatch is suspended from the trestles, attached by ropes to the two-wheel winch, and allowed to descend to the level where the compound can be deposited; the maneuver is controlled using the ropes – one of which makes it possible to open the hatch to release the compound. The detailed description of the finishing of the chest proves that Bélidor is describing a machine he has examined closely (probably the one used by Milet de Monville at Toulon).

"[...] the inside of the chest must be well finished and its joints well-caulked, and coated with tar from the outside, and closed with a sliding top, so that while descending, the water column thus put into motion removes not the slightest part of the concrete [...]."[42]

In his description of the accomplishments of Milet de Monville, Bélidor explains how to build a foundation in the sea using formwork – *encaissements* or coffers – with a maximum length of ten toises, to be closed at the ends using boards; after the pouring, other formwork sectors are assembled, and if the concrete has set too strongly, it will be necessary to "undergo extractions so that the new masonry binds better with the old: if, on the contrary, it does not solidify, it will flow into the elongated encasement on its own and will naturally bind with the concrete that is thrown on top."[43] Again thanks to concrete it is possible, according to Bélidor's advice, to create a solid substructure, or *massif*, for the piers of bridges when the riverbed is unstable, pouring the compound to be beaten with a rammer inside an enclosure formed by piles;[44] or it is possible, again for concrete foundations, to construct many pyramid-shaped chests whose facades are equilateral triangles, using boards from the demolition of ships, to fill with conglomerate and lower into the water for the making of a foundation[45] (the proposed system reminds us of the solutions described in 17th-century treatises, especially by Viviani). "The way to use these chests is to fill them with concrete mortar mixed with large stones, then close them and immerse them immediately in the water [...]."[46]

"Do we not have reason to be surprised that a custom practiced with such success in antiquity has been maintained only on the Mediterranean coast?" Bélidor wonders, regarding the concrete masonry with pozzolana or similar hydraulic powders, done by pouring, and preferable "to dry founda-

[38] *Ibid.*, p. 180.
[39] *Ibid.*, p. 181.
[40] *Ibid.*
[41] *Ibid.*, p. 185.
[42] *Ibid.*
[43] *Ibid.*, p. 188.
[44] *Ibid.*, p. 208.
[45] *Ibid.*, p. 209. *Cf.* plate X, figg. 14 and 16.
[46] *Ibid.*, p. 209. *Cf.* plate X, figg. 14-16.
[47] *Ibid.*, p. 187.
[48] See, Bergier, *Histoire des grands chemins de l'empire romain* cit. For several recipes for the composition of cement with lime, sand and brick powder, see, *ibid.*, p. 128. The construction systems based on cement described by Bergier are also referred to by Gautier (Henri Gautier, *Traité de la construction des chemins, où il est parlé de ceux des Romains, & de ceux des Modernes*, Paris, A. Cailleau, $1721^2$, p. 6).
[49] "Also they advised," Smeaton writes regarding the decisions of the committee, "the Core

tions made in chests and then plunged to the bottom of the sea" – recalling the recipe of Vitruvius for concrete.[47] All the premises exist, at this point, in the contemporary construction of major works of engineering, for the success of the rediscovery of what had been the ancient Roman method of construction, on which, until the publication of *Architecture Hydraulique*, very few indications circulate in manuals and in the archaeological literature, with only rare and quite generic images – apart from significant exceptions like the 17th-century book by Nicolas Bergier, *Histoire des grands chemins de l'empire romain*, reprinted at the start of the 1700s,[48] or *Antio e sue antichità* by Fontana.

## Etheridge's caissons and Barrow lime cement for the port of Ramsgate

The experiences involved in the construction of Westminster Bridge under the guidance of Labelye represent a crucial chapter in the training of certain technicians later assigned important projects of hydraulic engineering to be done in England. One of the protagonists of the construction of the machinery for that bridge, Etheridge, becomes the creator of major hydraulic works at the start of the 1750s.

The construction of the moles of the port of Ramsgate, facing Dunkerque, is one of the most controversial endeavours of English hydraulic engineering. The Westminster Bridge worksite and the technicians involved provide a reference point for the committee set up to evaluate the port project. Starting in 1736, Labelye himself is consulted to decide on the location of the port – his project dates back to 1737-38. The carpenter of Labelye's bridge in London, King, takes part in the discussions on the port in 1744, and perhaps it is following his death that his main collaborator, Etheridge, is called in to work on the project of the port of Ramsgate. After 1744 the projects follow one another, proposed by various engineers and sea captains, for the construction of the two moles, one in wood (project by Robert Brooke), the other in stone (project by William Ockenden). Construction begins in 1750. The masonry works, done with large blocks of stone, are entrusted to the contractor Thomas Preston. After observation, in 1751, of the lack of strength of the masonry under construction for the pier of Ockenden made with lime obtained by firing seashells – a product used above all in Holland – the committee issues technical recommendations, including that of using a mortar based on Dutch *trass*.[49] This is the context in which, in 1751, Etheridge finally intervenes, with the proposal of making the two pier arms with floating caissons, in keeping with the method used for the piers of the Westminster Bridge – he also proposes a way to flatten the seabed. Etheridge is appointed to supervise the work, alongside Preston.

The nature of the hydraulic mortar for the masonry of the piers is examined by Preston, following the failure of the trials using seashell lime. English builders had also made use of this type of lime since the time of Wren, but in the construction of masonry in the sea for the port of Ramsgate the product

to be lessened; the chalk to be well beaten, and mixed with gravel; not with mortar, as then practiced; and all the Ashler to be set with Tarras mortar, and the backing mortar to be mixed with brickdust or sea-coal ashes" (John Smeaton, *An historical report on Ramsgate harbour, written by order of, and addressed to the Trustees*, London, 1791², p. 17). *Cf.* Robert B. Matkin, *The Construction of Ramsgate Harbour*, in, "The Newcomen Society for the study of the history of Engineering and Technlology," vol. XLVIII, 1976-77, pp. 53-72; *Id.*, *The construction of Ramsgate Harbour*, in, Adrian Jarvis, *Port and Harbour Engineering*, Aldershot, Brookfield USA, Singapore, Sydney, Ashgate Variorum, 1998, pp. 125-44.

appears to be unsuitable, even when mixed with Dutch *trass*. "I was [...] informed," Smeaton writes, "that a part of the work at Ramsgate Pier had been done with this kind of lime, but was afterwards obliged to be taken up, on its dissolving quality in sea water being discovered."[50] Probably the process of dissolving in the sea at Ramsgate is the same one observed by Smeaton in experiments on seashell lime mortar: "On being put into water, after it was set, it did not dissolve, but did not acquire an additional hardness, on the contrary by degrees it macerated and dissolved, not internally, but gradually from the surface inwards[...]."[51] Preston's contribution concentrates precisely on the choice of the lime for the piers of Ramsgate.

"Mr. Preston, before the appointment of Mr. Etheridge," Smeaton reports, "not being satisfied with the use of shell lime under water, which was in part used at Ramsgate Harbour, recommended to the Trustees the trial of various other limestones, viz. Aberthaw, Barrow, and that of Boulogne, of which cargoes had been ordered; but their merits had not then been sufficiently investigated. Mr. Etheridge unfortunately had adopted shell lime for waterworks; and accounted that of Maidstone and St. Vincent's Rocks at Bristol Hot Wells as preferable to those above mentioned."[52]

To making of the lime for use in the masonry of the piers, blue stones are chosen, taken from the quarries of the village of Barrow-upon-Soar, transported to Ramsgate and then calcinated there. This is the type of lime recommended by Preston. For centuries the lime obtained at Barrow had been known for its hydraulic properties, so much so that it was even exported to Holland. The teams of masons charged with making the hydraulic lime for the construction of the moles of Ramsgate come from Barrow-upon-Soar.[53] In September 1752 construction is completed of 68 feet of the stone arm of the mole. That same year George Semple (1700 circa - 82), engaged in the construction of a bridge, visits the Ramsgate worksite, where he meets Etheridge and Preston – he admires the ability with which the "great Blocks of Portland Stone" are placed in the water, and gathers explanations on the foundation system.[54]

## Essex Bridge in Dublin, by Semple, and the concrete

The use of hydraulic mortars with high resistance obtained by calcination of local stones, like that from Barrow, brings quality to the construction of the most important English works of engineering. Precisely Barrow lime, after the experience at the port of Ramsgate, is recommended by Smeaton to make the mortar used in the masonry of the beacon of Spurn Point, at Spurn Head, facing the port of Grimsby – "or any other lime of equal quality for water-works," as he specifies in 1770.[55] Lime from Barrow continues to be recommended through-

[50] Smeaton, *A Narrative of the Building* cit., p. 106.
[51] *Ibid.*, p. 106.
[52] Smeaton, *An historical report on Ramsgate harbour* cit., pp. 20-21.
[53] *Cf.* John Britton, *The Beauties of England and Wales; or Delineations Topographical, Historical, and Descriptive, of Each County*, London, T. Maiden, 1807, vol. IX, p. 370.
[54] George Semple, *A Treatise on Building in Water*, Dublin, J.A. Husband, 1776, p. 28. *Cf.* Ruddock, *op. cit.*, pp. 38-44.
[55] *Cf. Additional remarks and observations, touching the construction of the Light-houses upon the Spurn Point, as approved of by the Honourable Corporation of Trinity-House, Deptford Strond, London*, 9 May 1770, in, *John Smeaton Reports of the late John Smeaton, F. R. S., made on various occasions, in the course of his employment as civil engineer*, London, Longman, Hurst, Rees, Orme, Brown, 1812, vol. I, (pp. 260-69), p. 261. Other parts of the lighthouse are to be made with a mixture of Barrow lime and pozzolana (*ibid.*, p. 264).
[56] *Cf.* William Marshall, *The rural economy of the Midland counties*, London, G. Nicol, 1790, p. 28.

out the 1700s to make the cement to use in cisterns and other hydraulic works.[56]

For the type of construction comparable to pouring, or the "boulder walls," one significant example is the exceptional masonry of gravel embedded in mortar made with Aberthaw lime, in the arch of the Pont y tu Prydd (Pontypridd) built in 1755-56 by William Edwards (1719-89), across the River Taff in Glamorganshire. The bridge is made with a single arch and three cylindrical holes at each end. It was constructed after the failure of a bridge with piers built in 1746 by Edwards himself, swept away by high water in 1748, and following the collapse of a second bridge with a single arch during construction in 1748, again based on his design. Aberthaw lime, like that of Barrow, continues to be successfully used, later, for important works of engineering.

But in the vast series of achievements connected with concrete, an outstanding role is played by the works done by Semple for the reconstruction of the Essex Bridge on the Liffey in Dublin. An Irish architect and builder, first commissioned in 1751 to repair, with a wooden structure, a breach in a deck sector spanning two arches in the bridge built in 1676-87. Semple constructs a new stone bridge, erected a few years later on the same site, with four pylons, in 1753-55. It would be demolished in 1872. The purchase in London by Semple, in 1751, of various architecture treatises – those of Alberti, Palladio, Scamozzi and Serlio, studied for the rules laid out for the construction of bridges – is done with the hope of also finding indications on the way of building an enclosure to make the foundations of the pylons in a dry place – the "inclosure" or "cofferdam," corresponding to the French *batardeau* and already seen in England among the proposals for the construction of Westminster Bridge.

Though the design of Semple's bridge is based on that of Westminster, the Essex Bridge opens a chapter in English hydraulic construction using the

**193**. William Edwards, bridge on the Taff River, Pontypridd, in Glamorganshire, 1755-56 (photograph by Bara-Koukoug, 2011).

**194**. George Semple, Essex Bridge, on the Liffey River, Dublin (G. Semple, *A Treatise on Building in Water*, Dublin, J.A. Husband, 1776).

ancient method, while applying the cofferdam system as opposed to the caissons of Labelye. It is built despite the doubts expressed by Etheridge and Preston during Semple's visit to Ramsgate. At the same time, Semple is looking for the ideal components for the making of the cement, which he sees as Dutch *trass*, or, as an alternative, the hydraulic lime from Leigh, near Warrington.

"I shall not relate," Semple recalls regarding his trip to London that began on 20 May 1752, "the various opinions of several judicious persons, whom I consulted concerning my inclosure, or coffer-dam; but in brief, I purchased more books in hopes of meeting with something concerning it, as I found I could not acquire any practical knowledge from professed artists. I purchased two chain-pumps, three double-screw, six jacks of different sizes, two drudging-engines, and several sorts of Tarrass, and went to Leigh near Warrington, to procure some light-lime (which as I was informed, would set under water as well as Tarrass) picked up a few labourers at Liverpool, and on the 23d of June returned to Dublin."[57]

Only in 1753 does Semple find concrete examples for the construction of the cofferdam when he gains possession, thanks to a bookseller in Dublin, of the four volumes by Bélidor of *Architecture Hydraulique* and a drawing of the cofferdam used for the reconstruction of the bridge at Orléans. Precisely in the volume of Bélidor published in 1753, Semple reads the description of the main foundation methods, drawing inspiration directly from what was done on the worksite for the bridge of Orléans. The development of the masonry for the bridge in Dublin happens against the backdrop of the discoveries made during the excavation of the bed of the Liffey at the piers, when a solid conglomerate is found that Semple studies to understand its composition. This marks the beginning of his research on the ancient methods of concrete.

"[...] we found under the inner part of it," he recalls after having dismantled the foundation grille of the third pier, "a smooth floor of clean sharp gravel, hard and firm under our feet; but when we thought to go on with filling the floats with it, to carry it away, as we had done in the bed of the river between the piers, we found it very hard, and as the shovel could not move it we tried the pick; but when we had got a little further into it, the pick made little impression upon it, and when we got further into it, we found that we had no other way to break it, but by undermining it, and then break it off in pieces with the largest sledge we had. The middle of it was an actual petrification of about a foot thick, but not so hard in the bottom as in the top. Some of the piers were much more petrified than others, and particularly this, as being better guarded by the base of the effigy, and having imbibed the petrifying qualities that soaked from it. The grit or grain of it greatly resembled that of mill-stone; but the colour was just the same with the bed of gravel it lay upon. Where the gravel partook of mud, the pier that lay on it was not near so hard, and those piers did not petrify at all that lay on beds that were not gravelly."[58]

The discovery of this conglomerate in the foundations of the old bridge, confirmed by similar masonry solutions found by Semple in 1752, when he is commissioned to repair another structure in Dublin, the Ormond Bridge, begun in 1684, leads Semple to investigate the fabrication of hydraulic mortars with the same characteristics of those found in the foundations of the two bridges. The architecture treatises, probably also that of Vitruvius, supply him with fundamental information for the preparation of the compound and its deployment.

[57] Semple, *op. cit.*, p. 29.
[58] *Ibid.*, p. 40.

Semple reads the considerations on hydraulic masonry that, as Scamozzi recommends, must be fabricated "with strong Puzzoli, Terrass, or Cement", and laid on the riverbed "by proper machines and cables."[59] The pages of the treatises consulted by Semple also include explanations on masonry poured inside masonry enclosures or wooden formwork. From the *Quattro Libri dell'Architettura*, which he reads in an English translation (probably that by Ware), he underlines the passages of Palladio on "walls called *reimpitura*, or *coffer-work*" that were "made by the Ancients, by taking planks laid edgeways, according to the thickness of the walls, filling the void or space between them, with cement and all sorts of small stones mingled together, and continued after this manner from course to course [...]."[60] But it is the reading of the *De re ædificatoria* of Alberti, again in an English translation, that shows Semple the vast range of applications of conglomerate for foundations of different kinds, to be poured into "cases" – corresponding to the "coffers" indicated by the translator of the treatise of Palladio.[61] Semple knows about Alberti's recommendation to create the first course of foundations with large stones and then to lay two outer walls, "filling (or stuffing as he calls it) the inside with small stones, and lime-liquid," making sure to select, for the filler, "small stones," because they join "more easily, and knit better with the cement than large ones."[62] Semple's interest in types of masonry done with formwork is confirmed by the fact that he also carefully reads the passages on the method, again described by Alberti, for construction of walls using wooden formwork in which to place earth mixed with aggregates and pressed.[63]

Together with the practical advice and observations of Fontana, Shaw, Gilij, Zendrini, Milet de Monville, Bélidor, De Cormontaingne, Pottard, Boffrand or Nollet on poured construction, what Semple discovers and brings back into current use by reading the Italian treatises represents another fundamental episode in the gradual rise of the construction system of concrete and formwork.

Semple's memoir of his own experiences with mortar constitutes an essential document for awareness of the persistence across the centuries of Roman masonry techniques, still practiced by masons even in England. The narration of the masons on the types of masonry simply confirms what Semple has learned from the treatises of Alberti and Palladio on stone-faced cement core walls:

"I have from my childhood, been well acquainted with the nature of lime and sand made in mortar, of all sorts, that have been used in buildings in these countries, and tried numerous experiments with them; on which, together with what I have observed and learned from old experienced workmen, during the course of upwards of sixty years, I think, I can safely affirm, that good mortar, that is, mortar made of pure and well-burnt lime-stone, and properly made up with sharp, clean sand, free from any sort of earth, loam or mud, will within some considerable time actually petrify, and as it were, turn to the consistence of a stone. I remember I had one of my remarks from an old *Scotch* mason, which I shall give you in his own identical words, that is, *when a hundred years are past and gane, then gude mortar is grown to a stain*; (or a stone). My father (who was a workman about the year 1675) often told me, and my own repeated observations convince me, that the methods masons practiced in former times, in building churches, abbeys, castles or other sumptuous edifices in this country, was to this effect. After they laid the out-side courses with large stones, laid on the flat in swimming beds of mortar, they hearted their walls with their spawls and smallest stones, and as they laid them in, they poured

[59] *Ibid.*, p. 15.
[60] *Ibid.*, p. 76.
[61] *Ibid.*, p. 77.
[62] *Ibid.*, p. 78.
[63] *Ibid.*

195-196. George Semple, Essex Bridge, on the Liffey River, Dublin (G. Semple, *A Treatise on Building in Water*, Dublin, J.A. Husband, 1776).

in plenty of boiling grout, or hot lime-liquid among them, so as to incorporate them together, as if it were with melted lead, whereby the heat of it exhausted the moisture of the out-side mortar, and united most firmly both it and the stones, and filled every pore (which as the masons term it) set, that is, grew hard immediately, and this method was taught to our ancient masons, by the *Romish clergy* that came to plant *christianity* in these countries, and I affirm, that in many of such old buildings, I have seen the mortar, as it were, run together and harder to break than the stone were."[64]

Semple believes that there are local "lime-stones" with which to make mortars for hydraulic masonry with better characteristics than those usually obtained using *trass*. To make this compound, he proposes a recipe for calcination of the rock – "roach-lime" – to then be pulverized and sifted, to obtain the hydraulic mortar. "This, I can assure you will not only stand as well, but is really preferable to any Tarrass."[65] Semple's awareness of the possibility of obtaining artificial products with characteristics that even surpass those of volcanic materials is of the utmost importance to reconstruct the genealogy of the experimentation that leads up to the fabrication of artificial pozzolana.

Semple's research provides other cases of poured construction after observations such as those of Dance on the faced and filled masonry in the piers of the medieval London Bridge. In February 1753 Sample begins work on his cofferdam to block a sector of the river, for the construction of the bridge completed in May 1755 – the piers rest on a continuous foundation, or "thorough foundation," as advised by Alberti.

"The method we took in laying the foundation," explains Semple, "was this: after we had drove the oak piles down to the rock [...], we spread a plentiful coat of roach-lime and sharp gravel over the ground, and laid a course of large flat stones, and filled and hearted them in close about the pile for about

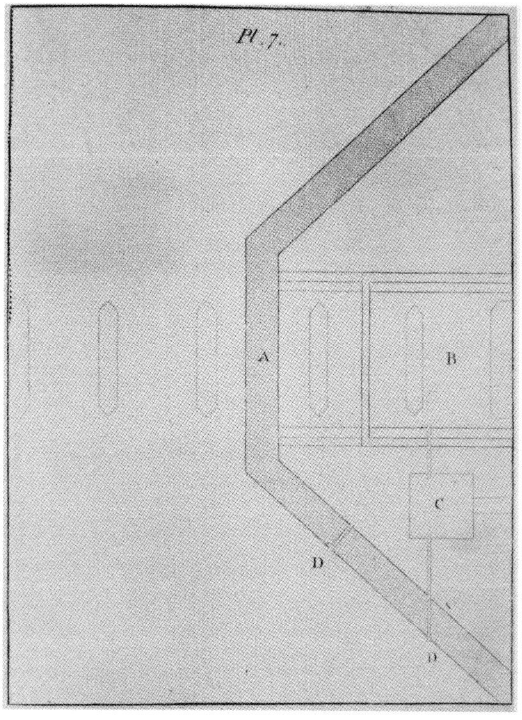


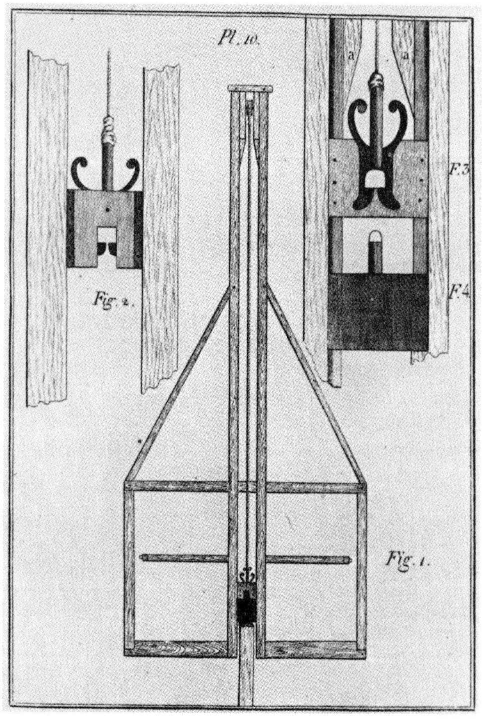

a foot high; then we covered that course with another plentiful coat of the dry grout (i.e. the roach-lime and sharp gravel) and the next course and all the rest was laid with mortar after the usual manner, and each course was grouted above; and so we went on till we came to the level for the caps that were laid over each row of piles, and then wrought up and levelled them, on which we laid three beams stretching the whole length of the pier from sterling to sterling, and filled up about them with the masonry; but the thorough foundation that lay under the arches, was laid and securely bedded, and many of the stones were rather pitched upon their ends and wrought close together, so that there is a substantial stone floor for the bed of the river between the piers, on which we spread a course or stratum of about a foot thick of sharp coarse gravel mixed with new and old lime-rubbish, and we covered that with a plentiful bed of such coarse gravelly stuff as we could collect within the dams, which together, made on an average about 2 1/2 feet thick [...]."[66]

## The "murazzi" of Temanza: pozzolana and Istria stone in conflict

While construction continues along the Venice waterfront of *murazzi in pozzolana* using the technique and methods invented by Zendrini, in the late 1740s and early 1750s the logic of the system begins to be reconsidered, again due to the need (already evident in 1737-38) to save materials, due to the difficulties involved in getting pozzolana and large blocks of Istria stone to the worksites. In 1747, when an inspection is made of the works done by Rossi at Caroman, between the position numbers 69 and 71, started in 1742 and now at a length of 458 feet, the experts write of a work "worthy of a prince."[67] But at the same time, the problem of the scarcity of pozzolana arises – only 4 or 5 cartloads remain in the public warehouses.[68] Between December 1747 and August 1748 in Venice the pozzolana is owned by Girardi, and the experts turn to him to purchase the cartloads needed for the continuation of the work, at a cost of 36 lire per cartload.[69] Girardi continues to supply certain quantities of pozzolana in the years to follow,[70] flanked by Carlo Soardi (the pozzolana is still the material taken from the quarry of San Paolo),[71] though the main supplier is still the company Antonetti e Figli. In order to improve the pozzolana supply, the expert Gabriel Marcello emphasizes the need to reinforce the public storage facilities with secure walls and doors, to prevent theft of the precious material.[72]

The shortage of pozzolana leads the experts to propose other solutions to proceed with the construction of the levees, "because the lack of pozzolana cannot cause suspension of such a necessary, useful and deliberated work," as Marcello and Lucchese remark.[73] Both insist on the urgency of constructing a *zoccolo a pozzolana* as the foundation of the levee along the coast, even if it means postponing the masonry construction of the upper parts. Marcello

64 *Ibid.*, p. 79.
65 *Ibid.*, p. 81.
66 *Ibid.*, p. 49.
67 See, Gabriel Marcello, report, 29 May 1747, Savi ed Esecutori alle Acque, 5, Archivio di Stato di Venezia, and Matteo Lucchese, Tommaso Temanza, "Misure dello Scarpone in Caroman sul Lido di Pelestrina fra li N. 69, e 71," [1747], *ibid.*
68 Gabriel Marcello, report, 29 May 1747, *ibid.*
69 "Terminazioni," 20 December 1747, 13 March 1748 and 17 April 1748, Savi ed Esecutori alle Acque, 444, Archivio di Stato di Venezia.
70 "Terminazione," 9 June 1750, Savi ed Esecutori alle Acque, 444, Archivio di Stato di Venezia.
71 "Terminazione," December 1753, *ibid.*
72 Gabriel Marcello, report, 1 June 1748, Savi ed Esecutori alle Acque, 5, Archivio di Stato di Venezia.
73 Gabriel Marcello, report, 7 September 1747, *ibid.*

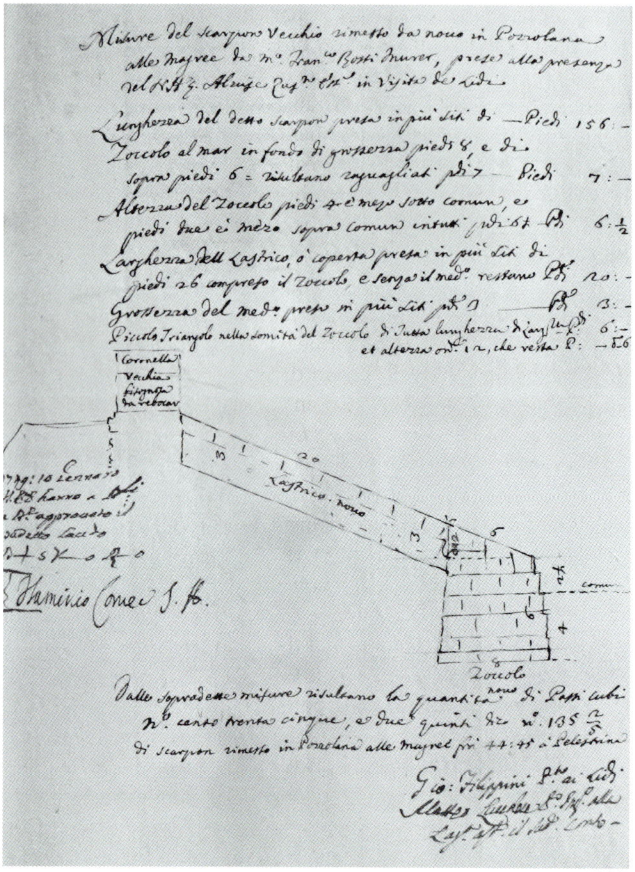


197. Giovanni Filippini, Matteo Lucchese, drawing of the section of a "murazzo," Venice, 1749 (Savi ed Esecutori alle Acque, 5, Archivio di Stato di Venezia).

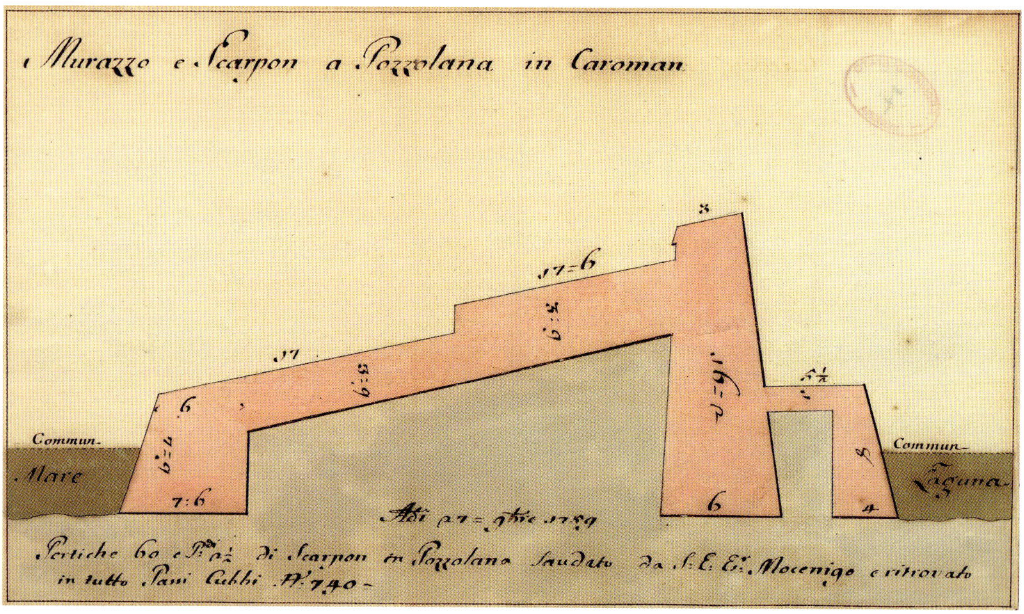


198. "Murazzo e scarpon a Pozzolana in Caroman," Venice (Savi ed Esecutori alle Acque, Archivio di Stato di Venezia).

proposes varying the profile of the sea wall in different stretches, to save materials.[74]

Inspections of the experimental *murazzi* of Tirali confirm the fundamental role of the *zoccolo a pozzolana* introduced by Zendrini. On this subject, what Alvise Giustinian writes in 1748 is revealing: "Immediately upwind of another similar pozzolana work at the Magre, there is another old work of large blocks of Istria stone, formed starting in the year 1722, but without a foundation, i.e. zoccolo, and without the pozzolana binder, so that the sea, digging at its foot, and entering there [...], removes the support of the stones, which fall down to the foot [...]. This protection, therefore, should have been rebuilt in pozzolana, and it was not, though it was suggested many other times, due to the scarcity of the pozzolana; now however, as there is a store due to the delivery made by Girardi to Chioggia, the time has come to do it."[75]

During the inspections of 1748 certain damage is noticed in the pozzolana work. First Lanciotto Zenier observes, in March, damage caused by ice on the pozzolana masonry, and the need to order annual repairs, to be assigned to Rossi, consisting in the filling of the damaged joints with the same pozzolana mortar.[76] In October Giustinian confirms having "seen in some parts surface damage caused by ice in the seams, so I believe it is necessary to make an agreement with the contractor Rossi, to fill them up again on an annual basis with pozzolana."[77] The inspection conducted by Ferigo Renier at the end of December 1749 to assess the damage caused by storms permits him to report on the good state of conservation of the first pozzolana works, and "some more evident damage" to the work at Capitello.[78] The causes of the damage to the *murazzo* of Capitello are attributed to the fact that since it was a sort of prototype, "very small" stones were used, and of "less consistency than those used later with better results," as well as to the lack of reinforcement of the *zoccolo*. Like his colleagues, Renier recognizes that the dislocations are partially caused by the lack of adequate "topping off" of the joints.[79] So in May 1750, as proposed by Filippini and Temanza, Rossi is hired, for a three-year period, to perform during summer the necessary pozzolana repairs of the *murazzi*.[80] At the end of the contract, though some experts assert that ice causes no damage to the levees,[81] the Savi ed Esecutori alle Acque renew Rossi's assignment – the compensation for this type of work is increased in 1758, to reflect the greater length of the segment of sea walls requiring maintenance.[82]

Appointed proto in November 1750, Temanza reopens the technical debate on the correct use of pozzolana in the "new method." Pozzolana, which had become so famous following the construction of the levees, is now (after the death of Zendrini, in 1747) seen for the first time as not the only ingredient to contribute to the solidity of the new structure.

[74] Gabriel Marcello, report, 1 June 1748, *ibid.* "I have observed that when that coast at that place is defended not just at the bottom, but also to quite a good height, all with marble in pozzolana on the same side as the sea; and for half its width all with marble, likewise in pozzolana, also in the middle of all its surface; and with all the base, also on the side of the lagoon, quite deep and all in marble, again with pozzolana, that protection can be said to be guaranteed; and that any additional elevation above that embankment is merely, in my way of thinking, an embellishment, with no enhancement of protection. So it is done at the other places, for the height of the wall of these real works has to also defend the land behind them from the spray of the sea."

[75] Alvise Giustinian, report, 4 October 1748, *ibid.*

[76] Lanciotto Zenier, report, 28 March 1748, *ibid.*

[77] Alvise Giustinian, report, 4 October 1748, *ibid.*

[78] Ferigo Renier, report, 1 January 1750, *ibid.*

[79] *Ibid.*

[80] "Terminazione," 24 May 1750, Savi ed Esecutori alle Acque, 444, Archivio di Stato di Venezia.

[81] Ferigo Renier, Michel Zen, report, 8 May 1753, Savi ed Esecutori alle Acque, 6, Archivio di Stato di Venezia. "Here it is appropriate to speak with great praise of these fabrications," we read about the *murazzi a pozzolana*, "both with respect to the idea, and with respect to Mastro Rossi, who was their builder. They could not be better bound, but they require retouching every year because of ice, which is the only thing that can bring damage to the operations themselves; and that they be protected (as we said) by groyones at the end of their sides."

[82] "Terminazione," 19 August 1758, Savi ed Esecutori alle Acque, 413, Archivio di Stato di Venezia.

In the report of 31 March 1751 written following an inspection of the works on the shorelines, Temanza notices the defects seen by Renier in the first *murazzo* at Capitello, which he too considers a prototype for the testing of the "new method" with pozzolana, later perfected in the construction of other sea wall segments.

"[...] that great work, from the foundations to the summit," Temanza writes, "has felt no detrimental effects; even though its figure is not the best suited for the needs. Since this was the first breakwater of this type to be built, almost like a trial for the works in pozzolana, it is little wonder that it did not achieve that best figure, which later, thanks to experience, has been given to other works. The damages are thus in the upper part of the sea wall, and consist in the breakage of the last course of stones that form its edge."[83]

The causes of the damage, visible above all at the upper "edge" of the *murazzo*, are attributed not to the "weakening of the pozzolana" but to the size of the stones – "too small in bulk" – and to their "friable" nature, making them "easily cracked and broken."[84]

At the start of June 1751 Temanza makes another shoreline inspection. His report written on 7 June contains a veiled critical assessment on the way pozzolana has been and continues to be used for the *murazzi*. Temanza's entire reasoning is based on the force of resistance to the impact of the sea exerted by the breakwater of stones, and therefore the benefits of the method introduced in Venetian practice by Tirali, to the point where the first two barriers without pozzolana become, in his view, the true prototypes of the system in the execution phase, because they have proven, over the years, the effectiveness of a method whose shortcomings were caused only by the yielding of the foundations at their base. In spite of the usual commendation of pozzolana included in the report, Temanza does not see pozzolana cement as the fundamental binder of construction with large stones, because it was not used correctly, in the way applied by the ancient Romans, to bind pieces of stone, gravel or rubble into a single block. The fact that the great builders, contemporaries of Temanza, know about the ideal size of aggregates in concrete is demonstrated by the statements written by Bélidor or by Semple on the size of the stones to embed in the mortar to obtain a conglomerate in the Roman way. In the *murazzi* the use of pozzolana cement as a binder of large stones, and furthermore stones that are not porous, as Temanza himself emphasizes, can only seem illogical to a connoisseur of the construction described by Vitruvius, and a brilliant pupil of Poleni. It is no coincidence that Temanza, from *De Architectura*, cites the phrase on the potential of pozzolana written in the second book, at the chapter entitled *De pulvere puteolano*. Temanza expresses himself in the report of 7 June in just these terms:

"One work only in pozzolana did I see done, and this at Caroman, between the positions numbered 71 and 73. The precise site is the one where previously there stood a great mass of large Istria stones, arranged almost in a heap, and without lime mortar; but of such force that it was able to resist the aggression of the sea for nearly half a century. The fact that the stones were not bound together reduced their resistance, and gradually their composition was dissolved. Therefore the deliberations of Your Excellencies were provident, to provide that site with the usual scarpone a pozzolana[85] in continuity with the neighboring breakwaters, using those same materials that had formed the original mass. A

[83] Tommaso Temanza, "Relazione Temanza circa il murazzo à pozzolana sopra il lido di Malamocco del giorno primo aprile 1751," 31 March 1751, Savi ed Esecutori alle Acque, 76, Archivio di Stato di Venezia.

[84] *Ibid*.

[85] The "*scarpone a pozzolana*" is synonymous of "*murazzo a pozzolana*."

large portion of this foundation has been made to date; and it may be hoped that most of the job will be done by next winter. [...]. The usefulness explained heretofore of placing large and very heavy stones at the outer parts of the murazzi should in no way alter the positive opinion deservedly conceived by all regarding pozzolana. Since through the virtue and force of pozzolana *etiam moles que construuntur in [mari, sub] aqua solidescunt*, as Vitruvius has left written. Nevertheless, mixed with lime, it does not have such a grip that when sturdily placed it holds the stones together; so that the very frequent strong blows of the sea do not cause them to come apart. And much more in our case, because the Istria stones are not very porous, and as a result they are not suited to the penetration of the very many minute particles, which when the stones are highly burdened could keep stone tenaciously clinging to stone, and united. The bulk of the stones, their great weight, added to the virtue and force of the pozzolana, can make those masses indissoluble, to great public benefit, enhancing the localities where they are erected. In fact, the pozzolana works built to date in Caroman excite the curiosity of strangers to come and see them, and one day they will be seen as one of the wonders of the Dominante."[86]

Temanza recommends limiting the use of large Istria stones in the sea walls to the *zoccolo* towards the sea and the edge towards the Lagoon. In the light of Temanza's observations, the experiment Zendrini had attempted at Capitello, with a caisson filled with pebbles, rubbles and pozzolana, according to the method practiced in the ports of the Mediterranean and observed by him at Viareggio, takes on new meaning. That experiment, which turned out to be a failure, was done precisely with the assistance of Temanza. While Zendrini's subsequent statements tend to conceal the system originally envisioned – caissons filled with shards and pebbles, embedded in pozzolana cement – now Temanza seems to offer a glimpse of what should have been the true system of construction in the Roman manner – the logical one for the use of pozzolana cement.

After 1751 Temanza and the other technicians involved in the construction of the *murazzi* attempt to follow the practical indications of Vitruvius on the use of pozzolana, and therefore to distinguish, in the masonry structure of the *murazzi*, between the parts with pozzolana cement and those with large Istria stones. On this subject Temanza's stance is significant, stated in the report of 21 February 1759, with the agreement of Lucchese and Tommaso Scalfarotto, regarding the future method of construction of the *murazzi*.[87] Once again, the starting point for the reasoning on the pozzolana works are the two boulders, or "*massi*," built by Tirali at Caroman: "The great depth of these breaches, and the promptness required to close them, as well as the lack of a sand which when mixed with lime would harden in the water, as does pozzolana, led him to propose placing stone upon stone, without any mixture that would fill the gaps between one and the next. Actually it was useless to make those works in this way as they employed vast stones of ten or more thousands each, which though not connected have validly resisted the immense force of the sea [...]."[88] Then the observations and principles expressed by Temanza in 1751 are cited, to confirm the validity of the construction of the *murazzi*. Pozzolana and large stones are again presented as ingredients in opposition in the masonry structure and the strength of the murazzi, as proven at this point by a long trial period, and once again attributed to the stones of large size, those utilized by Tirali.

[86] Tommaso Temanza, 7 June 1751, *ibid*.
[87] Tommaso Temanza, Matteo Lucchese, Tommaso Scalfuroto, report, 21 February 1759, *ibid*.
[88] *Ibid*.

"Then he proceeded to reason on the value of pozzolana," we read about the report by Temanza in 1751, "and its lack of penetration into those stones, as they were not porous, so that they could not bind together. From this a general theory began to emerge, that in the external parts of the embankment [*scarpone*], more than elsewhere, the specific weight of the stones is important to keep them still, immobile, standing up to the continuous, very violent blows of the sea. This principle, first with the examinations made by this very rigorous Magistrate, and then with experience, was acknowledged as very useful. And thus the custom began of using large, heavy stones in the embankments. The two boulders [*massi*] built by Tiralli at Caroman, which stood up to the sea for so many years by their weight alone, the experience of many years, and the adverse cases, have always confirmed this above opinion in us. Therefore we repute it is necessary to use large stones of four or more thousands of weight, in the whole frontal zoccolo, in the step at the middle of the embankment, and in the edge of the parapet towards the Lagoon. Without stones of such bulk, the work would be weak, and feeble."[89]

There is no reference, in the conclusion of Temanza, Lucchese and Scalfarotto, to pozzolana cement as an ingredient that gives strength to the parts of the *murazzo* most impacted by the waves. "The pozzolana embankments [*scarponi a pozzolana*] are works built for the preservation and the perpetuity of the Lagoon, as well as the beauty of the Dominante. But one cannot build for posterity except by exceeding in terms of solidity, as with learning one admires in the works of the Romans," is their ambiguous statement on the new method of Zendrini, and perhaps the result of discussions among the *proti* on deployment of pozzolana without following the principles of Roman construction.[90]

Apart from the correct use of pozzolana, it is clear that there are now the premises, in the discussions of the technicians working for the Savi e Esecutori alle Acque launched by Temanza on the way of deploying pozzolana cement in the *murazzi*, for an apologia of the magnificence of Roman construction, enriched by what until this point had not been sufficiently documented: the technical details of ancient construction.

## Pozzolana for construction along the Adriatic according to the methods of Zendrini or Vanvitelli

For the foundations of moles along the Adriatic coast, from Rimini to Pesaro, two models of hydraulic construction are followed, with variations, in Ancona and Venice, those of Vanvitelli and Zendrini, with at times a more direct reference to the ancient Roman construction method of the wall with a cement nucleus. On the various worksites pozzolana confirms its role as the fundamental component for mortars. So the shipping traffic to transport pozzolana from the port of Civitavecchia along the Adriatic coast intensifies halfway through the 1700s, to supply not only the works of Vanvitelli and Zendrini, but also those of their followers.

The work on the port of Rimini offers another opportunity to experiment with pozzolana cement construction. After the failed masonry experiment in 1732 and following the recommendations of experts like Zendrini and Eustachio Manfredi (1674-1739) pozzolana is used again starting in

[89] *Ibid*.

[90] *Ibid*. The condition of the shores of Malamocco, Pellestrina and Chioggia, at the end of the 1760s, is documented by a drawing by Scalfurotto (Grillo, *op. cit*., pp. 100-01, ASV, SEA, Lidi, drawing 46; the drawing is from July 1771).

1735, and imported by Enrico Rigazzi and Alessandro Guidoni,[91] specifically for the works at the port done based on a design by Vanvitelli and under the supervision of Giovan Battista Banderati (a small fort and a pier in masonry instead of wooden pile structures). To build the mole, a circular pile structure is raised, or an *incassatura*,[92] around the site of the foundation; then wooden poles are inserted along the footprint of the pier, and masonry is built using methods based on the ancient Roman procedure: large stones for the outer walls and a poured core, with pebbles embedded in pozzolana mortar.

For the construction of the mole of the port of Pesaro two solutions are proposed, based on the results achieved in Ancona and Rimini. The solution derived from the Rimini model calls for laying, over the foundation piles, "partially rugged blocks of marble, but made flat on the lower part, the upper part, and at the head."[93] The fact that the masonry is still based on Roman structures with a cement core can be gleaned from the description of the works: "The first layer of the marble blocks has a height of two Venetian feet, and once the marble blocks are placed lengthwise, work begins for the inside with pozzolana, stones and rubble poured as filler up to the level of the aforesaid first layer of stone, and then the second layer of stone is made, making sure that every two rows the marble pieces are connected by a third, placing between each two pieces of marble another positioned lengthwise, with the end cut to form a dovetail, and thus proceeding, also with the said fillers, one continues to the total determined height of the mole."[94] The fact that the stones have to be clamped with dovetail connections confirms the circulation of technical solutions known to the experts of hydraulic construction – just consider what Tirali had proposed for the levees of the Venetian Lagoon. To decide which model to implement for the mole of Pesaro, from 1755 to 1767 various experts were consulted, from Giuseppe Belomo to Nicola Michetti, Angelo Porzi, Romoaldo Bertaglia, Claudio Storace, Lanico Ircadi, Pio Fantoni, from different cities – Genoa, Naples, Ancona, Rimini, Rome. Most of the experts are critical of the model of Rimini, due to the lack of connection between the foundation piles and the masonry, and favor the model of Ancona, also the choice of the master builder of the works of Vanvitelli, Belomo. What Storace has to say against the mode of fabrication of the pier of Rimini is significant: "their layer of blocks of marble, on the ends, or on the surface of the piles, and with conglomerate filler to be inserted under said marble; for many reasons is a totally false way of fabrication, and it can never be sufficiently disapproved by those who have practical experience in works done under water."[95] The masonry works with pozzolana done at the port of Pesaro set the example for the works of extension of the pier done during the 1800s at the same port (on that occasion reference is still made, as a masonry model, to the sea walls of the Venetian Lagoon).[96]

## Magnificence of Roman construction, according to Piranesi

Only against the backdrop of the important construction achievements in progress in Italy, in Ancona or Venice, and of the inventions of the *proti* and sci-

[91] For the documents, see, Alessandro Serpieri, *Il porto di Rimini, dalle origini a oggi, tra storia e cronaca*, Rimini, Luisè, 2004.

[92] Cit. in, *Pareri, e direzioni originali di diversi ingegnieri, ed architetti di mare consultati dall'E.mo signor cardinale Gio. Francesco Stoppani legato di Pesaro, sopra la maniera più stabile da fabbricarsi il nuovo molo di pietra d'Istria all'imboccatura del porto di Pesaro l'anno 1755*, in, *Del porto di Rimino, lettera di un Riminese ad un'amico di Roma, coll'appendice di documenti*, Roma, Bernabò e Lazzarini, 1768, p. 29.

[93] *Ibid.*

[94] *Ibid.*

[95] *Ibid.*, p. 30.

[96] *Cf.* Maurizio Brighenti, "Memoria sul porto di Rimini" (*23.a Sessione ordinaria. 24 maggio 1860*, in, *Rendiconto delle sessioni dell'Accademia delle scienze dell'Istituto di Bologna. Anno Accademico 1858-1859*, Bologna, Tipi Arcivescovili, 1859, (pp. 66-71), p. 68).

entists of Venice, from Tirali to Zendrini, Poleni, Lucchese and Temanza, or of the engineers of the Papal State under Vanvitelli, is it possible to fully grasp the meaning of the work of one of the greatest visionary archaeologists, and the one with the greatest passion for Roman methods of construction: Piranesi.

Knowledge of an international level of Roman construction, and of the kind that is being applied in contemporary hydraulic works, is not gained through the circulation of the observations of travelers. The international fame of Roman construction and the pozzolana cement technique can be attributed to the four volumes of etchings of Piranesi, dedicated to Roman antiquities, entitled *Le Antichità Romane* and published in Rome in 1756.[97] In the views prior to the *Antichità Romane* Piranesi had simply paid tribute to the ruins and palaces of Rome, without documenting their technical details, as can be seen in the *Varie Vedute di Roma Antica e Moderna* published in Rome from 1741 to 1748, or *Alcune Vedute di Archi Trionfali ed altri monumenti inalzati da Romani parte de quali se veggono in Roma e parte per l'Italia*, published in 1748, again in Rome.

In the views with Roman ruins, artists of the generations prior to Piranesi, up to his contemporaries, depict temples, triumphal arches or pyramids only as assemblages of stone blocks, without any reference to the most typical construction system of Roman architecture – see, as indicative examples, the views by Marco Ricci (1676-1730), Giovanni Paolo Pannini (1691-1765), Giovanni Antonio Canal better known as Canaletto (1697-1768), Gianfranco Costa (1711-72), certain fantastic ruins depicted by Giuseppe Galli Bibiena (1696-1757), some dating back to about 1740 and published in *Architetture, e Prospettive*, or the later ones inserted in the views of Antonio Visentini (1688-1782). Rarely does the ruin become a pretext for the technical depiction of materials and structures, as demonstrated by a direct forerunner of the collections of drawings of Piranesi, the ten volumes by Giuseppe Vasi (1710-82), entitled *Magnificenze di Roma antica e moderna* and published in Rome between 1746 and 1761 circa, an homage to conventional veduta prints. Exceptions do exist, like the depictions of Hendrik van Lint (1684-1763), in which it is possible to make out, in the view of the mausoleum known as Tor de' Schiavi, on the Via Prenestina, the imprints left by amphorae embedded in the pouring of the roof of the destroyed portico.

It is only with Piranesi that the veduta genre converges with the tradition of technical drafting of architects, and ruins are observed and drawn no longer with the eyes of Canaletto or Vasi, but with those of Sangallo or Rusconi, of Lucchese or Temanza. Piranesi has the ability to find the most effective graphic representation to emphasize, in the European debate, the orientation in progress since the first decades of the 1700s towards grand hydraulic construction systems such as those documented in the treatises of the engineers, from that of Gautier to those of Bélidor, all the way to the exploits of the Lazzaretto and the pier of Ancona, and the *murazzi* of Venice.

The etchings of the *Antichità Romane* offer useful cutaways to understand the various construction systems, and they are designed to illustrate all the technical characteristics and materials of the Roman construction machine, precisely like the drawings of Sangallo and Rusconi, of Bélidor, or the drawings of the sea walls prepared by the *proti*. The fact that plates are even prepared on the foundation systems, like those of the Theater of Marcellus, of the

[97] Giovanni Battista Piranesi, *Le Antichità Romane*, Roma, Bouchard e Gravier, 1756, 4 vols.

**199**. Giuseppe Galli Bibiena, "A large funeral hall in ruins" (C.D. Denison, M.N. Rosenfeld, S. Wiles, ed., *Exploring Rome: Piranesi and His Contemporaries*, Cambridge-Massachusetts, London, 1993).

**200**. Giovanni Battista Piranesi, "Veduta degli avanzi del Tempio della Pace," Rome (G.B. Piranesi, *Varie Vedute di Roma Antica e Moderna*, Roma, 1741-48).

**201**. Hendrik van Lint, Tor de' Schiavi, on the Via Prenestina, near Rome (Palazzo Braschi, Museo di Roma, Rome).

Tom. IV
XXXII
L
I
H
D
C
B
Veduta di una parte de' fondamenti del Teatro di Marcello

**202**. Giovanni Battista Piranesi, "Veduta di una parte de' fondamenti del Teatro Marcello," Rome (G.B. Piranesi, *Le Antichità Romane*, Roma, Bouchard and Gravier, 1756, vol. IV).

mausoleum and bridges of Hadrian, Fabricius, and Cestius, the Porticus Octaviæ or the Temple of Portunus demonstrates, in an emblematic way, the originality of the vision of Piranesi and his familiarity with a territory like that of Venice, where the crucial technical and constructive question had to do with foundations, as seen in the pozzolana levees. The fact that Piranesi's choice of subjects overlooked by other veduta makers in the *Antichità Romane* is guided by the images of the powerful Venetian protective constructions is confirmed by plates like the view showing part of the foundations of the Theater of Marcellus and the view of the foundations of the Mausoleum of Hadrian[98]. Everything – from the choice of the vantage point from above on the substructures with piles, to the very form of the structure, with the great blocks of stone and the nucleus in pozzolana cement – contributes to render magnificent not so much the "foundations of the Theater of Marcellus" as, instead, the feat in progress along the shores of the Venetian Republic, the *murazzi*, and to prove, precisely with views of Rome, the intrinsic ancient and Roman character repeatedly underlined by Zendrini, Lucchese and Temanza. It is also worth noting, to emphasize the timely technical pertinence of certain views of Piranesi, that the view of the foundations of the Theater of Marcellus shows a system that coincides almost perfectly with that of the mole of Rimini.

To understand the reasons behind the rise of a technical vision of the Roman ruins in the *Antichità Romane*, we should remember that Piranesi is the son of the Venetian stonecutter Angelo and of Laura Lucchese, sister of the *proto* Matteo, a collaborator of Zendrini and Temanza. Piranesi also works for the Magistrato alle Acque, under his uncle (Lucchese himself, who teaches him the basics of technical drafting), and then for Giovanni Scalfarotto (circa 1690-1764), and it is thanks to this appointment that he has direct knowledge of the construction of the *murazzi* and of pozzolana cement. This is why many of the etchings Piranesi makes of the pozzolana works of ancient Rome can be seen as secret commemorations of the achievements of Zendrini, Temanza and Lucchese at the Venetian Lagoon.

His devotion to construction is confirmed by the fact that Piranesi dedicates the *Prima Parte di Architetture, e prospettive* to Nicola Giobbe, one of the most important building contractors in Rome (involved in the construction of the Fountain of Trevi and other worksites, under Salvi). In Giobbe's home, Piranesi can make use of a vast library of books on architecture, including manuals of construction and civil and military engineering, like those of Rusconi, Amico, Capra, Viola, Guarini, Bartoli, Branca, Zonca and Carlo Fontana, as well as a collection of architectural drawings.[99] Precisely at Giobbe's home, Piranesi meets Salvi along with one of the architects most deeply involved in continuing the Roman constructive tradition of pozzolana

[98] *Veduta di una parte de' fondamenti del Teatro Marcello*, in, Giovanni Battista Piranesi, *Le Antichità Romane* cit., vol. IV, plate XXXII; *Veduta del sotterraneo del fondamento del Mausoleo, che fu eretto da Elio Adriano*, *ibid.*, plate VI.

[99] See, *Nicola Giobbe. Testament extrait de l'inventaire après décès*, in, Georges Bunel, *Recherches sur les débouts de Piranèse à Rome: les Frères Pagliarini et Nicala Giobbe*, in, Georges Bunel, ed., *Piranèse et les français, colloque tenu à la Villa Médicis, 12-14 mai 1976*, Roma, Edizioni dell'Elefante, pp. 77-146. To confirm the role played by technique in the drawings of Piranesi we can also mention that the publisher Niccolò and Marco Pagliarini, in the same year 1743, released the *Prima Parte di Architetture e Prospettive* of Piranesi, and the short essay by Zabaglia on carpentry for centering (*Castelli, e Ponti di Maestro Niccola Zabaglia, con alcune ingegnose pratiche, e con la descrizione del trasporto dell'obelisco Vaticano, e di altri del cavaliere Domenico Fontana*, Roma, Pagliarini, 1743). *Cf.* Georges Brunel, *Recherches sur les débuts de Piranèse à Rome: les frères Pagliarini et Nicola Giobbe*, in, Bunel, *op. cit.*, (pp. 77-146), p. 79.

Tom. IV
SEZIONE TRASVERSALE DE' CVNEI INFERIORI DEL TEATRO DI MARCELLO
XXXI
Fig. I.
Fig. II.
Fig. I. A. Muro divisorio dell' andito dell' Ordine Equestre, che rimane dalla parte opposta. B. Via de' Senatori, che passava sotto lo stesso andito. C. Muri con volta che reggevano la scala loro soprapposta, la quale dal detto andito conduceva ai sedili. D. Cloache. E. Fondamenti. Fig. II. A. Sezione dell' andito dell' Ordine Equestre. B. Porta con iscala appartenente al vomitorio dello stess.º Ordine. C. Archi di tevoloni per corroborazione della fabbrica.

Tom. IV
VIII
Palmi
Romani
Spaccato degli Speroni, i quali servono di contrafforti al gran Fondamento del Maus.º d'Elio Adriano. A Parte del Bassamento quadrato, su cui posa la gran Mole. B Palizzate sotto agli Speroni, e al gran Fondam.to C Parte angolare della Pianta del gran Fondamento, dimostrato in alzata nelle Tavole precedenti.

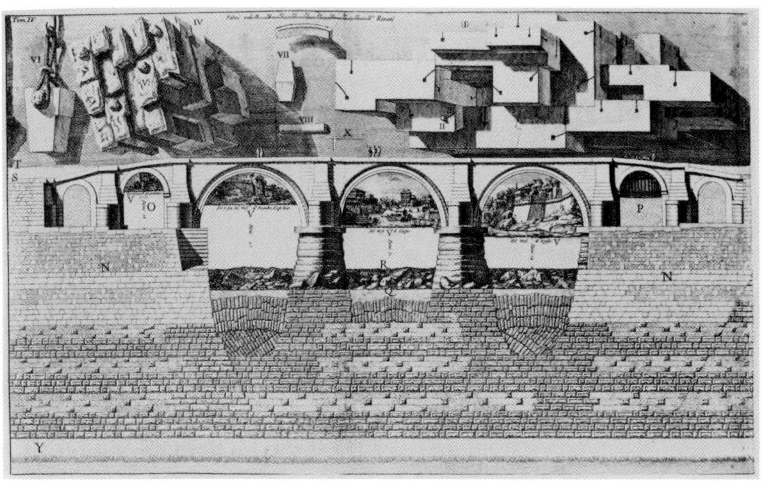
Tom. IV

**203**. Giovanni Battista Piranesi, "Sezione trasversale de' cunei inferiori del Teatro Marcello," Rome (G.B. Piranesi, *Le Antichità Romane*, Roma, Bouchard and Gravier, 1756, vol. IV).

**204**. Giovanni Battista Piranesi, "Spaccato degli Speroni, i quali servono di contrafforti al gran Fondamento del Maus.o d'Elio Adriano," Rome (G.B. Piranesi, *Le Antichità Romane*, Roma, Bouchard and Gravier, 1756, vol. IV).

**205**. Giovanni Battista Piranesi, hypothetical section of the foundations of the Bridge and Castle of Sant'Angelo, Rome (G.B. Piranesi, *Le Antichità Romane*, Roma, Bouchard and Gravier, 1756, vol. IV).

cement in Italy, Vanvitelli, whom he truly admires. It is significant that in the letter of presentation of the *Prima Parte di Architetture, e prospettive* Piranesi mentions the works by Vanvitelli, the "Port and Lazzaretto of Ancona just completed,"[100] captured in the views of Vasi; this very work must have appeared to him as emblematic of epic Roman construction in poured concrete.

In the campaign of surveys and engravings systematically carried out by the workshop of Piranesi in approximately 1750 to prepare for the publication of the *Antichità Romane* we can see a precise desire to document, with commemorative intent, the particular qualities of Roman construction, its techniques and materials. We can even state that precisely the etchings of Piranesi interpret, in a sublime way, that impassioned rediscovery of Roman construction that culminates in the invention of a cement with similar physical and mechanical characteristics (though the silence of historians on this subject is disconcerting).[101] It is possible to identify, in the sequence of the *Antichità Romane*, a crescendo of the role of construction details from the first to the last volume, where the depictions even take on overtones of engineering. Piranesi illustrates details with the same precision found in the drawings for the *murazzi a pozzolana* of Temanza and Lucchese; that the drawings of his collection become essential documents to understand the Roman construction systems is no coincidence, at least until the time of Rondelet.

In almost all the plates of the *Antichità Romane* the state of ruin or the refined cutaway design make the construction system immediately and eloquently visible. The presence of many close-up drawings and legends, though terse, is fundamental for the success of Piranesi's cultural project.

From the whole of the etchings of the *Antichità Romane*, the technique of concrete poured between various types of outer walls, and to make different structures, proves to be prevalent in Roman construction. The terms used by Piranesi to illustrate this technique are not always the ones applied in the treatises derived from Vitruvius. The core formed by the pouring is called *opera incerta* or *riempitura* – the first term coming from the Latin *opus incertum*, the second from the writings of Palladio.[102] When the core takes on great size, Piranesi uses the term *masso*, or mass, the same one applied by the *proti* to indicate the murazzi of Tirali. The *masso*, Piranesi explains regarding "a tomb on the ancient Appian Way at Vigna Buonamici," is "made of courses of flint shards with lime, and pozzolana," and was once covered by *travertini*, now "detached from the mass, and taken elsewhere."[103] Another great *masso* can be seen in the plan of the Mausoleum of Cecilia Metella.[104]

The *opera incerta*, or "filler," Piranesi explains, is a compound of fragments, of "any sort of stone chips,"[105] embedded in pozzolana-base mortar: "From inside

[100] Giovanni Battista Piranesi, *Reveritissimo Sig. Sig. Padrone Collendissimo*, 18 July 1743, in, *Piranesi. Early architectural fantasies. A Catalogue Raisonné of his Etching*, Washington, Chicago London, National Gallery of Art, University of Chicago Press, 1986, pp. 96-97.

[101] The historical reasons that led to an exclusively ideological interpretation of the work of Piranesi have been documented from Vincent Scully Jr. to Manfredo Tafuri and pupils. For the importance of Piranesi in the history of construction techniques in the 1700s (though limited to his definition of "opera saracinesca"), see for example what is written in, Daniela Esposito, *Tecniche costruttive murarie medievali: murature 'a tufelli' in area romana*, Roma, L'Erma di Bretschneider, 1998, pp. 58-67.

[102] For the definition "maniera riempiuta" see, Palladio, *op. cit.*, p. 13.

[103] *Pianta d'un Sepolcro situato sull'antica Via Appia alla Vigna Buonamici* (Piranesi, *Le Antichità Romane* cit., vol. II, plate XXXVI). See also *Spaccato del Sepolcro antecedente* (*ibid.*, plate XXXVII).

[104] *Pianta del Mausoleo di Cecilia Metella [...]*, *ibid.*, vol. III, plate XLIX.

[105] *Tavolozza triangolare martellinata delle mura d'Aureliano [...]*, *ibid.*, vol. I, plate VIII, fig. I.

**206-207**. Giovanni Battista Piranesi, "Pianta d'un Sepolcro sull'antica Via Appia vicino alla Vigna Buonamici e Spaccato del Sepolcro antecedente," near Rome (G.B. Piranesi, *Le Antichità Romane*, Roma, Bouchard and Gravier, 1756, vol. II).

**208**. Giovanni Battista Piranesi, "Pianta di un Sepolcro sull'antica Via Appia nella Vigna Buonamici [...]," near Rome (G.B. Piranesi, *Le Antichità Romane*, Rome, Bouchard and Gravier, 1756, vol. III).

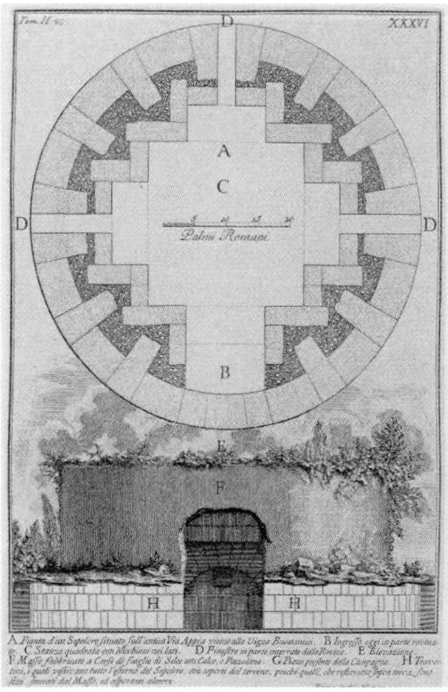

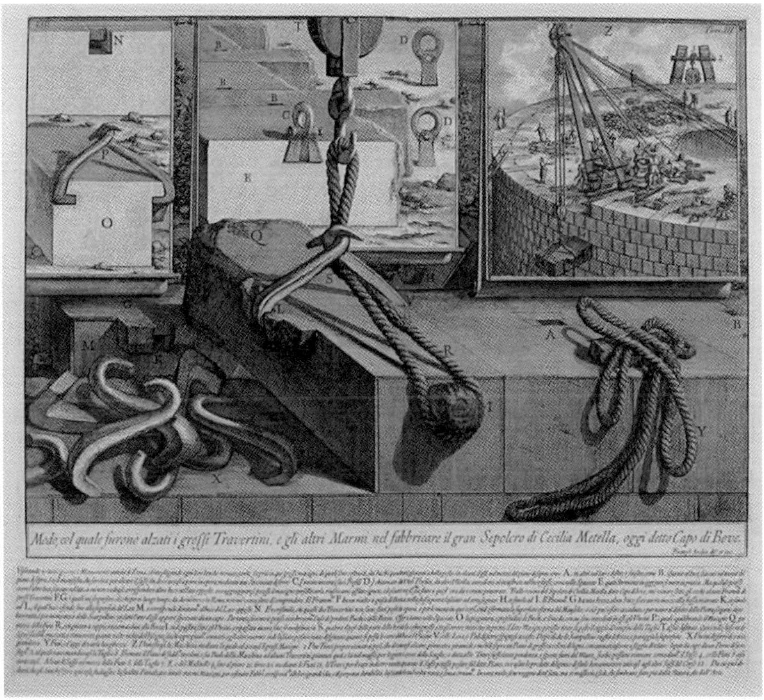


**209**. Giovanni Battista Piranesi, "Modo, col quale furono alzati i grossi Travertini, e gli altri Marmi nel fabbricare il gran Sepolcro di Ceciclia Metella, oggi detto Capo di Bove," near Rome (G.B. Piranesi, *Le Antichità Romane*, Roma, Bouchard and Gravier, 1756, vol. III).

**210**. Giovanni Battista Piranesi, "Pianta del Sepolcro di Alessandro Severo, situato fuori di Porta S. Giovanni [...]," Rome (G.B. Piranesi, *Le Antichità Romane*, Roma, Bouchard and Gravier, 1756, vol. II).

**211**. Giovanni Battista Piranesi, "Pianta del Mausoleo di Cecilia Metella [...]," near Rome (G.B. Piranesi, *Le Antichità Romane*, Roma, Bouchard and Gravier, 1756, vol. III).

it is filled with pozzolana lime, bits of stone and pieces of brick."[106] The walls built in *opera incerta* are usually depicted with outer surfaces in "squared blocks, sawed diagonally to form the triangular bricks."[107] The cross-sections of the great *massi* show the stratification of *opera incerta* regularly alternating with masonry rows. To bond together the two outer faces and the *opera incerta*, courses of special bricks, called "tegoloni,"[108] are laid horizontally in the thickness of the wall, as Piranesi himself explains: "walls constructed [...] of triangular bricks; in the middle, of filler alternating every four palms with horizontal courses of bricks."[109] Regarding the etching of the "remains of Hadrian's mausoleum and bridge," Piranesi describes the construction process: "here in perspective are shown the external courses of Peperino and Travertine, and how they are filled inside with layers of cement filler, or 'opera incerta', composed of broken stones placed in order, and well packed with lime and pozzolana, alternating every so often with large Peperino stones, to connect said courses."[110] In a detail of the view of the Porticus Octaviæ, Piranesi shows the horizontal connections in the poured wall better than in other views: "demonstration of the construction of the wall of the portico, at every four palm's of height there was an equal course of large tiles that bound the external panels with the internal filler."[111]

The structural strength of the conglomerate is demonstrated in the *Veduta di un gran Masso, Avanzo del Sepolcro della Famiglia de' Metelli sulla Via Appia*, where the cement core is reduced to a single central support point of a mass with a large overhang on all four sides.[112] Behind the powerful buttresses in blocks of peperino that form the stone masonry of the foundations of the Theater of Marcellus, Piranesi depicts the cement nucleus with facing in *opus reticolatum* of what he calls "internal foundations of opera incerta."[113]

Foundations, pylons, walls, vaults and domes are all constructive features the etchings of Piranesi demonstrate can be made with the Roman system of facing and filler, or *riempitura*. The drawing of the dome of the Pantheon, with the structure featuring series of brick arches, intersected and filled, is a fundamental contribution to the knowledge of Roman concrete construction of vaulted structures.

Almost as if to summon the ghosts of the artificial stone of ancient Egypt, Piranesi shows that a pyramid can also be erected with *riempitura*, as demonstrated by the plan of the Pyramid of Cestius: "mass composed of chips of flint, pozzolana, and lime."[114]

Piranesi also explains the compound known as *lastrico*, destined to attract increasing attention on the part of travelers. From his survey, it appears that the *lastrico* was used by Roman builders to cover conduits and to seal the stones in aqueducts, or to form pavements in baths or tombs.[115] The drawing of the

[106] *Mattoni quadrati, segati a linee diagonali per formare Mattoni triangolari [...], ibid.*, vol. III, plate V.
[107] *Ibid.*
[108] *Tavolozza triangolare martellinata delle mura d'Aureliano [...], ibid.*, vol. I, plate VIII, fig. I.
[109] *Pianta del Mausoleo di Sant'Elena Madre di Costantino, ibid.*, vol. III, plate XVI.
[110] *Avanzi del Mausoleo, e del Ponte d'Elio Adriano [...], ibid.*, vol. IV, plate VII.
[111] *Veduta di uno de' fianchi della facciata del portico di Ottavia, ibid.*, vol. IV, plate XLI.
[112] *Veduta di un gran Masso, Avanzo del Sepolcro della Famiglia de' Metelli sulla Via Appia, ibid.*, vol. III, plate XV.
[113] *Veduta di una parte de' fondamenti del Teatro Marcello, ibid.*, vol. IV, plate XXXII.
[114] *Pianta della Piramide di Caio Cestio presso la Porta di S. Paolo, ibid.*, vol. III, plate XLIII.
[115] *Cf. Pezzo di tufo [...], ibid.*, vol. I, plate IX, fig. I; *Spaccato del Condotto di Caracalla [...], ibid.*, vol. I, plate XIX, fig. II. In the *Pianta* e *Spaccato del Sepolcro di Alessandro Severo*, the powerful of the "masso" is shown with a drawing intended to convey the particular nature of the poured construction – "built of crushed stone."

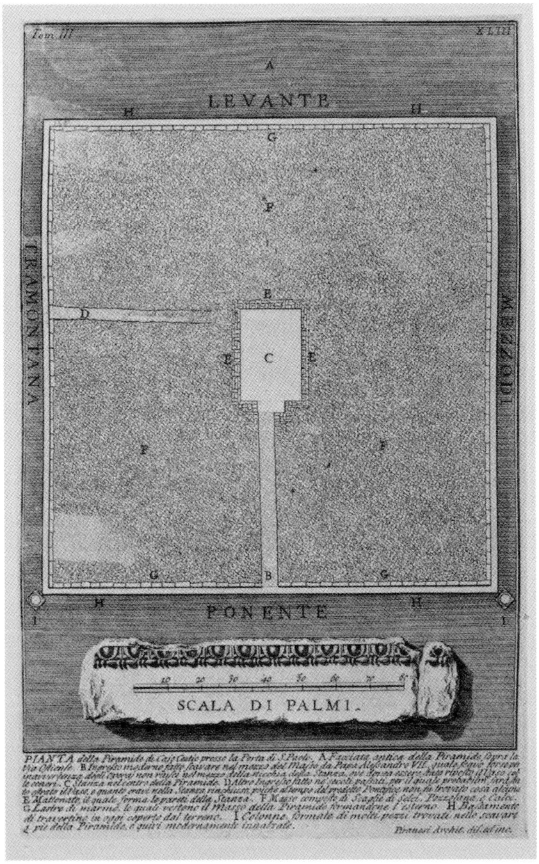

**212-213**. Giovanni Battista Piranesi, "Pianta della Piramide di Cajo Cestio presso la Porta di S. Paolo [...] e Spaccato della Piramide di Cajo Cestio," Rome (G.B. Piranesi, *Le Antichità Romane*, Roma, Bouchard and Gravier, 1756, vol. III).

**214**. Hubert Robert, Pyramides, 1760 circa (Northampton, Smith College Museum of Art).

Tom III. V
E
B
C
A

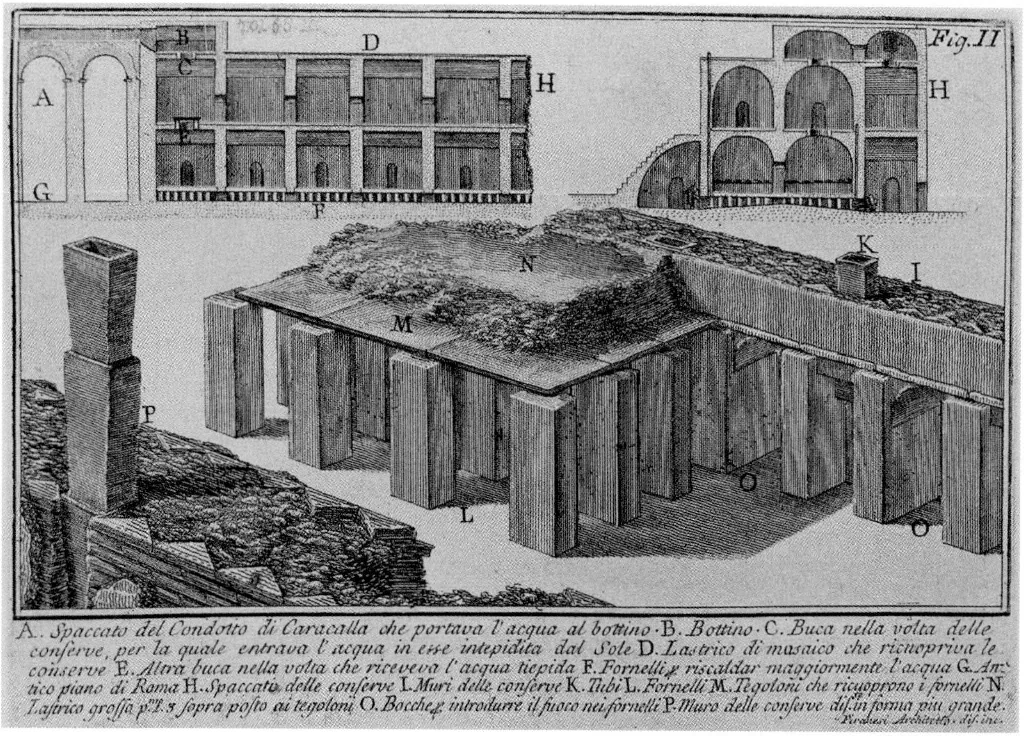
Fig. II
A
B
C
D
E
F
G
H
K
I
N
M
P
L
O
A. Spaccato del Condotto di Caracalla che portava l' acqua al bottino · B. Bottino · C. Buca nella volta delle
conserve, per la quale entrava l' acqua in esse intepidita dal Sole D. Lastrico di masaico che ricopriva le
conserve E. Altra buca nella volta che riceveva l' acqua tiepida F. Fornelli p. riscaldar maggiormente l' acqua G. An-
tico piano di Roma H. Spaccato delle conserve I. Muri delle conserve K. Tubi L. Fornelli M. Tegoloni che ricuoprono i fornelli N.
Lastrico grosso p.mi 3 sopra posto ai tegoloni O. Bocche p. introdurre il fuoco nei fornelli P. Muro delle conserve dis. in forma piu grande.
Piranesi Architetto dis. inc.

**215**. Giovanni Battista Piranesi, "Mattoni quadrati segati a linee diagonali per formare i Mattoni triangolari [...]" (G.B. Piranesi, *Le Antichità Romane*, Roma, Bouchard e Gravier, 1756, vol. III).

**216**. Giovanni Battista Piranesi, "Veduta degli Avanzi di Fabbrica magnifica sepolcrale co' sue Rovine la quale si vede vicina a Torre de' Schiavi [...]," near Rome (G.B. Piranesi, *Le Antichità Romane*, Roma, Bouchard and Gravier, 1756, vol. II).

**217**. Giovanni Battista Piranesi, "Spaccato del Condotto di Caracalla che portava l'acqua al bottino [...]," Rome (G.B. Piranesi, *Le Antichità Romane*, Roma, Bouchard and Gravier, 1756, vol. I).

aqueduct, *Avanzo del Condotto dell'Acqua Alsietina*, showing ruined walls, with the "lastrico composed of broken pots," the channel of the conduit "made with opera incerta, and clad on the outside and inside with opera reticolata,"[116] can be seen in relation to the major work of engineering undertaken by Vanvitelli in 1753: the Carolino Aqueduct. Vanvitelli himself offers a description of his aqueduct that contains echoes of those of Piranesi for the aqueducts of ancient Rome: "to prevent this very clear water from losing anything of its original purity, with the most precise diligence the whole inside of the conduit is fabricated and coated with a strong cement composed of lime mortar, lapilli and pozzolana."[117]

The working of the *lastrico* can be such as to create a clean surface, as Piranesi observes in the paving of the tomb of Alexander Severus: "lastrico composed of minute chips, bricks and lime, covered by another smoother layer of lastrico composed of the same material, but much finer."[118] A similar compound is documented as the screed of the roadbed of the Appian Way. "Bed of well-packed ground," the legend reports, "beaten with poles before spreading the large filler, one palm in height, resembling the lastrico composed of pozzolana lime and chips of flint, with the chert placed with force on top."[119]

The fact that the *Antichità Romane* include a number of plates on Roman bridges, with details on the foundations, piers and centering, makes Piranesi's anthology useful as documentation for engineers working on such structures.[120] A fundamental contribution of the collection has to do precisely with the centering system used by the Romans to build bridges. The drawings of the Cestius bridge in Rome, whose reconstruction dates back to the 4th century AD, illustrate the role of the corbels at the bases of the arches and the overhanging course at the top of the foundation pylons, as anchors for the structures of the centering.[121] The major European bridges of the time probably influence Piranesi's interest in this particular aspect of the construction.

Piranesi is aware of the fact that his compendium marks a turning point of veduta depiction and, at the same time, of the very idea of antiquity that is emerging through a range of different international publications. In the short introduction for scholars, *Presentazione agli studiosi della antichità romane*, in the first volume of the *Antichità Romane*, he complains of the lack of studies on Roman construction that would be able to "re-establish its importance and the most solid way of building." The statement projects his etchings into a contemporary dimension, almost as if they were plates of examples for future construction, and perhaps it reflects concerns that Piranesi may have discussed with builders like Lucchese, Temanza, Giobbe or Vanvitelli. Some of the criteria of representation developed to make construction sequences understandable are the same as those applied in famous construction manuals towards the end of the 1700s and in the early 1800s – including that of Rondelet.

"I have therefore depicted in the present volumes with all possible care and precision the above-said ruins," Piranesi explains regarding the drawing

[116] *Avanzo del Condotto dell'Acqua Alsietina*, *ibid.*, vol. I, plate XII, fig. 1.
[117] Luigi Vanvitelli, *Vita dell'architetto Luigi Vanvitelli*, Napoli, A. Trani, 1823, p. 48.
[118] *Pianta del Sepolcro di Alessandro Severo*, in, Piranesi, *Le Antichità Romane* cit., vol. II, plate XXXI, and, *Spaccato del Sepolcro di Alessandro Severo*, *ibid.*, plate XXXII.
[119] *Veduta dell'antica Via Appia*, *ibid.*, vol. III, plate VII.
[120] See, among the other plates, for the comprehension of the masonry core of "opera incerta," *Veduta dell'avanzo di una Pila del Ponte Trionfale*, *ibid.*, vol. IV, plate XIII; *Spaccato del Ponte Fabrizio, detto de' Capi*, *ibid.*, plate XIX (here we find an indication of the "courses of opera incerta"); *Elevazione del Ponte Ferrato*, *ibid.*, plate XXIII. In *Profilo del Tempio della Fortuna Virile* the vaults of the base are poured (*ibid.*, plate L, fig. II). In the *Sezione del serraglio delle fiere fabbricato da Domiziano* we see a powerful poured mass placed on the ground: "Walls of opera incerta, that stand in the territory of that hill, and that inside the structure are covered by tablets" (*ibid.*, plate LIV).
[121] *Cf.* Lynne C. Lancaster, *Concrete Vaulted Construction in Imperial Rome. Innovations in context*, New York, Cambridge University Press, 2005, p. 33.

**218**. Giovanni Battista Piranesi, "Veduta dell'avanzo di una delle Pile del Ponte Trionfale composta di travertini, e tufi, i quali rinchiudono, e maravigliosamente legano l'interno lavoro de corsi d'opera incerta [...]," Rome (G.B. Piranesi, *Le Antichità Romane*, Roma, Bouchard and Gravier, 1756, vol. IV).

**219**. Giovanni Battista Piranesi, "Elevazione del Ponte Ferrato [...]," Rome (G.B. Piranesi, *Le Antichità Romane*, Roma, Bouchard and Gravier, 1756, vol. IV).

**220**. Giovanni Battista Piranesi, "Spaccato del Ponte Fabrizio [...]," Rome (G.B. Piranesi, *Le Antichità Romane*, Roma, Bouchard and Gravier, 1756, vol. IV).

criteria adopted in the *Antichità Romane*, "representing many of them not only in their outer appearance, but also in terms of plan, and the interior: distinguishing their members by means of sections and profiles: and indicating their materials, and at times the manner of their construction, in keeping with what I have been able to portray during the course of many years of tireless observation and research [...]."[122]

The collections of etchings produced by Piranesi after the *Antichità Romane* like *Le Rovine del Castello dell'Acqua Giulia situato in Roma presso S. Eusebio, e falsamente detto dell'Acqua Marcia*, published in 1761, and the two published in 1762, *Il Campo Marzio dell'Antica Roma* and *Di due spelonche ornate dagli antichi alla Riva del Lago Albano*, reflect the tenets of the new orientation, though without achieving the same precision of technical detail.[123] After the publications of the *Antichità Romane*, no other veduta artist would venture into such technical detail as in the etchings of Piranesi, not even the abbot Ridolfino Venuti (1705-63), editor of the volumes of the *Accurata, e succinta descrizione topografica delle antichità di Roma*, published in 1763, though he does share Piranesi's interest in works of hydraulic engineering like the Cloaca Maxima.[124]

[122] In the rare construction sites initiated by Piranesi in Rome pozzolana is used to make the mortar, as is customary in that city. For example, to prepare for the visit, on 13 October 1766, of Clement XIII to the worksite of the church of Santa Maria del Priorato, a coating is prepared with pozzolana, spread in front and in back of the church, and on the garden, to create a sort of pavement (Giovanni Battista Piranesi, "Lavori fatti al muro di fianco ove fa prospetto alla Piazza avanti la Chiesa dalla parte dell'Orto del Convento dei P.P. de S. Sabina," 10 April 1767, in, Pierluigi Panza, *Piranesi architetto*, Milano, Guerini Studio, 1998, p. 170).

[123] *Il Campo Marzio dell'Antica Roma, Opera di G.B. Piranesi socio della reale società degli antiquari di Londra*, Roma, 1762.

[124] Ridolfino Venuti, *Accurata, e succinta descrizione topografica delle antichità di Roma*, Roma, G.B. Bernabò, G. Lazzarini, 1763, vol. I, pp. 52-53.

## Roman nostalgia, from Adam to Clérisseau

Restoring the qualities of Roman architecture means first of all staging the sequences of impressive vaulted spaces bordered by those powerful masonry structures only concrete could make possible, as Alberti had already intuited. Also thanks to the etchings of Piranesi and his imitators, architects rediscover the Roman qualities, and they attempt to replicate them in contemporary works. But for most of these architects the question is not constructive; it is not related to the erection of massive structures. No architecture attempts to repeat the feat carried out much earlier, in about 1466, at the Palazzo San Marco, in Rome, where a barrel vault was built in concrete with centering and formwork. Instead, while inventors, military men, engineers, naturalists and chemists attempt busily to discover the secret that made it possible to erect the great *massi* depicted by Piranesi – the composition of the cement and its modes of application – the young artists and architects who visit Rome in the 1750s and 1760s as part of their training, and sometimes even come into contact with Piranesi, consider the "magnificence" of Roman architecture as a question of the charm of the ruin in the landscape, and of the perception of sequences of spaces characterized by decorative stucco facings. Nothing could be further from Piranesi's intuition on the true nature of Roman architecture than what is seen and drawn by his young European admirers.

What William Chambers (1723-96) writes about the "celebrated Italian Artist" sounds like revenge against Piranesi for not having been able to curry his favor during his stay in Rome, from 1750 to 1755: "[he] knew little of construction or calculation, yet less of the contrivance of habitable structures, or the modes of carrying real works into execution [...]."[125] Chambers is one of the travelers initiated in the cult of Roman ruins. It is precisely the evoking of one of these ruins that he imagines in his project for a fantastic mausoleum for the Prince of Wales, designed in February 1752, when he is in Rome. The thickness of the masonry shown in the drawing are those of poured Roman walls and domes – a sort of critical reworking of the tomb of Cecilia Metella – though the pouring technique does not seem to have been indicated by Chambers as a fundamental feature of Roman construction. "My intention was to imitate a Roman antiquity built of brick with an incrustation of stone," he writes, regarding another one of his projects, the "Ruined Triumphal Arch for Kew" in 1759.[126]

In the drawings of two acquaintances of Chambers, Charles-Louis Clérisseau (1721-1820) and Robert Adam (1728-92), the reference to Roman construction does not go beyond the generic indication of a "mass," meaning powerful, thick masonry. Their drawings of Hadrian's Villa at Tivoli,[127] of the so-called nymphaeum of Egeria in Rome,[128] or the notebook of drawings of ruins in the Roman countryside, attributed to Adam,[129] confirm the rise of a new way of looking at Roman architecture brought about by Piranesi, with an increasing

[125] William Chambers, *Treatise on the Decorative Part of Civil Architecture*, London, 1791, p. 10.
[126] William Chambers, *Plans, Elevations, Sections and Perspective Views of the Gardens and Buildings at Kew in Surrey, the Seat of Her Royal Highness the Princess Dowager of Wales*, London, 1763 (cit. in, Thomas J. McCormick, *Charles-Louis Clérisseau and the Genesis of Neo-Classicism*, Cambridge-Massachusetts, MIT Press, 1990, p. 21).
[127] McCormick, *op. cit.*, pp. 40-41, figg. 31-34.
[128] The drawings of Robert Adam, *Grotto of Egeria* (Penicuik House, Midlothian), and of Clérisseau, *Vue de l'intérieur de la grotte de la nimphe Egérie à Rome* (Hermitage), could document a type of ruin Montano might have imagined restoring to its original state (for the drawings of Adam and Clérisseau see, McCormick, *op. cit.*, p. 46, figg. 37-38).
[129] *Cf.* Cara D. Denison, Myra Nan Rosenfeld, Stephanie Wiles, ed., *Exploring Rome: Piranesi and His Contemporaries*, Cambridge-Massachusetts, London, 1993, pp. 129-31. See, John Fleming, *Robert Adam and His Circle in Edinburgh & Rome*, London, J. Murray, 1962.

**221**. Giovanni Battista Piranesi, "Rovine d'una Galleria di Statue nella Villa Adriana," Tivoli (G.B. Piranesi, *Vedute di Roma*, Roma, 1757).

**222**. William Chambers, mausoleum for the Prince of Wales, 1752 (T.J. McCormick, *Charles-Louis Clérisseau and the Genesis of Neo-Classicism*, Cambridge-Massachusetts, MIT Press, 1990).

**223**. Robert Adam, view of ruins in the Roman countryside (C.D. Denison, M.N. Rosenfeld, S. Wiles, ed., *Exploring Rome: Piranesi and His Contemporaries*, Cambridge-Massachusetts, London, 1993).

focus on cement construction techniques (in 1755 Clérisseau and Adam explore several Roman ruins in the company of Piranesi himself).[130]

Some of the drawings by Adam of ruined vaults, like the ones he sees in Hadrian's Villa in 1756, with the masonry reduced to fragments suspended in the void and against the sky, demonstrate that he knows about the Roman technique, not in squared stone, but in cement and brick.[131] Like Chambers, Adam too imagines erecting a fake Roman ruin with the thickness of the "mass," in Hackfall Park, Yorkshire, around 1758.[132]

Clérisseau's view of a fantastic Roman hall and the so-called "room of the ruins" in the convent of Trinità dei Monti in Rome, painted in approximately 1766 for the mathematician-monks Thomas Le Sueur and François Jacquier, bear witness to the spread of the taste for the construction methods of Roman masonry vaults clad in stucco. But Clérisseau, when he documents the construction details, depicts vaults or walls of brick, not of poured material, even if certain drawings of arched or vaulted monolithic structures cannot possibly represent anything else other than the *masso* repeatedly described by Piranesi. The fact that Adam sets aside the idea, developed with Clérisseau, of surveying the Roman ruins published by Antoine Babuty Desgodets (1653-1728) and making a collection of the drawings with the precise measurements, opting instead to study ruins like those of Hadrian's Villa and the Baths of Caracalla or of Diocletian (according to what he states in a letter in 1756) should probably be seen as confirmation of the charm exerted by the spatial configurations generated by Roman concrete, rather than by the relics of the architectural orders,[133] but without an associated interest in technical matters. From this standpoint, the drawings of Clérisseau and Adam represent emblematic cases of the rendering, in perspective, of the vast size of the dome of the Pantheon, and not its constructive system.[134]

The exercise of Clérisseau and Adam of imagining fantastic, grand Roman baths in their original state, as in the sketch by Adam of a "reconstruction based

[130] McCormick, *op. cit.*, pp. 28-29. With them, there is also the painter Laurent Pécheux (1729-1821).
[131] *Cf.* Mario Bevilacqua, Heather Hyde Minor, Fabio Barry, ed., *The Serpent and the Stylus. Essays on G. B. Piranesi*, Ann Arbor, Michigan, University of Michigan Press, 2006, p. 220.
[132] *Cf.* Bevilacqua, Minor, Barry, *op. cit.*, p. 226.
[133] Robert Adam, letter to James Adam, 11 September 1756, cit. in, McCormick, *op. cit.*, p. 52.
[134] *Ibid.*, pp. 44-45, figg. 35-36.

**224**. Charles-Louis Clérisseau, ancient tomb near Pozzuoli (T.J. McCormick, *Charles-Louis Clérisseau and the Genesis of Neo-Classicism*, Cambridge-Massachusetts, MIT Press, 1990).

**225**. Robert Adam, "View of the Inside of the Vestibule," Diocletian's Palace, Split (R. Adam, *Ruins of the Palace of the Emperor Diocletian at Spalatro in Dalmatia*, London, 1764).

**226**. Charles-Louis Clérisseau, "Room of the ruins," Convent of Trinita dei Monti, Rome, 1766 circa (T.J. McCormick, *Charles-Louis Clérisseau and the Genesis of Neo-Classicism*, Cambridge-Massachusetts, MIT Press, 1990).

on the Roman baths," cannot help but raise questions about what constructive means would be necessary, in the contemporary era, to make that vision reality.[135]

Certain plates of the publication produced by Robert Adam, with the collaboration of Clérisseau, *Ruins of the Palace of Diocletian*, printed in 1764, show ruined Roman domes, of brick and poured material, where the large stones are only those of the fragments of trabeations and columns lying on the ground.[136]

Nevertheless, not one of Piranesi's imitators dares to publish views of construction details with the precision and force of image of those engraved by the Venetian visionary. And in fact neither Clérisseau nor the Adam brothers are involved in the rediscovery of the cement that had made it possible to erect the monuments they had seen in ruins. However, this rediscovery does takes place during the period in which they are leading figures in the growing vogue of antiquities, both in Europe and the United States. It is no coincidence that Adam does file a patent for a facing cement. The entire lesson of Rome and its ruins translates, then, into the theory of a sequence of rooms varying by their coverings and thus generating a movement, as Robert's brother James Adam eloquently notes (in 1760 he goes, with Clérisseau, to visit Naples, Pozzuoli and Baia).

"One thing which [conduces]," he writes in 1762, "greatly to the beauty and variety of movement in the section is the form of the vaults and domes. A proper mixture of domes, vaults and coved ceilings and flat soffits over rooms of various shapes and sizes are capable of forming such a beautiful variety as cannot fail to delight and charm the instructed spectator. A movement in the section is likewise derived from steps in a great circular or long room and also mention to be made of an attic occasioning interior movement."[137]

The possibilities of erecting Roman vaults are based on structural systems that were already established during the course of the 1600s – masonry in stone or brick, or wooden carpentry with an envelope of light materials applied and then plastered. It is significant that in 1758 Adam designs a Roman ruin for the

[135] For the drawing see, *ibid.*, p. 51, fig. 47.
[136] See, *View of the Inside of the Vestibule*, in, Robert Adam, *Ruins of the palace of the Emperor Diocletian at Spalatro in Dalmatia*, London, For the Author, 1764, plate XXIII.
[137] James Adam, notes for a treatise on the theory of architecture, Rome, 27 November 1762, cit. in, John Fleming, *Robert Adam and His Circle in Edinburgh & Rome*, London, J. Murray, 1962, p. 315.

**227**. Robert Adam, drawing of a ruin, property of Nathaniel Curzon, Kedleston, Derbyshire (T.J. McCormick, *Charles-Louis Clérisseau and the Genesis of Neo-Classicism*, Cambridge-Massachusetts, MIT Press, 1990).

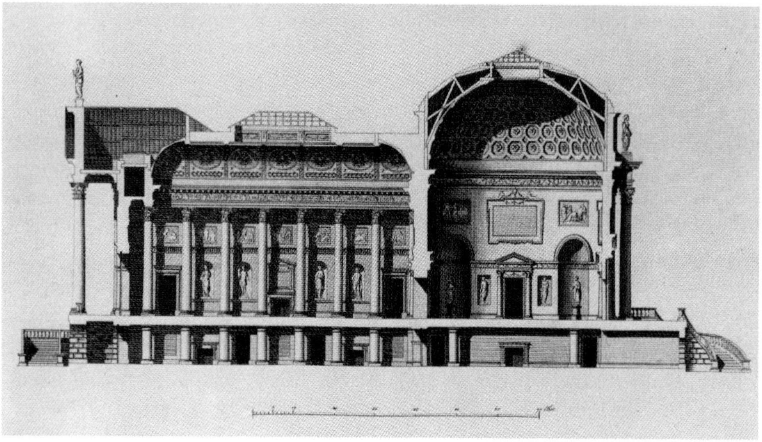

**228**. Robert Adam, property of Nathaniel Curzon, Kedleston, Derbyshire, 1760-65 (*The works of architecture of Robert and John Adam*, London, P. Emsly, 1774).

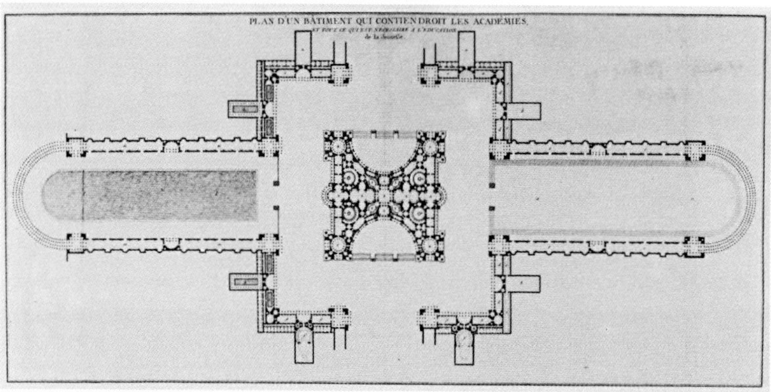


**229**. Marie-Joseph Peyre, "Plan of a building with academies, and all necessary for the education of youth" (M.-J. Peyre, *Oeuvres d'architectures, contenants différents projets d'édifices publics et particuliers, et plusieurs bâtiments*, Paris, 1765).

park belonging to Nathaniel Curzon at Kedleston, with vaults and domes whose masonry bulk would be typical of poured structures, similar to what he had seen at Hadrian's villa or in Rome; but in the villas he builds, the vaults and domes are reduced to mere thin shell, at times perhaps made with shaped and laid elements, and protected by carpentry supporting the roof (as demonstrated by the section of Kedleston Hall, and in particular the Saloon, in 1760).[138]

Just how far the idea of the Roman sequence of spaces could be developed in contemporary architecture, without any concern regarding the technique required to replicate it, can be seen in the projects of Marie-Joseph Peyre (1730-85), who also visited Rome from 1753 to 1756, an enthusiast of Hadrian's Villa and, peerlessly, of the Campus Martius of Piranesi.

Other artists are influenced by the vision of Piranesi of the constructive magnificence of Roman architecture, but are skeptical about the possibility of using poured materials to make vaults. This can be deduced from the many views of Roman ruins drawn by Hubert Robert (1733-1808), who lives in Rome from 1754 to 1765. At times Robert draws ruins in the company of Piranesi. In 1760 he accompanies the abbot Jean-Claude-Richard de Saint-Non (1727-91) on visits to the Roman ruins at Baia and Pozzuoli, where other masterpieces of pozzolana cement construction can be seen.[139] Even if the view of the ruins of the cement vaults of the Colosseum, *Washerwomen in the Ruins of the Colosseum*, dating back to the 1760s, belongs to the genre of the Piranesian *masso*, it must be stressed that Robert's vision of an architecture of Roman forms, one that can be built with French methods, is epitomized by the setting of the *Finding of the Laocoön*, painted in 1773. Here we find a grand space roofed with barrel vaults in cut stone with a regular geometric pattern, in keeping with the tradition of the *coupe des pierres* – a transfiguration of the perspective effects of the Scala Regia of Bernini in the Vatican.[140]

## Research of Delorme on the Roman aqueducts, or mobile mold

Though the veduta makers do not seem willing to understand and depict the original technical characteristics of Roman architecture, the construction details of bridges and aqueducts engraved by Piranesi do not remain isolated episodes for very long. The growing interest in hydraulic works prompts engineers to delve into Roman archaeology, often with the aim of documenting technical solutions that may be of use for contemporary construction. After Fontana and Piranesi, an important role in the discovery of the compound and the construction methods used by the ancient Romans for concrete masonry is played by Delorme, the engineer from Lyon behind the hydraulic works of the Canal de Givors, constructed to connect the Rhone and the Loire.

Perhaps stimulated precisely by the success of the views of the *Antichità Romane* of Piranesi, Delorme conducts his own major campaign of observation of the vestiges of Roman aqueducts around Lyon, and on 5 June 1759 he

[138] Bevilacqua, Minor, Barry, *op. cit.*, p. 228.

[139] Jean-Claude-Richard de Saint-Non, *Voyage pittoresque à Naples et en Sicilie*, Paris, Dufour et C.ie, 1829, vol. I. Saint-Non describes the terraces of houses in Naples and vicinity: "They are built, in fact, as can be seen, without a roof, covered only by flat roofs (terrasses) made out of this eternal pozzolana cement, whose property is to harden in water. [...]. These flat roofs are indeed extremely solid and stand up to heat as well as rain. They are extremely pleasant for Neapolitans who frequent them in order to cool off; for them these flat roofs are new gardens to decorate with trellises, vines and flowers" (*ibid.*, pp. 166-67).

[140] See, Hubert Robert, *Washerwomen in the Ruins of the Colosseum* (Brooklyn Museum, New York), and, Hubert Robert, *The Finding of the Laocoön* (Virginia Museum of Fine Arts, Richmond).

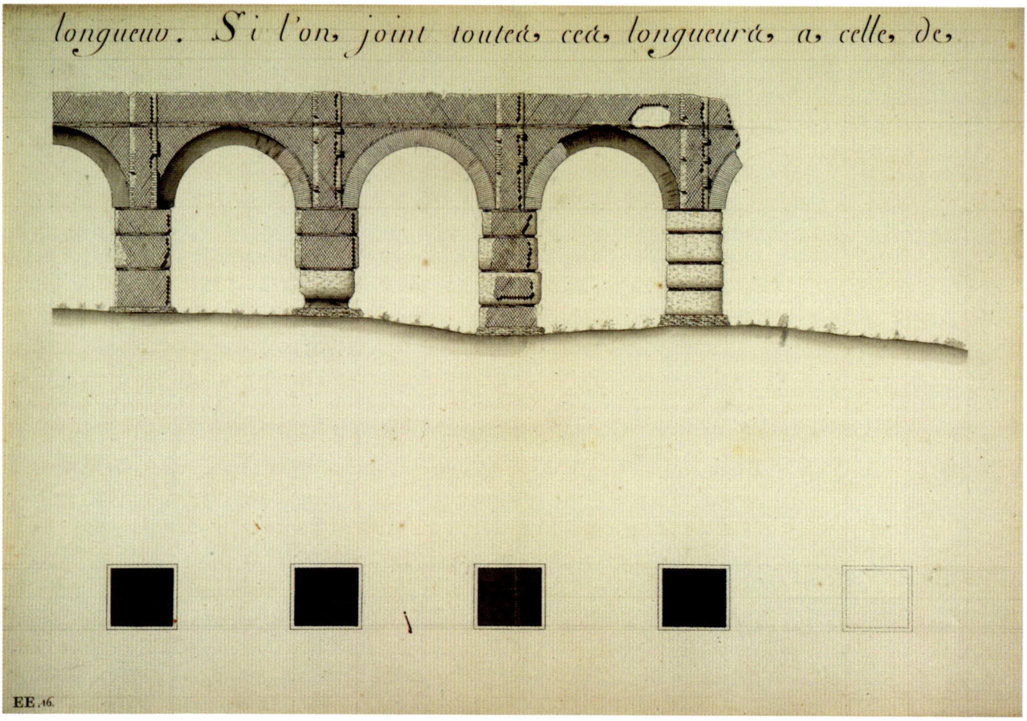


**230-232**. Guillaume-Marie Delorme, surveys of the Roman aqueduct near Lyon, 1759 (Académie des Sciences, Belles-Lettres et Arts, 39 Fi 98, 100 and 127, Archives Municipales, Lyon).

presents the results to the Académie des Sciences, Belles-Lettres et Arts of Lyon (the Aqueduct of Gier, known as "of the Pilat"; those of Brévenne at Ecully, Tassin-la-Demi-Lune, Yzeron at Craponne, and Plat de l'Air at Chaponost; as well as works like the Pont des Granges in the town of Saint-Maurice-sur-Dargoire). He continues to make surveys of the Roman aqueducts in the Lyon area at least until the 1760s, also with a publication in mind, never completed due to the death of its patron, Anne Claude Philippe de Tubières, Count of Caylus (1692-1765).

The publication of the first series of drawings, in 1760, entitled *Recherches sur les Aqueducs de Lyon, construits par les Romains*, contains useful information on the Roman type of conglomerate masonry. The fact that Delorme begins a series of experiments to determine the composition of Roman mortar just after this publication underscores the impact of this archaeological research on contemporary concrete construction.

"Speaking from experience," he writes, "one-third quicklime and two-thirds large-grained sand always make for an excellent concrete and thick mortar; and with the comparison I made of their mortar with the one of my experiments, it seems that they have observed the same proportions. Much less lime is needed when fine-grained sand is used; and this is why our workers prefer sand from the Rhone over sand from the Saône, which is coarser-grained and infinitely better."[141]

Delorme explains the type of masonry that recurs in the Roman aqueducts near Lyon. "This construction," he writes about the segment of aqueduct in the

[141] Guillaume-Marie Delorme, *Recherches sur les Aqueducs de Lyon, construits par les Romains*, Lyon, A. Delaroche, 1760, p. 33. Important references to Delorme's essay and the Roman construction systems for aqueducts are found in, Thomas Pownall, *Notices and Descriptions of Antiquities of the Provincia Romana of Gaul, New Provence, Languedoc, and Dauphine [...]*, London, J. Nichols, 1788, pp. 170 *et seq.*

village of Chaignon, "was made out of small rubble stones, from three to six inches thick, always lain on a mortar bed, that is completely sealed at the joints, and forms a complete and unbreakable unit. For, I observed that, with full knowledge of the facts, the large rubble was rejected for the entirety of this large construction. [...]. Using a good mortar contributes much to the solidity of their construction. It was made with large-grained sand from rivers [...]. One can see that the lime was not used sparingly, and it was made with proper calcareous stones. [...]. The cement used in the Aqueducts was composed of pulverized brick for the last layers of coating: for in the thickness of one and a half inches that they put on the side of the walls, broken pieces the size of peas were mixed in. The layer in the basin, six inches thick, contains walnut-sized and egg-sized pieces of brick. The mixture was made with newly slacked lime, without any other composition; unless it was diluted with wine or vinegar, which is not perceptible. Besides, what matters the most in making a good cement, like a good mortar, is that

233-234. "Architecture Maçonnerie". Figure 13 of plate II documents the system of coffers and filler already indicated by Palladio; plate IX show some solutions already explained by Bélidor (*Encyclopédie, ou Dictionnaire Raisonné des Sciences, des Arts et des Metiers. Recueil de planches, sur les sciences, les arts liberaux, et les arts méchaniques, avec leur explication*, Paris, Briasson, David, Le Breton, Durand, 1762).

both are well bonded, and that the mixture is made to exacting measures before being watered down. One can tell that this is the case when the color is uniform throughout the mixture. It is only then that one can start pouring a little water if it is for making mortar, and some wine or vinegar if it is for cement."[142]

*Recherches sur les Aqueducs de Lyon* also contains an important description of the processes of construction by pouring into formwork – coffer (*encaissements*) and molds (*moules*) are the terms used by Delorme. In this sense, Delorme's contribution completes what was written by Fontana in 1710 and by Bélidor in 1753 regarding formwork, and confirms that at this point engineers are increasingly interested in the question of the temporary wooden structure required for the pouring, going well beyond the hints on wooden formwork found in the descriptions of voyages made in the early decades of the 1700s in Mediterranean countries. Delorme's hypotheses on the construction process using formwork remain isolated in the specialized literature on Roman archaeology. Not even Piranesi ventures into the study of those temporary wooden structures that had left only imprints behind them, on the mass of the concrete. Delorme is one of the first to pose the crucial question of formwork in ancient Roman construction, with this concept emerging from his earlier studies of *pisé*, containing an accurate description of the *encaissements* in wood.

Starting with the criteria of construction in *pisé* using sliding molds, Delorme imagines a process of pouring used by the ancient Romans in which the courses of bricks passing through the concrete core also mark the phase of sliding of the mold, indicating the sequence of the pourings.

"These piers," he writes regarding the Roman aqueducts, "could only have been built using four-feet high mold, tied together by hoops or belts, like with our square wine tanks. This mold was necessary for forming the construction and maintaining it until it had solidified enough to support itself; for without this precaution the thresholds could not have held up against the force exerted by the wall facings. The molds were supported by props. Their supports were not fixed to the structure, as is ordinarily done with scaffolding. We would be able to recognize putlog holes if there were any. What's more, they would be impossible to make with walls linked in a network. If one were to build the pile without interruption, it would be necessary to have as many molds as the elevation required: the props would then be useless. The worker inside the mold did not need to raise, with even courses, more than one level at a time of the four sides of the rhombus-shaped pile, for which the mold would serve as a support, using plumb-bob and line. The mass of the pier was filled step by step as the courses of the walls were elevated; and after a four foot elevation, the whole structure was covered with two courses of these large bricks that bind the wall to the masonry piers, on which, by using a second mold, it was raised another four feet, and so on until the same level of the other piers was reached in order to receive the arcs that formed the bridge."[143]

By now, at the start of the 1760s, it becomes clear that some of the masonry masses in conglomerate admired by Piranesi were made thanks to temporary wooden structures, like those seen in Rome by Nollet or those still in use for *pisé*. So after Delorme's detailed description of the process of concrete construction using formwork, it should come as no surprise that the technique of wall construction with cement material poured into molds is documented by the *Encyclopédie* of Diderot and D'Alembert. In 1762 a plate of drawings in the *Encyclopédie* depicts a *mur de remplage* or filler wall, alluding to the term of Palladio, also used by Piranesi, of "filler."[144] The description of the construction method confirms the spread of hypotheses regarding formwork and the ancient Romans, such as the one outlined by Delorme, though the example of the concrete masonry poured into formwork seen on Lake Garda, comes directly from the treatise of Palladio.[145]

"The sixth way [to build] was [with] filler walls (fig. 13)," we read in the *Encyclopédie* about the *murs de remplage*, "by building a kind of chest as high as it was desired for the layers to be, made with planks held up by flying buttresses, filled with mortar, cement and all different forms and sizes of stone: layer by layer was built this way: and there are examples of walls built this way in Sirmion, on Lake Garda. There was yet another ancient way to build ramparts (fig. 14); by making walls four feet thick, six feet apart, linked together by transversal walls also six feet long, to make a type of chest of six square feet that was then filled with earth and stone."[146]

Another similar solution, also documented by the *Encyclopédie*, is that of the walls built for the foundations of bridge piers, where formworks are used, to be filled with oily soil.[147]

[142] Delorme, *Recherches sur les Aqueducs de Lyon* cit., pp. 31-34.
[143] *Ibid.*, pp. 41-43.
[144] *Cf.* Denis Diderot, Jean-Baptiste Le Rond d'Alembert, ed., *Encyclopédie, ou Dictionnaire Raisonné des Sciences, des Arts et des Métiers. Recueil de planches, sur les sciences, les arts liberaux, et les arts méchaniques, avec leur explication*, Paris, Briasson, David, Le Breton, Durand, 1762, p. 15, plate II, fig. 13.
[145] *Cf.* Palladio, *op. cit.*, p. 13.
[146] Jacques-Raymond Lucote, *Maçonnerie*, in, Denis Diderot, Jean-Baptiste Le Rond d'Alembert, ed., *Encyclopédie Ou Dictionnaire Raisonné Des Sciences, Des Arts et des métiers*, Neufchastel, S. Faulche & Compagnie, 1765, vol. IX, (pp. 803-36), p. 806.
[147] Denis Diderot, Jean-Baptiste Le Rond d'Alembert, ed., *Encyclopédie, ou Dictionnaire Raisonné des Sciences, des Art set des Métiers. Recueil de planches, sur les sciences, les arts liberaux, et les arts méchaniques, avec leur explication. Seconde livraison*, Paris, Briasson, David, Le Breton, Durand, 1763, pp. 4-5, plate XIX.

## Observations of Winckelmann on the "a sacco" masonry

After the publications of Piranesi and Delorme, and in the wake of growing interest in construction based on pouring materials into formwork, even believers in the myth of the superiority of ancient Greece cannot help but notice the qualities of Roman construction and its potential for the evolution of contemporary practices. A literary work that meets with great international success, containing painstaking descriptions of the construction processes of the ancient Romans, is *Anmerkungen über die Baukunst der Alten*, written by the greatest 18th-century German art historian, Winckelmann, and published in 1762 (the preface is dated Rome, 1 December 1760).[148] *Anmerkungen* should be interpreted against the backdrop of the volumes of the *Antichità Romane*, as the translation into literary form of the fascination with construction conveyed by Piranesi through his drafting.

The fact that *Anmerkungen* was written during Winckelmann's stay in Rome, beginning in 1755, and after his first trip to Naples, in 1758, is decisive for an understanding of the many technical details on the Roman construction method, studied directly from ruins and contemporary worksites, and related in the book by an author trained in a university setting, who prior to reaching Italy knew of the ancient world only from literary sources. Winckelmann's interest in contemporary construction is demonstrated by his observations on the quarries of Rome, where he notes that tuff is still quarried, but in small pieces to use "for foundations, vaults and masonry filler."[149]

At the start of *Anmerkungen*, Winckelmann provides impassioned reasoning on the construction materials of the ancients, significantly putting the accent on tuff and pozzolana, on their use in ancient and contemporary construction, and the various terms in use in Naples and Rome – *rapillo*, *lapillo* or *peperino* – precisely when research is being conducted on volcanoes and their products at an international scale by naturalists and builders. It is no coincidence that Winckelmann's pages on volcanic products are often mentioned by German mineralogists and volcanologists during the course of the 1770s. And it is important, in this sense, that already in the mid-1760s Winckelmann is in contact with one of the greatest volcanologists and experts on pozzolana, Desmarest.

The pages on pozzolana offer an initial and significant overview of the various types found in Naples and Rome, also with remarks on their exportation. Winckelmann is so fascinated by the Roman construction method that he even raises the hypothesis that the ancient Romans discovered vaults as a result of the potential of pozzolana and cement.

"The third sort of the materials, namely lime mortar," he writes, "was prepared by the ancient Romans, as in the present as well, with pozzolana; an earth then known by the same name, that is pulvis puteolanus, undoubtedly because it was discovered for the first time at Puteoli, now known as Pozzuolo, near Naples. Pozzolana is either blackish or reddish: the blackish version is ferruginous, heavier and dryer than the other; it is used mostly for construction under water, because when exposed to air it crumbles easily; the other is more earthy, and better for construction on land. The first kind is found around Naples, though not the second: but both are quarried in Rome, and its vicin-

[148] Johann Joachim Winckelmann, *Anmerkungen über die Baukunst der Alten*, Leipzig, J.G. Dyck, 1762.

[149] *Ibid.*, p. 4.

[150] Johann Joachim Winckelmann, *Osservazioni sull'architettura degli antichi*, in, Carlo Fea, ed., *Opere di G. G. Winckelmann, Prima edizione italiana completa*, Prato, F. Giachetti, 1836, vol.

ity, and are not found in any other part of Italy. Given all this, it should be noted that the ancients made little use of reddish pozzolana; when, instead, today it is more highly prized than the blackish one. Nor is pozzolana found in the countryside of Rome near the sea: so the ancients, who used it at Anzio, must have brought it from Naples, where it is obtained still today; since it cost less to bring it there by sea than to transport it by land from Rome. It is brought to Tuscany by sea as far as Leghorn, and it is also shipped to other places. Alberti, in his work on architecture, speaks of pozzolana as a thing of which he had no knowledge, except through the reports of others: it could not have been known to him in any other way, since he was Florentine. In other places, it is confused with *rapillo*. It appears that pozzolana has never been found in Greece, as Vitruvius observes: and precisely due to lack of this material, the Greeks could not make vaults with ease. This should not lead us to believe that they did not know the secret of making excellent lime; we have proof of this in the large water tank of Sparta made with gravel, which with the binder of lime make a body as hard as the stones themselves. The two kinds of pozzolana set equally hard, like stone; in fact the lime that is mixed in becomes harder than the stones themselves, binding them and uniting them; as can be seen in the ruins of constructions on the sea, and built in the sea itself at Pozzuolo, and at Baja, and in all that country, and at the port of Anzio, where the piers that formed and enclosed the port, and also the above-mentions structures, were made with bricks. Pozzolana was likewise used for the paving, in ancient times, of large and small roads in Rome, and its vicinity: a method that can still be observed today."[150]

Winckelmann goes into detail on construction with pozzolana cement, giving the pages of the *Anmerkungen* the character of a manual of the Roman art of construction. Winckelmann grasps the particular qualities of this construction system, with rubble mixed with "pozzolana lime" to be poured into cavities – the rubble masonry, coffer-work according to Ware, or *a sacco* as he writes, conserving the Italian term, leading to the German translation: *muldenweis*.[151] Thus he explains that once the trench has been dug for the foundations of a construction, alternating layers are poured of "Kalk mit Puzzolana"[152] and fragments of tuff until the cavity has been filled, obtaining masonry that sets in two days; then the walls are built, using more pozzolana cement than pebbles. Winckelmann does not, like Delorme, address the case of walls inside formwork. Rudolf Erich Raspe (1736-94) contributes to spread Winckelmann's observations in German and English scientific circles regarding *a sacco* masonry with pozzolana.[153]

The explanations included in the *Anmerkungen* on the construction of vaults using wooden centering go into greater detail, almost as if Winckelmann intends to make up for the silence of the literature on this type of temporary structure and on the processes of pouring concrete. Like Fontana, Nollet and Delorme, Winckelmann too pays attention to the imprints left by the wooden boards of the centering on Roman cement, as documents that help to reconstruct the process of deployment of the materials. After all Winckelmann, like Nollet before him, can observe not only the ruins, but also the worksites of contemporary Rome, where masons are building vaults in conglomerate in

VI, pp. 58-61; Winckelmann, *Anmerkungen* cit., pp. 6-7.
[151] Winckelmann, *Anmerkungen* cit., p. 9.
[152] *Ibid.*, p. 8.
[153] *Cf.* Rudolf Erich Raspe, *An account of some German volcanos, and their productions. With a new hypothesis of the prismatical basaltes*, London, L. Davis, 1776, p. 132.

235. Giovanni Battista Piranesi, "Veduta della Villa dell'E.mo Sig.r Card. Alessandro Albani fuori di Porta Salaria," Rome (G.B. Piranesi, *Vedute di Roma*, Roma, 1769).

the desired form, including some with very depressed arches. It is in these terms that Winckelmann outlines the construction phases of poured vaults in ancient Rome:

"When the framework of the vault had been formed by boards, lime with rubble of tuff, or brick, was poured, as in the construction of foundations; and this up to a certain height, which was nine palms in the Baths of Diocletian: after which, another layer of lime was placed to unify the surface of the vault. In this way, a small number of men could finish a large vault in just one day. This method of fabrication can be observed in the works from which the coating has fallen, as in ruined vaults, like those of the Colosseum, for example, or the Baths of Titus, Caracalla and Diocletian; and particularly in the considerable ruins of Hadrian's Villa, where the imprint of the framework boards can still be seen. This rapid method of constructing vaults is no longer used: today they are made by hand, though still by using tuff and pozzolana. The filling, until everything is packed to the plateau level, is even still done with the trough method *a sacco*[154], and pressed for one day, as the ancients did. By means of that lime, it is possible to give the vault any chosen form. In Rome vaults are still made that are almost flat, just barely perceptible as being curved. These vaults are left for some time with their framework, so they can solidify. While making their vaults extremely strong, the ancients also attempted to make them as light as possible; this was done in two different ways. The most ordinary manner was to fill the vaults with rubble from Mount Vesuvius, mostly pebbles and shards. Black ones are found near Viterbo, in a place where there are boiling hot springs, with water that hardens eggs in an instant. This place is known as Bollicame, a name that refers to the boiling. The subterranean fire, like the scoria found here, lead to the idea that this was once a volcano. But the scoria of Viterbo is not very good for making vaults, as it is too weak. This type of scoria can clearly be observed in ancient buildings, and was found in the Pantheon, when it was recently restored. Nevertheless, neither Vitruvius nor his commentators have spoken of this manner of constructing vaults; and he mentions the volcanic materials of Vesuvius only in passing. Since the nature of this mountain was little known to the ancients, they did not extensively seek to discover its phenomena."[155]

Winckelmann's observations are completed by descriptions of the way of making light vaults by using empty earthen pots around which to pour small

stones and pozzolana lime *a sacco*,[156] or to cover the rooms of houses, as can also be seen in the ruins of Pompeii: "they were made at times with crushed and split Greek reeds, known as reed vaults, for which Vitruvius and Palladio provide indications on the construction. The vaults were given their form with wood, and with boards, on which to attach the flattened reeds, which are generally longer in Italy than in Germany. On the reeds, scoria from Vesuvius was placed; the lime mixed with pozzolana was placed over it; then the last coat was made with crushed marble and plaster."[157]

During the writing of the *Anmerkungen*, Winckelmann is involved in the construction of a "casino," outside the Porta Salara, of the cardinal Alessandro Albani (1692-1779), for whom since 1758 he has been in charge of the collection of books, prints, drawings and antiquities. Albani's casino, designed by Carlo Marchionni, becomes an example of the construction of vaults with the Roman method thanks to the pithy description included in the *Anmerkungen*. The few words devoted by Winckelmann in the context of his reasoning on volcanic materials used in construction and on pozzolana cement give Albani's casino the aura of a model for contemporary construction using ancient methods.

"Vaults covered with such volcanic material are very common in Naples," he observes, regarding the products of Vesuvius, "but Cardinal Albani was the first, and until now the only one, to have something similar built in Rome. Here is the way with which to proceed in such a construction: after the framework has been made, the sides are filled [...] up to the plateau, or halfway along the vault. This plateau is then covered with the rubble and lime, which amalgamate together and consolidate, so that it is nearly impossible to destroy such a work."[158]

154 Winckelmann, *Anmerkungen* cit., p. 9.
155 Winckelmann, *Osservazioni* cit., pp. 63-65; Winckelmann, *Anmerkungen* cit., pp. 8-10.
156 Winckelmann, *Osservazioni* cit., p. 65.
157 *Ibid.*, p. 166.
158 Winckelmann, *Osservazioni* cit., p. 65; Winckelmann, *Anmerkungen* cit., p. 10.

# chapter five
# caissons and hydraulic mortars in the 1750s

## Foundations without pozzolana: evolution of the caisson, from Perronet to Tardif

Construction tools and worksite logic in the fabrication of masonry for foundations in water – on riverbeds to erect the piers of bridges, or in the sea to make breakwaters, levees or ports – can be seen to change at times reflecting the map of the trade routes of exportation of pozzolana from Civitavecchia. Until the second half of the 1700s, the foundation system using the caisson made famous by Labelye is perfected against the backdrop of the difficulty in obtaining pozzolana, the sole ingredient that can guarantee the construction of masonry structures that set in water. The development of the caisson, the study of the technical details to make it watertight, and the inventions to permit use even in deep water and on unstable beds, represent the agenda of carpenters and engineers in and around 1750, with the aim of constructing masonry whose mortar can set slowly on a worksite not immersed in the water of a river or sea.

The caissons in use at Italian ports, from Civitavecchia to Viareggio to Ancona, permit the transport of masonry to the point where it is to be sunk, and are true permanent formwork in the pouring of concrete with pebbles embedded in pozzolana mortar. The caissons that become part of building practice in continental Europe, often like great floating machines, continue to transport another type of masonry, made with courses of stones clamped by metal and bound with a small amount of hydraulic mortar. In definitive terms, they constitute an evolution of the cofferdam – the masonry is not placed against the sides of the caisson, as in the Italian cases. The publication in 1751 of *A Description of Westminster Bridge,* in which Labelye explains the construction process using caissons for the piers on the bed of the Thames, contributes to the international spread of his foundation method. Labelye's model has a direct influence on the evolutions of the caisson in France, as it had developed at the start of the century.

The technical questions of foundations of bridge piers are discussed systematically by the Assemblée des Inspecteurs Généraux des Ponts et Chaussées in the 1740s, with practical recommendations and observations proposed by Pitrou, Boffrand, Bayeux and Jean Hupeau (1710-63). During one of these discussions in 1749, Louis de Règemorte (1709 circa - 1774) takes stock "on hydraulic construction works," concluding that the failure of the cofferdams should be attributed not to the lack of strength of the piles or sheet piles with the clay, but to the "erosion and loosening of the piles and the sheet piles."[1] But in the late 1740s and early 1750s, the engineer Perronet of the Ponts et Chaussées plays a fundamental role in the evolution of foundation systems. He is the one who makes the drawings and models of the system that permits the evolution of the caisson used for the Westminster Bridge, already envisioned by Labelye himself. Since 1739 Labelye has stated, though without venturing to offer a possible technical solution, that if the bed of the Thames was too sandy he would have inserted, as was customary, a large number of piles, and that in order to position the bottom grille of his caisson horizontally, he would have had the piles sawed off all at the same height. Etheridge's saw, a special machine, is developed at the worksite of Westminster Bridge that would make Labelye's hypothesis possible. It allows for the cutting of the support piles of the wooden reinforcements required for the construction of the arches of the bridge.[2] Perronet is in contact with Labelye, his "compatriot" and "close friend;"[3] it is to him that Labelye bequeaths the drawings and models of Westminster Bridge. Perronet possesses the description of the construction of the first pier of the bridge and the caisson system, reported in *A short account of the methods made use of in laying the foundation of the piers of Westminster-Bridge*, which he has translated into French by De Montigny.[4] Labelye himself provides Perronet with the drawing and the model of Etheridge's saw.[5]

On 18 May 1748 Perronet describes his own "proposed method to build the foundation of bridges and other structures that are constructed in water on piers without having to resort to neither cofferdams nor dewatering." This report forms the basis for many of the outstanding French projects for bridge-pier foundations in the 1750s. Perronet, during the 1750s and 1760s, continues to refine and correct details of the system outlined in 1748.

The stimulus for the invention of the foundation system is the yielding of the arch of Westminster Bridge, which demonstrates that Labelye's caisson can be successfully used in place of the cofferdam only when the underlying land is solid. From one of Perronet's notes, we learn that it is Labelye himself who tells the former about the causes of the "settlement" of 11 inches of the bridge pier, namely the presence of a sandy layer in the riverbed, and about the

[1] See, "Précis des assemblées des Ponts et Chaussées, tenues chez M. Trudaine, pour l'examen de differents projets," Ms 233, vol. XIII, Ecole des Ponts et Chaussées, Paris. Various ways of constructing the piles for the foundation of bridges are examined; Boffrand proposes putting *palplanches* between the first rows of piles (minutes of the meeting of 18 June 1747, *ibid.*).

[2] For the drawing see, William Etheridge, "A perspective view of the Engine made use of for sawing off, under water, the piles which help'd to support the centers for turning the arches of Westminster Bridge [...]," 1 May 1745, Ms 2170, Ecole des Ponts et Chaussées, Paris.

[3] Pierre-Charles Lesage, note, Ms 2170, Ecole des Ponts et Chaussées, Paris.

[4] *Cf.* "Pont de Westminster," Ms 2516, Ecole des Ponts et Chaussées, Paris.

[5] Lesage, note cit.

[6] Jean-Rodolphe Perronet, sheet of notes, "tassement de l'une des piles du Pont de Westminster," 9 February 1749, Ms 2516, Ecole des Ponts et Chaussées, Paris.

implemented fixes.[6] Perronet's method addresses the problem of bridge foundations on yielding terrain. The theoretical case of application of his method is that of the pier of a bridge of the size of those of Blois and Westminster; certain construction phases described by Perronet, like the creation of an enclosure of piles in the water corresponding to the footprint of the pier, follow indications made by Labelye in *A short account* A short account. But Perronet relies on the usual grille of beams with the piles inserted in the compartments. Not having the cofferdam available, to level the protruding ends of the piles, he describes a special saw, operated by 28 men, to be placed in the water – a variation of Etheridge's machine. "The purpose of this saw," Perronet explains, "is to cut the piles in each rank, the sheet piles, and the poles of the grille structure one after the other, each time flush with the top of the grille, on which the approximately 250 pound blade rests and slides along its flat surface."[7] For a detailed description of the functioning of the saw, he announces that another report will be prepared. Though calling for the usual clay to seal the compartments of the grille, a correction to the report shows reconsideration in order to obtain a more durable construction, made with a mortar composed of quicklime, coarse gravel, and gravel. According to Perronet's plan, the construction of the masonry would happen in two phases: first the laying of squared stones on a special platform, or "upper scaffolding"; then its lowering onto the grille using a device, or "mobile iron carriage," to rest it on precise points thanks to metal brackets previously attached for the purpose on the grille (complex operations are indicated for the pouring of the mortar and the completion of the masonry). The idea of building masonry outside of the water and then lowering it comes to him from the reading of the *Idea dell'architettura universale*, where Scamozzi describes the construction of the bridge of Ostia at the time of the Emperor Claudius.[8]

Perronet writes down the names of the persons to whom his report is sent in 1748 – including De Ville, De Montigny, De Voglie.[9] On 13 September 1750 he presents the construction methods of the foundations of the piers of Westminster Bridge[10] to the Assemblée des Inspecteurs Généraux des Ponts et Chaussées, while on 27 December he proposes his own project, documented by drawings and a model, to build bridges without cofferdams nor dewatering; Trudaine advises Pollart (or Pollard) to use Perronet's method for the piers of the bridge of Mantes.[11]

Perronet's report on his own foundation method reaches the hands of Bélidor, who copies it into the second volume of *Architecture Hydraulique*, published in 1753 – "this particular chapter that I received from Mr. Peronet [*sic*]."[12] In March 1751, Bélidor borrows from Perronet the manuscript of the French translation of *A short account* in order to insert the technical information on the caisson of Labelye in *Architecture Hydraulique*.[13] It is precisely thanks to Bélidor's chapter, entitled *Section Seconde où l'on décrit de*

[7] Jean-Rodolphe Perronet, "Moyen proposé pour fonder sur pilotis les ponts et les autres ouvrages à construire dans l'eau, sans etre obligé de faire des batard'eaux ni des épuisements," 18 May 1748, Ms 2102, Ecole des Ponts et Chaussées, Paris.
[8] Perronet himself recalls what was written by Scamozzi (*cf.* Perronet, *Mémoire sur les différentes méthodes* cit., p. 143).
[9] Jean-Rodolphe Perronet, "Moyen proposé pour fonder sur pilotis les ponts et les autres ouvrages à construire dans l'eau, sans etre obligé de faire des batard'eaux ni des épuisements," 18 May 1748, Ms 2102, Ecole des Ponts et Chaussées, Paris.
[10] Minutes of the meeting of 13 September 1750, in, "Extrait du Journal des assemblées des Ponts et Chaussées, tenues chez Mr Trudaine, depuis 1748 jusqu'en 1774," Ms 233, vol. XIII, Ecole des Ponts et Chaussées, Paris.
[11] Minutes of the meeting of 27 December 1750, *ibid.*
[12] Bélidor, *Architecture Hydraulique. Seconde Partie* cit., vol. II, 1753, pp. 460-63; also see the drawings at plate LX.
[13] Bernard Forest de Bélidor, letter to Jean-Rodolphe Perronet, 4 March 1751, Ms 2516, Ecole des Ponts et Chaussées, Paris.

*quelle maniere on a fondé par encaissement les piles du Pont de Westminster*, that Labelye's caisson emerges from the context of the construction of Westminster Bridge and joins the ranks of the large structures for the foundation of hydraulic works, becoming the model for the fantastic floating structures created in the 1770s and 1780s.[14] Explanations and drawings by Bélidor are based on the documents of Labelye and the translation of De Montagny made available by Perronet.

Towards 1751 Perronet shows Bélidor the model of his own machine for sawing the foundation piles.[15] In *Architecture Hydraulique*, Bélidor only mentions Perronet's model, without explaining the functioning of the machine, because in his view it has not been sufficiently put into practice. "I would have gladly included it in this volume, had it been proven on the large scale," he explains.[16] But Perronet's machine is described at the end of *Architecture Hydraulique*, classified as a "new method of sawing off the pile heads under water" and accompanied by drawings that show how it is derived from Etheridge's machine.[17] The explanations show that Perronet's saw is designed to simultaneously cut, at the deck level, multiple tops of piers inserted between the beams of the wooden grille for the foundation of the pier of a bridge, in the case of underlying terrain that is not very solid – this corresponds to Perronet's report in 1748. Over this grille with the leveled piles, it would be possible to position a caisson like that of Labelye, or – as Perronet suggests – to lower stone masonry already laid on a special platform, on the worksite but out of the water.[18]

Also in the second volume of *Architecture Hydraulique*, after having described Labelye's caisson system and before mentioning the machine of Perronet, Bélidor proposes an alternative method for foundations with piles and masonry, which goes on to also influence English engineers like Semple. Again in this case, the novelty with respect to Labelye's foundation method with the caisson and that of the machines of Perronet consists in the use of concrete and the ease of its deployment, demonstrating the singular position of Bélidor in the panorama of French technical culture at mid-century – a true disciple of Milet de Monville and of the systems in use along the Mediterranean coasts.

To construct the foundations in concrete, Bélidor explains that once the piles have been inserted and once an enclosure of boards has been made, stones are thrown in, beaten with "rammers," followed by pouring of "concrete mixed with stones," also beaten with "rammers," and finally a layer of concrete; then the block is left to harden for several months. "One sees," says his comment, implying criticism of the mechanical contrivance of Perronet's method, "that by following this approach, there is no need to shorten the piles, nor must they be sunk relative to the longitudinal or transverse ground beams of the grillage [...]."[19] So as opposed to the type of refinement of hydraulic construction pursued by Labelye and Perronet, increasingly mechanized like clockwork, Bélidor suggests the construction practices based on those of the ancient Romans, and on the use of concrete.

The grille, the platform, the caisson and the horizontal saw are studied by the engineers of the Ponts et Chaussées thanks to the urgings of Perronet,

[14] Bélidor, *Architecture Hydraulique. Seconde Partie* cit., vol. II, 1753, pp. 198-206.
[15] *Ibid.*, p. 208.
[16] *Ibid.*, p. 209.
[17] "Dessein pour l'intelligence d'une nouvelle maniere de receper les pilotis au fond de l'eau et de fonder les piles des ponts sans faire des batardeaux" (*ibid.*, plate LX, p. 480).
[18] *Ibid.*, pp. 459-62.
[19] *Ibid.*, p. 208.

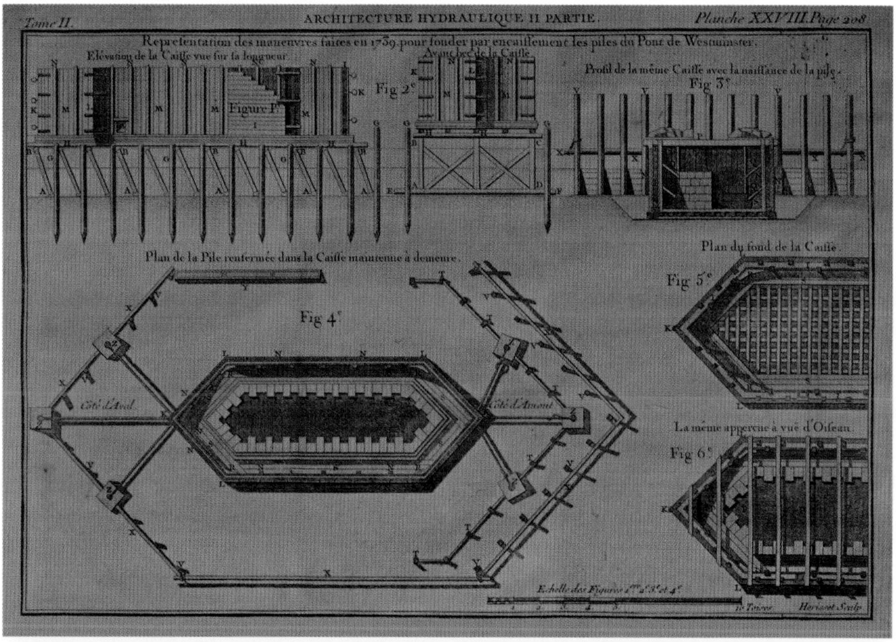


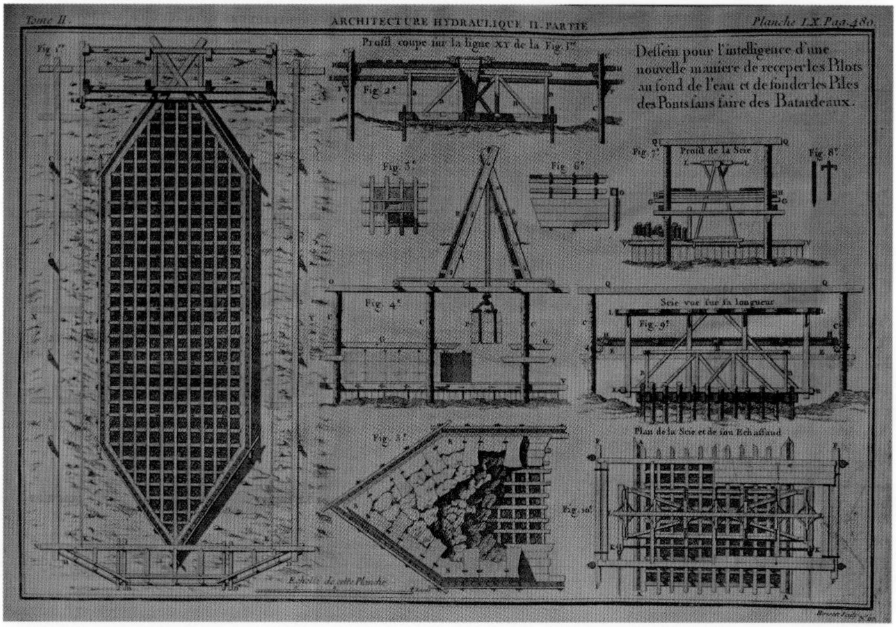


**236**. Bernard Forest de Bélidor, "Representation of work carried out in 1739 for the founding by caissons of the piles of the Westminster Bridge" (B.F. de Bélidor, *Architecture Hydraulique. Seconde Partie, qui comprend l'Art de diriger les eaux de la Mer & des Rivieres à l'avantage de la défense des Places, du Commerce, & de l'Agriculture*, Paris, Ch.-A. Jombert, vol. II, 1753).

**237**. Bernard Forest de Bélidor, machine for sawing multiple piles under water at one time, "new technique for trimming underwater pilotis," corresponding to the descriptions of Perronet (B.F. de Bélidor, *Architecture Hydraulique. Seconde Partie, qui comprend l'Art de diriger les eaux de la Mer & des Rivieres à l'avantage de la défense des Places, du Commerce, & de l'Agriculture*, Paris, Ch.-A. Jombert, vol. II, 1753).

238. Jean-Rodolphe Perronet, "New technique to made the cofferdams" 1754 (Ms 2531, Ecole des Ponts et Chaussées, Paris).

in the context of research on systems of foundation with piles, but without cofferdams nor dewatering.[20]

The draft of a report on systems of foundation without cofferdams nor dewatering, which later becomes part of the report presented to the Académie Royale des Sciences of Paris on 13 November 1765, is written by Perronet in 1753, in the context of the annual prize organized by the city of Besançon. Now Perronet imagines a masonry construction process that is different from the one described in 1748. His foundation system is still composed of a grille and piles inserted between the beams to consolidate the terrain, with the piles sawed off by a machine with a triangular "frame"; but over the grille, he now advises placing a caisson derived from that of Labelye, with removable sides, in which to make the masonry.[21] Perronet specifies the technical conditions regarding the masonry that lead him to study the question of the foundations. The method of Labelye that he has perfected is necessary when it is impossible to obtain suitable products for the making of concrete, with which otherwise it would be easy to construct foundations with piles and loose stones under water: "this construction demands, above all, excellent mortar. To this end, it must be composed [...] of pozzolana, cinders from Boulogne, or Dutch trass, and cement, and the lime must be made of shells, of marble, or of marl."

A drawing from 1754 documents Perronet's design for another foundation method, using an enclosure with sides formed by poles of different lengths, equipped with tips reinforced with iron to insert in the ground[22]. This may be the drawing referred to in the proposal made in March 1755 by Perronet to the Assemblée des Inspecteurs Généraux des Ponts et Chaussées, for a cofferdam foundation system: "I presented a model of two commonly used cofferdams

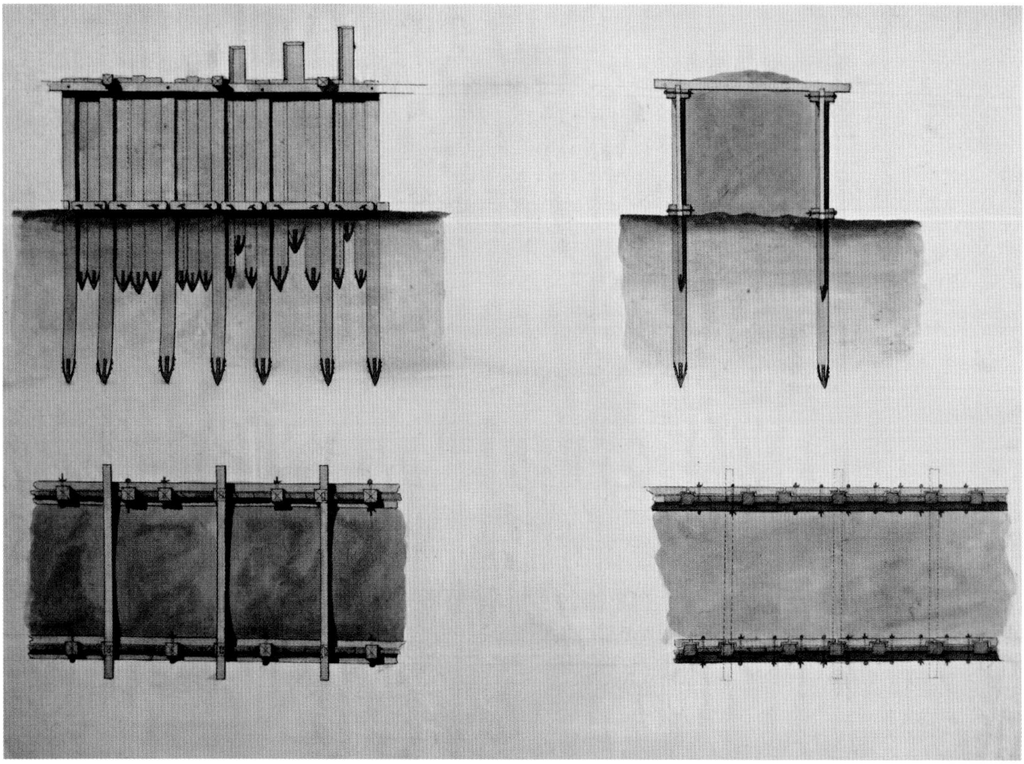

and the one that I propose, with moveable crossbars below and between the piles, to hold the sheet piles in a slot located closer to their extremity. Mr. De Règemorte did not approve of this design; several engineers thought it could be useful for sandy terrain, which remains to be proven."[23]

During that same year, 1755, Perronet finds the right conditions to attempt an initial application of the method outlined in 1748. The strong currents of the high waters of the Ain River make the construction of cofferdams problematic for the bridge at Chazey, on the road from Lyon to Geneva, under the supervision of Jean Montant André Le Ragois de Saint-André.[24] De Saint-André consults the second volume of *Architecture Hydraulique* to learn about the features of Perronet's method.[25] In April and May, Perronet sends De Saint-André the drawings and models of the foundation grille, of the saw for the piles and the platform on which to construct the masonry before lowering it into the water, specifying that he has never put the system into actual practice.[26] After receiving the drawings and models, De Saint-André develops the project and the related models to test the functioning, as advised by Perronet,[27] and on 30 December 1755 he prepares the "cost plan and detailed estimate for the construction works for the abutment block of the bridge at Chazey near Bresse according the new method put forward by Mr. Perronet [...] to avoid the expenses related to cofferdams and dewatering."[28] Besides the "grille following the model of Mr. Perronet," he describes a type of saw with a crank, operated by two men in a supine position. A note indicates that for the first arch of the bridge it has not been possible to use Perronet's saw: "we made this out of iron, which was easier to maneuver, but can only saw one sheet pile or pile at a time. In the other pouring we will follow that of Mr. Perronet's in individual

[20] Starting in 1750, proposals are advanced by Perronet, on 27 December 1750, Dié Gendrier (1705-91), an engineer at Bourges, 21 January 1753, Tardif on 27 January 1754, and Gendrier on 5 January 1755 (minutes of the meetings, in, "Extrait du Journal des assemblées des Ponts et Chaussées" cit.).

[21] [Jean-Rodolphe Perronet], "Projet pour concourir au prix des arts de la ville de Besançon pour l'année 1753," 23 June 1753, Ms 2102, Ecole des Ponts et Chaussées, Paris. See also, "Moyen proposé pour fonder sur pilotis les ponts et les autres ouvrages à construire dans l'eau, sans être obligé de faire des batardeaux ny des epuisemens," *ibid.* In this area of studies on foundation systems, there is also the project by Perronet documented in the drawing "Nouvelle maniere de faire des batardeaux faite par le Sr. Perronet en 1754," a variant of a system in any case also documented by Gautier in his treatise (Ms 2531, Ecole des Ponts et Chaussées, Paris). On the question of the cofferdams there are also contributions by the engineers Jean-Joseph Bochet de Coluel (? - 1778) and Jean-Charles Philibert Trudaine de Montigny (1733-77). See, "Memoires sur les epaisseurs des batardeaux. N.a Voir ce qui a été dit sur cela par Ms. Montigny et Coluel a l'occasion des projets proposés aux sous inspecteurs en 1748," Ms 2531, Ecole des Ponts et Chaussées, Paris.

[22] "Nouvelle maniere de faire des batardeaux faite par le Sr. Perronet en 1754," Ms 2531, Ecole des Ponts et Chaussées, Paris.

[23] Minutes of the meeting of 16 March 1755, in, "Extrait du Journal des assemblées des Ponts et Chaussées" cit.

[24] Perronet, *Mémoire sur les différentes méthodes* cit., p. 146. During digging in the river and on other nearby sites layers were found of pebbles "surrounded by a kind of concrete as hard as the most perfectly crafted mortar; the concrete parts are found in blocks of 7 to 8 feet in diameter and are difficult to pierce even with a iron probe." "The stones from the Ain River have also a characteristic of making excellent lime, especially when care is taken to burn it with charcoal extracted from the forest," De Saint-André writes (André Le Ragois de Saint-André, "Memoire et détail des ouvrages faits pour la fondation des culées du pont de Chazey sur la riviere d'Ain, Route de Lyon à Genève par la Bresse et le Bugey, le tout construit en maçonnerie de pierre de taille etablies sur pilotis, grillage et platteforme suivant la nouvelle méthode de M. Perronet [...] afin d'eviter la depense des batard'eaux et les epuisements presqu'impossible dans un torrent tel que celui du d'Ain," 30 November 1756, Ms 2102, Ecole des Ponts et Chaussées, Paris). In his memoir, De Saint-André uses the word *"bletton"* as a substitute for *"béton."*

[25] *Cf.* André Le Ragois de Saint-André, letter to Jean-Rodolphe Perronet, 1 April 1755, Ms 2102, Ecole des Ponts et Chaussées, Paris.

[26] Jean-Rodolphe Perronet, letter to André Le Ragois de Saint-André, 29 April 1755, copy, *ibid.* In May, when he is in Lyon, De Saint-André receives the model of Perronet's machine he intends building on the bridge worksite (André Le Ragois de Saint-André, letter to Jean-Rodolphe Perronet, 25 May 1755, *ibid.*).

[27] André Le Ragois de Saint-André, letter to Jean-Rodolphe Perronet, 29 September 1755, *ibid.*

[28] André Le Ragois de Saint-André, "Toisé et detail estimatif des ouvrages faits pour la fondation de la culée du pont de Chazey du coté de Bresse suivant la nouvelle methode de Mr Perronet [...] afin d'eviter les depenses des batardeaux et les epuisemens [...]," 30 December 1755, *ibid.*

cases."[29] For the other arch, De Saint-André uses Perronet's saw, which however he considers "very cumbersome to maneuver, especially during flooding."[30] The construction of the masonry of the piers is done with the devices envisioned by Perronet of the platform and the machine for lowering the masonry into the water. The masonry cladding is in squared stones, connected with iron fasteners; "its inside is lined with one or several courses of broken stones or ashlar, good lime mortar, and cement."[31]

The cost estimate done by Perronet for the construction of the cofferdams of certain bridges in the early decades of the 1700s demonstrates how expensive the system was considered to be, as well as being not very effective.[32] "The foundations of bridges and other masonry structures in water," he would state in the 1760s, "must be executed carefully and solidly; the approach commonly followed to this end using cofferdams and dewatering is very slow and costly, and nothing is less certain than its success. The often unavoidable flaws in the construction of the cofferdams, infiltrations and all too abundant sources, and flooding can delay, topple and backfill parts of the construction work; these kinds of events are far too common and sometimes lead to the loss of the work of an entire campaign."[33]

In the meantime, for the bridge on the Allier at Moulins, built in 1753-63, Règemorte creates an exceptional foundation in continuous masonry on wooden decks, a "foundation slab" like the one built one century earlier by François Blondel, which crosses all the piers of the bridge to form a stable base on the sandy, shifting natural base (the special boarding permits the use of cofferdams, preventing infiltration of water through the sand of the bed during the construction of the masonry).[34]

Starting precisely with the proven limits of the cofferdam, and against the backdrop of the success of the caisson of Labelye and the method of Perronet, another foundation system for deep waters and swampy, unstable land is presented to the Assemblée des Inspecteurs Généraux des Ponts in 1754 and published in 1757 with the title *Nouvelle methode d'encaissement*. The author is Jean-Marie-René Tardif (1716-66), engineer of the Ponts & Chaussées de la Généralité d'Amiens and son-in-law of a great builder of bridges, Pitrou.[35] Tardif's system consists of a cofferdam to be constructed out of water and sunk like a caisson. The particularity of what Tardif himself defines in multiple ways – "*encaissement*," "*casson*" and "*batardeau*" – lies in the possibility of also making foundations in deep water, thanks to sides devised so that they can be raised at will, according to the depth of the terrain suitable for the foundation (though a type of extensible sides had already appeared at the start of the 1700s).[36]

"It consists," Tardif explains, "of the construction of a caisson, which is nothing other than a cofferdam in one piece, composed of two circumferential frameworks, one placed within the other, separated by a gap of approximately two-and-a-half feet and connected by a sloped plane formed at the bottom of the interior structure. This cofferdam, built at the site of the foundation and

[29] "Toisé et detail estimatif des ouvrages faits pour la fondation de la culée du pont de Chazey du coté de Bresse suivant la nouvelle methode de Mr Perronet [...] afin d'eviter les depenses des batardeaux et les epuisemens [...]," 30 December 1755, *ibid.*

[30] André Le Ragois de Saint-André, "Memoire et détail des ouvrages faits pour la fondation des culées du pont de Chazey sur la riviere d'Ain, Route de Lyon à Genève par la Bresse et le Bugey, le tout construit en maçonnerie de pierre de taille etablies sur pilotis, grillage et platteforme suivant la nouvelle méthode de M. Perronet [...] afin d'eviter la depense des batard'eaux et les epuisements presqu'impossible dans un torrent tel que celui du d'Ain," 30 November 1756, *ibid.*

[31] Perronet, *Mémoire sur les différentes méthodes* cit., p. 146.

[32] [Jean-Rodolphe Perronet], "Extrait des prix aux quels sont revenus les Batardeaux et les epuisemens de differens ponts," [1753 circa], Ms 2102, Ecole des Ponts et Chaussées, Paris.

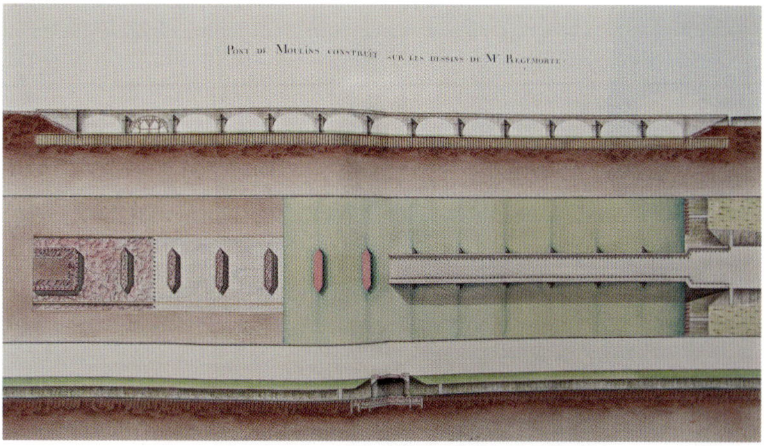

**239**. Louis de Règemorte, bridge on the Allier River, Moulins, 1753-63 (Ms Fol. 9/1, Ecole des Ponts et Chaussées, Paris).

loaded, between both structures, with masonry or clay soil, sinks into the ground under its own weight and by dredging which is performed at its center and once it has reached the tuff or the solid ground, it allows to easily and solidly erect the structure under construction."

To fill the interspace in the sides, which are like true formwork, Tardif describes a masonry that can be poured with machines like the ones used at Toulon by Milet de Monville for concrete, and described by Bélidor. Tardif uses the expression *mortier de béton* (concrete mortar) to indicate hydraulic masonry made with pozzolana.[37] The fact that Tardif's invention of the special bottomless caisson happens precisely due to the lack of concrete, attempting to improve on the formwork with respect to the system used by Milet de Monville and described by Bélidor, is confirmed by the statement of Tardif himself, where he mentions pozzolana: "May it be added that concrete mortar, composed of one third quicklime and two thirds pozzolana, the only kind adapted for this particular way of laying foundations, can only be used in places where these components are available."[38] "Besides," Tardif writes, further demonstrating how much this method is relevant in the absence of pozzolana concrete, "the use of concrete mortar is not necessary for the solidity of foundations built using caissons. It suffices that the mortar used for its masonry be composed of lime and sand. The carpentry within which it is enclosed protects it from all forms of degradation."[39]

The various applications proposed by Tardif for his new method call for the combination of multiple caissons to create foundation structures for bridges or pentagonal forts to be built in the sea (true variations of the Lazzaretto of Vanvitelli, to be built this way if pozzolana is not available). The transport of the caissons using a flotation system with barrels becomes a source of inspiration for the grand structures of the last decades of the century for the construction of foundations in the sea with concrete masonry.

[33] Perronet, *Mémoire sur les différentes méthodes* cit., p. 139.

[34] On the foundation system of the bridge see, Louis de Règemortes, *Description du nouveau pont de pierre construit sur la rivière d'Allier à Moulins*, Paris, Lottin, 1771.

[35] Tardif supervises the publication of the works of Pitrou (Robert Pitrou, *Recueil de ses différents projets d'architecture, de charpente, et autres*, Paris, 1756).

[36] Jean-Marie-René Tardif, *Nouvelle methode d'encaissement, pour fonder facilement et solidement à telle profondeur qu'il sera nécessaire dans les Riviers, dans les Marais, dans la Mer, à proximité des Côtes, & généralement dans les Terreins Sabloneux ou Vaseux*, Paris, H.D. Chaubert, 1757, pp. 4 r-v.

[37] *Ibid.*, p. 3.

[38] *Ibid.*

[39] *Ibid.*

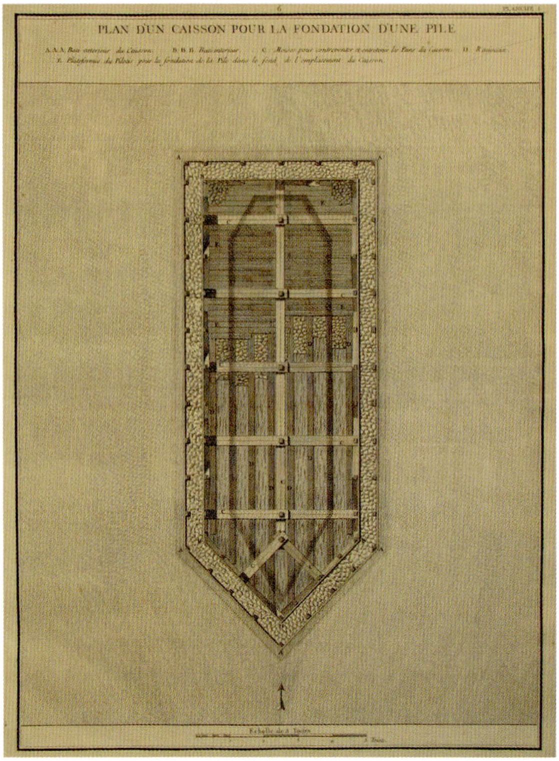


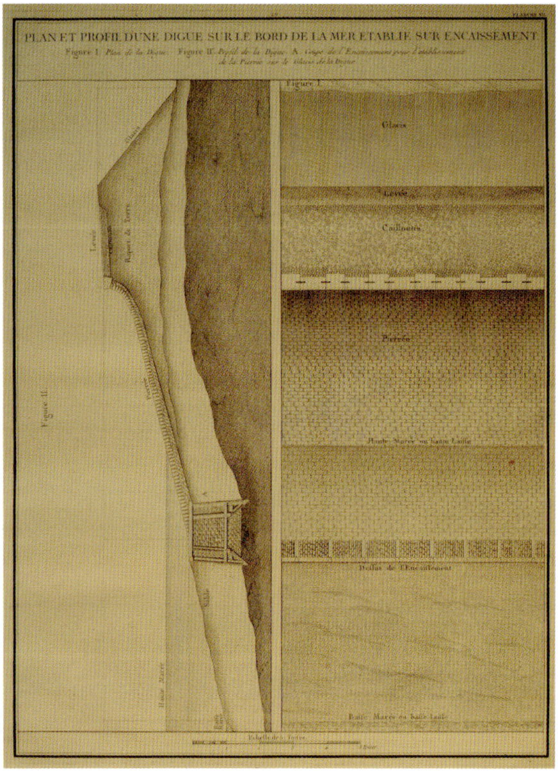


**240-242**. Jean-Marie-René Tardif, drawings of a "new method of the use of caissons." "Plan of a caisson for the fondation of a pier"; "plan and profile of a dike on the seashore built with caissons"; "profils of a caisson for the fondation of a pier" (J.-M.-R. Tardif, *Nouvelle methode d'encaissement, pour fonder facilement et solidement à telle profondeur qu'il sera nécessaire dans les Riviers, dans les Marais, dans la Mer, à proximité des Côtes, & généralement dans les Terreins Sabloneux ou Vaseux*, Paris, H.D. Chaubert, 1757).

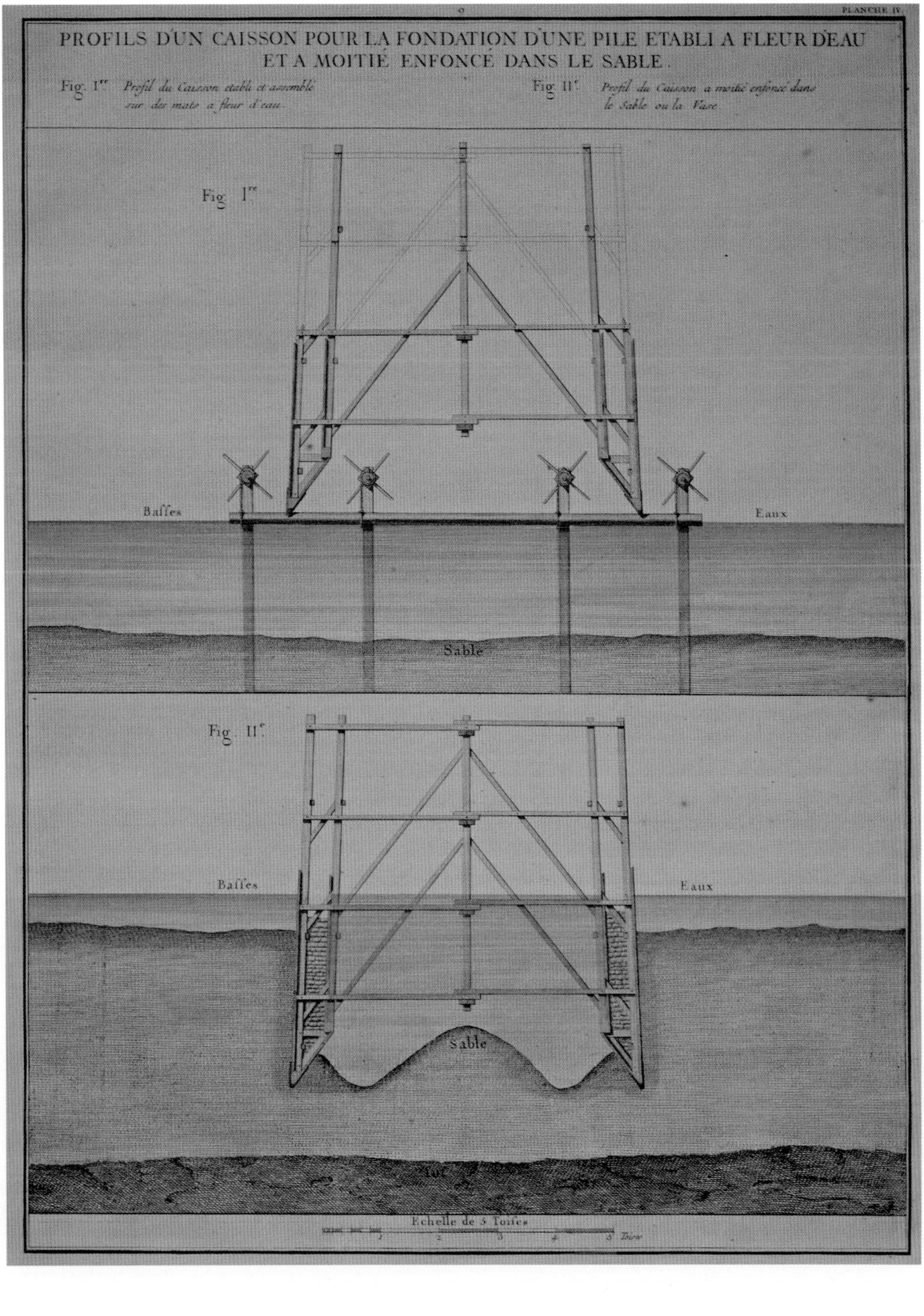
PLANCHE IV.
PROFILS D'UN CAISSON POUR LA FONDATION D'UNE PILE ETABLI A FLEUR D'EAU
ET A MOITIÉ ENFONCÉ DANS LE SABLE.
Fig. I.re Profil du Caisson etabli et assemblé sur des mats a fleur d'eau.
Fig. II.e Profil du Caisson a moitié enfoncé dans le Sable ou la Vase.
Fig. I.re
Basses
Eaux
Sable
Fig. II.e
Basses
Eaux
Sable
Echelle de 5 Toises
1
2
3
4
5 Toises

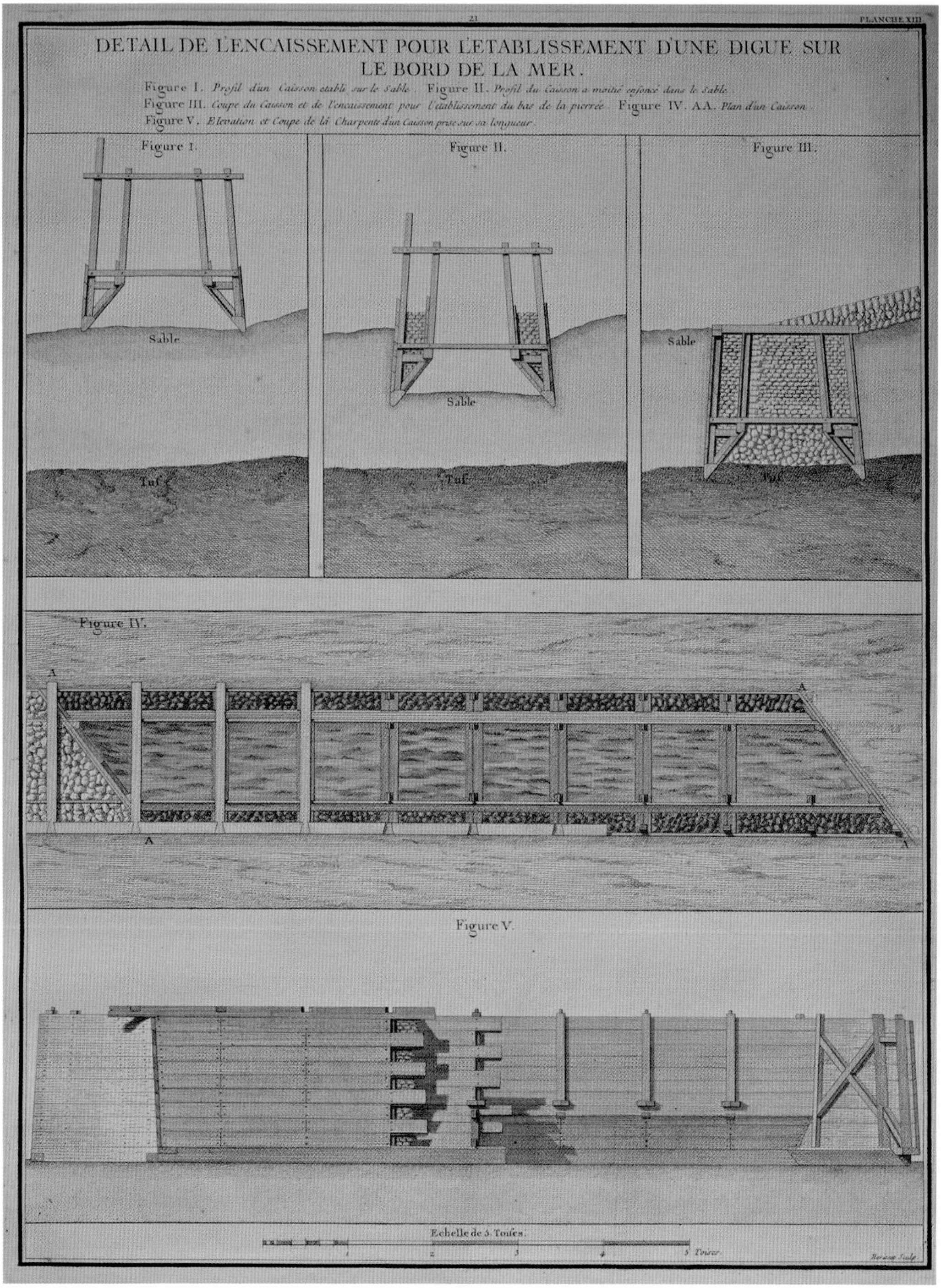


**243-244**. Jean-Marie-René Tardif, drawings of a "new method of the use of caissons." "Detail of a caisson for the construction of a dike on the seashore"; "detail of a fondation made with caisson for a fortification wall" (J.-M.-R. Tardif, *Nouvelle methode d'encaissement, pour fonder facilement et solidement à telle profondeur qu'il sera nécessaire dans les Riviers, dans les Marais, dans la Mer, à proximité des Côtes, & généralement dans les Terreins Sabloneux ou Vaseux*, Paris, H.D. Chaubert, 1757).

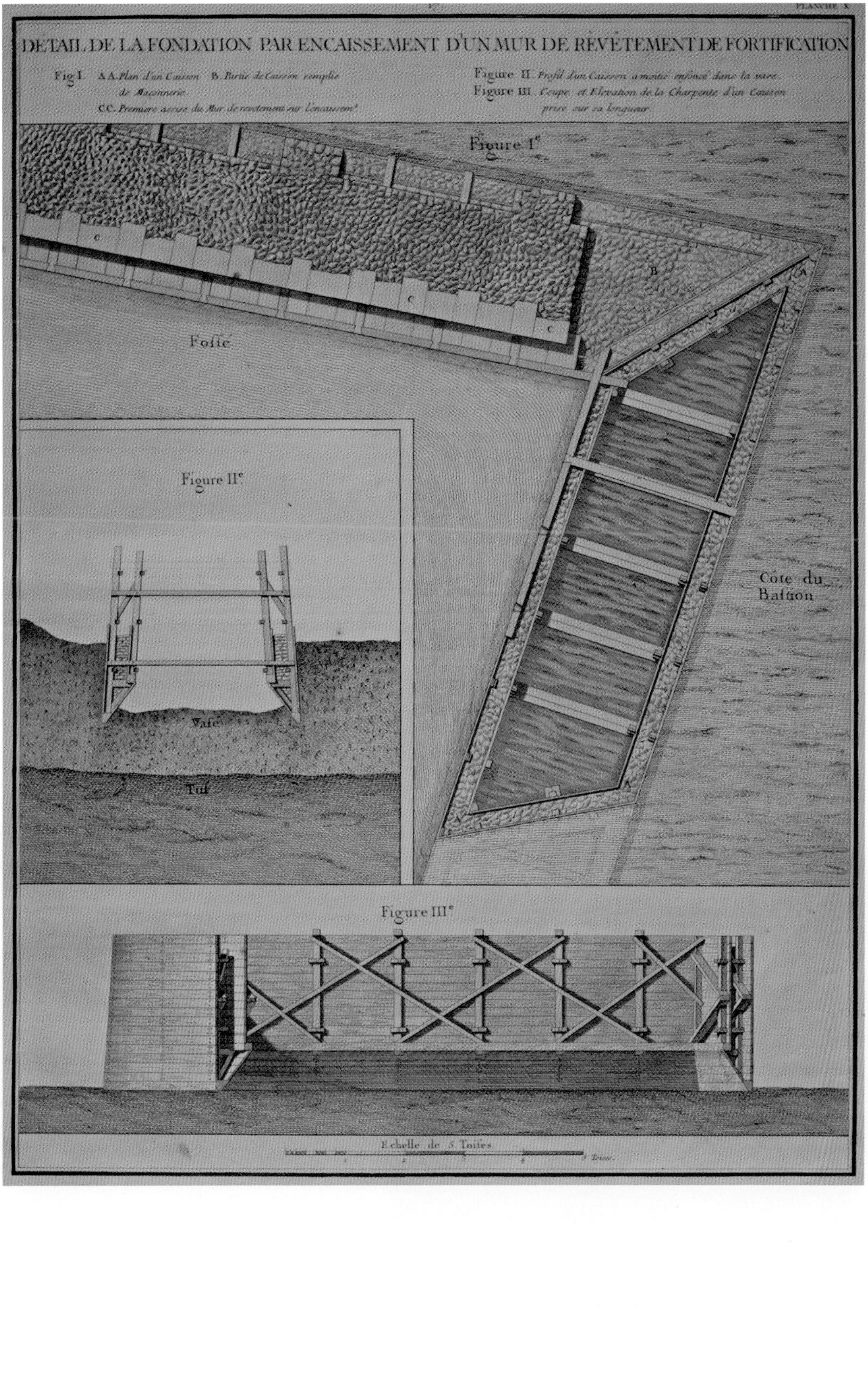
PLANCHE X
DETAIL DE LA FONDATION PAR ENCAISSEMENT D'UN MUR DE REVÊTEMENT DE FORTIFICATION
Fig I. A A. Plan d'un Caisson B. Partie de Caisson remplie de Maçonnerie
C C. Premiere assise du Mur de revetement sur l'encaissem.t
Figure II. Profil d'un Caisson à moitié enfoncé dans la vase.
Figure III. Coupe et Elevation de la Charpente d'un Caisson prise sur sa longueur
Figure I.e
Fossé
Côte du Bastion
Figure II.e
Vase
Tuf
Figure III.e
Echelle de 5 Toises

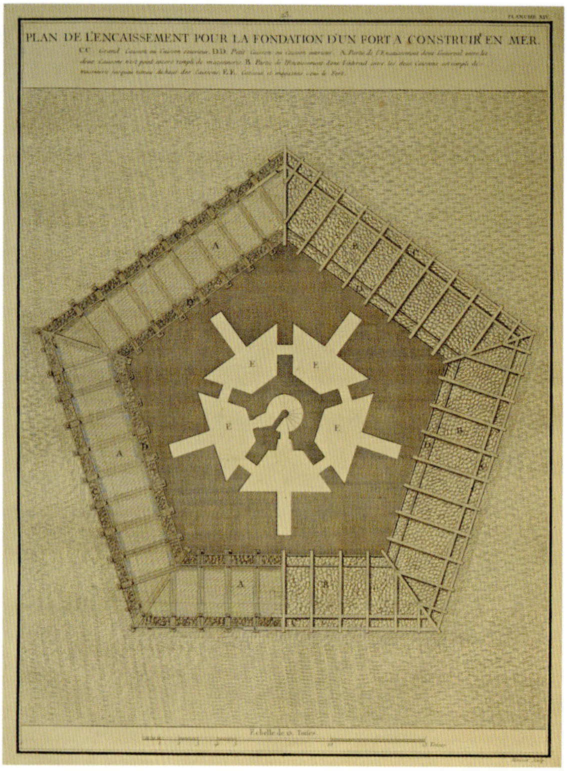


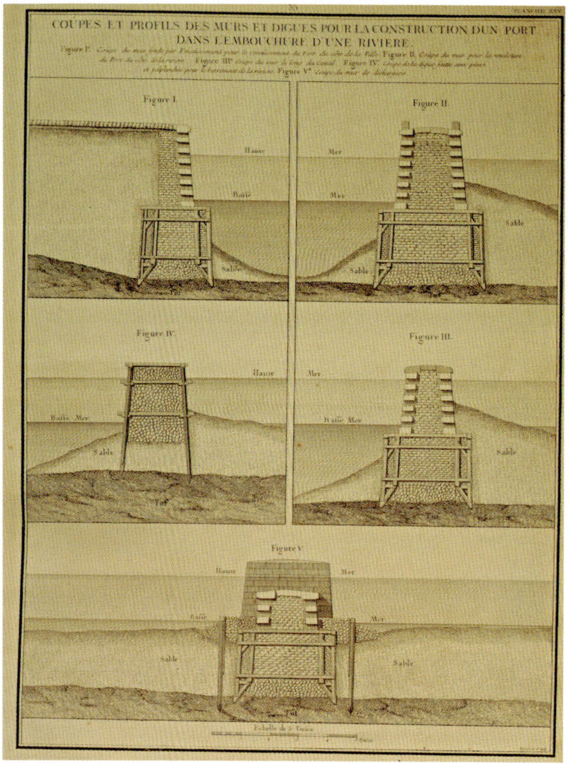


**245-247**. Jean-Marie-René Tardif, drawings of a "new method of the use of caissons." "Plan of the caissons for the fondation of a fort built in the sea"; "sections and profiles of walls and dikes for the contruction of a port in the mouth of a river"; "section of a fort built in the sea and of the caissons of its fondation" (J.-M.-R. Tardif, *Nouvelle methode d'encaissement, pour fonder facilement et solidement à telle profondeur qu'il sera nécessaire dans les Riviers, dans les Marais, dans la Mer, à proximité des Côtes, & généralement dans les Terreins Sabloneux ou Vaseux*, Paris, H.D. Chaubert, 1757).

20
PLANCHE XX.
COUPE DU FORT CONSTRUIT EN MER ET DE L'ENCAISSEMENT DE SA FONDATION
PRISE SUR LA LIGNE A.Y. DE LA PLANCHE XV.
A Profil du grand Caisson ou du Caisson exterieur
B Profil du petit Caisson ou du Caisson interieur
haute Marée
baſſe Marée
A
B
C
D
Sable
Tuf
Echelle de 15 Toises
15 Toises

### Caissons, saw and hydraulic mortar for the bridge at Saumur

After Perronet's projects for foundations without cofferdams, after its publicizing by Bélidor and the experimentation done by De Saint-André, and after the caisson-cofferdam project of Tardif, comes the first important application, in 1757, for the construction of the second pier of the bridge on the large branch of the Loire, at Saumur, of Labelye's caisson deployed with the techniques outlined by Perronet.

After having tested the terrain in 1752 and obtained approval, in 1753, of the project for the construction of a bridge with twelve arches at Saumur, the engineer Jean-Baptiste Bentivoglio, known as De Voglie (1723/24-1777),[40] supervises the work, assisted by Louis-Alexandre de Cessart (1719-1806). The foundation of the first pier, begun only in 1756 (the first stone is laid on 5 October) and completed in 1757, demonstrates that the great depth of the water makes the cofferdam solution unfeasible, both due to the difficulty in finding wood of sufficient length, and due to the lack of strength of the sides.[41] When De Voglie is making the foundations with cofferdam in the Loire, at the end of 1756, the bridge of De Saint-André at Chazey is examined by technicians of the Ponts et Chaussées to assess the results of the work and potential improvements to Perronet's system. At the start of December Perronet receives all the documents and models made by De Saint-André to construct the bridge with the new method.[42] On 16 December Perronet submits to the Assemblée des Inspecteurs Généraux des Ponts et Chaussées the model he has asked De Saint-André to make, to see the "way in which he laid the foundations for one of the abutments of the bridge at Chazey."[43] It is against the backdrop of these experiences that De Voglie is contacted by Perronet to inform him of the application of the foundation method used at Chazey.[44] So De Voglie starts to develop the project of the foundation of the second pier of the bridge at Saumur based on the method of the caisson of Labelye modified by Perronet, and with the experiences of De Saint-André in mind. Perronet himself sends along the report of De Saint-

**248**. Plan of Saumur, with the bridges on the Loire started in 1756 (F.14.10200(5)2, Archives Nationales, Paris).

**249**. Jean-Baptiste De Voglie, bridge on the Loire, at Saumur, 1752-65 (F.14.10200(5)3, Archives Nationales, Paris).

**250**. Jean-Baptiste De Voglie, bridge on the Loire, at Saumur, 1752-65. Detail of the drawing with the pier laid in 1756, "avec épuisemens" (right), and the one laid in 1758, "sans épuisemens," with caisson (DG 3200, Ecole des Ponts et Chaussées, Paris).

André on the construction of the bridge at Chazey. De Voglie expresses a series of doubts regarding the depth of the foundation in the ground, the thickness of the masonry and the way of joining the stones used by De Saint-André, and he does not think it will be possible to duplicate, at Saumur, the saw of De Saint-André with the two operators lying down, given the fact that a bridge pier on the Loire requires about 180 piles.[45]

[40] *Cf.* De Voglie, in, François-Pierre H. Tarbé de Saint-Hardouin, *Notices Biographiques sur les ingénieurs des ponts et chaussées, depuis la création du Corps, en 1716, jusqu'à nos jours*, Paris, Baudry et C.ie, 1884, pp. 33-34.

[41] The difficulties in installing the piles and boards of the batardeau emerge since 1752 (Jean-Baptiste de Voglie, letter to Jean-Rodolphe Perronet, 29 July 1752, Ms 2087, Ecole des Ponts et Chaussées, Paris).

[42] De Saint-André sends Perronet "all of the machines that I had executed at the bridge at Chazey following the approach you invented to lay foundations under water in the absence of cofferdams with a grillage, piles and a platform" (André Le Ragois de Saint-André, letter to Jean-Rodolphe Perronet, 3 December 1756, Ms 2102, Ecole des Ponts et Chaussées, Paris). This is the "model of the method put into practice at the Pont de Chazey," accompanied by "8 pieces of stonework, one of which is fitted" with the "large saw" using "bent bolts," iron studs, and drawings of the scaffolding (André Le Ragois de Saint-André, letter to Boury, 6 September 1756, *ibid.*).

[43] Minutes of the meeting of 16 December 1756, in, "Extrait du Journal des assemblées des Ponts et Chaussées" cit.

[44] *Cf.* Jean Hupeau, "Observations sur la nouvelle methode de fonder les ponts sans batardeaux ny epuisements, proposée par M. Perronet et executée par M. de St. André au Pont de Chazey en Bresse, sur la riviere du D'Ain," 4 February 1757, Ms 2102, Ecole des Ponts et Chaussées, Paris; Mathieu Bayeux, "Observations sur la nouvelle methode des fonder les ponts sans battardeaux ny epuisemens proposées par M. Perronet et executée par M. de St. André au pont de Chazey en Bresse sur la riviere d'Ain," 22 February 1757, *ibid.*

[45] Jean-Baptiste de Voglie, letter to Jean-Rodolphe Perronet, 5 January 1757, *ibid.*

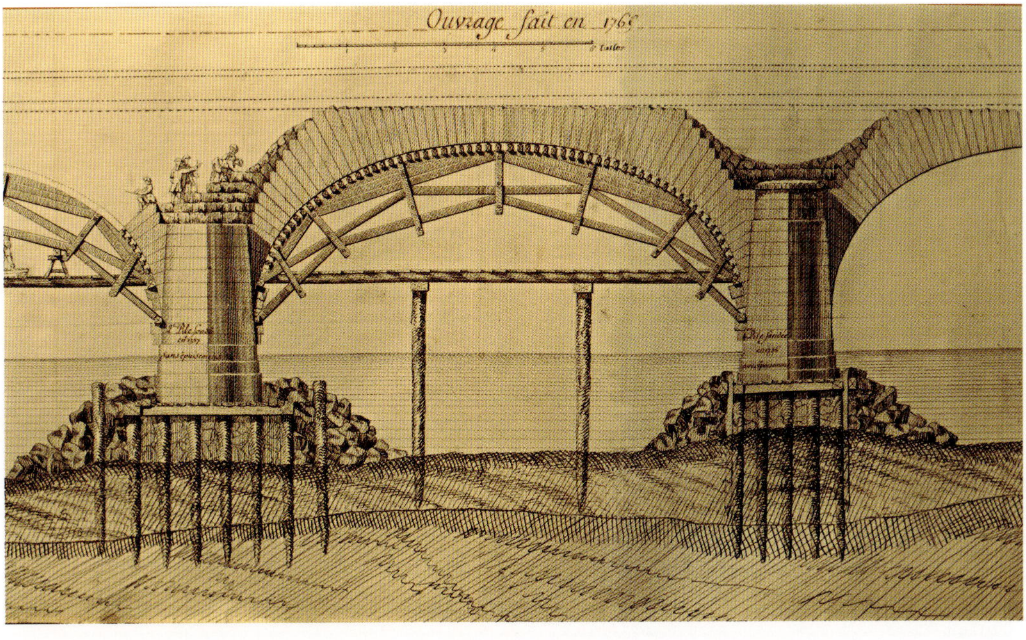

In April 1757 De Voglie prepares the project for the foundation of the second pier, following the method of Labelye perfected by Perronet.[46] The foundation supporting the caisson is formed by a series of long poles circled by iron chains, with gaps filled by stones and earth. The caisson, built by the carpenter Quincé, is to measure 48 x 20 feet by 8 feet in height, with the ends shaped as isosceles triangles, and well-sealed boards – "care must be taken to caulk and tar all of the joints and line them with foam and plaster."[47] The construction of the masonry happens inside the caisson that gradually sinks as the courses are laid, until it touches the plane of the foundation. The masons have a platform, specially made around the caisson. The pier has facings made of stone of Champigny, and the blocks are anchored with metal clamps, while the nucleus is filled with stones and lime mortar. For the mortar De Voglie makes use of Doué lime, perhaps following the advice of the contractor for the masonry works, Alexandre Cailleau.

On 21 September 1757 the first caisson with masonry is sunk onto the pile foundation; the work on the bridge is completed in October 1765, under the supervision of François-Michel Lecreulx.[48]

On 19 February 1758 De Voglie presents to the Assemblée des Inspecteurs Généraux des Ponts et Chaussées the models "of how, the previous year, he had erected one of the piles of the Saumur Bridge six feet below the low flow line using a Labelye caisson without cofferdams nor dewatering," to which Perronet adds the comment, "as I had proposed in the report that Mr. Bélidor printed in his *Architecture Hydraulique*."[49] The caisson of De Voglie is not the only factor that influences the developments of foundation systems in the sea on the part of French engineers in the second half of the century. The improvements to the saw designed by Perronet and built by De Saint-André to cut the underwater piles to the same height make it possible for the saw built at Saumur to be used at great depths, making it a model for the French school of engineering until the early 1800s, starting with its publication in the plates of the *Encyclopédie* of Diderot and D'Alembert. The saw used at Saumur is built with the collaboration of the local locksmith Gamory, known as l'Angoumois, and the help of De Cessart.

If Guettard, Nollet and Soufflot pay increasingly systematic attention to the stones and mortars of ancient Roman construction, it is because in France the search for a binder capable of setting masonry underwater, like that of the Romans, is the order of the day for major engineering works. Precisely on the Saumur worksite, De Montigny observes the use in the fall of 1752 of a stone

[46] Inside the caisson the masonry will be done above ground, up to 6 feet and 9 inches, using two cranes installed at the edge (Jean-Baptiste de Voglie, "Mémoire pour servir à la fondation d'une des piles du pont de Saumur sur la Riviere de Loire sans Batardeaux ny Epuisements," 17 April 1757, Ms 44, Ecole des Ponts et Chaussées, Paris). *Cf.* also, Jean-Baptiste de Voglie, letter to Jean-Rodolphe Perronet, 14 March 1762, Ms 1831, Ecole des Ponts et Chaussées, Paris (the letter is published in, Eugène-Jean-Marie Vignon, *Études historiques sur l'administration des voies publiques en France aux Dix-septième et Dix-huitième siècles*, Paris, Dunod, 1862, vol. II, p. 335).

[47] Jean-Baptiste de Voglie, "Mémoire pour servir à la fondation d'une des piles du pont de Saumur sur la Riviere de Loire sans Batardeaux ny Epuisements," 17 April 1757, Ms 44, Ecole des Ponts et Chaussées, Paris. On the bridge of Saumur see, Louis-Victor Dubois d'Arneuville, ed., *Description des travaux hydrauliques de Louis-Alexandre de Cessart*, Paris, Baudouin, 1806, vol. I, pp. 47-224. See also, "Journal des opérations faites pour la construction du pont de pierre de Saumur sur le grand bras de la rivière de Loire," Ms 22-30, Bibliothèque Municipale, Saumur.

[48] Perronet, *Mémoire sur les différentes méthodes* cit., pp. 146-47.

[49] Minutes of the meeting of 19 February 1758, in, "Extrait du Journal des assemblées des Ponts et Chaussées" cit.

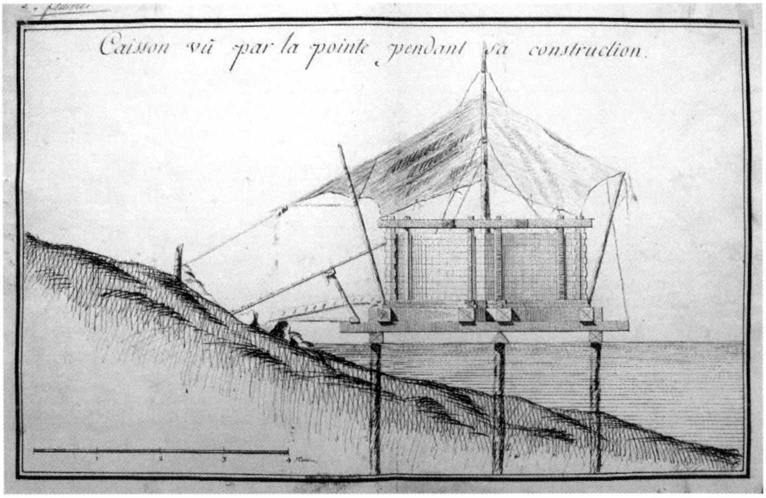


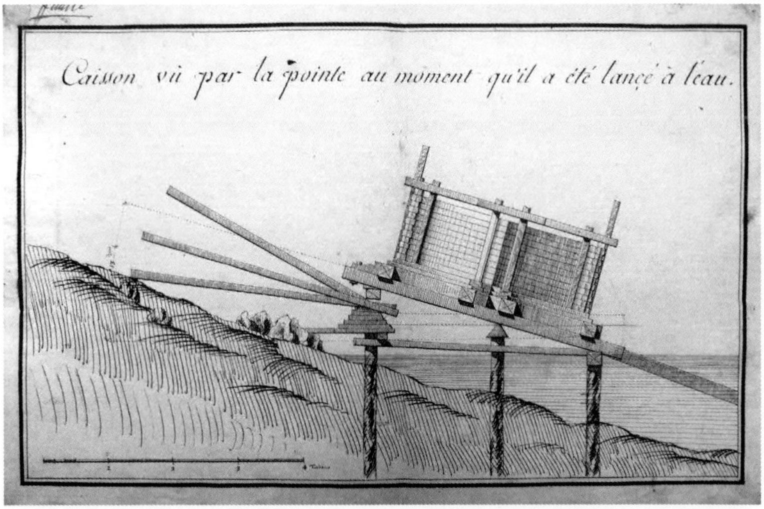


**251-252**. Jean-Baptiste De Voglie, bridge on the Loire, at Saumur, 1752-65. Caisson (Ms 2533, Ecole des Ponts et Chaussées, Paris).

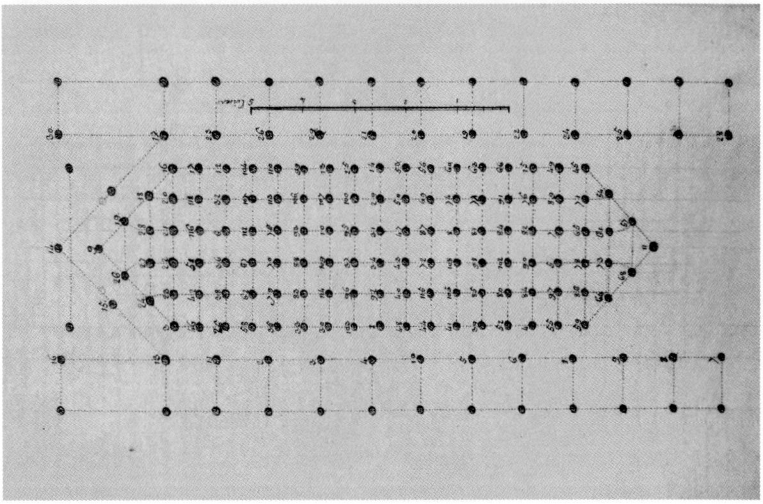

**253**. Jean-Baptiste De Voglie, bridge on the Loire, at Saumur, 1752-65. Jean Hupeau, drawing of the piles of a pier, 1757 (Ms 2537, Ecole des Ponts et Chaussées, Paris).

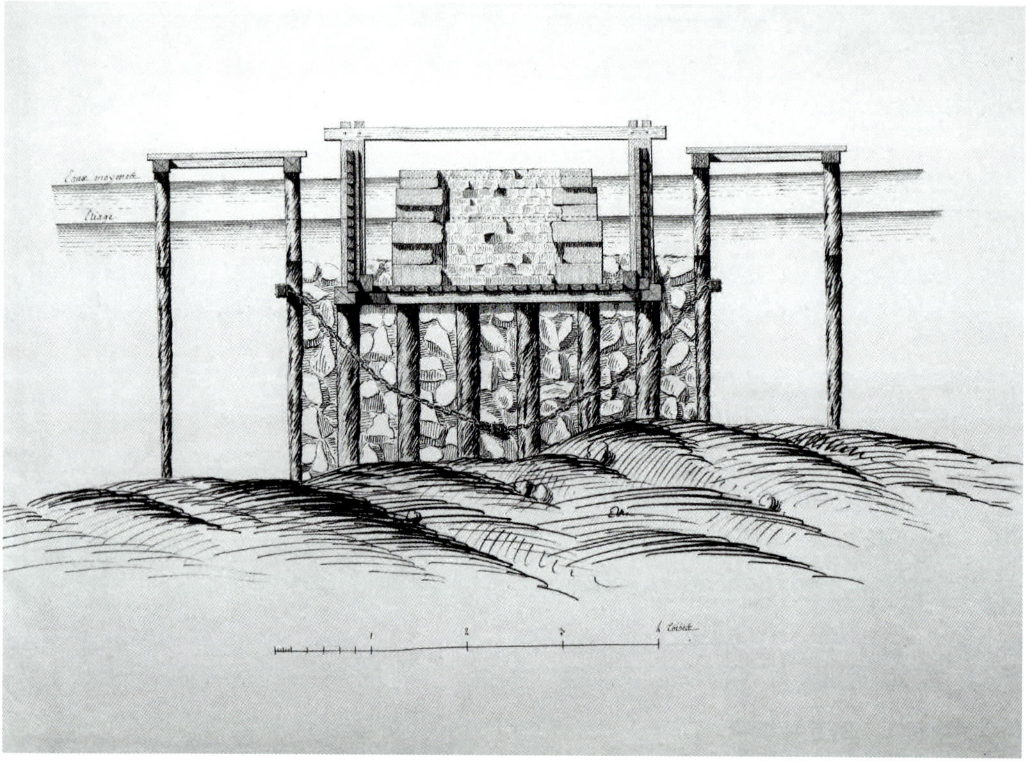

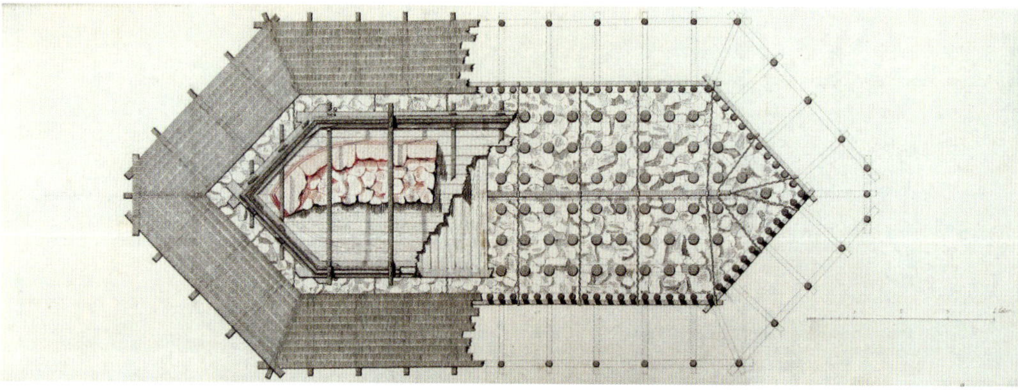

**254-255**. Jean-Baptiste De Voglie, bridge on the Loire, at Saumur, 1752-65. Caisson with masonry (Ms 2533, Ecole des Ponts et Chaussées, Paris).

**256**. Jean-Baptiste De Voglie, bridge on the Loire, at Saumur, 1752-65. Saw for cutting piles under water (P. Patte, *Mémoires sur les objets les plus importans de l'architecture*, Paris, B. Rozet, 1769).

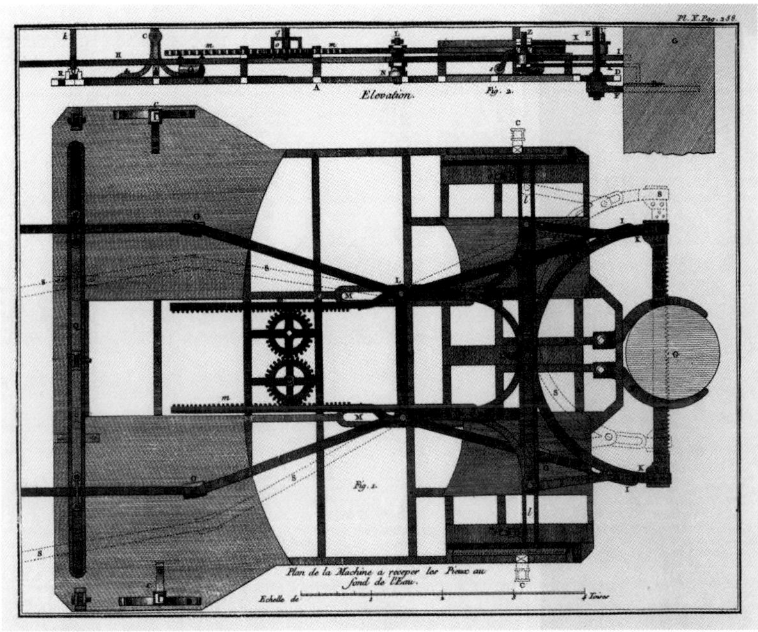


"of a dark shade of gray," taken from quarries at Doué, and with the characteristic of producing a lime that sets "in twenty four hours in such a way that a knife cannot penetrate it."[50] De Montigny delivers a memo in which he states that this "lime should be of more widespread and general use" – but it is only during the course of the 1800s that the lime produced with the clayey calcareous soil from Lias, from the deposit between Doué and Montreuil-Bellay, becomes one of the most widespread in France for hydraulic works.

"Among the fossils from Douay that I was shown at Saumur," De Montigny writes, "the one that appeared to deserve most attention is a limestone of most extraordinary quality. The lime derived from this rock is dark gray; I have seen some of this lime be slaked at Saumur; it takes a long time to heat and does not swell at all, but absorbs much heat as it is slaked. The lime does not slake well when exposed to air, and can therefore be preserved more easily than lime from white rock. But its most unique feature that makes it extremely useful is that, by covering it with water once it is slaked, it hardens within twenty-four hours to the point that a knife can no longer penetrate it; this fact was certified to me by Mr. Voglie and Mr. Caillot [*sic*: Cailleau], a contractor at the Saumur Bridge who uses it daily for underwater construction."[51]

The overview of the main systems for hydraulic foundations outlined by Perronet in the 1760s, after the experimentation of De Saint-André and De Voglie, confirms the opposition between different construction systems that took form in 1753 with *Architecture Hydraulique*: on the one hand, the construction in poured concrete; on the other, the increasingly sophisticated equipment for operating in water with traditional criteria, made possible precisely by special machinery, from the caisson to the saw for cutting the piles.

[50] Etienne Mignot de Montigny, "Voyage dans l'Orleanais, le Blesois, la Tourraine, l'Anjou, et la Bretagne, fait en 1752, depuis le 9 septembre jusqu'au 23 octobre, par de Montigny," Ms 2840, Bibliothèque Mazarine, Paris.

[51] *Ibid.*

Pl. XI
Opération de la Machine à receper les Pieux au fond de l'Eau.
Patte Del.

**257**. Jean-Baptiste De Voglie, bridge on the Loire, at Saumur, 1752-65. Saw for cutting piles under water (P. Patte, *Mémoires sur les objets les plus importans de l'architecture*, Paris, B. Rozet, 1769).

Perronet, too, before introducing his own machine for the cutting of the piles, cannot help but appreciate, like Bélidor before him, the simplicity and ease of construction in water using only concrete. He takes significant passages from *Architecture Hydraulique* to describe the process of construction of "foundations made of a masonry referred to as concrete," done with pozzolana, Dutch *trass* or "cinders from Boulogne."[52]

"[...] the foundation is made of a masonry called concrete," Perronet writes, "which is composed of pozzolana, Dutch trass, or cinders from Boulogne; one of these compounds is blended with quicklime and broken stone or small pebbles, laid in layers alternating with other layers of medium sized rocks; this concrete masonry is lowered to the seabed after having removed the mud or other soft materials that can be found there, and small chests that can open to the bottom are used to lower the masonry, to place it as evenly as possible on the surface on which the foundation is to be erected, and the whole structure is built atop these large footings until it reaches the surface of the sea and quickly forms a mass of great solidity. This method is used successfully in Mediterranean ports, and the pier built in 1748 in Toulon in the new Darse was constructed in the way I just explained. In the architecture of Vitruvius, one can see that the Romans were aware of this type of construction and used it for moles and jetties, except that they did not lower the masonry in cases; they threw them into the water after having surrounded the foundation under construction with a line of piles and sheet piles when the depth of the sea allowed for it."[53]

Though he praises the Roman system, Perronet points out the difficulties involved in its implementation, due to the need to obtain pozzolana: "This type of construction works very well, but it is not always the case that pozzolana, or an equivalent material, is readily available; common mortars cannot replace it and do not bestow the masonry with the solidity required of these kinds of structures."[54]

## The hydraulic mortars of Smeaton for the Eddystone lighthouse

Following the fundamental experiences of Semple on the worksite for the Essex Bridge, the search for a cement-based compound having hydraulic properties in England resumes and is developed on a scientific basis by the most famous English engineer, Smeaton. A voyage to Holland in 1755 takes on the symbolic value of a prologue to the experiences of Smeaton with this type of compound. Certain segments of the journey are dictated by the need to visit hydraulic works, like the piers of Dunkerque described in the treatises of Bélidor, which Smeaton mentions in his travel diary.[55] When he finds himself at Boesinghe, he admires the "sas or lock" he has already seen in *Architecture Hydraulique* by Bélidor, which he describes as one of the masterpieces of "hydraulick architecture" seen during his journey. He also admires its "small stones, and not very square hewn, but the whole seemed so well cemented together that there did not appear the least defect."[56] "I made it my business to enquire, as particularly as I could, concerning the Tarras," he recalls in the book in which he describes the porous, grayish stone from Germany he has

[52] Perronet, *Mémoire sur les différentes méthodes* cit., pp. 143-44.
[53] *Ibid.*
[54] *Ibid.*, p. 144.
[55] *Cf.* Arthur Titley, *John Smeaton's Diary of his journey to the Low Countries*, 1755, The Newcomen Society, 1938, pp. 4 and 7.
[56] *Ibid.*, p. 7.

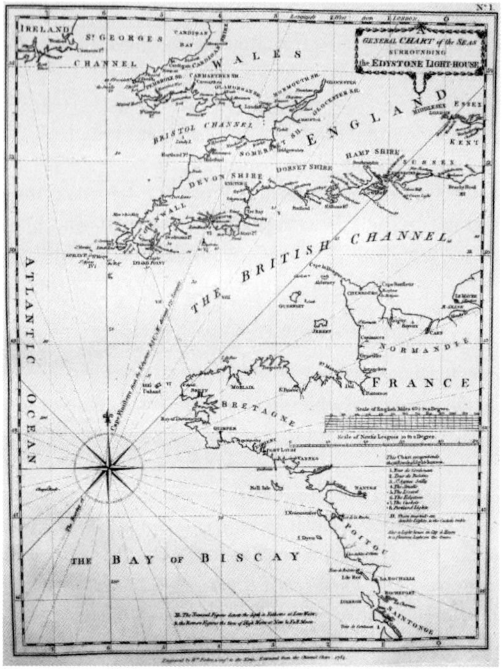


**258**. "General chart of the seas surrounding the Edystone Light-House" (J. Smeaton, *A Narrative of the Building and a Description of the Construction of the Edystone Lighthouse with Stone*, London, G. Nicol, 1791).

seen in Holland, the mills that grind it into powder and the other processes to which it is subjected before its use in hydraulic masonry.[57] So while Chambers and the Adam brothers study the Roman ruins and the way of translating their sequences of spaces and decorations into contemporary architecture, focusing only on questions of facing mortar, Smeaton opens the English chapter of scientific research on cement fabricated in the Roman manner, by looking to Holland. Like the engineers in the wake of Perronet, he too wants to identify a mortar to use for works in the sea or in rivers.

Smeaton finds an opportunity to study the development of a hydraulic mortar when the Royal Society of London selects him as the engineer capable of undertaking the reconstruction of the lighthouse on the cliff known as Eddystone, near the bay of Plymouth on the Channel, following the destruction by fire on 2 December 1755 of the wooden lighthouse built in 1706-09 by John Rudyard, or Rudyerd (1650-1718 circa). While Rudyerd's idea was to resist the force of violent storms with the elasticity of a wooden structure, Smeaton, on the other hand, aims to achieve a masonry structure of maximum stiffness, thus finding himself having to study all the necessary techniques to transform stone masonry, laid in courses, into an ideal monolith. This is why his research is largely concerned with the connections between the blocks of stone in a single course and between superimposed courses, which Smeaton makes by means of wedges inserted in the joints, based on the art of carpenters – "joggle" is his definition[58] – and a cut of the

[57] Smeaton, *A Narrative of the Building* cit., p. 111.
[58] *Ibid.*, p. 149.
[59] *Ibid.*, p. 43.
[60] *Ibid.*, p. 41.
[61] *Ibid.*, pp. 53-54.
[62] *Ibid.*, p. 40.
[63] *Ibid.*, p. 66. *Cf.* George Godwin, *Prize Essay upon the Nature and Proprties of Concrete, and its application to Construction up to the present period*, in, "Transactions of the Institute of British Architects of London. Sessions 1835-36," vol. I, part I, London, 1836, (pp. 1-35), p. 12. The examples of medieval use of concrete are still known in the 1800s. "In Rochester Castle the staircases are composed of it; the under sides, or soffites, show to this day the marks of the boards which sustained it till it was set" (*Of Foundation. On Land*, in, *Encyclopædia Britannica, or Dictionary of Arts, Sciences,*

stones like the one illustrated by Bélidor in the case of Cherbourg or the one seen by Smeaton himself in the stones of the sidewalks of London.[59] The conoid shape of his new lighthouse and the structural role of the various connection slabs of the round wall also contribute to make a column of maximum strength against storms - a "conical form," as Smeaton calls it, which makes the lighthouse similar to the trunk of an oak tree.[60] The whole construction system is studied, stone by stone, row by row, using a wooden model mounted on a model of the cliff, which Smeaton makes derived from precise surveying (using with a tool he constructs himself, based on what Alberti describes in *De Statua*).[61]

Through the documents communicated to him at the start of 1756, Smeaton discovers the expedient, used by Rudyerd, of the wooden barrier to protect the mortar of the foundation from sea water during the setting, "in case the cement was not of the very best kind possible for water-works."[62] It is precisely this detail, fundamental for the construction of the foundations and the new lighthouse entirely in stone, that triggers Smeaton's research on the composition of a mortar capable of setting in water. This research begins in the spring of 1756, when Smeaton travels through the various counties, around Plymouth or London, to observe the nature of the different rocks or sea pebbles, and that of the mortars used for the masonry of houses. On one of these trips, towards May, he visits the ruins of the walls of Corfe Castle, built in the Roman manner, with small stones held together by a solid cement, and writes a description destined to be repeatedly quoted until the mid-1800s in English texts on cement:

"I found its solidity, as usual in ancient buildings, did not consist in having been built with large hewn stones throughout, for the interior filling of the walls was with rough rubble, and fragments of the quarries, the interstices being entirely filled up with mortar, that undoubtedly had originally been fluid, and in that state poured in and from the nature of the component matter, as well as time, the whole mass had become strongly cemented together. I found that the mortar was composed of lime (doubtless originally of a good quality), with a considerable admixture of sharp sand and pebbles."[63]

The question of the type of mortar for the masonry of the Eddystone lighthouse becomes timely only in December 1756, when the worksite is already in progress, when the rock of the cliff has already been prepared for the first rows of stones, and when the stone has already arrived from Portland; at this point it becomes urgent, as Smeaton himself remarks, "to have a cement the most perfect that was possible, to resist the extreme violence of the sea."[64] So in the winter of 1756-57 Smeaton conducts a series of experiments in Plymouth with various compounds, to find the formula for a cement and a mortar with hydraulic characteristics and such strength as to ensure rapid bonding of stones even during sea storms. This is the beginning of that scientific research which in England will lead to the discovery of Roman cement and subsequent cement-based products.

The aim of the experiments of Smeaton is to ascertain, through repeated trials, the rules followed by masons in the preparation of the cement-based

*and General Literature*, Edinburgh, A. and Ch. Black, 1860, vol. XX, p. 721, reprinted in, Arthur Ashpitel, ed., *Treatise on architecture including the arts of construction, building, stone-masonry, arch, carpentry, roof, joinery, and strength of materials*, Edinburgh, A. and Ch. Black, 1867, p. 178).

[64] Smeaton, *A Narrative of the Building* cit., p. 97. In July 1756 Smeaton organizes, in Plymouth, the work team, with Josias Jessop, appointed as his assistant, and Thomas Richardson and William Hill, two master masons, each heading the two teams of masons at the orders of Jessop (starting in 1757 the stonecutter William Tyrrell, who had prepared the Portland stone for the Westminster Bridge, is also involved in the work).

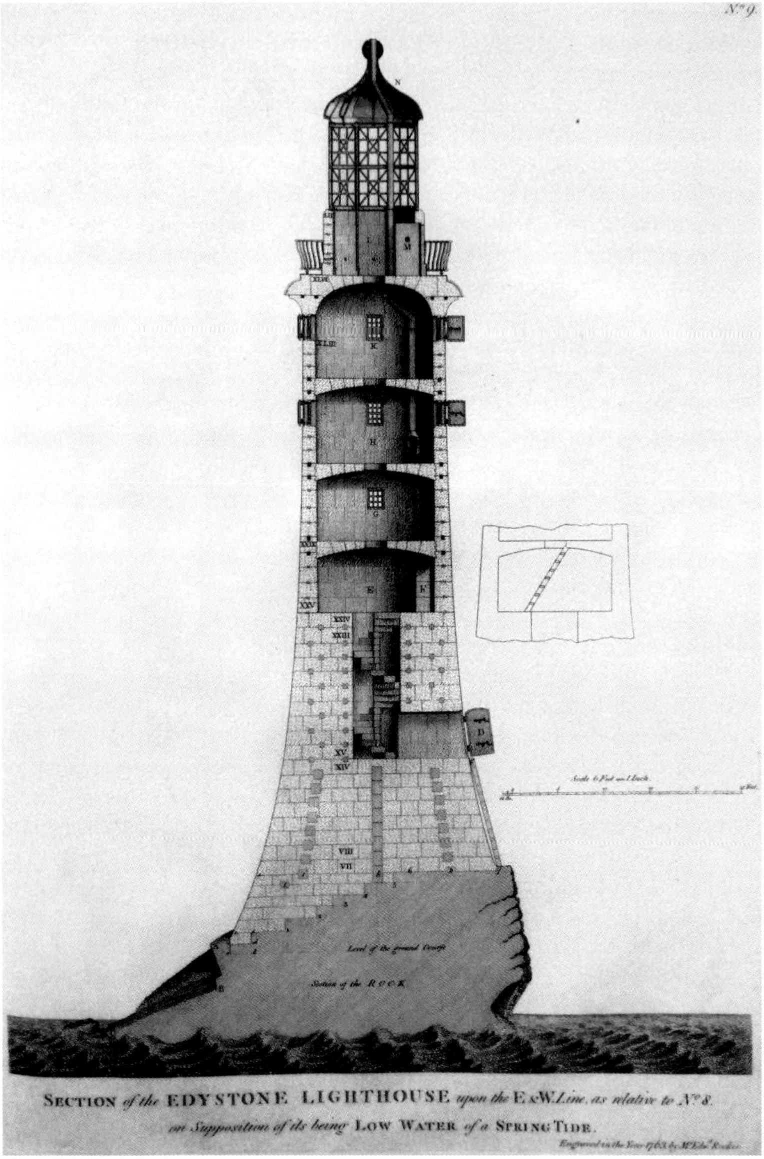

**259**, **261**. John Smeaton, Eddystone lighthouse, Plymouth Sound, 1756-59 (J. Smeaton, *A Narrative of the Building and a Description of the Construction of the Edystone Lighthouse with Stone*, London, G. Nicol, 1791).

**260**. John Smeaton, Eddystone lighthouse, Plymouth Sound, 1756-59. Remains of the lighthouse (M. Palmer, *Eddystone. The Finger of Light*, Rendlesham, Palmridge Publishing, 1998).

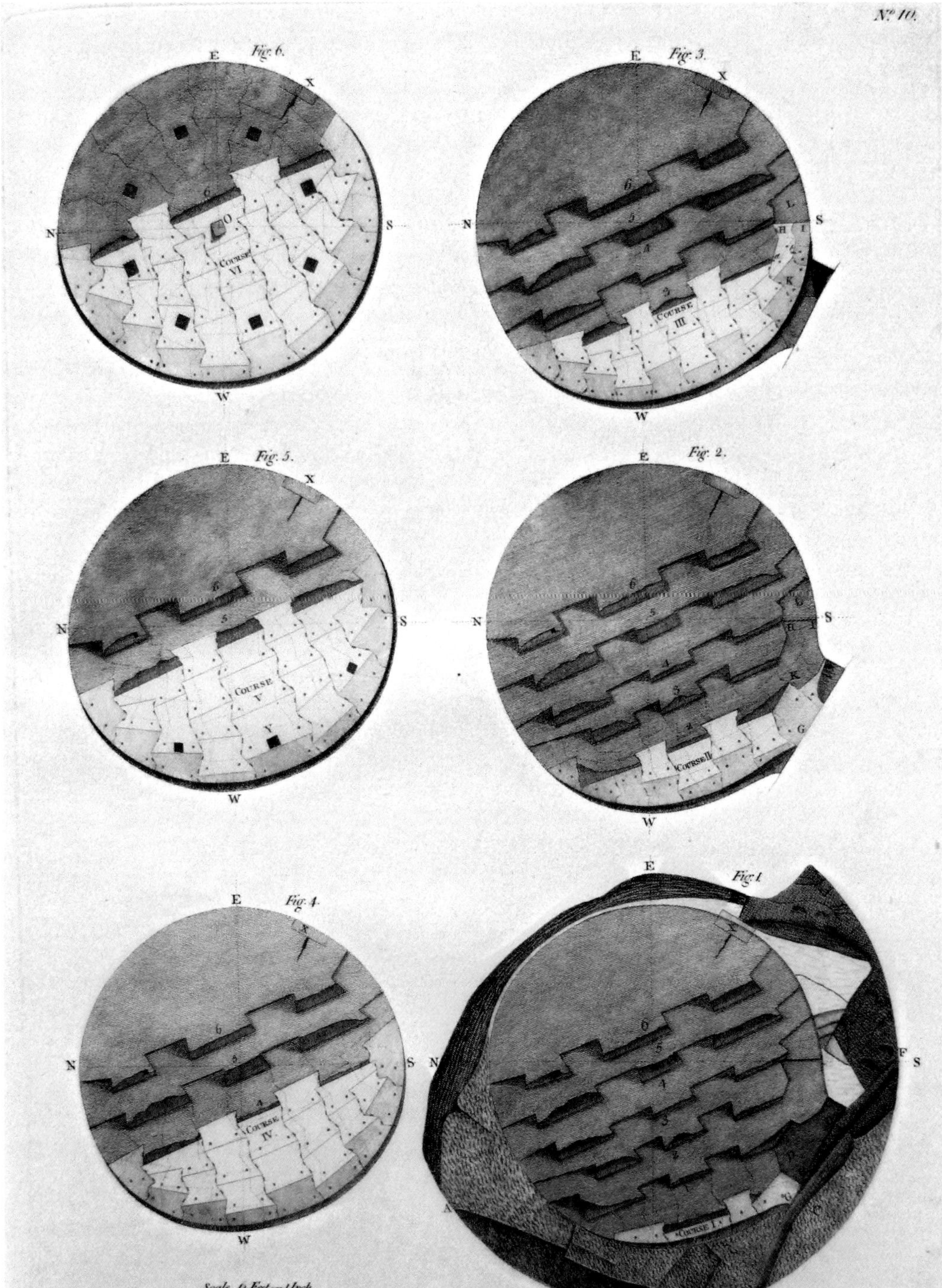


PLANS *of the* ROCK *after being cut: & prepared to receive the* STONE BUILDING.

*Shewing the Six* FOUNDATION COURSES.

compounds, passed down from generation to generation (he reports on discussions, on many occasions, with the mason). The idea of subjecting the various compounds to testing probably comes to him from what Bélidor has written on the trials conducted by Milet de Monville to establish the modes of setting in water of the various concretes – Smeaton himself mentions *Architecture Hydraulique* regarding his experiments.[65] We should not overlook the fact that during his stays in Plymouth, Smeaton was a guest at Notte Street, in the home of William Cookworthy (1705-80), a chemist and mineralogist, inventor and man of letters known for his research in different fields, and an owner of several patents. Probably it is precisely Cookworthy who helps to orient Smeaton towards the experimental tests currently used in scientific practice, which he also applies in his laboratory (to identify the process of fabrication of the compound of Chinese porcelain, for example).[66]

When he begins the experiments, Smeaton knows that "two measures of quenched or flaked lime, in the dry powder, mixed with one measure of Dutch Tarras, and both very well beat together to the consistence of a paste, using as little water as possible, was the common composition, generally used in the construction of the best water-work both in stone and brick [...]";[67] but he does not know about the effects on the properties of the mortar, with equal proportions of the compound, of the nature of the calcined stones,[68] nor does he know if it is true, as the masons claim, that fresh water, not salt water, should be used to prepare the mortar, or what will be the effects of the degree of calcination of the stones on the nature of the final compound.

The various compounds tested by Smeaton are shaped into spheres two inches in diameter, which are then pressed to form disks, or "flat pots," to immerse in water. "[...] what happened to the ball in this state was the criterion by which I judged of the validity of the composition for our purposes," he says.[69] The extremely small size of the test samples confirms a type of experimentation as conducted in scientific laboratories, like that of Cookworthy, and not on construction sites, as in the experiences of Milet de Monville.

To control the amounts of substances in the compounds, Smeaton uses the unit of measure of a small box in which he mixes lime, water and "tarras, or other gross matter, gradually, working it after each addition till it was become tough,"[70] reaching the conclusion that the proportions currently used by masons produce a mortar that is effective for conventional masonry, but not for masonry used in water. Then he establishes, still through experimentation, the correct proportion for a compound with hydraulic properties: "Balls even of the common composition of Tarras mortar (two parts flaked lime to one of tarras) I found would not always stand this test but balls made up with equal measures of lime and tarras would general stand it."[71]

The testing with the balls show him that unlike what is asserted by the masons, "the strength of the lime must depend upon some other quality, than of the hardness of the stone," and that the use of fresh or salt water does not produce apparent differences in the days immediately following the making of the compound, though after two or three months of immersion the compound made with sea water performs better[72] (in the construction of the

[65] *Ibid.*, p. 105, note.
[66] *Cf.* A. Douglas Selleck, *Cookworthy 1705-80 and his circle*, Plymouth, Press Roman, 1978.
[67] Smeaton, *A Narrative of the Building* cit., p. 102.
[68] "I found it indeed commonly asserted by Masons, that the harder or stronger the Limestone was, the stronger would be the Lime, but whether this maxim chiefly regarded the usual composition of lime and sand in common buildings [...] or whether it held good also in Tarras Mortar, did not appear [...]" (*ibid.*, p. 103).

lighthouse, this experimentation permits him to decide to mix the materials transported to the cliff directly with sea water).

Another series of trials contributes to determine to what extent the type of calcareous stone influences the quality of the lime. These tests are done starting with the calcination of Aberthaw stone, considered the best for the job, and are then compared with the performance of lime samples produced with stones from the vicinity of Plymouth, demonstrating that the difference in hardness between the two cases, already after 24 hours, "was very remarkable": "the composition of two measures of Aberthaw to one of Tarras, considerably exceeded in hardness that of common lime and Tarras, in equal parts [;] the composition of Aberthaw and Tarras in equal parts was still considerably harder, and this difference was the more apparent, the longer the composition were kept."[73]

The trials conducted to answer questions regarding the multiple firing of a product – "whether Tarras mortar, after having been once well beaten, becomes better by being repeatedly beaten over again?" – establish that "the composition of equal quantities of Aberthaw lime and tarras seemed adapted in the best manner possible to the purpose."[74]

Other hydraulic compounds used on various English worksites are subjected to the ball experiments, like shell lime, i.e. cockle, or "other shells burnt" that is believed to produce an excellent mortar, regarding which Smeaton is aware of the use made by Wren for Saint Paul's Cathedral.[75] The observation of the surface decomposition of the compound when kept immersed at length in water leads him to express a critical view: "I concluded it totally unfit for our use."[76] He also conducts trials on plaster of Paris, with the aim of improving its performance: "I tried many experiments to improve its qualities by admixture, but did not find that either sand or tarras was of any use."[77]

Following these experiments and with the fundamental help of Cookworthy, Smeaton studies in greater depth the composition of Aberthaw stone, which has turned out to be the most effective stone for producing hydraulic mortar. The results of Cookworthy's chemical analysis lead Smeaton to the conviction, which is fundamental for subsequent research on artificial cement, that "the most pure limestone was not the best for making mortar, especially for building in water."[78] "[...] for some reason or other," he adds, "when a limestone is intimately mixed with a proportion of clay, which by burning is converted into brick, it is made to act more strongly as a cement. This suggested to me the idea, that an admixture of clay in the composition of limestone, when treated as above specified, might be the most certain index of the validity of a limestone for aquatic building."[79]

Smeaton's testing could not help but consider the effects of the addition to the compound of those powders universally thought to be very effective for hydraulic masonry: Dutch *tarras* and pozzolana. While he is aware of the very widespread use in England of the *tarras* imported from Holland, whose efficacy he was able to observe firsthand during his trip to the Netherlands, he learns of the properties of pozzolana only by reading Bélidor's treatise.[80] As chance would have it, in Plymouth Smeaton runs across that exceptional shipment of pozzolana that had been brought there by a merchant with an eye on selling it for

69 *Ibid.*, p. 104.
70 *Ibid.*
71 *Ibid.*
72 *Ibid.*, p. 105.
73 *Ibid.*
74 *Ibid.*, pp. 105-06.
75 *Ibid.*, p. 106.
76 *Ibid.*
77 *Ibid.*
78 *Ibid.*, p. 108.
79 *Ibid.*
80 *Ibid.*, p. 106.

the construction of the Westminster Bridge. But then the merchant wants to sell the material, and Smeaton purchases it at the same price as Dutch *tarras*.[81] Again, the analysis of the composition of the pozzolana is done with Cookworthy (it is the chemist who shows him that the pozzolana contains iron).[82] Smeaton's enthusiasm about the results achieved with compounds made with Italian pozzolana mixed with Aberthaw lime shines through in his words:

"I perceived that in every state of it, if made into a mortar with Aberthaw lime, it exceeded in hardness any of the compositions commonly used in dry work, and in wet and dry, or wholly wet, was far superior to any I had ever seen or experienced, for, when the balls were immersed in water, they appeared to acquire hardness progressively, which progress, while they were out of the water and became dry, was indeed suspended, but yet they retained the hardness they had there acquired. With respect to those balls that were kept constantly under water, they did not seem inclined to undergo any change in form, only to acquire hardness gradually, insomuch that I did not doubt but to make a cement that would equal the best merchantable Portland stone in solidity and durability."[83]

Having recognized the volcanic origin of *tarras* and pozzolana, Smeaton conducts other tests on similar products, from pumice to coal cinders, bricks to tile dust, reaching the fundamental conclusion that they "all are possessed of an absorbent property, which caused the mortar made with them to set somewhat more quickly, than when made up with sand only [...]."[84]

For the mortar of the lower levels of the masonry of the Eddystone lighthouse, Smeaton decides to use precisely the combination that proves most effective in the trials, that of two parts pozzolana and two of Aberthaw lime – the lime is sent to Plymouth in sealed casks – and he points out that he is the first to invent this blend: "I apprehend, unless they [the ancient Romans] have the Lyas limestone in Italy [a term for the layer of stone of the Aberthaw type], it was reserved to the Edystone, to have those two materials first combined, and of consequence, for far as yet appeared, the perfect composition of water mortar was now first ascertained to be a proper mixture of blue lyas lime and puzzolana."[85]

In June 1757 construction of the lighthouse resumes with the laying of the first course of stones over the rock foundations. The mortar is protected by applying a layer of plaster of Paris to the joints, "to prevent the immediate wash of the sea upon the mortar."[86] "I also observed," he writes in his diary, on 8 August 1757, "that as in mending the pointings we had in some place made trial of Dutch tarras as well as puzzolana, interchangeably, the puzzolana, for hard service, was evidently superior to the tarras and some particular joints had proved so difficult, that I was obliged to try other expedients, the best of which was to clop oakum very small, and beat it up along with the mortar."[87]

Completed in October 1759, the Eddystone lighthouse becomes the first major English work to be made with hydraulic masonry using pozzolana. It is acclaimed by the English themselves in the decades to follow as the beginning of the history of contemporary cement, though actually it is but one chapter of the narrative that is unfolding during the course of the 1700s, one of the

[81] *Ibid.*, p. 109. According to an Italian observer at Plymouth on 21 August 1760, "to make that lighthouse sturdy, so it would not be damaged by the sea, that sand known to the Romans as pozzolana was brought from Rome [...]" (*Lettere familiari di Giuseppe Baretti a' suoi tre fratelli Filippo, Giovanni e Amedeo*, Milano, Società Tipografica de' Classici Italiani, vol. III, 1839, p. 21).
On the spread of knowledge of pozzolana in England, see, *An Account of the Manners and Customs of Italy; with observations on the mistakes of some travellers with regard to that country, by Jospeph Baretti, vol. I, the second edition, corrected, with notes and an appendix added, in answer to Samuel Sharp*, London, T. Davies, 1769, p. 22.

many beginnings, alongside those of Fontana, Vanvitelli, Zendrini, Milet de Monville or Bélidor, and certainly a crucial one, above all for having created, after the attempts of Semple, the premises for English construction culture to conduct experimental research on concrete based on testing and analysis, rather than the recipes of builders alone. Nevertheless, the results of the many experiments of Smeaton on mortars are not made public until 1791.

Following an inspection of the lighthouse, on 10 August 1766 Smeaton writes a report in which he also indicates the condition of the cement of the joints between the stones: "The cement every where below high-water mark, and from there to six feet above it, was perfectly entire, as left by the trowel, and seems changed to the colour and consistence of iron-stone. From thence to ten or twelve feet above high-water mark, the cement appears here and there to be somewhat impoverished in the joints, or fretted, and I observed that this was more so on the East side than on the West above the parts last described, the cement appears perfectly flush in the joints, quite to the top, as left by the trowel, being no ways altered in colour or consistence."[88] When he returns to inspect the lighthouse in August 1787, he notices that the "the cement in the upper works was in most places sensibly corroded."[89]

Smeaton continues to make works with hydraulic mortars similar to the one used for Eddystone, when in 1760 he does works to improve the navigability of the Calder River, even using the lime left over from the Eddystone worksite, well conserved in sealed casks.

## The worksite of the Blackfriars Bridge in London by Mylne

Another work of engineering takes its place in the process of European recognition of Roman magnificence spurred by the impassioned propaganda of Piranesi. From 1760 to 1769 the Scottish architect and engineer Robert Mylne (1733-1811) makes the Blackfriars Bridge, in London, using the system of Labelye of the caisson foundations implemented for the Westminster Bridge, and a type of masonry partially based on that used by Smeaton for the lighthouse of Eddystone.

Mylne likes to describe himself as "a scholar of Piranesi,"[90] since during his stay in Rome, from 1755 to 1758, in the company of his brother William, an apprentice mason, he took drawing lessons from Piranesi himself (the Mylnes reach Rome after passing through Toulon to visit the naval shipyard, disembarking at Civitavecchia).[91] When Mylne arrives in Rome he has already completed an apprenticeship in the workshop of the carpenter Daniel Wright, and he is also probably well aware of masonry techniques (he comes from a family of architects and builders). It may be no coincidence that Piranesi studies the structure of the foundations of the bridges of ancient Rome precisely in the mid-1750s, when his workshop often hosts a young Scotsman interested and expert in construction techniques (the most important plates on the construction of bridges are published in the fourth volume of the *Antichità Romane*). During his stay in Rome, Mylne studies the system of the aqueducts,[92]

[82] Smeaton, *A Narrative of the Building* cit., p. 111.
[83] *Ibid.*, p. 109.
[84] *Ibid.*, p. 111.
[85] *Ibid.*
[86] *Ibid.*, p. 125.
[87] *Ibid.*, p. 130.
[88] *Ibid.*, p. 179.
[89] *Ibid.*, p. 181.
[90] Cit. in, John Wilton-Ely, *Giovanni Battista Piranesi. Vision und Werk*, München, Hirmer Verlag, 1978, p. 145, note 112.
[91] Christopher Gotch, *The Missing Years of Robert Mylne*, in, "The Architectural Review," vol. CX, September 1951, no. 657, (pp. 179-82), p. 179.
[92] *Cf.* Albert Edward Richardson, *Robert Mylne, Architect and Engineer, 1733 to 1811*, London, B.T. Batsford Ltd, 1955, p. 14.

and he probably has in-depth knowledge of pozzolana cement and its hydraulic qualities, since once he returns home he has pozzolana shipped to him in England.[93] Confirmation of Mylne's interest in hydraulic construction can also be seen in the fact that one of the stops on his journey back to England is Holland and the port of Rotterdam, "a country so new and so much connected with the mechanical part of our trade," as he writes.[94]

As soon as he has returned to London, in October 1759 he submits a project in the competition for the new bridge over the Thames at Blackfriars, alongside other competitors like Smeaton, Chambers, George Dance the Elder, John Gwynn (1713-86) and Isaac Ware (1704-66). Mylne wins, in February 1760, after having demonstrated the structural logic of his elliptical arches (using the example of the Santa Trinita bridge in Florence), countering the claims in favor of the semicircular arches of Johnson – the same Johnson who takes part in the debate on the natural or artificial stone of Stonehenge.[95] To defend the elliptical arches, a booklet is published in 1760, entitled *Observation on Bridge Building and the Several Plans offered for a New Bridge*. Behind the pseudonym of Publicus, indicated as the name of the author of the booklet, lurks Mylne himself[96] (the pseudonym confirms that Mylne has also learned the literary methods of Piranesi).

*Observation on Bridge Building* contains the explanation of the complex structural machine created by Mylne so that the cutting of the stones, their assembly, the poured filler and the inverted arches create a system so as to construct, in the masonry of each span and each pier, a skeleton to stiffen the structure against static strains. This conception of a lightened masonry mass ribbed by a skeleton can only come from Mylne's knowledge of Roman masonry, including poured works like the cupolas crossed by brick ties. But it is above all in the drawing *Spaccato del Ponte Fabrizio, detto de' Capi* by Piranesi that we can find the direct source of inspiration for the system proposed by Mylne.[97] The drawing of the skeleton for the Blackfriars Bridge can even be considered the result of the overturning of the foundations with inverted arches drawn by Piranesi. Precisely the fantastic section of the Ponte Fabrizio, with the arch and foundation that create a single circumference line, resurfaces in the original project by Mylne, in which the line of the

[93] *Cf.* Robert Ward, *The Man Who Buried Nelson: The Surprising Life of Robert Mylne*, London, Tempus, 2007, p. 103.
[94] Cit. in, Ruddock, *op. cit.*, pp. 63-79.
[95] The letters of Johnson on Mylne's project are published in, *Considerations on the Plans offered for the Construction of Black-Friars Bridge in Three Letters, to the Printer of the Gazetteer, in, Arthur Murphy, The works of Samuel Johnson, LL.D.*, London, L. Hansard, 1801, vol. II, pp. 375-83.
[96] *Observations on Bridge Building and the Several Plans offered for a New Bridge [...]*, London, J. Townsend, 1760.
[97] *Spaccato del Ponte Fabrizio, detto de' Capi, in, Piranesi, Le Antichità Romane* cit., vol. IV, plate XIX.

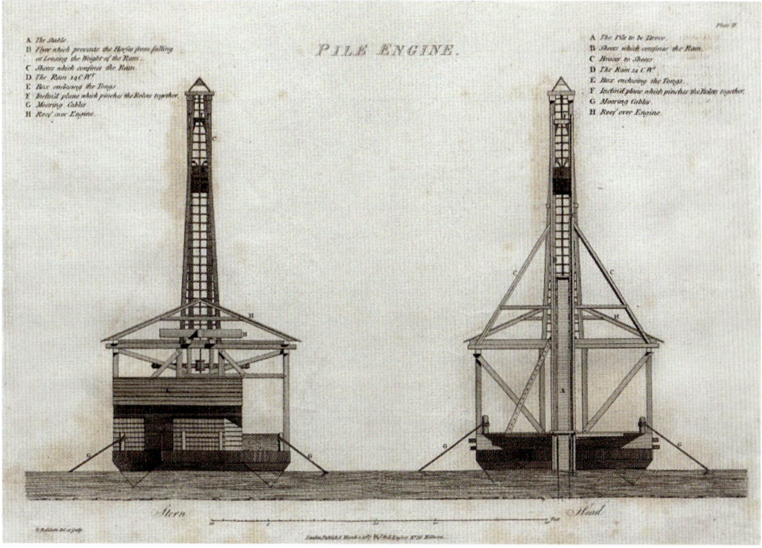


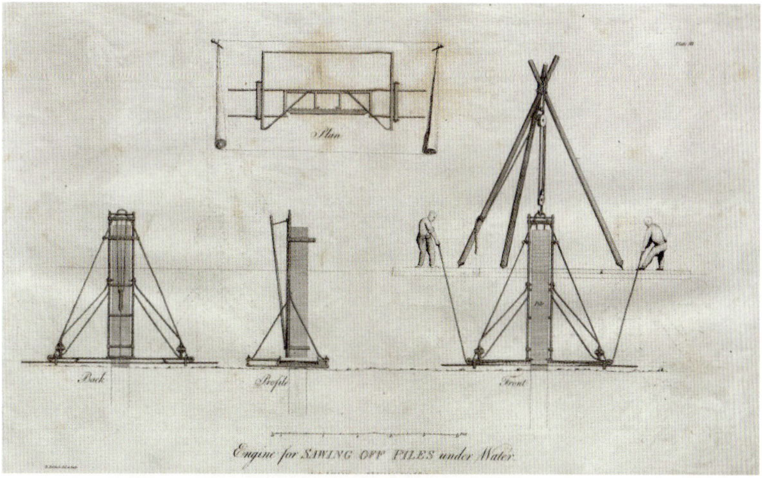


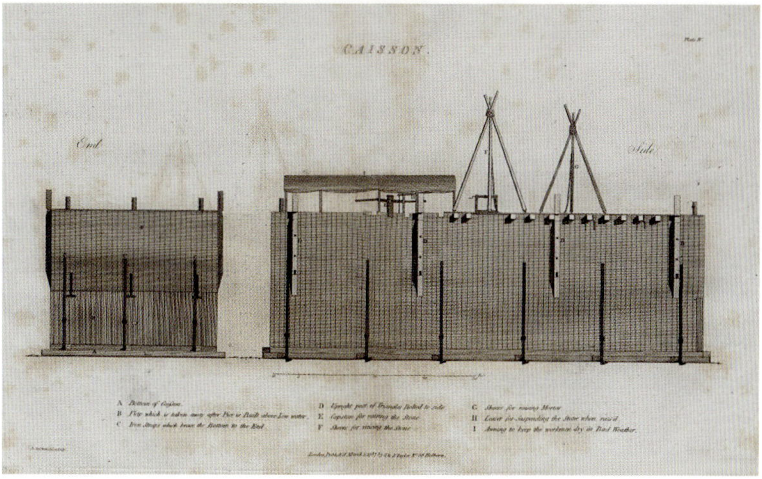


**262-265**. Robert Mylne, Blackfriars Bridge, London, 1760-69. Plan and elevation; "Pile Engine"; "Engine for sawing off piles under Water"; "Caisson" (R. Baldwin, *Plans of the machines and centering used in erecting Blackfriars' Bridge*, London, I. and J., Taylor, 1787).

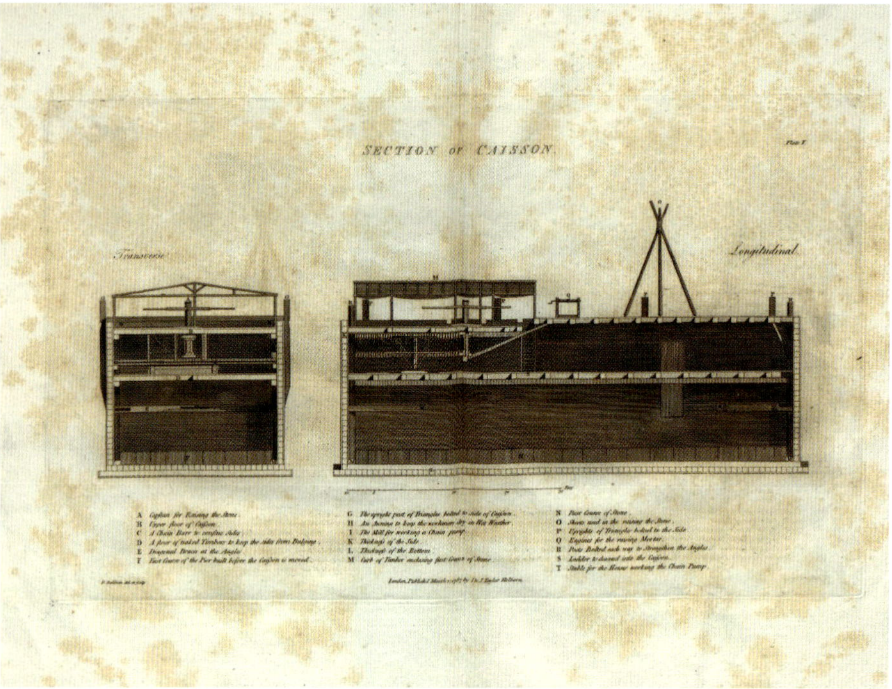
SECTION OF CAISSON.
Plate V
Transverse
Longitudinal
A Capstan for Raising the Stone.
B Upper floor of Caisson.
G The upright part of Triangles bolted to side of Caisson.
I The Mill for working a Chain pump.
K Thickness of the Side.
L Thickness of the Bottom.
N First Course of Stone.
Q Engines for the raising Mortar.
S Ladder to descend into the Caisson.
T Stable for the Horse working the Chain Pump.

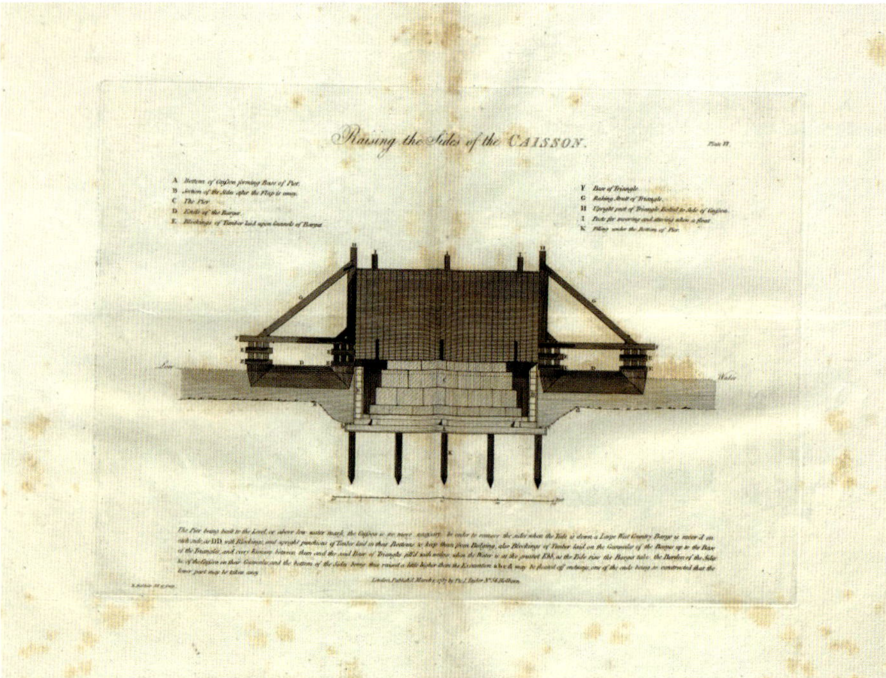
Raising the Sides of the CAISSON.
Plate VI
Low
Water

**266-267**. Robert Mylne, Blackfriars Bridge, London, 1760-69. "Section of Caisson"; "Raising the sides of the Caisson" (R. Baldwin, *Plans of the machines and centering used in erecting Blackfriars' Bridge*, London, I. and J., Taylor, 1787).

arches bends, extending all the way to the sides of the pylons. The initial idea later abandoned in the construction of consolidating the stones with special anchors, again in stone, inserted in the joints – "joggles," as they are called by Mylne[98] – can be a demonstration of continuity not just with the solution of Smeaton for the stones of the Eddystone lighthouse, but also with certain techniques of Roman construction that emerge in the etchings of Piranesi of the *Antichità Romane* (consider, for example, the shape of the wedges of the arch shown in the "cutaway" of the "Tomb on the ancient Appian Way,"[99] or the interlocks between blocks in the foundations of the Theater of Marcellus, though Mylne's system is closer to the one documented by Cesariano for the relieving arches in a trabeation and to that used by De l'Orme for his flat arch).

While for the piers of the bridge Dance or Smeaton had foreseen use of an "inclosure or cofferdam,"[100] for the construction of the Blackfriars Bridge, begun in 1760 the masonry is implemented with the caisson system invented by Labelye, but now improved. For this, Mylne is assisted by Robert Baldwin. "The method of building the caissons," we read in the booklet of Publicus, "within high water-mark, and yet above it, upon frames that fall to a level, where the side may be floated and fixed to them. The method of making the sides of the caisson higher than high water-mark, and splitting them in the middle to be carried away, after they have preserved the pier dry from the laying the first course of stone, till it is as high as high water-mark."[101]

Once the area has been closed off, as Labelye had done, the bed of the Thames is excavated until the right terrain for the sinking of the piles is found. Then Mylne proceeds according to the method recommended by Perronet and put into practice by De Voglie: he prepares a wooden grille with the piles inserted, making up for the defect that arose during the construction of the Westminster Bridge – Mylne too, like De Voglie, uses a saw to level the piles, based on that of Etheridge. The caisson to construct the masonry of the piers is equipped with a base with three layers of wooden boards reinforced by iron hooks, and with removable sides (27 feet high, as opposed to the 16 feet of Labelye's caisson, requiring a complex system of carpentry, also with metal reinforcements).[102] After the construction of the first two piers (the caisson of the second pier is put into the water on 20 May 1762),[103] the system is modified so that the bottom of the caisson reaches the position of the pier with the first layer of masonry already laid, prior to sinking. In general, the sides are stripped after the construction and reused to form other caissons – they are held in position by means of two barges.

The piers are built according to the stone skeleton system that forms the basis of the entire structural machinery of the bridge. The skeleton is composed of blocks of Portland stone, laid to form quadrangular compartments – some blocks are hewn square with inclined joints for a better grip on the stones of the outer courses assembled like flat arches (the Portland stone of the facings decays rapidly). The compartments are filled with alternating layers of the same stone, but with random sizes, and of poured stone rubble.[104] The

[98] *Ibid.*, p. 28.
[99] Piranesi, *Le Antichità Romane* cit., vol. II, plate XXXVII.
[100] *Observations on Bridge Building* cit., p. 41.
[101] *Ibid.*, p. 30.
[102] For documentation of the systems used by Mylne see, Robert Baldwin, *Plans of the machines and centering used in erecting Blackfriars' Bridge*, London, I. and J., Taylor, 1787.
[103] From the diary of Mylne, cit. in, Richardson, *op. cit.*, p. 57.
[104] See the drawing in, Ruddock, *op. cit.*, p. 77, fig. 70.

pier in the *Veduta dell'avanzo di una delle Pile del Ponte Trionfale*, engraved by Piranesi, also seems to be constructed with a similar skeleton that cannot, however, be found elsewhere in the piers of English bridges in the 1700s. Mylne's system is also completely different from the model indicated by Gautier of the piers with well-laid rows, with stones of the same height both in the nucleus and on the faces.[105] Confirmation of Mylne's personal knowledge of Roman construction can also be found in his projects after that of the Blackfriars Bridge, where he makes use of pozzolana (in the case of the bridge in London he writes about "very hard mortar").[106]

The fact that the Blackfriars Bridge is conceived as a first monumental European manifestation of the rebirth of the magnificence of Roman construction championed by Piranesi is demonstrated by Mylne's request to the great master to engrave the view of the worksite at the moment of the closure of the first arch – the arch is completed in October 1764, but the correspondence between Piranesi and Mylne for the etching dates back to 1760, while the etching is dated August 1764.[107] For Piranesi this is a demonstration of what the state of the ruins depicted in his *Antichità Romane* rendered visible of Roman construction, namely the fundamental role of the worksite and its phases. Not only does Piranesi's view of the Blackfriars Bridge transform, thanks to the perspective, the elliptical arches into Roman ones; it also immortalizes the worksite of the bridge, with all the innovative centerings, easily knocked down thanks to a special system of wedges at the base. From the contemporary worksite in the Roman manner, Piranesi shows the scaffolding with a force that is quite different from that of the etchings published in the book by Nicola Zabaglia (1664-1750), focusing precisely on frameworks and centerings.[108]

[105] Gautier, *Traité des ponts* cit., p. 235.
[106] From Mylne's diary, 27 May 1765, cit. in, Richardson, *op. cit.*, p. 71. For direct references to pozzolana see: 30 June 1773 (p. 95); 18 November 1774, expenses for the Newcastle Bridge (p. 100); and 12 August 1775 (p. 103).
[107] See, Piranesi, letter to Robert Mylne, 11 November 1764 in, Christopher Gotch, *The Missing Years of Robert Mylne*, in, "The Architectural Review," vol. CX, September 1951, no. 657, (pp. 179-82), p. 182.
The view engraved by Edward Rooker (1724-74), showing the worksite of the Blackfriars Bridge in July 1766, is published following the act of Parliament of 31 December 1766, and made on the basis of that of Piranesi.
[108] See, Nicola Zabaglia, *Castelli e ponti di maestro Niccola Zabaglia con alcune ingegnose pratiche, e con la descrizione del transporto dell'obelisco vaticano e di altri del cavaliere Domenico Fontana*, Roma, N. e M. Pagliarini, 1743.
[109] To understand the value and impact of the drawing by Piranesi it will suffice to compare it with the one published in 1761 by Temanza for the Roman bridge in Rimini – a longitudinal section that offers but a glimpse of a structural nucleus different from the stone masonry of the outer surfaces (in the description Temanza makes no mention of the technique of concrete). *Cf.* Tommaso Temanza, *Delle antichità di Rimino, libri due*, Venezia, G. Pasquali, 1761, p. 16.
[110] *Observations on Bridge Building* cit., pp. 18-19.

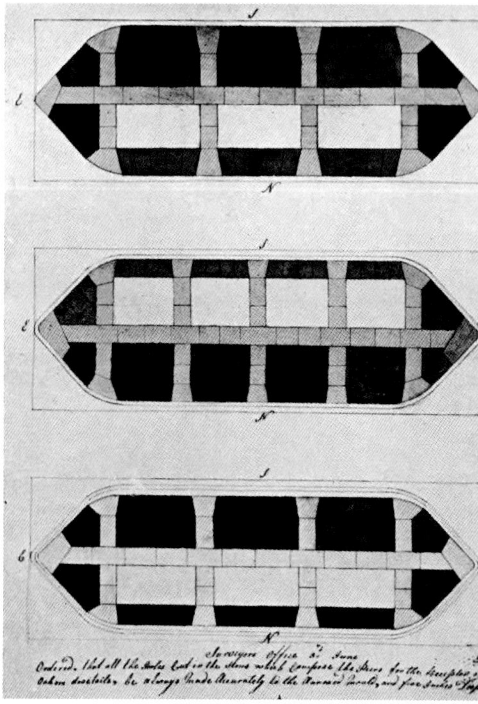

**268**. Robert Mylne, Blackfriars Bridge, London, 1760-69. Drawing of the masonry of the piers (A.E. Richardson, *Robert Mylne, Architect and Engineer, 1733 to 1811*, London, B.T. Batsford, 1955).

**269**. Giovanni Battista Piranesi, "A View of the Intended Bridge at Blackfriars London in August MDCCLXIIII, by Robert Mylne Architect, Engraved by Piranesi at Rome" (J. Wilton-Ely, *The mind and art of Giovanni Battista Piranesi*, London, Thames and Hudson, 1988)

In the etching Piranesi depicts, behind the arc and the part of the wall in wedges of perfectly squared stones, the start of a masonry stratification of small stones, in keeping with the Roman concrete technique. The layers of this masonry, which have the height of the rows of stones, rest on a stone course Piranesi depicts penetrating into the vault, with the function of an ashlar that grips the stone system of the facing to the poured system of the core. Perhaps this detail is intended to demonstrate the existence of a skeleton and of a "mass" in the body of the construction of the bridge.[109]

The discussion of Publicus of one of the projects submitted for the competition for the Blackfriars Bridge, that of Dance, contains telling indications on the method of foundation with concrete in the Roman manner, done with pozzolana, and on how the use of wooden compartments in which to construct the masonry in water is an alternative precisely needed due to the lack of such a concrete with hydraulic properties.

"The manner in which he proposes to lay the foundation of the piers," Publicus observes, "he says, is no chimera; because it is an invention of Vitruvius. It is a lamentable case that all the inventions of great men come under a disrepute, by being made use of by people less knowing. Vitruvius recommends in building a mole, to make an inclosure all round, and pump the water out, so as to make it dry; and this only in such places of the Mediterranean sea, where no Puozzollo [*sic*] earth could be found. Mr. Dance might have observed, that Vitruvius recommends this in a sea, where there is no rise and fall of tides. Here in the river Thames there is one of 16 feet, so that there would be an immense weight of water upon the outside of this inclosure."[110]

A. Ostium, sive Emissarium Cloacae Maximae in Tiberim. B Directio Cloacae Forum versus. C Reliquiae Pulcri Litoris ex topho. D Triplex Cloacae ex albano lapide arcuatio, quatenus extenditur latitudo Pulcri Litoris, videlicet a C usq. ad E. F Demonstratio eiusdem triplicis arcuationis. G Anterides Cloacae
A
B
C
D
E
F
G

# chapter six
# pozzolanas and new cement compounds, from cronstedt to loriot

## The nature of pozzolana according to Cronstedt

The fact that Smeaton conducts his analyses of compounds for the fabrication of hydraulic concrete in collaboration with the chemist and pharmacist Cookworthy is emblematic of the change in progress halfway through the 1700s in that chapter of the art of construction that has to do with mortars. After the initial scientific experiments on lime conducted at the start of the century, the chemists, especially in Sweden, enter the scene in search of the nature of pozzolana, alongside the volcanologists, who study the various products found in different geographical areas close to extinct volcanoes. The contribution of the chemists aims to identify the components of pozzolana. The background of the chemical analyses of volcanic products is the construction of important hydraulic works, and the costs and timing involved in transporting pozzolana from Italy to Sweden. The major hydraulic works of Swedish ports and canals are done first by Christopher Polhem (1661-1751), then by his son Gabriel (1700-72), and later by Daniel af Thunberg (1712-88), a pupil of Polhem.

Thanks to his voyage of 1695-96, Polhem has personal knowledge of cement manufacture as practiced in the Netherlands and Western Germany, which he helps spread through Sweden by his writings and contributions; he also recommends which materials mixed with lime and sand can produce resistant cements, insisting on the need for innovative methods of artificially producing building materials.[1]

The port of Karlskrona is the most demanding worksite with respect to pozzolana. After the initial work, based on a project by Christopher Polhem from 1714 to 1724, it is decided to expand the naval shipyard under the supervision of Gabriel, with work done in 1757, in keeping with a plan that calls for thirty new docks arranged according to a traditional design, laying the

**270-271**. Port of Karlskrona. Plans with drydocks, 1771 (Cartes et Plans, GE SH 18 PF 11 DIV 8 P 7, and GE SH 18 PF 11 DIV 8 P 8, Bibliothèque Nationale de France, Paris).

groundwork for what was to become, at the end of the century, the most important military port in Sweden. Thunberg supervises the work on the port, which does not enter its decisive phase until the 1770s. The material used to wall up the beds of the basins is pozzolana imported from Italy. It is no coincidence that an expert volcanologist like Johann Jakob Ferber (1743-90) takes an interest, when he is in Stockholm, in the works made with pozzolana at the port of Karlskrona, as demonstrated by a letter he writes to Von Linné.[2]

But the worksite that launches advanced experimentation in Europe to find artificial products that can function as alternatives to Italian pozzolana and Dutch *trass* is the project to make the rapids of Trollhättan navigable, along the Göta älv River. The construction of a navigable route to join the Baltic and North Seas, crossing the Trollhättan rapids, based on the model of the Canal du Midi, is a project that fits into the initiatives already begun in the 1600s, with the construction of canals and locks, some entrusted to Dutch technicians. The project by Christopher Polhem, dating back to 1717 and approved in 1718, calls for making the Trollhättan rapids navigable and thus creating a river connection from Göteborg, through the Vättern Lake, all the way to Norrköping. Work begins immediately, but is not continued until 1748 – Gabriel Polhem is also involved. The construction of the dam in the last waterfall to create a large basin happens under the guidance of Wodman,[3] using caissons filled with stones and connected together by iron brackets.[4] But in September 1755 the force of the water laden with lumber destroys the dam and the work is suspended.

It is not until 1757 that Thunberg proposes, in collaboration with Samuel Sohlberg (1708-85), a cut in the rocks to bypass the Trollhättan rapids, returning to the grand project of the connection of two seas. In the meantime, in 1768 work is completed on the Brinkebergskulle locks in the Karlsgraf canal, as designed by Thunberg.[5]

It is no coincidence, if we take the constructive exploits at Karlskrona and Trollhättan into account, that the decisive chapter of the scientific research on the nature of Italian pozzolana, to determine the characteristics that make it ideal for hydraulic mortar and to therefore be able to start production of an artificial substitute, happens precisely in Sweden, where the cost of shipping the material from Civitavecchia is higher. The chemist and mineralogist

[1] *Cf.* Nils Cronstedt, *Polhem's Contribution to the Art of Building*, in, Torsten Althin, ed., *Christopher Polhem, the father of Swedish Technology*, Hartford, Trustees of Trinity College, 1963, (pp. 183-226), p. 212-14. For some general indications on how to recognize the best mortar, in the context of reasoning on construction materials, see, Gabriel Polhem, *Rede von der Benuzung der verschiedenen Baustoffenin, gehalten vor der Königl. Akademie der Wissenschafften zu Stokholm, [....] den 5 Winterm 1760*, in, Gottlieb Sigmund Gruner, *Auserlesene Sammlung zum Vortheil der Staatswirthschaft, der Naturfosrchung, und des Feldbaues*, Bern, E. Thurneysen, 1769, (pp. 189-205) pp. 203-04. *Cf.* Nils Cronstedt, *Polhem's Contribution to the Art of Building*, in, Torsten Althin, ed., *Christopher Polhem, the father of Swedish Technology*, Hartford, Trustees of Trinity College, 1963, (pp. 183-226), p. 223.

[2] Johann Jacob Ferber, letter to Carl Nilsson Linnaeus (Von Linné), 12 May 1771, cit. in, Theodor Magnus Fries, Johan Markus Hulth, *Bref och skrifvelser af och till Carl von Linné*, Uppsala Universitet, Aktiebolaget Ljus, 1912, p. 73. For the works see, Johan Fellers, ed., *Essais de batir sous l'eau mis en oeuvre par Mr. Thunberg, Chevalier de l'Ordre de Wasa* [...], Stockholm, 1774.

[3] *Cf.* Charles Colville Frankland, *Narrative of a visit to the courts of Russia and Sweden, in the Years 1830 and 1831*, London, H. Colburn, R. Bentley, 1832, vol. I, p. 50.

[4] *Cf.* Paolo Frisi, *Dei canali navigabili*, in, *Raccolta d'autori che trattano del moto dell'acque*, Firenze, Stamperia di Sua Altezza Reale, 1770, vol. VII, p. 562. See, in particular, Samuel Ebbe Bring, *Trollhätte kanals historia till 1844*, Stockholm, Kungliga Vattenfallsstyrelsen, 1911.

[5] *Cf.* Gustaf Lagerheim, *Notice sur les canaux de Trollhätta et de Gothie en Suède*, in, "Annales des Ponts et Chaussées. Mémoires et Documens," 1834, I semester, pp. 315-36.

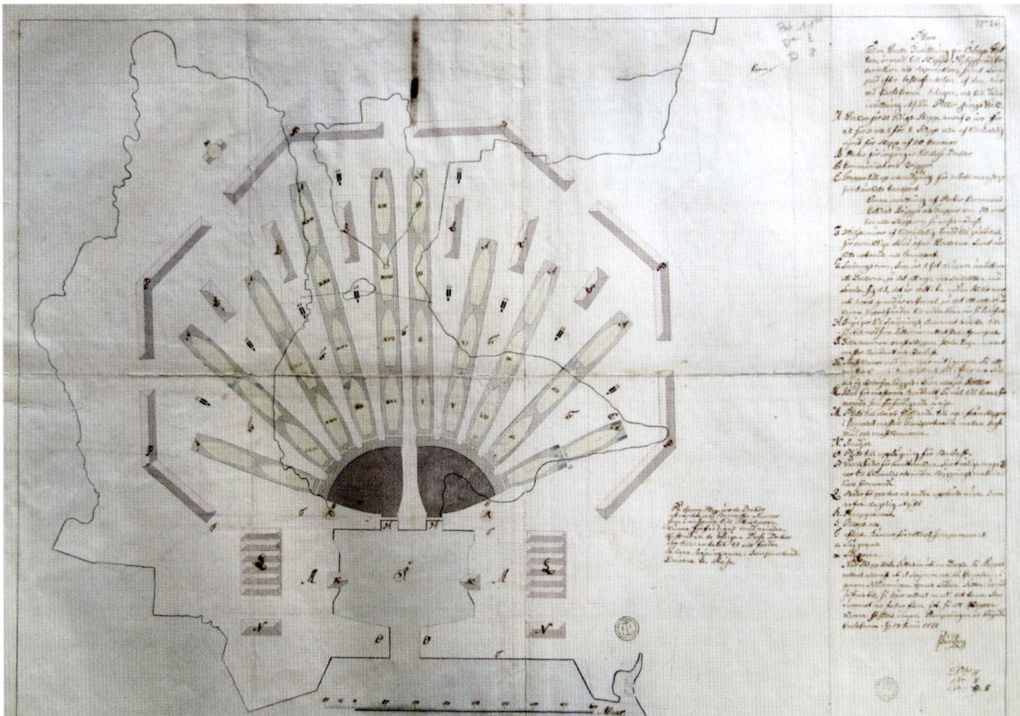

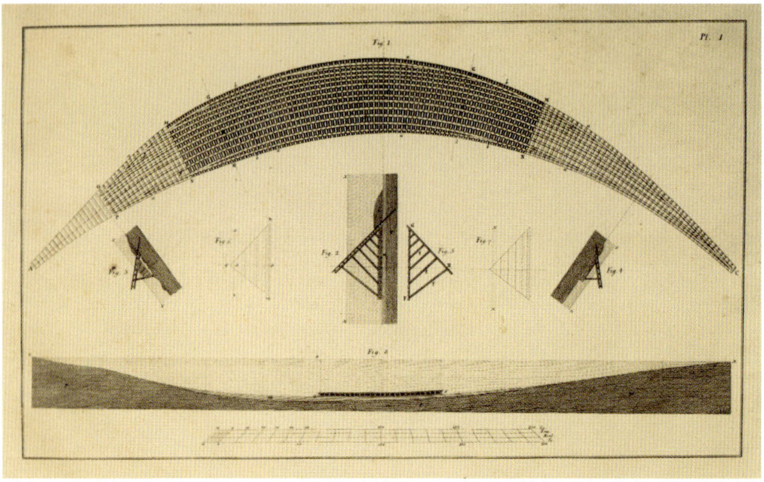


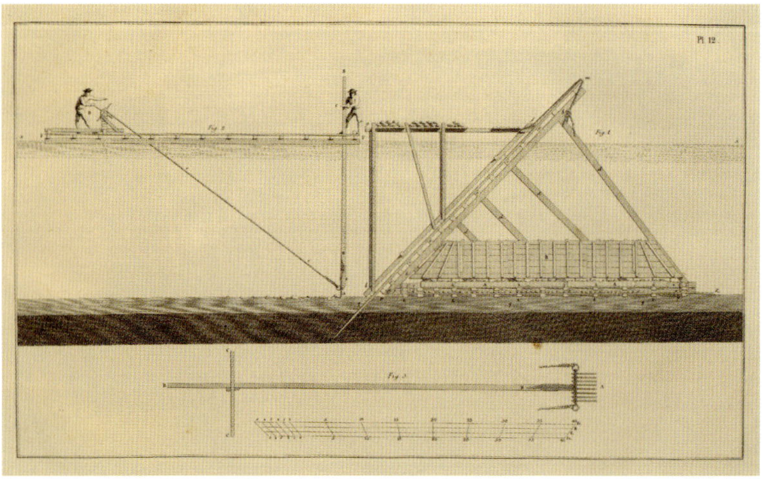


**272-275**. Port of Karlskrona.
View and defensive works (J. Fellers, ed., *Essais de batir sous l'eau mis en oeuvre par Mr. Thunberg, Chevalier de l'Ordre de Wasa, Directeur des ouvrages de Mechanique, à l'occasion des Nouvelles Formes qu'on construit a Carlscrona*, Stockholm, 1774).

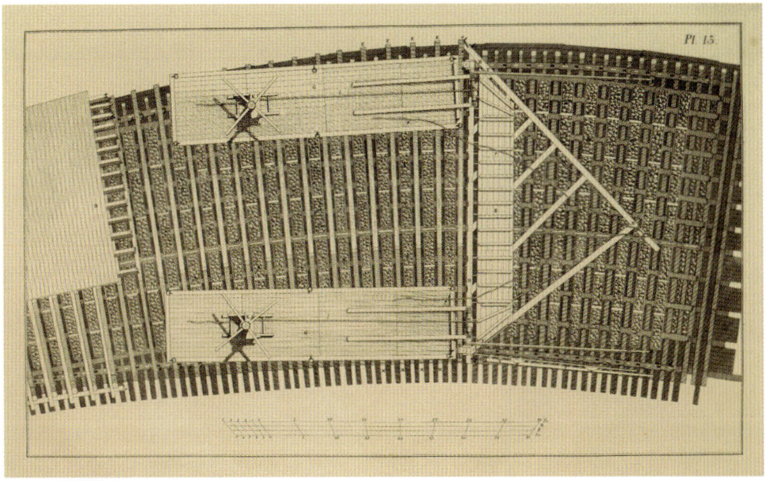

Axel Fredrik Cronstedt (1722-65), influenced by the work of Polhem and with a degree from the University of Uppsala, where he enrolled in 1738, is the one who attributes the particular properties of what he calls "terra pouzzolana," in the making of hydraulic cements, to the ferrous components found in the soil subjected to the action of volcanic fire.[6] Smeaton had reached similar conclusions (according to what he states in 1791). The analyses are conducted by Cronstedt using the method he was taught by his mentor at Uppsala, Brandt, while the classifications of natural products proposed by Wallerius provide the scientific support. Along with pozzolana, Cronstedt also examines a volcanic product in similarly widespread use for hydraulic constructions in the countries of Northern Europe, as an alternative to the Italian substance, namely Dutch *trass*.

"The *Terra Puzzolana* or *Terras* is a sandy marle," he writes in 1758, "mixed with iron and baked in a strong fire. Its effects, however, in the cement, may, perhaps, depend only on the iron which has been reduced into a particular substance by means of subterraneous fires, evident signs of which are observable in the places where it is obtained."[7]

Apart from the accuracy of the hypothesis, Cronstedt focuses the scientific research on the question of the identification of the component that makes pozzolana the most effective ingredient for the making of hydraulic mortars. Other scholars will extend this same question to the analysis of the various limes used by masons and endowed with properties similar to those of pozzolana, continuing – where the limes are concerned – the line of study begun in the 1740s by Duhamel du Monceau, Malouin, Macquer and also Wallerius. After the contribution of Cronstedt, the issue is to discover, through chemical analysis, the precise components of pozzolana and of hydraulic limes, in order to proceed with their artificial production and to improve their performance. The first renowned artificial pozzolana is developed precisely at the Trollhättan worksite, at the start of the 1770s.

[6] *Cf.* Axel Frederic Cronsted, *Försök til mineralogie, eller mineralrikets upställning*, Stockholm, Wildiska, 1758; Axel Frederic Cronsted, *Versuch einer Mineralogie*, Kopenhagen, Rothenschen Buchhandlung, 1760, p. 45.

[7] Cronsted, *Versuch einer Mineralogie* cit., p. 45 (*Id.*, *An essay towards a system of mineralogy*, London, Ch. Dilly, vol. I, 1788, p. 72).

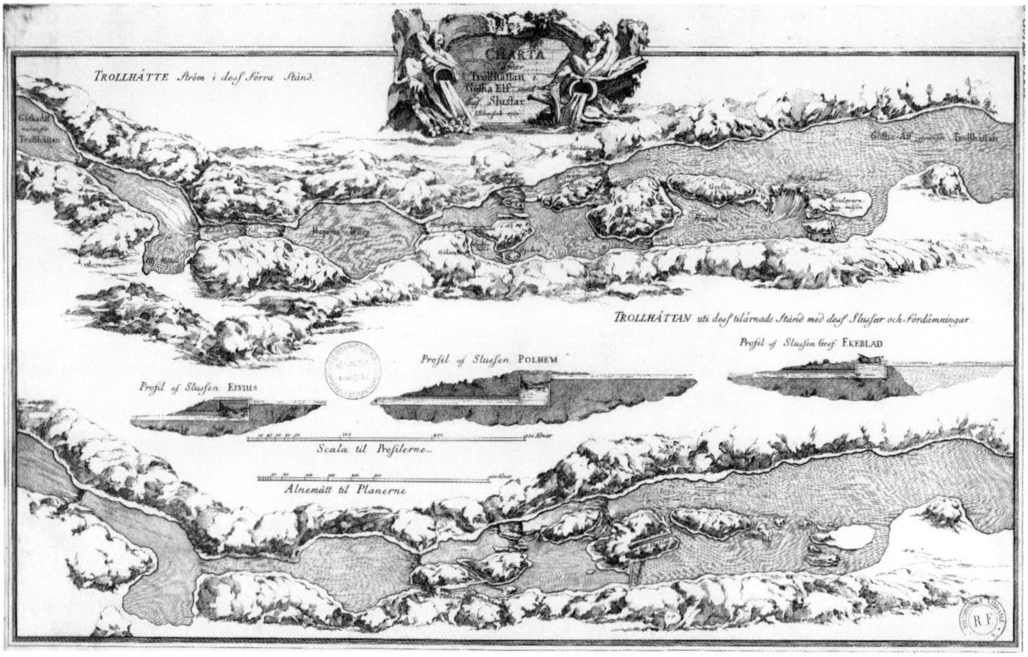


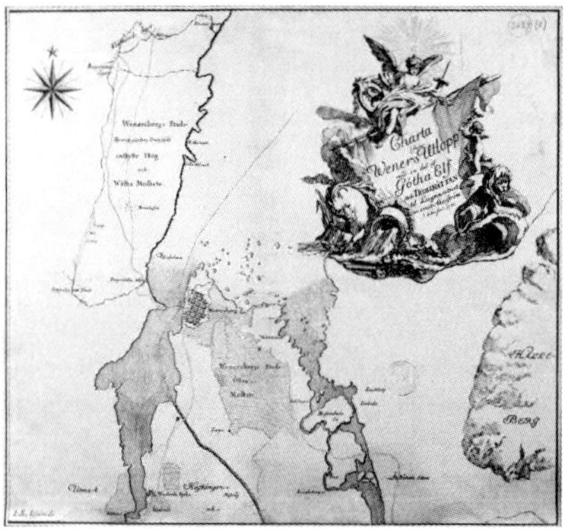

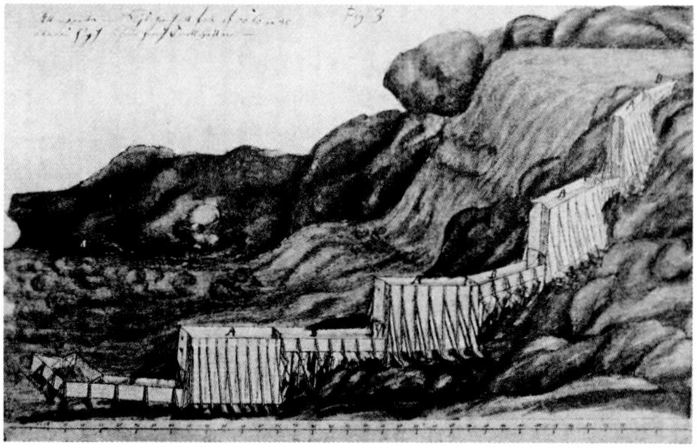

**276-278**. Trollhättan rapids, on the Göta älv River, 1750 circa (Cartes et Plans, GE DD-2987 (2902), 6E DD-3629 (2), Bibliothèque Nationale de France, Paris; S.E. Bring, *Trollhätte kanals historia till 1844*, Stockholm, Kungliga Vattenfallsstyrelsen, 1911).

In the context of German and, more generally, European culture, we should not overlook the observations on pozzolana made by the Italian Giovanni Antonio Scopoli (1723-88), a naturalist and physician, professor of chemistry, mineralogy and metallurgy at the Mining Academy of Schemnitz (now Banská Štiavnica) – in 1777 he becomes a professor at the University of Pavia.[8] Scopoli analyzes a series of different lavas by firing them in a furnace, and reveals how to make improvements through the addition of a certain amount of fired, pulverized clay to the cements.[9]

## Discovery of the first artificial products

A breakthrough in the research on artificial products for mortars comes in the 1750s, not through the work of chemists, but through that of experimenters of different types who are seeking an alternative to the importing of volcanic products with hydraulic properties. The fact that products are made and presented as surrogates of natural volcanic materials constitutes a crucial chapter for the production of artificial pozzolana, one that takes place in the 1770s. In England and Holland we can find the first attempts to produce such pozzolanic materials.

In March 1750 Isaac Dent, in London, files a patent for "making tarrass used in brick and stone work for building bridges and wharfs."[10] The patent, significantly, has to do with the construction of hydraulic works for which Dutch *trass* or products with similar characteristics would have to be imported. The statement, right from the title of the patent, that the method makes it possible to produce "tarrass" indicates the economic context of European trade in volcanic products for cement and the national ambition to be liberated from the constraints of importation.

"Take freestone," we read in the specifications of Dent, "and burnt it in a kiln or furnace till it becomes like a body of fire, then take it out and cool it, and then grind it in a mill till it becomes a fine powder, and then skreen or sift it through a fine skreen or sieve to separate the fine stuff from the hard lumps which the mill cannot easily break, and then the same becomes terras; or (where such freestone is scarce, or becomes too dear to answer the end, and cannot be had at a moderate expense), mix fine light loam or clay with water, and work it up into a substance, and lay it out to dry and harden in the sun and wind till it becomes hard almost like a stone, then burn it in a kiln or furnace, and grind, skreen, or sift in the very like manner as above directed when it is made with freestone."[11]

Semple's assertion regarding the possibility of obtaining artificial products with characteristics similar to those of *trass* comes against the backdrop of his personal experiences and those of researchers like Dent.[12]

Even more explicit than that of Dent is the ambition of a Dutch inventor to free the producers in his country from the need to import *tuffstein* from Germany. It is no coincidence that this inventor is a mechanic residing in Dordrech, the town best known in Holland for the production of *trass*. Records

[8] The studies of Scopoli are referenced, for example, by Christian Ludwig Ziegler (*Id.*, *Fortsetzung des aufsatzes von dem tarras, der puzzolane und deren substitutionsarten*, in, "Hannoverisches Magazin," XI, 1 March 1773, no. 18, coll. 279-88).

[9] *Cf.* Giuseppe Gioeni, *Saggio di litologia Vesuviana*, Napoli, 1791, p. 44, note.

[10] *Cf.* Bennet Woodcroft, ed., *Subject-matter index of patent of invention, from March 2, 1617 [...], to October 1*, 1852, Part. I, London, G.E. Eyre, W. Spottiswoode, 1857, p. 108. Dent's patent is number 654.

[11] *A.D. 1750, March 13. N° 654. Dent, Isaac, A method of making terras*, in, Bennet Woodcroft, ed., *Patents for inventions: abridgments of specifications relating to bridges, viaducts, and aqueducts, A.D. 1750-1866*, London, G.E. Eyre, W. Spottiswoode, 1868, p. 1.

[12] Semple, *op. cit.*, p. 81.

report on the experiments of the mechanic from Dordrech, conducted around 1750 to identify a substitute for German tuff.

"This ingenious mechanic tried an experiment of a substitute for this Hessian stone, which he found to answer to admiration. He then went to the owner of the Hessian Mountain, and told him he could make as good terras without his stone, but offered to take all his stone at about the then present price. The stone-merchant, well knowing there was not stone in Holland or any of the Seven Provinces, refused his offer, and even made another advance on his stone. The mechanic returned to Dort, made terras of the substitute which he found at home, and obtained a patent of the states for fifteen years for no other terras to be used in any of the Dutch provinces; he sold it cheaper, it was proved to be better, and the mechanic made an immense fortune. This substitute is to be found in every maritime country; and the author of this work is able to point out the whole process of making it equal to the Dutch. To make terras properly for use, as a water-cement, mix fresh-burnt lime, not flacked too much (and if stone-lime the better), and to every bushel of real Dutch terras add two of lime, as little water as possible, and well beat it with beaters, on a wood or stone floor."[13]

It is probable that the reports on the patents of Dent and of the mechanic from Dordrech, or on the trials conducted by Semple for the artificial fabrication of *trass*, indicate the existence of more detailed research conducted in various European localities to avoid the need to import natural powders. Those attempts do not remain isolated for long, given the fact that in December 1763 Thomas Manlove obtains a similar patent in London, again focused on the "art of making tarrass."[14]

## Journeys of naturalists to study pozzolana in Italy

In the passage from the 1750s to the 1760s, two schools of thought emerge on voyages to Italy to study architecture: on the one hand, there are young architects interested in ruins, where they find inspiration for new ways of composing spaces; on the other, there are the naturalists, scientists from different disciplines, as well as curious scholars of all kinds, who want to closely observe the materials and processes of fabrication that can be useful to the evolution of the art of construction. The latter group pays better attention to anonymous construction as they discover the fact that masons know already what could be considered as the secrets of the Roman constructive traditions. It is significant that in the unfolding of these two ways of observing Italy, Soufflot assumes the role of authority for the scientists, confirming the constructive intelligence of an architect who, not coincidentally, becomes a protagonist in the experimentation on Roman cement.

After the discoveries of Guettard in Auvergne, the question of lava and its use in construction becomes one of the themes discussed by scientists and travelers who cross Italy and visit the areas of Rome and Naples. La Condamine, who stays in Italy in 1755-56 (accompanied by the architect Moreau), is interested in Italian lavas: knows about the hypotheses of Guettard regarding similarities between the lavas of Auvergne and those of the zone of Naples;

[13] Entry *Cæmentum*, in, John Wilkes, ed., *Encyclopædia Londinensis, or, Universal dictionary of arts, sciences, and literature*, London, Encyclopædia, 1817, vol. XV, p. 455.

[14] *Cf.* Bennet Woodcroft, ed., *Subject-matter index of patent of invention, from March 2, 1617 [...], to October 1, 1852, Part. I*, London, G.E. Eyre, W. Spottiswoode, 1857, p. 108. Manlove's patent is number 802.

[15] Charles-Marie de La Condamine, *Extrait d'un journal de Voyage en Italie*,

observes the various uses of lavas, including for the construction of road pavements between Rome and Naples; asserts the volcanic origin of pozzolana – though not commenting on, like others, its use in cements; and confirms that the considerations of Soufflot have become a source of information on lavas in construction in the vicinity of Rome and Naples.

"A different opinion, no less decisive in this matter," he writes, "is that of Mr. Soufflot, the royal building inspector. After having read this report after its publication, I learned from Mr. Soufflot himself that during his last trip to Italy in 1752 [*sic*], he traveled to Capo-di-Bove, one mile from Rome, to the quarry from which the pavement used in this city is extracted, and that he realized that this supposed stone was no different from the lava that paves the streets of Naples. It could be extracted from a place nearer to Rome than Capo-di-Bove, since I saw pure lava when I was leaving Rome, and very near to the gate of San Sebastian, on the way from Frascati to Rome. It is surprising that a fact that is so easy to verify could to this day seem a paradox and is not more generally known."[15]

Among the scholars of phenomena of Italian volcanism we find the Swiss Guillaume-Antoine Deluc (1729-1812), physicist and geologist, and his brother Jean-André Deluc (1727-1817), naturalist, who in 1756 and 1757 visit Vesuvius and Aetna, collecting a fine series of samples of volcanic materials, for which Guillaume-Antoine publishes a *catalogue raisonné*.

In the trip through Italy done in 1758, Pierre-Jean Grosley (1718-85) describes, during a stay in Naples, the houses of the city and the technical characteristics of the mortars that make their flat roofs possible. The fact that in his reasoning on Italian techniques he also mentions the Louvre or the Astronomical Observatory of Paris for their flat roofs offers a glimpse of how certain 17th-century works are destined to become focal points for 18th-century experimentation on cement in France.

"These houses, built with a sort of very light tuff," Grosley writes on the buildings of Naples, "that rise up to seven or eight stories high, are all completed as platforms, with neither a roof nor a cover. This type of construction makes them look not like the Louvre, the Bourbon Palace, or the Paris Observatory, but rather like a house that has lost its rooftop to the flames. [...] this type of construction reflects not the climate, it rains as much as anywhere else, but rather the convenience of pozzolana supplied by all of the surroundings of Naples. This material, composed partly of metal, partly of small crystals that are rough the touch, mixed with lime-marble or shells, constitute a plaster that is hardened, not destroyed, by water. Once it has been spread out on a platform, it is even necessary to keep it under water for a few days, so that the lime, by incorporating the water, becomes able to resist the sun, which, without taking this precaution, would transform it back into dust. If, instead of cement, this sand had been used to bind the rocks used to pave the platform of the Paris Observatory, it can be assumed that this magnificent monument from the reign of Louis XIV would have lasted for more than a century, in the condition it was in when it left the hands of Mr. Perrault."[16]

in, "Histoire de l'Académie Royale des Sciences, Année MDCCLVII, avec les Mémoires de Mathématique & de Physique, pour la même Année," Paris, Imprimerie Royale, 1762, (pp. 336-410), p. 382.

16 *Nouveaux Mémoires sur l'Italie et sur les Italiens, par deux gentilshommes suédois*, Londres, J. Nourse, vol. III, 1764, pp. 68-69. Grosley travels with Belly, a shopkeeper.

**279**. View of the Vesuvius (J.-C.-R. de Saint-Non, *Voyage pittoresque à Naples et en Sicilie*, Paris, Dufour and C.ie, 1829).

Also in the wake of the discoveries of Guettard, the botanist Auguste-Denis Fougeroux de Bondaroy (1732-89), nephew of Duhamel du Monceau, examines lavas during his stay in Naples in 1763 - *Sur le Vésuve* is the essay in which he sums up his main observations. Unlike La Condamine, Fougeroux de Bondaroy proposes a classification in lavas and stones of the various volcanic materials, with observations on their applications in construction, and pays closer attention to the materials used to make mortar, like the *lapilli* – which "are used in mortars to build the terraces of houses in this city and its surroundings"[17] – and, in particular, pozzolana.

"At Mount Vesuvius one also finds the type of rock that, when broken into small pieces, is known throughout Italy by the name of *pozzolana*. Pozzolana, which, as we know draws its name from Pouzzoles, where it can be found in large quantity, and where it is loaded onto structures to transport it, is, as is well known, widely used for maritime structures and masonry built under water. One glance at pozzolana and the soil it is found on is enough to reveal that it is the product of underground fires; it is a stone, burnt and broken into small pieces, that can be added to mortar like our cement: it is found where ever volcanoes once existed; it soaks up humidity with a greater force than our bricks or burnt and beaten tiles because it has suffered a greater degree of fire, and for a longer duration. It is known that colcothar, for this unique reason, is superior to our ordinary cement for terraces and masonry structures. Agricola believed that cement is bituminous, sulfurous, and aluminous.[18] Pliny, Vitruvius, De l'Orme already cited it as excellent for the production of mortar for buildings. [...]. It is known that genuine pozzolana can be found in Rome and near Civita-Vecchia, etc. and I have no doubt that, following the Apennine Mountains, one can find a similar burnt soil even further away [...]."[19]

[17] Auguste-Denis Fougeroux de Bondaroy, *Sur le Vésuve*, in, "Histoire de l'Académie Royale des Sciences, Année MDCCLXVI, avec les Mémoires de Mathématique & de Physique, pour la même Année," Paris, Imprimerie Royale, 1769, (pp. 70-99), p. 83.

[18] "[...] *read*: Agricola believes this cement to be made with pozzolana," written in, *Fautes à corriger*, in, "Histoire de l'Académie Royale des Sciences. Année MDCCLXVII, avec les Mémoires de Mathématique & de Physique," Paris, 1770, p.n.n.

[19] Fougeroux de Bondaroy, *op. cit.*, pp. 85-86.

[20] Entry *Pozzolane*, in, Jacques-Christophe Valmont de Bomare, *Dictionnaire raisonné*

**280**. Pietro Antoniani, eruption of Vesuvius in 1767 (C. Knight, *Hamilton a Napoli. Cultura, svaghi, civiltà di una grande capitale europea*, Napoli, Electa Napoli, 2003).

Fougeroux de Bondaroy is among those who take up the chemical analyses of Cronstedt on lavic materials, confirming the presence of "molten metallic materials."

In this same period other recipes for Roman cement are reported by Jacques-Christophe Valmont de Bomare (1731-1807), trained as a surgeon, an expert on metals, as well as on their deposits and foundries for their fabrication. He has extensive knowledge of the volcanoes of Iceland and is a naturalist and friend of Georges-Louis Leclerc, Count of Buffon (1707-88). Valmont de Bomare describes the particular nature of pozzolana in his *Dictionnaire Raisoné Universel d'Histoire Naturelle*, published in 1764 – a short text destined to become a point of reference when the invention of a cement is discussed the following year.

"This name is given," he writes about pozzolana, "to a sort of sand that is found on the territory of Pouzzol, a town in Italy, near Naples; it can also be found on Guadeloupe, Martinique, and on l'Isle de France. Pozzolana is to be considered a mixture of sandy, earthy, and ferrous parts, hardened, bound, and attached together, up to the size of a pea and dried by underground fires. This kind of sand is reddish brown and of a crusty or gravelly form. It is successfully used to cement the stones of moles and structures built in maritime sites and even in the sea: one third of lime is added to it: it is then spread out in a very large amount of water, and used immediately after; for it is characterized by the fact that it hardens as quickly as calcinated and fused gypsum."[20]

To confirm the growing interest in ancient Roman construction, we can recall that even the astronomer Joseph-Jérôme Le Français de Lalande (1732-1807), traveling in Italy in 1765-66, mentions, when writing about Vesuvius, "pozzolana, which makes the most highly esteemed Roman cement for construction in water;"[21] and observes the flat roofs of the houses.[22]

*Universel d'Histoire Naturelle*, Paris, Didot le Jeune, vol. IV, 1764, pp. 495-96.

[21] Joseph-Jérôme Le Français de Lalande, *Voyage d'un François en Italie, fait dans les années 1765 & 1766 [...]. Nouvelle édition corrigée & considérablement augmentée par un Savant très-distingué, qui a parcouru cette charmante partie de l'Europe l'année 1767*, Yverdon, 1770, vol. VIII, p. 56.

[22] "One of the remarkable things that can be observed in Naples is the lastrico, or cememt, with which the terraces and the tops of houses are covered [...]" (Joseph-Jérôme Le Français de Lalande, *Voyage en Italie*, Yverdon, 1787, vol. V, p. 488).

Another voyage in Italy of importance for the observations of stones used for construction, especially in ancient times, in the context of the research on French pozzolanas, is the trip in 1762 of the abbot and professor of Eloquence at the University of Paris, Antoine Guénée (1717-1803),[23] in the company of a certain D'Aubreuil. The report on the voyage is published in 1768 with the title *Sur la Minéralogie de l'Italie*, edited by Guettard, who had introduced his friend Guénée to the study of geology during travels in France. Guettard combines the notes of the two travelers with the considerations of Nollet, La Condamine and Grosley.[24] Like La Condamine, Guettard also recalls the observations of Soufflot, confirming the crucial importance of the travels in Italy of the architect for the evolution of knowledge of geology and construction.[25]

The pages in which Guettard refers to the observations made by Guénée and D'Aubreuil in Naples and vicinity contain many considerations on lavic stones, which by now are the materials most assiduously studied during voyages in Italy, after the discoveries in Auvergne. Information is also reported on the various pozzolana quarries – for example, that of *Grottoni*, near Rome, where "a very red pozzolana is extracted."[26]

"When approaching Aversa, and even more so Naples," Guettard writes, "the soil becomes lighter, as in the areas around Pozzuoli, Baia, Cuma, etc., as well as at Portici, Pompeii, Statia. It is probably mixed with a large amount of ashes and other pulverized volcanic materials [...]. The stones in the catacombs at Naples and those at Cuma and others are of a kind of tender and whitish-yellow rubble. Naples is built with these stones and with the marble that is extracted nearby. The entire Capo di Monte mountain and all of those surrounding the city appear to be made up of this type of rubble. When going from Naples to Pozzuoli, one comes across a hill made up of a blueish rock that gallery slaves cut for public works, especially for the pavement used in the city. This rock is a kind of lava. One part of the land belonging to Pozzuoli, Baia and Cuma is made up of small-grained pozzolana, light pumice the color of ash, cinders (sometimes in their natural state of division), occasionally slightly gathered in a mass; the remainder is made up of white rocks, which constitute the mountain of Cuma. Above Cuma, one comes across old pools or water reservoirs dug into the same stone."[27]

To demonstrate the growing success of pozzolana, we can note that in the new edition in 1762 of the treatise of Liger, *La nouvelle maison rustique*, an explanation is provided regarding "cement or Pouzzol."[28]

## Observations on Vicentine pozzolanas, from Arduino to Fortis

Italian scientists systematically search for pozzolanas in the inland territories of the Venetian Republic only in the 1760s, following the discoveries made in France by Guettard and his followers. Though is worth remembering that the technicians working for the Magistrato alle Acque knew of the existence of so-called Vicenza pozzolana as early as the start of the 1720s.

[23] See, *Paul Guiton et l'Italie des voyageurs au XVIIIe siècle: son projet de bibliographie critique des voyageurs français en Italie: deux manuscrits (R 9707 et R 9705) de la Bibliothèque municipale de Grenoble,* Gilles Bertrand, 1999.

[24] See, Jean-Etienne Guettard, *Neuvième mémoire. Sur la Minéralogie de l'Italie*, in, *Id.*, *Mémoires sur différentes parties des sciences et arts*, Paris, L. Prault, 1768, vol. I, pp. 347-439.

[25] *Ibid.*, p. 379, note.

[26] *Ibid.*, p. 386.

[27] *Ibid.*, p. 367.

[28] Louis Liger, *La nouvelle maison rustique, ou économie rurale, pratique et générale de tous les biens de campagne*, Paris, Knapen, 1762$^8$, p. 26.

[29] Ezio Vaccari, *Giovanni Arduino (1714-1795). Il contributo di uno scienziato veneto al dibattito settecentesco sulle scienze della Terra*, Firenze, Olschki, 1993, pp. 102 *et seq*.

[30] *Lettera Seconda di Giovanni Arduino veronese-vicentino, Sopra varie sue Osservazioni fatte in diverse parti del Territorio di Vicenza,*

The protagonist of the first important observations on volcanic materials near Vicenza is Giovanni Arduino (1714-95), a Venetian scientist specializing in geology and mineralogy, who in October 1740 made notes on volcanic substances during a trip from Vicenza to Recoaro, along the Agno Valley, in the company of Antonio Vallisnieri, or Vallisneri Junior (1708-77), a naturalist of the University of Padua.[29] It may be no coincidence that the interest of naturalists in Vicentine pozzolanas emerges at the start of the 1740s, if we consider the fact that in 1738, based on the proposals of Filippini, comparison trials were conducted at the Venetian Lagoon to test local pozzolanas against those imported from Civitavecchia to make the cement of the sea walls.

In the 1750s and 1760s some fundamental essays are published on Vicenza pozzolana. In 1759 Arduino writes a letter "on his various Observations made in different parts of the Territory of Vicenza, and elsewhere, regarding Theory of the Earth and Mineralogy," where he discusses "ancient volcanoes" and the "true volcanic pumices, and the gravel of pure vitrified iron" he has gathered "in the mountains of Fara and of Lugo," and that constitute, in his view, the "demonstration that our mountains of stones and black tuff, and of other colors, owe their form to igneous force."[30]

In 1764 the Augustinian abbot Alberto Fortis (1741-1803), a disciple of Arduino, studies the effects of extinct volcanoes in the Vicenza area, on which he reports in a letter sent to Arduino and published in 1765 by the "Giornale d'Italia."[31] Also in 1765, Arduino observes geological phenomena of volcanic origin in Val d'Alpone, near Verona, and asks Father Tommaso Riccardi to inform him about the "variety of the effects" caused by Vesuvius, and to provide him with rock samples of the Neapolitan volcano for the purpose of "comparison with our lavas, pumices, glasses, sands, ashes, and volcanic talcs," in order to evaluate "the degrees of similarity."[32] In 1769 Arduino is still actively investigating volcanic products from the Vicenza area, also discussing them with Antonio Zanon (1696-1770), an agronomist from Udine,[33] and with Fortis.[34]

The observations of Arduino and Fortis lay the groundwork for experiments on Vicentine pozzolanas conducted in the 1780s by the group of scientists of the Studio of Padua gathered around Fortis, and by the *proti* of the Magistrato alle Acque, with the aim of assessing their potential for use in the making of cement for the construction of the sea walls of Venice, as an alternative to the pozzolana imported from Civitavecchia.

## The edict of 1763 on the pozzolana trade at Civitavecchia

Until the 1770s the discoveries of deposits of pozzolana in Auvergne and in locations around Vicenza remain relegated to the studies of naturalists. The pozzolana trade is not impacted by the new map of resources, and continues to have its center in Civitavecchia for shipping to the countries of the Mediterranean, while shipments to the northern seas of volcanic products similar to

*ed altrove, appartenenti alla Teoria Terrestre, ed alla Mineralogia*, Vicenza, 30 March 1759, cit. in, Ezio Vaccari, *Giovanni Arduino (1714-1795). Il contributo di uno scienziato veneto al dibattito settecentesco sulle scienze della Terra*, Firenze, Olschki, 1993, pp. 179-80.

[31] *Descrizione Fisica de' Colli di Montegalda, luogo del territorio Vicentino, addrizzata dal Dottissimo P. Alberto Fortis Agostiniano al Chiariss. Sig. Giovanni Arduino*, in, "Giornale d'Italia," vol. I, 1765, pp. 83-87.

[32] Cit. in, Vaccari, *op. cit.*, p. 209.

[33] Giovanni Arduino, *Effetti di antichissimi estinti vulcani: e altri fenomeni, e prodotti fossili osservati da Giovanni Arduino nei monti della Villa di Chiampo, ed in altri luoghi del territorio di Vicenza e di quello di Verona, da esso riferiti con lettera al chiarissimo signor Antonio Zanon*, Milocco, 1775.

[34] See, Vaccari, *op. cit.*, p. 212.

pozzolana continue to come from ports along the Rhine, where German tuff is sold, and from Dutch ports, from where *trass* is exported.

From the early 1760s the port of Civitavecchia reports a significant increase in pozzolana shipments, leading to economic measures aimed at bringing more proceeds into the coffers of the Vatican. While during the entire first half of the 1700s the export of pozzolana was subject only to taxes to be paid to the consuls of the various nations involved, in 1763 the Papal State issues an edict, prepared by the Commissioner General of the Papal Navy, Pasquale Acquaviva (1718-88), imposing a duty on ships using the port of Civitavecchia. The tax consists in the payment of a guard to oversee loading operations of pozzolana in the port. Though the tax is applied equally to all those who buy pozzolana in Civitavecchia, the measure clearly effects above all France, the leading importer of pozzolana for many years. So it comes as no surprise that the French consul at Civitavecchia has harsh words for the edict of Acquaviva.

"The other tax concerns Pozzolana," is his comment. "It is a red soil, found in the vicinity of Rome, specific for building foundations in water, and that the arsenals in France and the Canal de Languedoc use extensively for their construction and repair work. Now, in this port, this precious earth has become a staple trade good, and it is a good like any other, valued at what could be referred to as an exorbitant price. This does not keep our captains from Languedoc from purchasing this soil in times of peace, and occasionally the Messrs the Intendants of the royal Arsenals place orders of this kind. Your Excellency may be surprised to hear that the tax that has just been imposed on pozzolana is nothing other than the wages for a guard, whose task it is to ensure that no one throws this merchandise into the sea. Can one imagine, Sire, anything more extravagant? Under the false pretext of filling the harbor, they impose on foreign nations the tax of a guard, whose duty it is to keep the owners of the ships from throwing their wares into the sea? This circumstance reveals the spirit behind this new decree, that is, without giving much attention to how, to try to reward an individual at the expense of foreign nations. It is true that all are included in the decree under consideration, but regarding pozzolana, the French will pay the largest portion of the standard tax."[35]

The new tax is applied, in June 1763, to the cargo of pozzolana taken on by three "skippers from Languedoc" that bring it to France to "build a basin near Toulon."[36] Over the years to follow pozzolana is exported to French ports where important works of construction are being done in the sea. Marseille and Agde are the destinations of different shipments, in May and June 1767, with loads of 110, 249, 130, 120, 103, 153, 123 and 120 cartloads of pozzolona, in French ships coming from Agde, Cette (now Sète), Genoa and Trieste.[37] The English and the Dutch continue to refrain from importing significant quantities of pozzolana, while in Sweden pozzolana arrives in large amounts above all in the years around 1770, to supply major hydraulic worksites. Major Italian destinations for pozzolana shipped from Civitavecchia in the 1760s include the worksites of Vanvitelli in Ancona. Venetian, Swedish,

[35] Jean-Baptiste-François Vidau, 28 March 1763, in, "Correspondence Consulaire, Civitavecchia, 1744-1774," AE, B1.374, Archives Nationales, Paris.
[36] Jean-Baptiste-François Vidau, 20 June 1763, *ibid*.
[37] "Suite de l'Etat des batiments françois qu'on fait commerce a Civitavecchia pendant les trois mois d'avril, may, et juin de l'an 1767," *ibid*.
[38] "Questions join a la lettre de M. Vidau du 3 Mars 1766. Reponces," *ibid*. "Pozzolana earth from Rome is another object [as the wool that is exported], but the miserable quality of the goods is a good demonstration of what should be done with it," we read in this same document.
[39] Johann Joachim Winckelmann, *Frammento di un nuovo lavoro relativo alle Osservazioni sull'architettura*

Danish and Hamburg-based ships do not trade directly with the Papal State, and come to Civitavecchia to be "chartered to transport Roman pozzolanic soil from here to Ancona, where they are still working on completing the new branch of the port and its lighthouse."[38] Some of Winckelmann's notes on the pozzolana trade at the port of Civitavecchia confirm the size of the shipping to Ancona: "Roman pozzolana is loaded at Civita Vecchia onto foreign ships as ballast, after having been sifted in Rome, because in the port it is not possible to build without this earth. For the edifice presently being built and for the enlargement of the port of Ancona every year seventeen thousand cartloads of pozzolana are shipped, and those boats are obliged to sail entirely around the southern part of Italy."[39] A dispatch in 1767 of the Venetian consul on activities at Civitavecchia informs of the arrival of a Swedish ship in "this port to take on red Pozzolana of Rome for Ancona," after "having deposited in Corsica its cargo of Jesuits."[40] Besides Ancona, other Italian ports continue to be supplied with pozzolana from Civitavecchia – ships bring it, for example, to Leghorn, in journeys that last eight days.[41] Pozzolana exports are also recorded at the port of Naples, from which boats loaded with the precious material include vessels bound for Barletta.[42]

"Yellow pozzolana can be found in the papal state in the vicinity of Rome and in other parts of Italy," writes one of the leading European mineralogists. "It is transported to Civitavecchia, from where it is sent to Sweden, to France, to Holland, and more generally, to half of Europe. There is black pozzolana on the Vesuvius, but the best variant of this color exists in the vicinity of Naples, where it is extracted from the Torre dell'Annuzia, the gray one from Pozzuoli, the yellowish-brown from everywhere; the gray one is as fine as the finest of them."[43]

## Cement and mold, according to Marsy

The question of the construction of walls with formwork and of cement made with the ancient methods continue to be timely at the end of the 1750s, in the wake of the reports of various observers in the early decades of the century. Certain accounts of travels in distant lands, along the Mediterranean and in the Far East, contribute to spread the myth of cement made in the ancient way, in keeping with different traditions. It is in the area of research on conglomerates that certain descriptions supplied by travelers in the Mediterranean become important sources for the reconstruction of the criteria in use to make concrete works.

One of the most significant reports for knowledge of the system of pouring cement into formwork is that of the Jesuit abbot and man of letters François-Marie de Marsy (1714-63), written on the basis of literary sources in his monumental *Histoire moderne des Chinois, des Japonais, des Indiens, des Persans, des Turcs, des Russiens*. Though some of Marsy's observations replicate what has already been reported by Shaw, it is worth noting, to understand the success of the descriptions provided in the *Histoire moderne*, that one of the most famous discoverers of what was considered the true recipe of

*degli antichi, tratto dal manoscritto di Winckelmann pubblicato per la prima volta nella nuova edizione di Dresda*, 1762-1768, in, Carlo Fea, ed., *Opere di G. G. Winckelmann, Prima edizione italiana completa*, Prato, F. Giachetti, 1836, vol. VI, p. 232.

40 Domenico Natale, 3 August 1767, Archivio Proprio Ambasciata Roma, 155, Archivio di Stato di Venezia.

41 *Cf. Livorno, 10 agosto*, in, "Gazzetta Toscana," 1768, no. 33, p. 144.

42 Paolo Macry, *Mercato e società nel Regno di Napoli: commercio del grano e politica economica nel Settecento*, Napoli, Guida, 1974, p. 77.

43 Johann Jacob Ferber, *Lettres sur la minéralogie et sur divers autres objets de l'histoire naturelle de l'Italie, écrites par Mr. Ferber à Mr. Le Chev. De Born. Ouvrage traduit de l'allemand, enrichi de notes & d'observations faites sur les lieux, par le B. De Dietrich*, Strasbourg, Bauer & Treuttel, 1776, p. 180.

Roman cement in the course of the 1770s, Polycarpe de La Faye, makes reference in great detail to the writings of Marsy on construction done with cement and formwork.[44]

Like Delorme, Marsy too is interested in the construction of earthen walls, discussing this system with reference to the houses of the Moguls, or Moghuls, in India.[45] But the interest of the inventors of cement is piqued above all by the description of the construction practices still in use in certain cities along the Mediterranean Sea. Marsy says that in these cities the cement has the same solidity and strength of that of the ancients, and that it is composed of three parts lime, in powder, two parts ordinary ash, and one part fine sand. The process of working these three materials consists in continuous beating for three days and three nights, gradually adding water and oil, until the compound takes on a certain consistency. "This mortar is used to build arches, cisterns and terraces."[46] The construction of walls in Morocco is done with a compound of ordinary earth, clay, sand and cement, which is poured into formwork, or "wooden molds that resemble bottomless troughs," three feet wide by seven feet long, formed by large boards reinforced with iron clamps.[47] The compound is packed into the formwork by means of large piles. Marsy specifies that the most solid walls are made with this system, but with a compound based only on cement and sand. He also explains the fabrication of the "artificial stone," one over the next, making them solid thanks to continuous beating.

Marsy is also the author of a dictionary of terms of art and architecture containing the entries *Beton* and *Ciment*. The former is defined as a "type of mortar that is thrown into foundations and that becomes extremely hard";[48] while for the latter he writes: "Cement is commonly made with crushed bricks that are mixed with lime. The best cement is the powder from Pozzuoli. A very hard cement can also be made using bricks, glass, stone-coal, well washed granitic sand, iron chippings that fall from anvils, and with quicklime that has been thoroughly crushed."[49]

### Quicklime, chippings, mold: Loriot and Roman cement

"Piranesi would be hard-pressed to convince anyone that the Romans did not borrow anything from Greek architecture, and they had masters of this art from their very foundation," De Marigny remarks on the letter sent to him by Piranesi on 16 February 1762. In that letter Piranesi informs him of the publication of some of his collections of etchings, including *Della Magnificenza ed Architettura de' Romani*, printed in 1761, and the *Campo Marzio dell'Antica Roma*, printed in 1762, all devoted to the understanding and the "apologia" of Roman architecture, also through "the most precise details of everything that is most reasonable and worthy of being chosen in this art-form," as Piranesi himself writes.[50] Piranesi continues to engrave plates of construction details that demonstrate the

[44] Polycarpe de La Faye, *Mémoire pour servir de suite aux Recherches sur la préparation que les romains donnoient à la chaux dont ils se servoient pour leurs constructions, & sur la composition & l'emploi de leurs mortiers*, Paris, Imprimerie Royale, 1778, pp. 12-14.

[45] François-Marie de Marsy, *Histoire moderne des Chinois, des Japonnois, des Indiens, des Persans, des Turcs, des Russiens, &c.*, Paris, Saillant, 1758, vol. IV, p. 228.

[46] François-Marie de Marsy, *Histoire moderne des Chinois, des Japonais, des Indiens, des Persans, des Turcs, des Russiens, pour servir de suite à l'"Histoire ancienne" de M. Rollin*, Paris, Desaint & Saillant, 1762, vol. X, p. 440.

[47] *Ibid.*, p. 441.

[48] François-Marie de Marsy, *Dictionnaire abregé de peinture et d'architecture*, Paris, 1746, vol. I, p. 70.

[49] *Ibid.*, pp. 124-25.

[50] Cara D. Denison, Myra Nan Rosenfeld, Stephanie Wiles, ed., *Exploring Rome: Piranesi and His Contemporaries*, Cambridge-Massachusetts, London, MIT press, 1993, p. 108, note 60.

**281**. Giovanni Battista Piranesi, "Ostium, sive Emissarium Cloacae Maximae in Tiberim [...]," Rome (G.B. Piranesi, *Della Magnificenza ed architettura de' romani*, Roma, 1761).

**282**. Anonymous, view of the Cloaca Maxima, Rome (III.A.32, Ecole des Ponts et Chaussées, Paris).

potential of pouring of the *lastrico*, like the plates that show the tomb "falsely said to be of the Orazi, and Curazi" and the "water cistern," in the collection *Antichità d'Albano e di Castel Gandolfo* published in 1764.

One of the most famous 18th-century inventions of the art of building, which would mark the beginning in Europe of frenetic research on the fabrication and composition of cement, comes precisely against the backdrop of a theoretical debate in which the defenders of Roman architecture and those of its Greek counterpart clash violently to assert the superiority of one or the other, as forecast by De Marigny. The protagonists of this debate, involving personalities of European culture and philosophy, are Piranesi and Pierre-Jean Mariette (1694-1774). The first strives, in *Della Magnificenza*, to demonstrate the functional and anti-decorative character of Roman architecture and its Etruscan, not Greek, roots,[51] while the second attacks Piranesi to assert the superiority of Greek architecture, in a letter published in 1764 on the pages of the "Gazette Littéraire de l'Europe" (in 1764 and 1765 Piranesi makes two responses to Mariette).[52]

The apodictic statements of Piranesi on the magnificence of the Romans are simply the prelude of archaeological research on the cement technique begun in the mid 1770s, in order to discover the secret of the Romans. In *Della Magnificenza ed Architettura de' Romani* Piranesi revives the timely character of the passage of Vitruvius on the "opera segnina" or *opus signinum*, describing the formula for a cement suitable to clad cisterns and the conduits of aqueducts. Piranesi translates the words of Vitruvius into Italian and concludes with a short quote from Pliny, again about cement. Vitruvius and Pliny are the authors most often mentioned by certain 18th-century inventors of cement considered to be "Roman."

"Here is what must be done in works of opus signinum," we read in the excerpt of Vitruvius published by Piranesi. "First the sand is prepared, which must be very pure and fine; the cements must be of flints crushed in such a way that the pieces to be used will not weigh more than one pound each. Very strong lime should be mixed with the sand, making sure that five parts of this should correspond to two parts of lime mortar. The cements will be blended in the vessel; and with this composition the walls of the ditch will be coated, and the material poured to the desired height; and then beaten with a mallet of wood bound by metal. And Pliny: What has art not invented! It has gone so far as to make use of broken shards to compose the work known as signinum, mixing with them lime mortar, so that it will last longer." "The crust of this on aqueducts can be seen to be more lasting than stones," Piranesi adds.[53] And he continues: "As can be seen, it was already known to those masons

[51] *Cf.* Giovanni Battista Piranesi, *Della Magnificenza ed architettura de' romani, 1761, introduction by Antonio Giuliani*, Milano, Polifilo, 1993.
[52] *Lettre de M. Mariette aux Auteurs de la Gazette Littéraire de l'Europe*, in, "Gazette Littéraire de l'Europe," vol. III, September-November 1764, pp. 232-47. Piranesi sets out to "praise the Romans" (*ibid.*, p. 232).
[53] Piranesi, *Della Magnificenza ed architettura de' romani* cit., p. XLVII.
[54] *Ibid.*, p. XCV.
[55] Charles Notoire, letter to Abel-François Poisson de Vandières, Marquis de Marigny, 16 June 1762, in, Anatole de Montaiglon, Jules Guiffrey, *Correspondance des Directeurs de l'Académie de France à Rome avec les surintendants des bâtiments*, vol. XI, 1754-1763, Paris, N. Charavay, 1901, pp. 425-26.
[56] On the life and works of Loriot see, "Précis exact et historique des découvertes et des dépenses faites par le Sr Loriot pour se rendre utile à ses concitoyens, depuis l'année 1742 qui etait la dix huitième de son age, jusqu'à ce jour," anonymous report sent on 30 June 1777 to Charles Claude Flahaut de La Billarderie, compte d'Angiviller (1730-1809), O.1.1294, 289, Archives Nationales, Paris. On the machines invented by Loriot see, "Etat des machines comprises dans le cabinet du feu S. Loriot, présenté au Roi," [1783], O.1.1294, 291, and

how to fabricate, with stones and with cements, since they had learned it from Noah, and from his sons, from which these methods originated."[54]

The experimenters with cement address the question of the superiority of the Greeks or the Romans without making theoretical declarations or statements of principle like those of Piranesi and Mariette, simply investigating the nature of the processes of manufacture and deployment of Roman cement. The fact that they are able to make their Roman cement into the basic technique for construction in the future definitely puts an end to the discussion on superiority, which continues for many decades but only in literary circles.

Against the background of the debate on the magnificence of the Romans, the French intensify their research on mortars, launched by Soufflot. Indeed, there are no traces of widespread knowledge and immediate practical consequences of the experiences of Smeaton on the use of hydraulic cement, put into practice for the construction of the Eddystone lighthouse.

The purchase by De Marigny, in 1762, of some collections of etchings and books by Piranesi, ordered after he has been informed of their publication by Piranesi himself (Robert is one of the intermediaries in the operation),[55] is an act that symbolizes the role Roman construction was to play in France. Further proof of the influence of Piranesi on contemporary construction comes from the fact that some collections of his etchings are among the books consulted by the engineers of the Ponts et Chaussées in Paris.

Only one study of an experimental character – one that starts with literary sources and ruins – would be able to discover what the general public and certain experts continue to consider to be the lost secret of the Romans: the nature of the compound used to make their cement. Such a study is attempted by Loriot, a workman-cum-inventor and brother of the architect Louis-Adam.[56] Loriot's experiments take place independent of the manufacture of the Roman cement based on pozzolana used by French engineers in various ports, from Toulon to Brest. It is probable that the French royal administration encourages his research from the outset, precisely in the hope of finding a product that could be made anywhere and that would eliminate the need to import pozzolana from Italy.

Loriot's first experiments on cement, interrupted due to several jobs to be done by order of the king, date back to 1764,[57] when for four and a half months[58] he travels through southern France with the aim of examining the specificities of cements in the ruins of Roman constructions.[59] When he begins this investigation the composition of the cement of the Romans is still wreathed in mystery and lore. It is interesting, for example, to see how Loriot learns that the red shade of the cement of the aqueduct of Arcier, near Besançon, is thought

O.1.1294, 297, Archives Nationales, Paris. More in general, see, "Dossier Loriot, 1773-1784," O.1.1294, 242-297, *ibid.* It is worth pointing out, also to confirm how hydraulic architecture involves the most famous inventors of cement in the 1700s, that in 1763 Loriot presents to the Assemblée des Inspecteurs Généraux des Ponts et Chaussées a machine "to hammer in piles with a drop pile hammer, different from the one by Wanloë [sic: Valoué]" (minutes of the meeting of 10 April 1763, in, "Extrait du Journal des assemblées des Ponts et Chaussées tenues chez mr. Trudaine, depuis 1748 jusqu'en 1774," Ms 233, vol. XIII, Ecole des Ponts et Chaussées, Paris).

57 *Cf. Mémoire sur une découverte dans l'art de bâtir faite par le Sr Loriot, Mécanicien, Pensionnaire du Roi; dans lequel l'on rend publique par ordre de sa Majesté la méthode de composer un Ciment ou Mortier propre à une infinité d'ouvrages, tant pour la construction, que pour la décoration*, Paris, M. Lambert, 1774, p. 26.

58 "Précis exact et historique des découvertes et des dépenses faites par le Sr Loriot pour se rendre utile à ses concitoyens, depuis l'année 1742 qui etait la dix huitième de son age, jusqu'à ce jour," anonymous report sent to D'Angiviller on 30 June 1777, O.1.1294, 289, Archives Nationales, Paris.

59 "Mr. Loriot, whose only goal was to find the composition of the ancient mortar or cement of the Romans spared no efforts to give as much rigour as possible to the discovery that he hoped to make; these motives pursuaded him to travel the southern provinces of the Kingdom to study more closely the precious remains of the Roman monuments that time had preserved" (*Mémoire sur une découverte dans l'art de bâtir faite par le Sr Loriot* cit., p. 14).

by the local peasants to be the result of the addition of ox blood to the compound.[60] During studies conducted at the port of Fréjus, with the goal of "taking advantage of the ruins in the ancient port, said to have been built by the Phocaeans and made more well-known by the Romans," he finds remains "of these old structures, and everything regarding construction using chippings," of Roman origin, and is able to observe the "toughness of their cement."[61]

That same year, in 1764, probably during the first days of December, Loriot writes a short memoir on the cement of the Romans, his "first memoir on the observation of roman construction [...]."[62] It constitutes the initial text of a reflection on Roman cement destined to expand after the series of experiments he conducts at the start of the 1770s on behalf of De Marigny. The writings will take the form of the famous *Mémoire sur une découverte dans l'art de bâtir faite par le Sr Loriot, Mécanicien, Pensionnaire du Roi; dans lequel l'on rend publique par ordre de sa Majesté la méthode de composer un Ciment ou Mortier propre à une infinité d'ouvrages, tant pour la construction, que pour la décoration*, published in 1774.[63]

The fact that Loriot declares his interest, in the report, in "construction" by means of "*cailloutages*," or chippings, observed in the ruins of the port of Fréjus, is a crucial indication of the construction mode he is seeking: the manufacture of conglomerate, of that massif also described by Piranesi. Apart from the port of Fréjus, the report mentions other ruins, some in generic terms - "the portions that still remain of ancient roman roads and other monuments" - others with precise indications of their location: "the aqueducts of Lyon, Besançon, and others." The indication of the aqueduct of Lyon may be a clue that Loriot has consulted the volume by Delorme on the *Recherches sur les Aqueducs de Lyon, construits par les Romains*, a fundamental work for knowledge of Roman construction methods. In the memoir of Loriot, the repeatedly evoked secret of the Romans is said to consist in the manipulation of materials to achieve a cement of rare toughness – "the secret of their cement." Loriot asserts that the characteristic of Roman cement lies in the use of quicklime mixed with pure sand and brick dust. It is the recipe of the *opus signinum* narrated by Vitruvius and Pliny, and made timely by Piranesi, though without specifying the nature of the lime. Loriot does not consider the various compounds described by Vitruvius and in architectural treatises; nor does he make any mention of the use of pozzolana or other volcanic products.

The discovery of what he believes is the method of manufacture of Roman cement opens up the possibility, according to Loriot, of updating the construction process of the ancient Romans through the pouring of the conglomerate into special wooden molds, or *encaissements*. Although construction by means of *cailloutages* (chippings) and *encaissements* (molds of wood) has already been explained by authors like Fontana or Bélidor, the modes of making of the formwork and the pouring described by Loriot lead us to believe that his treatise is based precisely on the essay by Delorme, also because to explain the formwork of the Romans Loriot calls on the case of the aqueduct studied in the *Recherches sur les Aqueducs de Lyon*.

60 *Ibid.*

61 [Antoine-Joseph Loriot], "Mémoire," Ms 2069, Ecole des Ponts et Chaussées, Paris. In the text, it is indicated that the report is presented to the Académie Royale des Sciences; "Sciences" is erased and replaced by "Architecture."

62 *Mémoire sur une découverte dans l'art de bâtir faite par le Sr Loriot* cit., p. 14. *Cf.* "Précis exact et historique des découvertes et des dépenses faites par le Sr Loriot" cit.

63 *Mémoire sur une découverte dans l'art de bâtir faite par le Sr Loriot* cit.

**283**. Map of the ancient port at Fréjus, 1756 (Cartes et Plans, GE D-15747, Bibliothèque Nationale de France, Paris).

**284**. Roman aqueduct, Fréjus (postcard, property of the author).

"First of all," we read in Loriot's memoir, "the Romans worked using mold [*encaissement*]. Their vertical and curved walls were built as if in a mold, using planks prepared to be placed one on top of the other as the wall grew in height, acting as a barrier to contain the materials in the required direction. This way the structure was set up between two planks, solidly raised and aligned with a plummet and a string, forming a coffer [*caisse*] fit to receive and contain any stone that they would have wanted to use for the construction. Regarding the foundations, it seems that the ditches, which reach down to the solid ground on which was to be built, acted as a coffer by channeling the almost always irregular thickness of the foundations, and it is only once the ground level was attained that they began to contain the flow in the molds [*encaisser*] in order to control the width of the foundations such that they would be adapted for the wall they were to carry.

Secondly, the Romans only used lime, washed sand, and crushed bricks to prepare their mortar or cement; but they were careful not to slake the lime

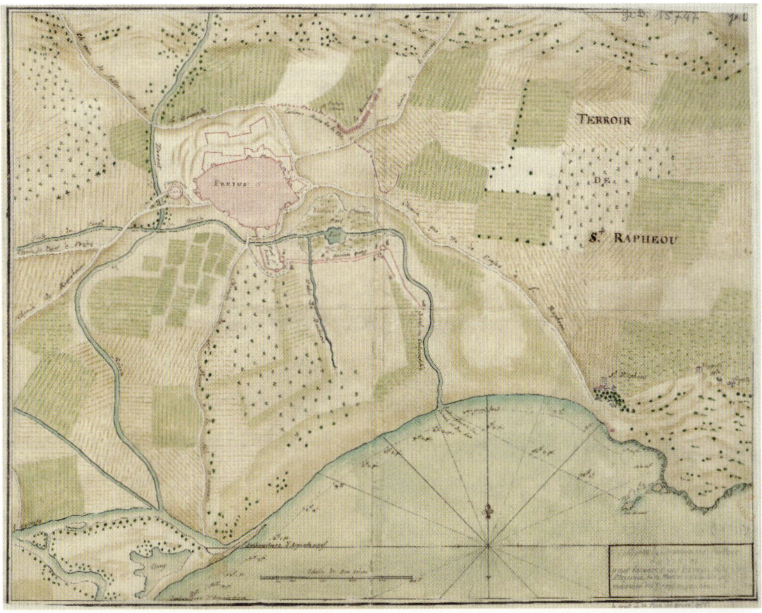

the way we do and let it sit as a liquid in holes in the soil for often up to several years before using it. In this state of decomposition, the lime is no different from a piece of soaked chalk, without strength or action, a greasy substance, only useful as a lubricant for sands, chippings and rubble, drawing them closer to each other without thereby acting on their substance, without incorporating itself into them and penetrating them like in Roman cement.

Why such a difference between our most recently built walls that fall to dust and that show no sign of adhesion between the various parts, while a hammer can only barely break free a few pieces of the finest chippings of the ancients? It is because the Romans prepared the amalgam at the very moment when the lime had all of its strength, it is because they only slaked it the instant that they used it, it is because they used it while it was hot, or boiling, so to speak. With the molds prepared, the workers arranged a bed of rubble and chippings to the mix, while others added crushed lime rocks into large troughs, into which they poured the water needed to melt it, mixed the lot and added washed sand, which they blended in this boiling liquid, along with crushed bricks, and they mixed everything until the blend was perfect. The cement was then immediately poured onto the bed of rubble and chippings, and since it was still very liquid and boiling hot, it perfectly filled the interstices so that the rubble and the pebbles were bathed, penetrated and fixed to the point that the whole became a single body. If the cement bed dominated the layer of rubble and chippings, more rubble and chippings were thrown on top, forming another layer, which in turn was covered by fresh cement. One can easily imagine the promptness and the neatness with which these structures that have lasted centuries were built. The walls required no other plaster. There were, however, sometimes two situations that called for the addition of something to this crude construction [*brute construction*], ornament on the one hand, in the form of a coating of external cladding, and on the other, a protective measure against the effect of water on the inside walls of an aqueduct that seemed to be coated using a special kind of cement. The external cladding that can be seen on the aqueduct of Lyon provides the most convincing proof of construction using a mold. It is nothing but rocks that are cut in square or diamond shapes as used for tiling, that are hardly more than two inches thick and seem to be embedded into the walls they clad. Their position in the field, an angle at the bottom, and their exact arrangement suggest that this could only have been accomplished using a form or mold, where they were first pressed against the coffer before lining it with rubble and chippings. The solidity and the precision with which they are embedded shows at the same time that this could only have been done using a liquid cement which fixed and puttied them in place so precisely in all directions that, despite their thinness, they can only be extracted by demolishing. Regarding the alleged plastering that forms the inside of the aqueducts and where a series of layers can be perceived, where the one that is in contact with the water is finer grained, while those that follow are progressively composed of coarser gravels and chippings: Mr. Loriot does not think that this was made by means of different plasters following the construction, rather, he believes that it was all made using the mold approach, the first cement was composed of finer sand, and it included a larger amount of crushed brick to become one with the boiling lime. This first layer of cement was applied, just one inch thick, on the planks of the mold and of the centering,

[64] [Loriot], "Mémoire" cit.
[65] Minutes of the meeting of 10 December 1764, in, Henry Lemonnier, *Procès-verbaux de l'Académie Royale d'architecture, 1671-1793,*

another layer composed of coarser sands and gravels and less crushed brick followed, and chippings were added in the end along with the rubble and the rocks, all bathed and penetrated by the same cement poured when hot. These various layers made a single wall mass, where it can easily be seen by the size of the sands and chippings that the materials used are not the same, but one does not see the kind of gaps that would be visible had the plaster been added afterwards. Furthermore, it is very remarkable that the interior layer does not chip off the way it does on anything that was plastered at a later time and was not fixed by the strength of the cement. There can be no doubt that a good, freshly made lime used following this method would have the same effect in our hands, no doubt either in the principles that the boiling lime water and the *gluten* that is the product of its decomposition penetrates the sands, chippings and the rocks with which everything is mixed together and that the bond creates a kind of petrification that time cannot destroy.

Everywhere rocks suited to make lime can be found, but they are not of the same quality. With this respect, the capital is very poorly furnished. There is only the kind that is made of a soft and porous rock, which, by the way, is half-wasted when it is loaded onto boats. Differences in the quality of the lime can lead to differences in that of the cement. The sand that is to be used must be washed, removing the soil pieces that are an obstacle to the bonding and the solidity because of the humidity that they confer whereas this kind of cement is intended to solidify like plaster and gypsum used to this end. Moreover, this construction method can be very advantageous even in situations where it seems necessary to build using ashlar and curiously, for both underground and outdoor fortifications."[64]

The Roman construction process described in the memoir of Loriot includes all the phases, from the selection of the components of the cement to the making of the compound, to the pouring in molds and the criteria to obtain wall finishes. Only certain passages of the essay of Fontana on Anzio, the treatises of Bélidor and the essay by Delorme on the aqueducts of Lyon contain similarly detailed indications. The isolation of the question of the manufacture of Roman cement from the various construction processes or from archaeological research, pointing to the manufacture as the foundation for contemporary construction, represents one of the most important characteristics of Loriot's memoir. From the technical viewpoint of the preparation of the compound, Loriot specifies the nature of the lime to use to imitate the strength of Roman cement: quicklime. From the Trévoux dictionary to the manuals of Gautier and Bélidor, to the essays of Tardif, Marsy and Delorme, quicklime is identified as an essential component for the manufacture of a strong cement.

The report is presented by Loriot to the Académie Royale d'Architecture on 10 December 1764 and described in the transcripts of the session as "an observational memoir" devoted to "Phocaean and Roman structures and their way of making and using mortar."[65] Perronet and Antoine-Michel Le Carpentier (1709-72) are charged by the Académie with the evaluation of Loriot's findings (though there are no known documents containing their assessments).[66] According to Loriot, the reaction of the members of the Académie Royale d'Architecture to the memoir is critical, in any case, not lastly due to the fact that Loriot makes no mention of all the technical literature on the question of Roman cement, starting with Vitruvius. Perhaps foreseeing a certain hostility

Paris, A. Colin, 1922, vol. VII, p. 190. On this same occasion Loriot brings the model of a hydraulic machine and that of a pile driving machine (*ibid.*).

[66] *Ibid.*

on the part of the academy members, prior to sending the memoir to the Académie Royale d'Architecture, Loriot has a copy sent to De Marigny. The intervention of De Marigny in favor of Loriot's research becomes decisive, from this point on, for the success, in France and abroad, of what is to become the cement of Loriot. De Marigny is not just the superintendent of royal constructions; he is also an expert on Roman cement, which he has personally been able to study during his trip to Italy under the guidance of Soufflot. Against the background of the criticisms advanced by the Académie Royale d'Architecture, his opinion takes on particular weight, expressed in the letter sent to Loriot on 13 February 1765. The letter contains useful information on the criteria of manufacture and deployment of the compound based on pozzolana and *lapilli* observed by De Marigny himself during his stay in Naples. One fundamental passage supports Loriot's hypothesis regarding the use of quicklime as a criterion for the making of the Roman compound. The fact that De Marigny suggests that Loriot, in the manufacture of the cement, make use of the processes of manipulation practiced in Naples, has a role in the development of the Loriot cement that rises to renown in the 1770s.

"[...] it deem it all the more worthy of your attention," De Marigny writes to Loriot, "that it seems that you very closely approach the method the Romans observed in their construction; it is certain that, throughout Italy, and mainly in Naples where they are the most solid, it is common practice to use lime only the moment that it is slaked. It is still hot when it is mixed with the *Rapillo*, which we replace rather imperfectly with sand. Once the mixture, to which pozzolana is almost always added, is made, it is gradually wetted and mixed until the mortar takes on the right consistence, which must be done as quickly as possible, so that the mortar retains a lot of heat when it is applied, and it may be to further increase and maintain this heat that care is taken to very frequently wet the structure with water to which quicklime is added repeatedly, because the mortar must be clear before it is used, and it may be a flaw that can only be remedied by the continuous wetting that I just spoke to you about. It would be good for you to know that the *Rapillo* that I mentioned seems to be nothing but rock that has been calcined by ancient volcanoes, and that is consequently very spongeous, very light and binds very intimately to boiling lime. From this observation, I conclude that the more the brick you intend to use is baked and the lighter it is, the closer your cement will come to being perfect. Perhaps it will become of greatest use for our reservoirs and basins if it is handled in the same way as in Naples. First, a layer of this mortar or cement, about twice as thick as it is intended to be in the end, is spread out on the surfaces that are to be made impervious to water. Once its has acquired enough consistence so that it will not give easily, we bring workers to site, and sitting on small stools, they use large mallets to beat this layer in a cascade of blows, all the while wetting it with water to which quicklime is added to the extent that is necessary, until they manage to reduce the thickness by a half, and once it is dry, no liquid can penetrate it."[67]

Based on the judgment expressed by De Marigny, Loriot prepares a second memoir which he sends to the Académie Royale d'Architecture, preceded by the transcription of De Marigny's letter, to add authority to his own statements. The pretext for the transcription is that the letter of De Marigny

[67] Abel-François Poisson de Vandières, marquis de Marigny, letter to Antoine-Joseph Loriot, 13 February 1765, Ms 2069, Ecole des Ponts et Chaussées, Paris. Loriot recalls the letter of De Marigny in the *Mémoire* of 1774 (*Mémoire sur une découverte dans l'art de bâtir*

contains useful "instructions" and "special manipulations" related to the making of cement in Naples.[68] Loriot also uses De Marigny's letter for the observations on the ways of making cement with quicklime in Italy, precisely because he believes that ancient Roman cement was made with that kind of lime.

In the second memoir, Loriot discusses the materials indicated by De Marigny as fundamental ingredients of the Neapolitan compound, lapilli and pozzolana. Loriot explains their geological characteristics and uses in building, though he simply refers, where pozzolana is concerned, what was written by Valmont de Bomare in the recently published *Dictionnaire*. According to Loriot, the observations of De Marigny on lapilli and pozzolana confirm the value of his own cement, given the fact that "thoroughly washed sand" takes the place of the lapilli, while "well-baked crushed brick" replaces the pozzolana, and that if worked with the Neapolitan method the result could produce similar technical performance.

"Mr. Loriot," we read in the "2e Mémoire," "has no doubt that a cement prepared following his method, but which would be poured thicker by replacing the *rapillo* with pumice, and to which the procedure of wetting and beating with a mallet would be applied, would produce the same effect as Neapolitan cement. But for the ordinary cement that he proposes for all kinds of structures, and curiously for aqueducts, canals, basins and massive terraced reservoirs for fortifications, vaults, casemates, and other underground structures even in aquatic environments, it requires neither watering nor beating; a roller whose weight could be increased at will would be enough to make the bed of the basins, canals, caves and other structures of this kind more even, and in this case, the roller would have to follow the workers without cooling the structure. Regarding the wetting, it only becomes necessary to pick up the construction work following an interruption and facilitate the bonding. It is considered that, following his approach, further watering is not needed, as on the one hand, his cement is served in a highly liquid state and at a time when the fusion of the lime is perfect, it offers a medium that exactly fills the gaps, and on the other hand, the chippings that Mr. Loriot uses are not susceptible to neither a higher nor a lower density due to the blows with the mallet, the structure takes on all of that to which it is susceptible by virtue of the weight of the materials by the cooling of the cement and the drying of the structure. He does not approve the use of all kinds of rubble in the structures he proposes with his cement, all porous and spongeous limestone rocks carry too much air and water, destructive agents that are entirely analogous to the same elements that attack the structure from the outside. To hold ground, the enemy must not have any intelligence. The most precise masonry and the most perfectly made finishing only concentrate the destructive agents, which in the end try to expand and cause destruction. Moreover, the high porosity of these rocks absorbs the principles of the lime which easily penetrate their surface without finding a binding site. But all fusible rocks and stones, hornstone, quartz, which at a glance resemble soft stone [*pierres molaires*] and hornstone [*quartz coriés* (*sic*)] can be used to an advantage with this cement, as well as rubble from refractory stones and frost-susceptible stone that are of no use in other applications. Due to their density, all of these bodies receive the imprint of the lime without absorbing it any more than the sands that are also fusible,

*faite par le Sr Loriot* cit., p. 17).

[68] [Antoine-Joseph Loriot], "2e Mémoire," 10 December 1764 [sic: February 1765], Ms 2069, Ecole des Ponts et Chaussées, Paris. The date of 10 December 1764 is apocryphal.

whose heat expresses analogous principles to petrification and the residue of the whole makes up no more than a truly petrified mass that is inaccessible to the penetrations of air and fire.

Mr. Loriot believes that he can, until he is given very detailed contrary examples, confidently support the claim that we have until now not been working following his principles, and that slaking the lime long before using it was a vulgar mistake that good authors, seduced by common practice, supported in their opinions, to the point of saying that, the longer it had been fused, the better it was. His method points towards the need to work with the mold method [*encaissement*], and the mold approach used by the Romans leads to the knowledge of their procedures, just as the debris of their works as well as plain physics allow to recognize the materials they used."[69]

The second memoir is presented to the Académie Royale d'Architecture on 25 February 1765, and submitted to Perronet for an opinion.[70] In March Loriot again turns to De Marigny, to submit to him the second memoir and the idea of a publication on his cement. Loriot suggests the possibility of a publication containing both memoirs, with the letter sent to him by De Marigny, in case of a negative opinion on the part of the Académie Royale d'Architecture. Loriot, also as a result of the criticism caused by his first memoir, fears another negative reception by the academy. Even later, the rise of his cement is punctuated by more or less open hostility on the part of certain architects of the Bâtiments du Roi and the Académie Royale d'Architecture. With Loriot's letter in March, the project of an essay on cement is born, which De Marigny encourages and finances, but only after having allowed Loriot to conduct experiments on various worksites, in his own holdings at Ménars and the royal buildings at Versailles and Paris.

"I am overly flattered," Loriot writes to De Marigny, "by the honor that you have graciously bestowed upon my memoir to not express to you my utmost recognition. It was presented to the Académie Royale d'architecture only a few days after I took the liberty of sending it to you, and I learnt that, upon its first reading, it caused some debate. Your instructive letter, Sir, encouraged me and I submitted a second memoir to which I took the liberty of attaching your letter. I have attached a copy of it here. I would be very flattered if this sign of my diligence could buy me the continuation of your powerful protection. I must expect a positive judgement from the *Académie*, but if this is not the case, would you disapprove, Sir, of my printing both memoirs as well as your letter, I request your permission, I have no doubt that their being public with your support would unite opinion in favor of the method that I propose, and that by developing the theory the practice would be sanctioned."[71]

The second memoir is also viewed negatively by the members of the Académie Royale d'Architecture, again due to Loriot's suggestion of the use of quicklime, as opposed to the specifications of ancient authors, especially Vitruvius and Pliny, cited by the members of the Académie. "If Sir Loriot is

[69] [Loriot], "2e Mémoire" cit.
[70] Minutes of the meeting of 25 February 1765, in, Lemonnier, *op. cit.*, vol. VII, p. 201.
[71] Antoine-Joseph Loriot, letter to Abel-François Poisson de Vandières, marquis de Marigny, 10 March 1765, O.1.1911/1,f. 17, Archives Nationales, Paris.
[72] *Mémoire sur une découverte dans l'art de bâtir faite par le Sr Loriot* cit., pp. 17-18. On the episode of the two reports presented to the Académie, see, *ibid.*, pp. 16-17.
[73] *Lettre de M. Patte Architecte de S.A.S. M. le Duc Regnant de Deux Ponts à M. Fréron, sur la construction vicieuse de la plûpart des bâtimens de Paris*, in, "L'Année Littéraire," vol. VI, 1767, pp. 53-64 (republished in, Patte, *Mémoires* cit., pp. 155-62).
[74] The manual contains only certain generic notions on the "sand and lime mortar," explaining that the process of drying is due to the "saline-

troubled by this lack of reception, he was not discouraged; he placed less blame on the *Académie* than on some of its members, whom he knew to be unfavorable, as well as on a quite natural and quite well-grounded hindrance, that it was accredited by the opinion of the two only ancient authors that dealt with this material. In fact, they did not fail to cite the testimony of Vitruvius and Pliny, primarily of the former, who, in his architecture, gave the highest praise to the chalk that had been slaked the longest ago."[72]

After the proposals of Loriot and the involvement of De Marigny, in the context of the Académie Royale d'Architecture, research on cement is conducted by Soufflot, but with a different orientation.

## Mortar without "abuses," from Patte to Baumé

In pursuit of the perfect recipe for making hydraulic mortar, in France other personalities, besides Loriot, continue to conduct research. The manufacture of mortar was one of the main technical concerns of Boffrand, and it is inherited by his pupil Pierre Patte (1723-1814). Patte sees the making of mortar and the use of the "plaster" to construct foundations, walls and vaults of buildings in Paris as one of the most dangerous and frequent "abuse" that threaten the solidity of those buildings. His critique expressed in 1767, in which he praises the methods of making mortar practiced by Boffrand and his simple machine for mixing lime and sand – developed on the worksite of the Hôpital des Enfants-Trouvés[73] – provokes the public contribution of the chemist and pharmacist Antoine Baumé (1728-1804), a friend of a great expert on lime, Macquer (they both teach courses in chemistry and pharmacology in their private school on Rue Saint-Denis, in Paris). For Baumé this marks the beginning of a focus on the chemical processes of mortars, and hydraulic mortars in particular, that is still not evident in his *Manuel de chymie* published in 1763 and reprinted in 1766.[74]

Following the considerations of Patte, Baumé explains the composition of "construction mortar" – "a mixture of quicklime, slaked by water and a vitrifiable soil"[75] – and discusses the materials and definition of cement: a "pulverized baked clay," that "is made of debris from tiles, bricks and stoneware trays."[76] Among the various types of cement, he indicates the one in use in Lorraine, composed of "claylike black lime" and very fine sand mixed with clay. "This mortar is so perfect that it is used with great success in garden basins," he writes, specifying that he has seen "joints made of this mortar that after ten years still bear their relief despite the fact that they are very often swept with a broom when the basins are cleaned [...]."[77] Then, more like a construction worker than a chemist, he proposes the practical recipe for the proportions of the perfect hydraulic mortar: "it takes approximately three parts of sand for one part of quicklime."[78] The indication of the nature of the lime demonstrates that the hypothesis of Loriot, though challenged by the members of the Académie Royale d'Architecture, is shared by other specialists. Baumé returns

earth material" that adheres to the "grains of sand" (Antoine Baumé, *Manuel de chymie, ou exposé des opérations de la chymie et de leurs produits*, Paris, P.-F. Didot, 1766², p. 78).

[75] Antoine Baumé, *Mortier à bâtir*, in, "L'Avantcoureur," 9 November 1767, no. 45, (pp. 707-09), p. 707. *Cf.* also, *Sur le Plâtre employé dans les Bâtimens*, in, "L'Avantcoureur," 2 November 1767, no. 44, pp. 692-94.

[76] Baumé, *Mortier à bâtir* cit., p. 708.

[77] *Ibid.*

[78] *Ibid.*, pp. 708-09. The recipe is no longer included in, Antoine Baumé, *Chymie expérimentale et raisonnée*, Paris, P.-F. Didot, 1773, vol. I (*cf.* p. 193).

to the subject of mortars, after having done a series of experiments on components and chemical reactions, in the essay *Chymie expérimentale et raisonnée* published in 1773, where he quotes the paragraphs on cement of the letter written in 1767 – but not the practical recipe.[79]

In his literary masterpiece *Mémoires sur les objets les plus importants de l'architecture*, written in 1768 and published in 1769, Patte goes back to the issues of the preparation of mortars and the techniques of Boffrand; it is here that he offers the definition of cement, also using the specifications of Baumé: "cement is a crushed tile which, having been mixed with lime instead of sand, makes an excellent mortar, which mainly finds its use in structures built in water, for terraces, and to build sandstone buildings. One must avoid mixing terra-cotta and brick, given that these materials, which do not have as strong a consistency as tile, which is baked to a higher degree, will necessarily alter its quality."[80] But, like Loriot or Baumé, Patte does not discuss the potential of pozzolana; he only mentions it regarding the construction of the foundations of a bridge, for which he advises using a "good mortar made of lime and cement or pozzolana."[81]

### Desmarest, D'Arcet and Soufflot, and the French pozzolanas

In spite of Loriot's Roman cement formula and recommendations like those of Patte and Baumé for the making of the perfect mortar, the question of pozzolana continues to be crucial for the construction of major hydraulic works. Without taking the need to identify a binder capable of permitting the construction in water of dikes or bridge piers, we cannot understand all the reasons behind the gradual turnaround in geological research on volcanic materials in France, after the discovery of Guettard.

Research on pozzolanas in Auvergne takes a new direction thanks to the work done in 1763 by Nicolas Desmarest (1725-1815), a mineralogist and member of the Académie Royale des Sciences of Paris. The fact that Desmarest holds a governmental position, assigned him by Trudaine, as inspector of factories to improve production, contributes to determine the practical character that sets French research on pozzolanas apart from this point on, still against the backdrop of large hydraulic works of engineering along the rivers and coasts of France. Unlike Guettard, Guénée and D'Aubreuil, Desmarest is particularly interested in pozzolana-based mortars, rather than stones in general. It is probable that from the outset Desmarest's geological interest in volcanic stones cannot be separated from a focus on the use of pozzolana in the making of mortar. In any case it is clear that after his explorations in Auvergne, Desmarest does not fail to inform, in different ways, through publications, memoirs and personal contact, builders, engineers and governmental organizations regarding the practical possibilities of the use of French pozzolanas in large constructed works.

Along the lines of the findings of Guettard, Desmarest identifies, in the volcanic deposits of Auvergne, a material similar to the Italian pozzolana, whose constitution is known to him thanks to samples he has examined. With the cooperation of the engineers François Pasumot (1733-1804), Jacques Dailley and

[79] *Cf.* Baumé, *Chymie expérimentale* cit., p. 193.
[80] Patte, *Mémoires* cit., p. 118. Patte warns that the Quai Pelletier has been built with this type of compound (*ibid.*, p. 218).
[81] *Ibid.*, p. 223.
[82] [Nicolas Desmarest], "Lettre sur la pozzolane, et particulierement sur celle d'Auvergne," 20 February 1779, B3, Archives de l'Académie Royale d'Architecture, Institut de France, Paris. The "Lettre sur la pozzolane" is published in, *Lettre à M. l'Abbé Bossut, de l'Académie des Sciences, sur les différentes sortes de Pozzolanes, & particulièrement sur celles qu'on peut tirer de l'Auvergne*, in, "Observations sur la Physique, sur l'Histoire naturelle et sur les Arts," vol. XIII, March 1779, pp. 192-204. The letter

Claude Pezet, from 1764 to 1766 surveys and observations are carried out for the making of a geological map of Auvergne, with particular focus on the area of Mont-Dore, again taking to completion work that had been begun by Guettard. In 1764, with a report on the "objects of natural history, of use for construction,"[82] Desmarest makes his initial discoveries known to the Minister of State Louis Phélypeaux de Saint-Florentin, Marquis de La Vrillière (1705-77), who intervenes to finance the research.

"Among these materials," Desmarest explains, recalling his proposal to the minister of a geological map of Auvergne, "I pointed out natural cements for different types of construction and known in Italy under the name pozzolanas. It is true that, at the time, I was only familiar with pozzolanas through their examination in small quantities, but these examinations were enough to determine their nature as porous, brittle volcanic slags, and to be struck by the similarities with the powdery materials that accompany the lava flows in Auvergne offered to me everywhere. Therefore, I announced to the Minister three kinds of natural cements that could take the place of pozzolana, and that could be extracted for a low cost from Auvergne by way of the Allier. [...]. At the same time, I proposed these three types of cements as such that can, alone, or mixed in certain proportions, be of use in structures that require solid mortars that are impervious to water and for which pozzolana has been used until now. I urged the Minister to consider how important it is to record on a map the huge amount of these volcanic sands that are dispersed in the surroundings of Clermont and along the banks of the Allier."[83]

Other lavic materials indicated by Desmarest to the minister for possible utilization in construction are the light "spongeous lavas," which in his view could be cut in the form of bricks to use in flat vaults, replacing the special light bricks manufactured in Paris by certain German workers, at a high cost.[84]

Precisely when discussion begins on the efficacy of Loriot's invention, the intervention of De La Vrillière allows Desmarest to visit Italy in 1765-66, in the entourage of the Duke Louis-Alexandre de La Rochefoucauld d'Enville (1743-92) and in the company of the draftsman Jean-Jacques de Boissieu (1736-1810). So halfway through the 1760s Loriot and Desmarest are the protagonists of the fundamental research on the two types of Roman cement, one based on a composition using quicklime, the other depending on pozzolana.

Desmarest's trip to Italy is financed precisely for the study of the qualities of pozzolanas and their application in construction, especially in Rome and Naples, and around the two cities, from the Bolsena Lake to Baia, in order to make a better selection of the lavas in the deposits of Auvergne. But Desmarest studies volcanic phenomena in Tuscany, Umbria, Latium, Campania and in the Colli Euganei, near Padua, of which he has learned through the literature of Arduino. The fact that Desmarest makes excursions through the Pre-Alps near Vicenza, from 13 to 17 April 1766,[85] in the company of Fortis, should not be underestimated for an understanding of the reasons behind the growing interest of Venetian scholars in volcanism and, above all, of their

of Desmarest is dated Paris, 20 January 1779. A note explains that the letter was read by Bossut to the Académie Royale d'Architecture on 1 February 1779 (*ibid.*, p. 198, note 1).

83 [Desmarest], "Lettre sur la pozzolane" cit.

84 *Ibid.* On the manufacture of special light bricks for flat vaults see, "Aux Messieurs de l'Académie Royale d'Architecture, assemblés le 13 novembre 1769," manuscript, B3, Archives de l'Académie Royale d'Architecture, Institut de France, Paris.

85 *Cf.* Luca Ciancio, *Autopsie della terra. Illuminismo e geologia in Alberto Fortis (1741-1803)*, Firenze, Olschki, 1995, pp. 124-25.

subsequent attempts to make use of local deposits for the production of a cement using local materials.

During the voyage in Italy Desmarest learns to distinguish the various types of pozzolana that are traded, and gains knowledge of the modes of manufacture of mortars and their uses in construction thanks to encounters with masons, architects and engineers; furthermore, he also visits ancient Roman works to assess the durability in time of the constructions made with pozzolana-based mortar. Thus, after the voyage of Soufflot, which marks the discovery of the potential of Italian pozzolana for building, the trip of Desmarest becomes a true scientific exploration whose goal is to imitate Italian construction with materials found in France.

"[...] I traveled through the volcanic Cantons of this beautiful country," Desmarest recalls regarding his trip to Italy, "observing, with utmost attention, the natural cements where ever they were to be found; and focused primarily on determining, in great precision, the nature of pozzolanas, the different states in which these precious materials can be found, and the uses for which they were best suited, in order to get a better idea of corresponding substances that the Auvergne could be hiding, and on the advantages that French architects would be able to draw from them. The result of observations and investigations I conducted on pozzolanas in particular in the surroundings of Naples and Rome are the following: 1st that they are a mixture of spongeous volcanic slags that are more or less vitrified, more or less brittle, more or less earthy and reduced to a certain degree of tenuity; 2nd that the different types or varieties of pozzolanas were the result of different raw materials that were altered or melted by fire, as well as of the various principles that are come together when brought in contact with the flame in the vast crucible of the volcanoes. As a consequence of these results, I was brought to distinguish several different types of natural cements. The first variety is a cluster of black slag, spongeous and vitrified, quite solid and reduced to granules of mediocre size. It is known in Naples under the same *Rapillo* or *Lapillo*. It closely resembles cinders or the residue from the combustion of charcoal. [...]. This volcanic sand, mixed with quicklime, slaked on the site, forms a mortar that is exclusively used for the terraces that the houses in Naples are covered with. It is also used to make tablets of various dimensions, which in a short time acquire a solidity that rivals that of ordinary stones in the surroundings of Paris. The second variety is a cluster of spongeous, brittle slag, of one or different colors: there are yellow, grey and red kinds. They are reduced to different degrees of tenuity [...]. To make ordinary mortar using this variety, one third quicklime is mixed to it, and in a short time, the mortar acquires a very strong solidity. The third variety is a debris of whitish pumices, in a powdery state. White pozzolana from the surroundings of Baia, of which several boatloads are shipped to Malta, and which according to word in Naples is thought to be the best kind, is of this type. The fourth variety is a

[86] [Desmarest], "Lettre sur la pozzolane" cit.

[87] *Mémoire sur l'origine et la nature du basalte à grandes colonnes polygones, déterminées par l'histoire naturelle de cette pierre, observée en Auvergne*, in, "Mémoire de l'Académie des Sciences," 1771 (1774), pp. 705-75. The map is entitled: "Carte d'une partie de l'Auvergne où sont figurés les courants de laves, dans lesquels se trouve le basalte en prismes, en boules, etc. pour servir à l'intelligence du mémoire de M. Desmarest sur ce basalte."

[88] Guillaume de La Haye, *Carte d'une partie de l'Auvergne (partie du Puy-de-Dôme) [...] levée par Pazumot et Dailley, 1774*. *Cf.* Kenneth L. Taylor, *New Light on Geological Mapping in Auvergne during the Eighteenth Century: The Pasumot-Desmarest Collaboration*, in, "Revue d'Histoire des Sciences," vol. XLVII, 1994, pp. 129-36. The map is published in 1774 with the title: "Carte d'une

cluster of baked clay, whitish, mainly spongeous. These soils are in the form of a dry and brittle dust, or in that of a tender stone, rubble: in this case it is the result of the union of the earthy molecules via a principle of ordinary calcareous infiltration. Thus, the terracotta form the basis of the tufa or rubble from Naples and of the *peperino* from the surroundings of Rome. [...]. The fifth type is a cluster of quite solid black granules that seem to be little chips of more or less compact lava, and in which only very little porosity can be discerned. These five types are sometimes found isolated, sometimes mixed together in different combinations, and in each case these igneous products are known and used in Italy by the name of pozzolana. Therefore, when one examines more closely the pozzolana from the catacombs at St. Sebastian of Capo di Bove, which is transported from Rome to Civita Vecchia for the shipments that the French, the Genovese and the Spanish come to make, one can easily see that it is a mixture of light slags (second type), whitish pumices (third type), porous, dry and brittle baked clay (fourth type) and chips of solid lava (fifth type) with white farinaceous tips; and that similar mixtures can be found in excavations conducted in the vicinity of Baia, of Pozzuoli, and of Naples to extract pozzolana. These same mixtures stand out everywhere on the steep shore of Lake Bolsena. In Italy, not only do they use these mixed pozzolanas to prepare mortar, but also the pure types that I believe to have distinguished and characterized above: and I never noticed that a specific choice was made; one variety is even replaced indistinctly by another if it can be found near the construction site, and the architects in Naples and Rome assured me that they had never observed any significant difference in the results. So, for example, the mortars that architectural ornaments are composed of in Rome and Naples are made with one part lime and two parts pozzolana: and whether one mixes the second, the third or the fourth type to the lime, or even the ground rubble, or the river sand, all of these different preparations take on the same consistency and bind equally well to the light layer of stucco that they are covered with."[86]

Once back in France, on the basis of observations made during his time in Italy, Desmarest probably orients the preparation of the geological map of Auvergne in such as way as to indicate the deposits of lavic materials similar to those used in Italian construction (he personally takes part in the campaign of surveying in September-October 1766). The map indicates five types of lavic materials: "ancient flows," "solid masses of ancient lava melted on the site," "modern flows," "basalt prisms in the flows," "basalt spheres in the flows."

Presented to the Académie Royale des Sciences in 1771 and published in 1774,[87] Desmarest's map, engraved by Guillaume de La Haye, becomes a geographical tool available to French builders to help them conduct experimental research on the manufacture of French cements according to the Roman method – the map indicates "the path and the limits of the materials that were melted and ejected by the volcanoes."[88] It lays the groundwork for

partie de l'Auvergne ou' sont figurés les courants de laves, dans lesquels se trouve le basalte en prismes, en boules, &c. pour servir à l'intelligence du Mémoire de M. Desmarest sur ce basalte. Levée par M.M. Pasumot et Dailley, Ingénieurs-Géog. du Roi" (*cf. Nicolas Desmarest, Mémoire sur l'origine & la nature du basalte à grandes colonnes polygones, déterminées par l'histoire naturelle de cette pierre, observée en Auvergne*, in, "Histoire de l'Académie Royale des Sciences. Année MDCCLXXI, avec les Mémoires de Mathématique & de Physique," Paris, Imprimerie Royale, 1774, pp. 705-75). *Cf.* Kenneth L. Taylor, *Geological travellers in Auvergne, 1751-1800*, in, Patrick N. Wyse Jackson, ed., *Four Centuries of Geological Travel: The Search for Knowledge on Foot, Bicycle, Sledge and Camel*, Geological Society, London, Special Publications, [2007], pp. 73-96; Kenneth L. Taylor, *The Beginnings of a Geological Naturalist: Desmarest, the Printed Word, and Nature*, in, "Earth Sciences History," vol. XX, 2001, no. 1, pp. 44-61. *Cf.* also, Antoine Vernière, *Les voyageurs et les naturalistes dans l'Auvergne et le Velay*, Clermont-Ferrand, G. Mont-Louis, 1900.

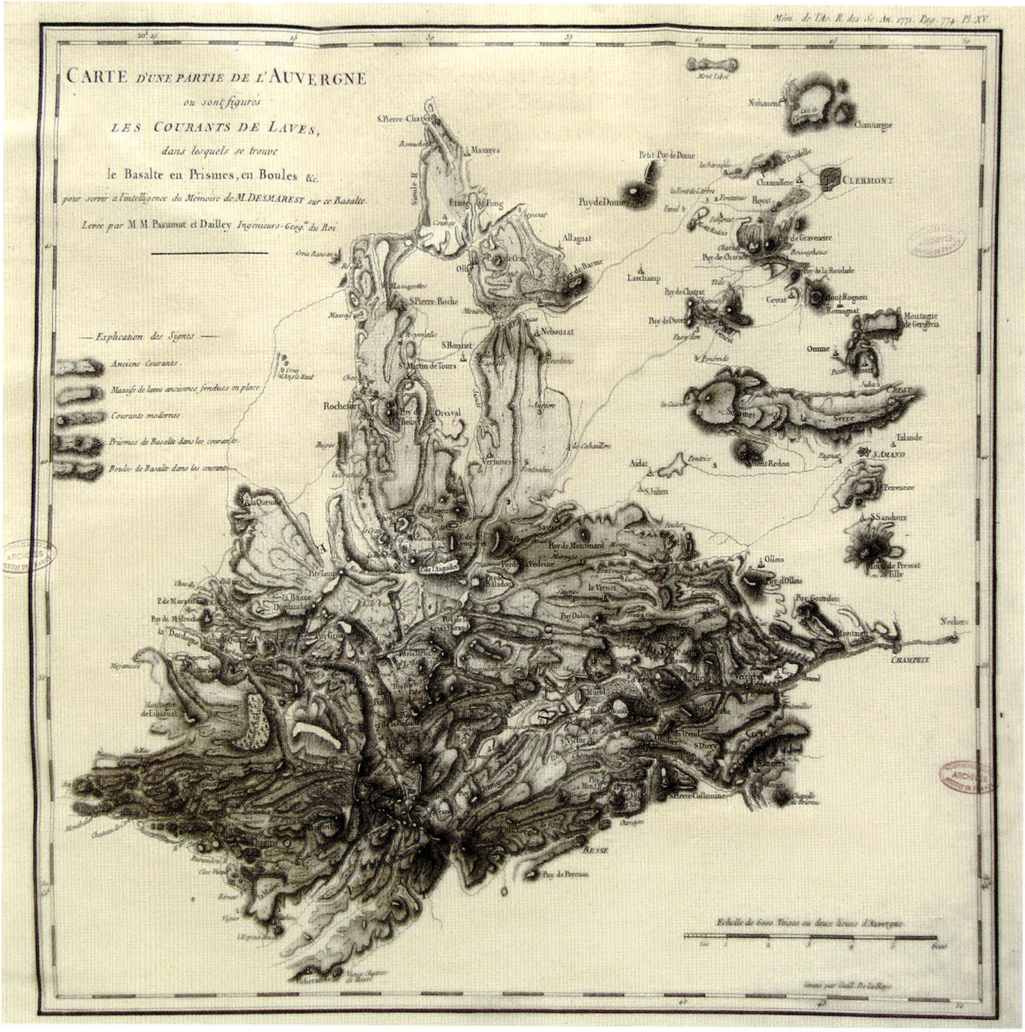


**285**. Nicolas Desmarest, "Map of a part of Auvergne with lava flows shown, in which one finds Basalt in the form of prisms, balls, &c." ("Histoire de l'Académie Royale des Sciences. Année MDCCLXXI, avec les Mémoires de Mathématique & de Physique," Paris, Imprimerie Royale, 1774).

the discovery of a natural hydraulic cement produced in France, suggesting an alternative to the one being studied at the same time by Loriot. Desmarest himself is aware of the potential of the pozzolanas of Auvergne in the evolution of French construction methods, as demonstrated by the fact that he immediately informs Trudaine of his discoveries, who among other things has also been the Intendent of Auvergne.

Thanks to Desmarest the pozzolanas of Auvergne join the ranks of the materials analyzed by chemists, confirming the interest caused by his discoveries and the practical potential of that volcanic product. The chemist Jean d'Arcet (1725-1801), destined to become director of the Manufacture de Sèvres, completes an extraordinary series of tests on different rocks and types of soil, starting with the criteria of analysis put into practice by Pott, but with the help of a much more powerful furnace, the one built by Louis-Léon-Félicité de Brancas, duke of Lauraguais, in 1766, to make porcelain similar to that of China. We should recall that in the porcelain venture Lauraguais is in contact with Guettard.

Desmarest offers D'Arcet various samples of the volcanic products of Auvergne, which are also subjected to the force of the furnace of Lauraguais. The results of the experiments with firing are submitted to the Académie Royale des Sciences of Paris on 16 and 28 May 1766, and lead to the conclusion that the stone of Auvergne has produced "a brown and transparent glass" and "could be identical to what the Italians refer to as their *Piperino*."[89] D'Arcet, in turn, has personal knowledge of the volcanic products mined at Andernack, where he went in the company of Lauraguais – "This stone so perfectly resembles that from Volvic near Clermont in Auvergne," he writes.[90]

After the exploration of Auvergne and the voyage to Italy, Desmarest continues to travel with the aim of expanding knowledge on pozzolanas and their various uses. In 1767 he identifies other deposits in France, in Limousin.[91] His statements regarding the possible use of this pozzolana confirm the interest in application of his research to the field of construction.

"The Romans," he writes, "recognized this natural cement and made use of it: for, in some places near the village of Chassenon there are remains of Roman buildings built with the most compact melted materials. They are bonded together very solidly with a mortar composed of lime and different types of burnt clay and slags that can be found dispersed on the surface of the volcanic terrain. As these pozzolanas could be of some use in the construction works commanded by the government to make the Charente River navigable upstream of Angoulême, and since this cluster is not far from this city, I deemed it necessary to inform the head engineers that are in charge of this work; they suggested to make a certain choice and use them above all for those constructions that are to be set in the water. Incidentally, these natural cements, transported to Angoulême, can also be used in some of our ports in the vicinity of the mouth of the Charente."[92]

[89] Jean d'Arcet, *Mémoire sur l'action du feu égal, violent, et continué pendant plusieurs jours sur un grand nombre de terres, de pierres & de chaux métalliques essayées pour la plupart telles qu'elles sortent du sein de la terre, lu à l'Académie Royale des Sciences les 16 & 28 Mai 1766*, Paris, P.G. Cavelier, 1766, p. 73.
[90] *Ibid.*, p. 76.
[91] *Lettre sur les différentes sortes de pozzolanes et particulièrement sur celles qu'on peut tirer de l'Auvergne, à M. l'Abbé Bossut, de l'Académie des Sciences [...]. Extraite du Journal de Physique, mois de mars 1779*, Paris, Clousier, 1779, p. 23.
[92] Nicolas Desmarest, *Lettre à M. l'Abbé Bossut, de l'Académie des Sciences, sur les différentes sortes de pozzolanes, & particulièrement sur celles qu'on peut tirer de l'Auvergne*, in, "Observations sur la physique, sur l'histoire naturelle et sur les arts," vol. XIII, March 1779, (pp. 192-204), p. 204.

The research on pozzolana of Auvergne has great impact, and Desmarest can write: "By means of these writings and the procedures that I have carried out to make the pozzolanas of Auvergne known, all of my observations on this important object are, since a long time, public in this province, and one can find educated people that can provide all the relevant information that one may need, either as an architect or as a naturalist."[93]

Nevertheless, based on the statements of Faujas de Saint-Fond, the discoveries of volcanic material in Auvergne do not seem to have attracted due attention in the world of construction, and their discussion remains relegated to the sphere of specialized publications: "When it was announced for the first time that there had been several volcanoes in Auvergne a very long time ago and that there, one could find pozzolana very similar to that from Italy, the imbeciles and the fools looked upon the worthy scientists that had made this useful discovery with pity: pozzolana was brought to Paris, but either it was poorly chosen, or more likely, natural history had not shone its light on architecture, and those that exercised this fine art were not as educated as they are now, afterwards pozzolana was only mentioned in some chemistry and natural history books [...]."[94]

But the remarks of Faujas de Saint-Fonds do not take the experiences of Soufflot into account, who precisely in the wake of the success of the research of Desmarest returns to become, though for a short time only, the protagonist of the pursuit of practical applications of pozzolana in construction (a role later played precisely by Faujas de Saint-Fonds). Though it cannot be demonstrated with documents, it is highly probable, given the clues, reports and the chronological sequence of the events, that Soufflot conducts his experiments based on the direct advice of Desmarest himself.

"Knowing about such precious materials," Desmarest recalls, "located in the center of France and dispersed along the banks of a navigable river, my first effort was to instruct the administration about it, in particular Mr. Trudaine. I also thought it necessary to point them out to Mr. Soufflot, who had himself observed pozzolana and its uses in Rome and Naples. This skilled architect was willing to conduct some tests with them. The Rapillo from Auvergne and the fecund variety mixed with lime gave him a mortar that solidified rather promptly and in that state appeared to be as solid and impervious to water as the terraces and tablets of Naples."[95]

First, in February 1768, Soufflot presents to the Académie Royale d'Architecture a brick cut from a block of lava - "a very light matter" – taken from the volcanoes of Auvergne.[96] The members of the Académie test the lava brick in water to see how it floats, and compare its weight with the light bricks of Bourgogne. In August Soufflot begins to conduct systematic studies to see if lavas, pozzolanas and lapilli taken precisely from the area explored by Desmarest, Mont-Dore, are of the same quality as those found at Vesuvius or near Naples. This is the hypothesis at the center of the explorations and observations of Guettard and Desmarest, in Auvergne and Italy, which Soufflot sets out to prove through practical experiences. One of Soufflot's application goals is to produce waterproof and durable slabs for terrace roofs, which are increasingly in vogue in residential and public buildings, and a mortar capable of

[93] *Ibid.*, pp. 12-13, note 1.
[94] Barthélemy Faujas de Saint-Fond, *Mémoire sur la manière de reconnoître les différentes especes de pouzzolane et de les employer dans les constructions sous l'eau et hors de l'eau, pour servir de Suite & de Supplément aux Recherches sur la Pouzzolane*, Amsterdam, Nyon, 1780, p. 51, note a. Probably Faujas de Saint-Fond refers precisely to the discoveries of Desmarest.
[95] [Desmarest], "Lettre sur la pozzolane" cit.
[96] Minutes of the meeting of 8 February 1768, in,

protecting the masonry of vaults from water seepage. Moreover, in May 1767 he had prepared a list of works to be done urgently on the terrace of the Paris Observatory, under pressure from De Marigny.[97] The story of Soufflot's attempts to test the potential of French pozzolana is narrated by Desmarest. From what he reports, it appears that Soufflot continues to be interested in the manufacture of light bricks for flat vaults, as suggested by Desmarest – the vaults might be the ones to be built in the church of Sainte-Geneviève in Paris.

"Following these initial successes," Desmarest reports, "Mr. Soufflot wished to conduct larger scale studies on pozzolanas from Auvergne; he employed Mr. Carré, contractor at the marble quarry at St. Léon, to deliver these different cements to Paris. In 1768, we saw two boats arrive at the Port de la Rapée, loaded with pozzolanas as well as porous lava suited to make light bricks [...] and even though these boats were shipwrecked, we were able to retrieve the bulk of the load from them, which Mr. Soufflot had deposited in the cellars of the Notre Dame; and I do not know the reasons that kept Mr. Soufflot from making use of them in the way he had suggested."[98]

Soufflot conducts experiments on volcanic products of Auvergne under the aegis of the Académie Royale d'Architecture, which records the results.

"[...] with the lapillo," we read in the minutes of the meeting of 5 September, "he made mortar composed of one third lime and two thirds of this lapillo, worked and crushed without water; he performed the same operation with ordinary cement; afterwards, he made two shallow troughs with these mortars by working them as they work the lapillo terraces in Naples. To this end, he first placed on a plank, which was surrounded by a three inch high rim, a layer two inches thick that he had beaten in every direction with a shovel blade, the following day he added a second layer, which he beat in the same manner and then flattened by beating. Following these procedures, he exposed both troughs to the frequent rains of the last month and the sun in his courtyard. After twenty six days, the trough made with the lapillo took on the consistency of the hardest Saint Leu stone. He removed it from the mold in a single piece, but the cement trough broke apart into several pieces, and did not, by far, have the same consistency: it had chapped in several places, while the other had been preserved undamaged without any loss of volume, as far as the Company had seen. Mr. Soufflot concluded from this test that using these materials, terraces as solid and robust as those of Naples could be made, and that they would be better suited to protect terraced vaults from the rain than ordinary cement."[99]

Soufflot makes use of pozzolana on the worksite opened in the Notre-Dame complex in Paris, where he has stocked volcanic material that arrived at Port de la Rapée. This can be gleaned from the printed version of the letter of Desmarest, where regarding the experiences of Soufflot it is written that "the only thing that was constructed with these materials is a door in the cloister of Notre Dame in Paris."[100] The fact that it is a door raises the question of the very nature of the structure envisioned by Soufflot – perhaps a concrete based on Auvergne pozzolana, reinforced by metal bars, like the stones Soufflot himself uses on the worksite of the church of Sainte-Geneviève?

Lemonnier, *op. cit.*, vol. VIII, 1924, pp. 5-6.

[97] *Cf.* Jean Mondain-Monval, *Soufflot, sa vie, son œuvre, son esthétique* (1713-1780), Paris, A. Lemerre, 1918, pp. 535-36, and pp. 294-95.

[98] [Desmarest], "Lettre sur la pozzolane" cit.

[99] Minutes of the meeting of 5 September 1768, in, Lemonnier, *op. cit.*, vol. VIII, 1924, p. 32.

[100] *Lettre sur les différentes sortes de pozzolanes* cit., p. 12.

Precisely when Soufflot is working on experiments with pozzolanas of Auvergne, Desmarest, driven by the same interest in pozzolana-based hydraulic mortars, makes a trip to Holland, in 1768, to observe the *trass* imported from Andernach. The contacts with D'Arcet, who was personally acquainted with the product of Andernach, may have been decisive to orient Desmarest towards German volcanic products processed in Holland. As in his voyage to Italy, Desmarest sets out to study the constructive qualities and methods of fabrication and use of cement.[101] His observations on Dutch *trass* are always made with an eye on identifying a similar material in Auvergne. Just as the volcanic material is mined at Andernach and marketed by means of barges on the Rhine, so the similar product from Auvergne could be transported, Desmarest claims, via L'Allier. Desmarest's considerations are all of a commercial nature, and become crucial for the future entrepreneurial activity of Faujas de Saint-Fond as a supplier of French pozzolanas.

At this point, the experiences of Desmarest and Soufflot on French pozzolanas and the experiences of Loriot on cement have created the conditions for the manufacture of Roman-type cements using French materials.

## Hamilton's collection of lavas from Vesuvius

The discoveries of Guettard, Desmarest and Soufflot on French lavic stones are the best known of a nascent branch of geology – volcanology – destined to become crucial precisely against the background of the experiments on the construction potential of pozzolana for hydraulic cements. While the French are busy studying the nature of pozzolana and trying to find similar materials far away from the famous Italian sites, and while Italy, Germany and Holland are linked by trade in volcanic products to use for construction, the English play a certain, though marginal, role in the evolution of scientific knowledge of lavic stones, thanks to the curiosity and activity of William Hamilton (1730-1803), a diplomat, antiquarian and collector of antique vases.[102] But the cultural background of Hamilton's first observations on lavas is not that of his French contemporaries, nor that of Smeaton; none of his phrases makes reference to the constructive potential of pozzolana, its use for the making of mortars in ancient Rome or on the worksites of Naples. Naturalists like Buffon and La Condamine are his favored counterparts, with whom he shares the

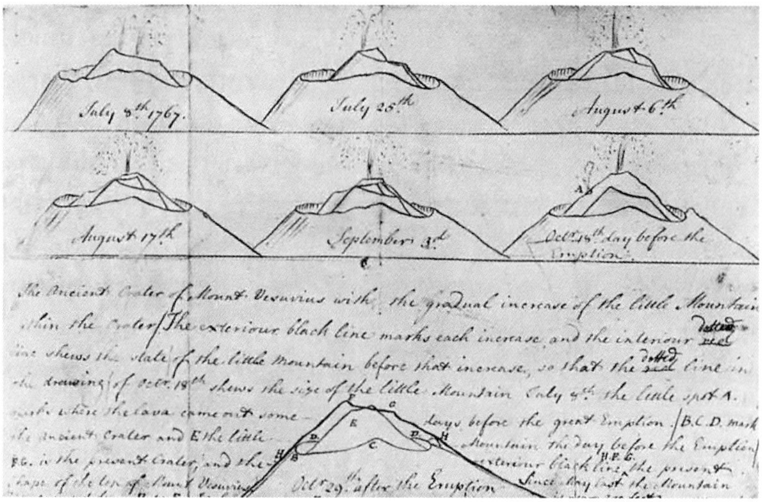


**286**. William Hamilton, sketches of the crater of Mount Vesuvius, during the eruption between May and December 1767 (I. Jenkins, K. Sloan, ed., *Vases & Volcanoes. Sir William Hamilton and his collection*, London, British Museum Press, 1996).

**287**. Pietro Fabris, samples of vulcanic products taken from the crater of Mount Vesuvius (C. Knight, *Hamilton a Napoli. Cultura, svaghi, civiltà di una grande capitale europea*, Napoli, Electa Napoli, 2003).

goal of discovering the geological reasons behind the formation of mountains.

From his arrival in Naples, in November 1764, and until October 1767, Hamilton observes the eruption phenomena of Vesuvius, takes notes and makes sketches of the volcano, and writes a series of observations in the form of letters he sends to the Royal Society of London, of which he became a member in 1766.[103] First published in "Philosophical Transactions," the letters converge together with others in two volumes, *Campi Phlegraei. Observations on the Volcanos of the Two Sicilies*, published in 1776 (the supplement, published in 1779, is mainly devoted to the description of the eruption of Vesuvius in August that same year). Hamilton also sends to London, to the Royal Society, samples of ash and lavic rocks of different kinds from the eruption of 1767, certain "very curious salts and sulphurs" found at Torre dell'Anziata, to be subjected to scientific study. "It is very extraordinary, that I cannot find, that any chemist here has ever been at the trouble of analyzing the productions of Vesuvius," is his comment.[104] In the spring of 1769 he visits Mount Aetna, and writes a report for the Royal Society. He continues to send lavic rock samples to the Royal Society in 1770.[105]

To document with drawings the various types of stones collected around Vesuvius, Hamilton relies on the collaboration of the painter Pietro Fabris (1740-92).[106] Lava, pumice and other volcanic slag are depicted on the shelves that house Hamilton's collection. The various types of lava documented by Fabris are also cut with a saw, made into blocks and polished to be displayed in special sorting cases. Perhaps the samples in Hamilton's collection are purchased directly in the shop of a Neapolitan – a trade that confirms the curiosity of European travelers regarding volcanic materials. "Complete collections of all the natural productions of Mount Vesuvius are now made and sold by

[101] *Cf. Lettre sur les différentes sortes de pozzolanes* cit., p. 15.
[102] See, John Thackray, *'The Modern Pliny'. Hamilton and Vesuvius*, in, Ian Jenkins, Kim Sloan, ed., *Vases & Volcanoes. Sir William Hamilton and his collection*, London, British Museum Press, 1996, pp. 65-74.
[103] Cfr. William Hamilton, letter to James Douglas, President of the Royal Society, 10 June 1766, in, William Hamilton, *Campi Phlegraei. Observations on the Volcanos of the Two Sicilies, as They have been communicated to the Royal Society of London. Observations sur les volcans des Deux Siciles, telles qu'elles ont été communiquées à la Societé Royale de Londres*, Naples, 1776, vol. I, p. 14.
[104] William Hamilton, letter to James Douglas, President of the Royal Society, 3 February 1767, in, Hamilton, *Campi Phlegraei* cit., p. 20.
[105] William Hamilton, letter to Matthew Maty, secretary of the Royal Society, 16 October 1770, in, Hamilton, *Campi Phlegraei* cit., pp. 52-89.
[106] Carlo Knight, *Hamilton a Napoli. Cultura, svaghi, civiltà di una grande capitale europea*, Napoli, Electa Napoli, 2003.

**288-290**. Pietro Fabris, volcanic products taken from the Fossa Grande at Resina; samples of minerals found by William Hamilton on Mount Vesuvius; samples of lava from Vesuvius (C. Knight, *Hamilton a Napoli. Cultura, svaghi, civiltà di una grande capitale europea*, Napoli, Electa Napoli, 2003).

Tommaso Valenziani at Naples opposite the King's Theatre," we read in the caption of a panel of drawings by Fabris.[107]

Fabris, in his activities for Hamilton, also documents certain quarries for the mining of tuff, near Casalnuovo, and the lava from the eruption of 1631 used to pave the streets of Naples.[108]

## Wark's cement patent

English architects and builders pay increasingly close attention to the development of compounds for wall facings in the ancient manner, so much so that the fashion ushered in by Inigo Jones flourishes anew, also in the wake of the voyages in the 1750s and 1760s made by the generation of the Chambers and the Adams. "Stucco" continues to be the widespread term in English to indicate facings made in the Italian way. Precisely during the course of the 1760s, certain significant improvements are made to the compound usually used for masonry coatings.

The first patent for a type of facing stucco, which would trigger a series of English patents on similar cementitious products, is filed on 27 July 1765 by the Reverend David Wark, of Haddington, Scotland, to coincide with the vogue of building brick walls with a cladding that imitates the look of stone. Like other similar compounds already in use in construction, Wark's is a mixture containing oil, because it is thought to be ideal to make waterproof protective facings, like a coat of oil spread over a painting.[109] "Composition or stone-paste, made with oils and other things, for covering walls, roofs, and

[107] Hamilton, *Campi Phlegraei* cit., plate XXXXIX.
[108] Hamilton, *Campi Phlegraei* cit., vol. II, plate VII.
[109] See patent number 834 of 1765, Scottish Record Office, TD77/142/4/22 (*cf.* Frank Kelsall, *Liardet versus Adam*, in, "Architectural History," vol. XXVII, 1984, pp. 118-26). *Cf.* also, Dana Arnold, *Reading Architectural History*, London, New York, Routledge, 2002, p. 65; Geoffrey Beard, *Decorative Plasterwork in Great Britain*, London, Phaidon, 1975, pp. 72-73; Roger L. Emerson, *The Philosphical Society of Edinburg 1748-1768*, in, "The British Journal for the History of Science," XIV, 1981, pp. 133-76.

domes, and for other purpose," we read in the title of Wark's patent.[110] Actually, it is then proven that this type of stucco, if applied on damp masonry, can be one of the causes of the detachment of facings.

After several constructions built by Wark – we know of two houses at York Buildings, made in 1765[111] – in 1766 Gwynn, an architect and engineer, recommending the new stucco for the facings of public buildings, describes its creative potential as a substitute for marble and other stones: "it is to be lamented that encouragement is not given to some ingenious person to find out a stucco or composition resembling stone, more durable than the common sort, and in which exterior ornaments might be easily wrought at a very easy expense."[112]

Wark's stucco patent becomes the property, starting in 1767, of the business of the Adam brothers, William Adam and Company. The fact that during their travels in Italy Robert and James Adam investigated decorative surfaces of ancient constructions plays a role in their decision to become the owners of products seen as having some of the physical characteristics of ancient building materials, and thus capable of producing a rich, ornamentation in the ancient manner at a limited cost. Actually their acquisition of the Wark stucco is simply the first in a series of other similar operations that lead the Adams to attempt to corner the market on stucco with an ancient appearance in England. William Adam and Company uses Wark stucco for the decorations and antique facings of the grand Adelphi Terrace residential complex, facing the Thames by means of an artificial platform with Piranesian underground spaces, a work that gains immediate renown – begun in 1768, the construction continues until 1772.[113] The fact that in 1776 the Adams state that the use of Wark stucco "had not been attended with the desired success"[114] offers a glimpse of the shortcomings of the compound.

## McCarthy, Davy, Pincot and the English artificial stone of the 1760s

The success of Wark's stucco coincides with the resumption of productive activity in the artificial stone field, applying terracotta-based techniques that appeared in England at the beginning of the century. One year after Wark files his patent, on 6 March 1766 Dennis McCarthy obtains a patent for a "composition or cement, called pietra-cotta."[115] McCarthy's product is of the type conceived by Holt and Langley, since it involves "a method of forming statues, &c., having the effect of the hardest stones."[116] The fact that the terms "cement" and "pietra-cotta" are used as synonyms indicates technical misconceptions that are widespread, at this point, in English construction culture regarding the nature of artificial stone. The works of McCarthy undoubtedly imitate those sculpted in stone and are produced by means of a compound to be shaped and fired in a furnace.

110 *Cf.* Bennet Woodcroft, ed., *Subject-matter index of patent of invention, from March 2, 1617 [...], to October 1, 1852, Part. I*, London, G.E. Eyre, W. Spottiswoode, 1857, p. 108.
111 *An Appeal to the Public on the Right of using Oil-cement or Composition for Stucco, &c., containing provisos in letters patent granted for inventions, and the provisos in the Act of Parliament for extending the term of the patent granted to John Liardet*, London, J. Hand, 1778, pp. 62-64.
112 John Gwynn, *London and Westminster Improved*, London, 1766, pp. 82-83.
113 *An Appeal to the Public* cit., pp. 62-64. *Cf.* Alistair Rowan, *After the Adelphi: forgotten years in the Adam brothers practice*, in, "Journal of the Royal Society of Arts," CXXII, September 1974, pp. 659-710.
114 Kelsall, *Liardet versus Adam* cit., p. 118.
115 *Cf.* Bennet Woodcroft, ed., *Subject-matter index of patent of invention, from March 2, 1617 [...], to October 1, 1852*, Part. I, London, G.E. Eyre, W. Spottiswoode, 1857, p. 108. McCarthy's patent is number 841.
116 James Patrick Muirhead, *The origin and progress of the mechanical inventions of James Watt*, London, J. Murray, 1854, vol. III, p. 220.
117 Wedgwood also invents "encaustic painting in various colours, in imitation of ancient etruscan and roman earthenware," to apply to his ceramic products (Josiah Wedgwood, patent n. 939, 16 November 1769, in, Bennet

The idea of the imitation of stone products is also behind the technical research on ceramics conducted since 1765 by Josiah Wedgwood (1730-95). In the evolution of English artificial stone, it is significant that Wedgwood becomes famous for the creation, developed with Thomas Bentley (1730-80), of an artificial Egyptian black basalt that is actually a hard ceramic composed of clay, iron-stone, ochre and manganese oxide. The Wedgwood factory makes vases, plates, teapots and other objects in this artificial black basalt, as opposed to the production based on the techniques of Holt and Langley, done for construction.[117] Nevertheless, the technical processes applied by Wedgwood fit into the subject area of artificial stone, so much so that in the last decades of the century several English observers point precisely to Wedgwood's experience as a way of conducting research on artificial cement to be obtained by firing special compounds.

The production of artificial stones for the construction industry, after the early work of Holt and Langley, and after McCarthy's patent, finds new stimulus thanks to the activity in 1767 of Daniel Pincot (?- 1792),[118] at his factory on Goulston Square, Whitechapel, later moved to Narrow Wall, and of George Davy, who opens a similar factory that same year, 1767, in the facility vacated on Goulston Square (his business closes in 1773).[119] Unlike his predecessors, Pincot does not file a patent for his compound (no patent exists for Davy's compound either). Whereas it is not known what objects were offered to the market by Davy's firm, the enterprise of Pincot immediately becomes famous for the production of statues, busts, vases, friezes, medallions, fireplaces, capitals and bases of columns. Some of these products are displayed, together with a copy of the Borghese Vase, at the Free Society of Artists exhibition held in London in 1767.[120]

Against the backdrop of the revival of artificial stone, reflected in the activity of the factories of Pincot, Davy and McCarthy, we see the spread of a taste for objects made in the antique fashion, in a climate like that of England where stone sculptures and architectural elements used to decorate villas and gardens rapidly decay due to weathering, and also have a high cost which limits their use. "[...] the weather in this island," Pincot writes, "has so destructive an effect on Portland, Bath, and other rock stone, and even marble itself, when exposed to the open air, that their surfaces are soon destroyed."[121]

The product Pincot puts on the market is made through processes of selection of clay, working and firing, based on what is done in China. Pincot claims that the material is much stronger and more durable than any kind of natural stone. The motivation to produce sculptural objects in artificial stone comes from the void left after the closing of the operation of Holt and Langley, from the revival of interest in ceramics produced in the ancient manner, aroused by the Wedgwood factory, and from the unsatisfactory technical results

Woodcroft, ed., *Subject-matter index of patents of invention, from March 2, 1617 [...], to October 1, 1852*, London, G. E. Eyre and W. Spottiswoode, 1867, p. 562). For the text, see, *Specification of the patent granted to Mr. Josiah Wedgwood, of Burslem, in the country of Stafford, potter to her Majesty; for his invention of ornamenting earthen and porcelain ware with an encaustic gold bronze, together with encaustic painting in various colours, dated Nov. 16, 1769*, in, *The repertory of arts and manufactures: consisting of original communications, specifications of patent inventions*, London, vol. VII, 1797, pp. 309-14.

[118] *Cf.* Daniel Pincot, *An Essay on the Origin, Nature, Uses, and Properties, of Artificial Stone: together with Some Observations upon common natural stone, clays, and burnt earths in general in which the durability of the latter is shewn to be equal, if not superior, to the hardest marbles, being the result of many experiments*, London, R. Hett, 1770.

[119] Davy's factory produces decorations in the ancient manner in artificial stone, probably for works by Robert Adam, like the Brentford Gateway, at Syon House, the residence of the earl of Northumberland (Kelly, *op. cit.*, p. 33).

[120] *Ibid.*, pp. 33 *et seq.*

[121] Pincot, *op. cit.*, p. 6.

achieved by the founder of the English tradition of this sort of product – Pincot considers Langley's product to be of better quality than that of Holt ("particularly some bustos, said to be tolerably executed," he opines).[122]

"But, it is evident from a considerable quantity of broken pieces now in my possession," Pincot writes, regarding fragments of works by Holt, "that such a work could not have held many years in great esteem without improvement, had its author lived; there being neither taste in the designs, nor neatness in the execution, though time has proved the materials durable in the severest trials. It is all covered on one side with an earthen ware, white glaze; and some of it is poorly painted with blue ornaments, baskets of flowers, &c."[123]

Pincot is aware that the English artificial stone, as it has been manufactured starting with Holt, differs from that produced by simple drying of a compound, since it is a ceramic product that therefore requires firing in a kiln. For this reason, he introduces a distinction between "artificial stone" and "burnt earth," though even he then classifies his own products as stones, since he wants to compete with works sculpted in natural stones or marble, and perhaps also to distinguish his output from that of Wedgwood. It is no coincidence that Pincot specifies that what is indicated as "artificial stone" among the ancients should actually be "formerly (and perhaps more properly) called, stone of baked or burnt earth."[124] Pincot pinpoints the origin of the misunderstanding of the meaning of artificial stone in the treatise of Holt, "where the name, artificial stone, is applied to burnt earths; for whenever this name hath occurred in other authors, it seems to mean those compositions which acquire hardness in the air only."[125]

By now, with Pincot, it is clear that English artificial stone is not a variation of the products based on a conglomerate, like the Italian *cantoni*. "There are various other methods of making artificial stone without burning," he specifies, "as by cementing their parts with gummous and divers other glutinous liquids, or by oily or resinous mixture; but the insufficiency of all these for any good purposes, renders them undeserving of any farther notice."[126] Perhaps precisely against the background of English production of artificial stone, from Holt to Pincot, the French inventors of Roman-method cement experiment, in the 1770s, with the application of the product for the making of sculptures and garden decorations.

The question of the existence of artificial stone in ancient civilizations, and that of Egypt in particular, though partially laid to rest, returns in a timely way with Pincot, who as an attentive reader of the treatise of Holt still sustains that thesis: "if I am not greatly mistaken, the English fall infinitely short of the skill of the ancient Egyptians; for I have long suspected that porphyry, granite, and other hard marbles, (no quarries of which can be found) are compositions. I could support this supposition by experiments, the account of which must at present be omitted."[127]

Pincot is the first producer of artificial stone to thoroughly discuss the question of the mold, and to indicate the advantages and drawbacks of this system of manufacture of sculptural objects. In his historical reconstruction of the invention of "burnt earth," starting from biblical times, he hypothesizes a process of working by direct shaping of the compound by the sculptor. The mold, in his view, would be a modern achievement – in any case he emphasizes that there exist no significant traces of the construction of molds in ancient sources. Based on his own experience, Pincot indicates the possible drawbacks of making certain figures of exceptional size with molds, in the moment in which the shaped clay is extracted from the mold to be placed in the kiln – he proposes breaking the figure into parts, and assembling the pieces after the firing. His reasoning also extends to the economic aspects of the production of artificial stone by means of molds, thanks to which the savings become more perceptible when the ornamentation of a piece is more complicated – he cites the example of a Corinthian capital, which is more economical than an Ionian, which in turn is more economical than a Doric capital, to be made according to his method.[128]

"But these disadvantages are counterbalanced by a property in the former of holding cement, with a firmness superior to the latter; and the joints may be made as fine as in any material whatever, so that if a subject be too long for its bulk, or too large for convenient management, it can be cut into several pieces, and so joined after burning, as to appear one entire stone. [...]. Again, as those productions are cast in moulds, and the price of the model and mould must fall on the casts that are made from it; so upon single subjects there can be no great saving of expense, unless modeled in the composition itself, which in some cases may be done with advantage: or else when the subject is so important, that a statuary would make a perfect model in clay, before he carves it in stone; then the making a mould from that clay, and so producing a cast, would save expence on a single subject. Once more: In plain masonry, such plinths, facies and shafts of columns, Tuscan bases and capitals, and things of like workmanship, there will be little or no saving."[129]

The success of manufactured products classified under the heading of "artificial stone" meets with the opposition of the stonecutters, who see it as a challenge to their artistic skills and to the quality of their works in stone, measured in terms of resistance to weather. In this sense it is significant that Pincot warns against the risk of bad publicity from the craftsmen specialized in sculptural works, who though they recognize the superiority of manufactured products with respect to natural stone, still advise clients to have works sculpted and to avoid the purchase of reproductions in artificial stone, which they dismiss as "a mere makeshift."[130]

When Pincot publishes, in 1770, the treatise *An Essay on the Origin, Nature, Uses, and Properties, of Artificial Stone*, the business of his factory is already associated with the person destined to make English artificial stone internationally famous: Eleanor Coade (1733-1821).

[122] *Ibid.*, p. 47.
[123] *Ibid.*
[124] *Ibid.*, p. 26.
[125] *Ibid.*, p. 46.
[126] *Ibid.*, p. 53.
[127] *Ibid.*, p. 45.
[128] *Ibid.*, p. 74.
[129] *Ibid.*, pp. 73-74.
[130] *Ibid.*, p. 78.

Fig. 2.a

# chapter seven
# cement works in ports

## The success of caissons in the technical literature

The evolution of the caisson for the foundations of bridge piers comes about, as demonstrated by De Saint-André, De Voglie and Mylne on the worksites of Chazey, Saumur and London, as a result of the perfecting of the saw used to even off the ends of the foundation piles according to the method of Perronet. The question of the origin of the invention of the saw is debated in the early months of 1761, starting with the report in "Mercure de France" in October on the assertions of the engineer Guillaume Viallet regarding Perronet's model and its making, attributed to De Cessart,[1] and the clarifications of Louis La Racine (1692-1763), published by the "Journal Encyclopédique" on 15 December, which attribute the invention to Perronet and the execution to De Voglie with the help of De Cessart.[2] After having read La Racine's article, on 14 March 1762 De Voglie writes to Perronet to specify the contributions of the various protagonists in the invention and construction of the saw.[3] The fact that Etheridge invented a similar saw is not mentioned by anyone – and in fact only Perronet has the drawing of that saw.

But beyond the knowledge made possible by Labelye's book on the Westminster Bridge or Bélidor's *Architecture Hydraulique*, and beyond the

[1] *Séance publique de la Société Littéraire de Châlons sur Marne*, in, "Mercure de France," vol. II, October 1761, pp. 163-69.
[2] La Racine, *Lettre adressée à un auteur de ce Journal, sur une nouvelle méthode de construire des ponts sur les grandes rivières, à Saumur ce 15 octobre 1761*, in, "Journal Encyclopédique," vol. VIII, 15 December 1761, pp. 107-18.
[3] Jean-Baptiste de Voglie, letter to Jean-Rodolphe Perronet, 14 March 1762, in, Eugène-Jean-Marie Vignon, *Études historiques sur l'administration des voies publiques en France aux Dix-septième et Dix-huitième siècles*, Paris, Dunod, 1862, vol. II, p. 335. See also, Jean-Rodolphe Perronet, letter to Jean-Baptiste de Voglie, 2 April 1762 (*ibid.*, pp. 336-37).

**291-293**. Caisson, machine for sawing piles under water used for the bridge at Saumur, and foundation for piers of bridge (D. Diderot, J.-B. D'Alembert, ed., *Encyclopédie, ou Dictionnaire Raisonné des Sciences, des Arts et des Métiers. Recueil de planches, sur les sciences, les arts libéraux et les arts méchaniques, avec leur explication, seconde livraison, en deux parties, Première partie*, Paris, Briasson, David, Le Breton, Durand, 1763).

discussions on the attribution of the invention of the saw for the piles, international renown is brought to the caisson with removable sides during the course of the 1760s by the *Encyclopédie* of Diderot and D'Alembert, and the contributions of Perronet and Patte.[4] The fact that the French technical literature puts the accent on the caisson version developed by De Voglie overshadows the importance of the prototype developed for the Westminster Bridge, so much so that the caisson with removable sides looks like a rediscovery of the French technical literature. At the same time, the English caissons, from the pioneering work of Labelye to the variants developed by Etheridge and Mylne, are not adequately discussed in the English technical writings.

In 1763 the *Encyclopédie* publishes plates on the machine to saw in water, *Machine à scier dans l'eau*, and the caisson for piles, *Grande Caisse pour les Piles*, documenting the works made for the construction of the bridge at Saumur.[5] The plate of the caisson includes all the technical details of its functioning, the turn-up walls, of horizontal boards held in position by posts with braces, the complex system of connection of the upper extremities of the posts by means of double members equipped with removable metal anchors, and the large threaded bolts that vertically cross the boards of the sides to guarantee a watertight seal.[6] But the drawings are made without the direct supervision of De Voglie and the entire illustration of the "new method" turns out to be "quite far from the truth and lacking in precision, as much as in the details as in the drawings," as De Voglie himself writes.[7] The same *Encyclopédie* also contains the drawing of the framework structure of a cofferdam with an interspace filled with "rich soil."[8]

The treatise *Recherches sur la construction la plus avantageuse des digues* by Charles Bossut and Viallet, honored in 1762 by the Académie Royale des Sciences, Inscriptions & Belles-Lettres of Toulouse, considered the continuation of the essays on hydraulic architecture of Bélidor and published in Paris in 1764, also contains useful information on what by now are the two types of caissons for construction in water, namely those with fixed or removable sides – the "construction with caissons or by *encaissement*" which the authors indicate as currently in use above all in the ports of the Mediterranean.[9] The same authors underline the precautions needed to ensure the connection of the masonry done in different successive caissons, and specify that the recent machines

[4] Perronet, *Mémoire sur les différentes méthodes* cit., p. 142.
[5] *Cf. Charpenterie*, in, Denis Diderot, Jean-Baptiste Le Rond d'Alembert, ed., *Encyclopédie, ou Dictionnaire Raisonné des Sciences, des Arts et des Métiers. Recueil de planches, sur les sciences, les arts libéraux et les arts méchaniques, avec leur explication, seconde livraison, en deux parties, Première partie*, Paris, Briasson, David, Le Breton, Durand, 1763, plate XXI, *Machine à scier dans l'eau*, and plate XXII, *Grande Caisse pour les Piles* (p. 5, for the legends).
[6] *Ibid.*, p. 5, legend of plate XXII.
[7] Jean-Baptiste de Voglie, letter to Jean-Rodolphe Perronet, 18 May 1771, Ms 2209, Ecole des Ponts et Chaussées, Paris. De Voglie responds to a letter sent to him by Perronet on 10 May to ask for the drawings of the "new method" of foundations at Saumur. De Voglie states that the *Encyclopédie* has already published "a description of the new method, but having been done, written and given to the Editors without my knowlege, it is quite far from the truth and lacking in precision, as much as in the details as in the drawings, which I could have provided" (*ibid.*). After the publication, the editors of the *Encyclopédie* had asked De Voglie for a description inserted in the *Supplement* to correct the errors of the drawings (*ibid.*). See, *Description des moyens mis en usage pour fonder sans batardeaux ni épuisemens les piles du pont de Saumur, sur le grand bras de la rivière de Loire, en 1757 & années suivantes*, at the entry *Ponts*, in, Denis Diderot, Jean-Baptiste Le Rond d'Alembert, ed., *Encyclopédie, ou Dictionnaire Raisonné des Sciences, des Arts et des Métiers*, Genève, Pellet, 1779, vol. XXVI, (pp. 711-91), pp. 730-42.
[8] *Cf. Charpenterie* cit., plate XIX, and p. 5.
[9] Charles Bossut, Guillaume Viallet, *Recherches sur la construction la plus avantageuse des digues*, Paris, Ch.-A. Jombert, 1764, p. 32.

Charpente, Grande Caisse pour les Piles.

Charpente, Machine à scier dans l'Eau.

Charpente Fondation de piles.

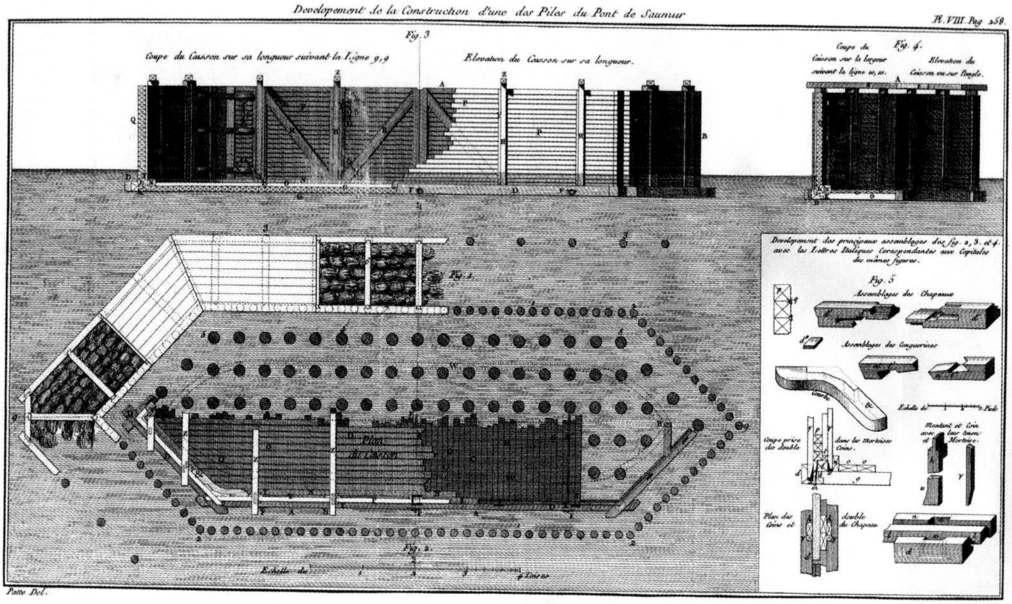


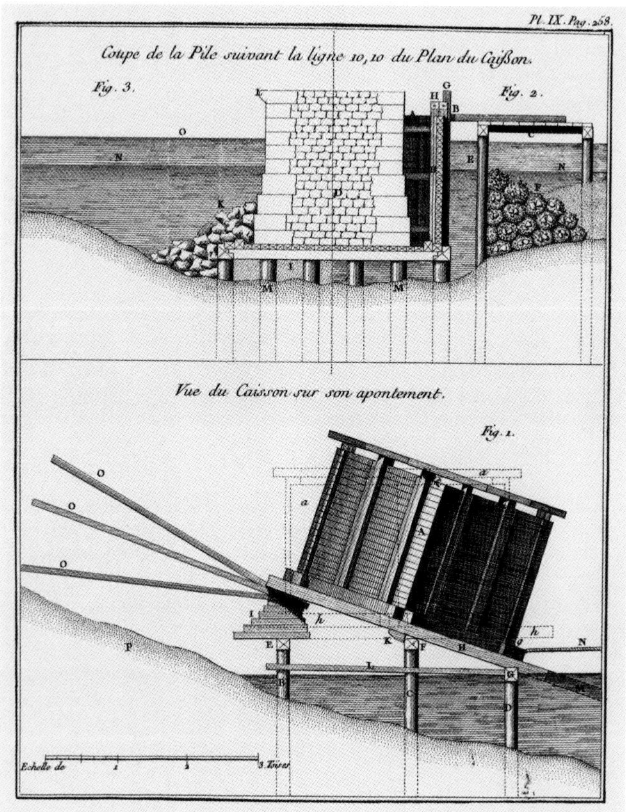


**294-295**. Pierre Patte, drawings of the caisson used for the construction of the piers of the bridge at Saumur (P. Patte, *Mémoires sur les objets les plus importans de l'architecture*, Paris, B. Rozet, 1769).

for cutting the ends of the foundation piles have offered "a new degree of perfection for the construction by *encaissement* [...]."[10]

Following the publication of the caisson in the *Encyclopédie* and against the backdrop of the work by Bossut and Viallet, Perronet returns to the notes he had made on multiple occasions on the foundation method based on that of Labelye, and on 13 November 1765 he presents a memoir to the Académie Royale des Sciences of Paris, later published in 1769, "on the different methods that have been used to make the foundations of masonry works in water, and principally on those that aim to eliminate cofferdams and dewatering in Bridge construction." It is here that Perronet again emphasizes the potential of pozzolana cement and indicates that without pozzolana it is necessary to perfect foundation systems derived from caissons, for which he reconstructs the genealogy from Labelye to De Voglie, also mentioning earlier examples, like the caissons described by Scamozzi.[11]

During the same year as the publication of Perronet's *Mémoire*, Patte pays tribute to the construction of the foundations of the piers of the bridge of Saumur in *Mémoires sur les objets les plus importans de l'architecture* (his essay is ready on 22 June 1768).[12] Patte sums up the various systems of building foundations in water, from the lost stones, poured "in a confused way, without binding," without mortar,[13] to the caissons of wood filled with masonry ("these caissons cannot properly set"),[14] to the wooden trellis that sink only after having built a masonry whose height exceeds that of the water,[15] to the cofferdams, all the way to the caissons.[16] The long description also based on direct accounts of Perronet and De Cessart and accompanied by four plates of drawings, of the works for the foundations of the bridge of Saumur, the caisson, its submersion and the special saw, summarizes the evolutionary process of the various foundation systems described by Patte.[17]

"The advantages of this new procedure," Patte concludes, "are tangible, both concerning economical savings, and regarding the elimination of an infinity of operations and other measures and difficulties related to the ordinary construction of bridges using cofferdams and dewatering. [...]. It is important to observe that, since the masonry is dry-built in the caisson, the mortars have time to set and develop the right consistency; in contrast, abiding to the conventional approach, the foundations are almost always raked out, given that there is a rush to cut the pumps to reduce the large costs they entail."[18]

After the publications of Diderot, D'Alembert, Perronet and Patte, the caisson becomes the prototype of those with the widest range of forms and uses invented by French engineers in the 1770s and 1780s, including students of Perronet, for hydraulic works destined to become fantastic masterpieces of construction technique, also with the use of pozzolana-based concrete.

The constructive culture of the Spanish military engineers continues to lead the hydraulic architecture for large harbor works. Their contribution is critical for the development of caissons, after the innovations introduced by Labelye and the variations proposed by French engineers. Thanks to the Spanish engineers, technical improvements made to caissons to transform them in a foundation system of bridge piles are applied to the original shape of the caisson used in the Mediterranean ports for the construction

[10] *Ibid.*, p. 33.
[11] Perronet, *Mémoire sur les différentes méthodes* cit., p. 142.
[12] *Cf.* Cochin, *Approbation*, in, Pierre Patte, *Mémoires sur les objets les plus importans de l'architecture*, Paris, B. Rozet, 1769, p. 374.
[13] *Ibid.*, p. 221.
[14] *Ibid.*, p. 223.
[15] *Ibid.*
[16] *Ibid.*, p. 227.
[17] *Ibid.*, pp. 230-58.
[18] *Ibid.*, pp. 251-52.

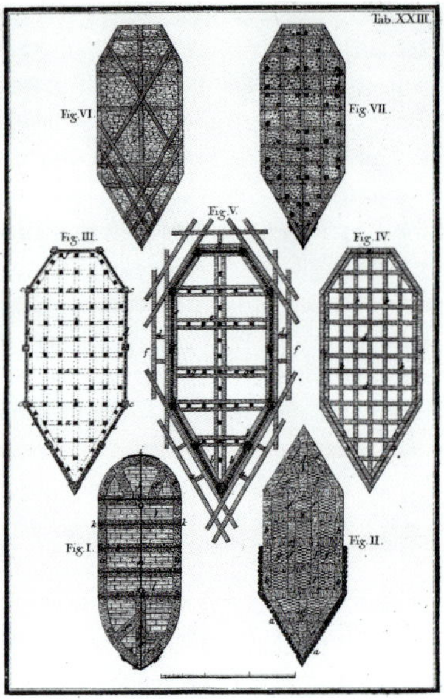


**296**. Foundation systems for the piers of bridges reported in the German treatises ([G. Walter], *Brücken-bau oder anweisung wie allerlei arten von brücken sowohl von holz als steinen nach dem desten regeln der zimmerkunst dauerhaft anzelegen sind [...]*, Augsburg, Beith, 1766).

**297-299**. Julián Sánchez Bort, project for the arsenal of El Ferrol, 1764-65. Plan, view of a caisson, and sections of masonry (Ministerio de Educación, Cultura y Deporte, Archivio General de Simancas, Mapas Planos y Dibujos, 04,092, 04,094, and 24,048).

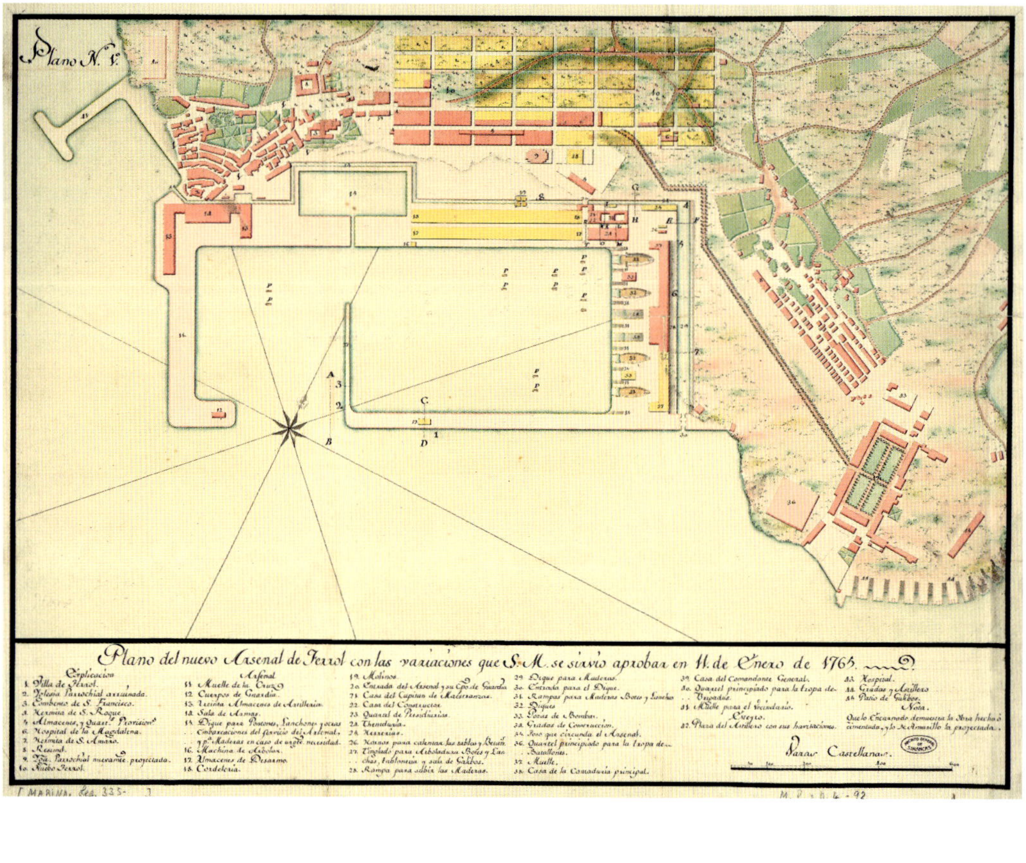

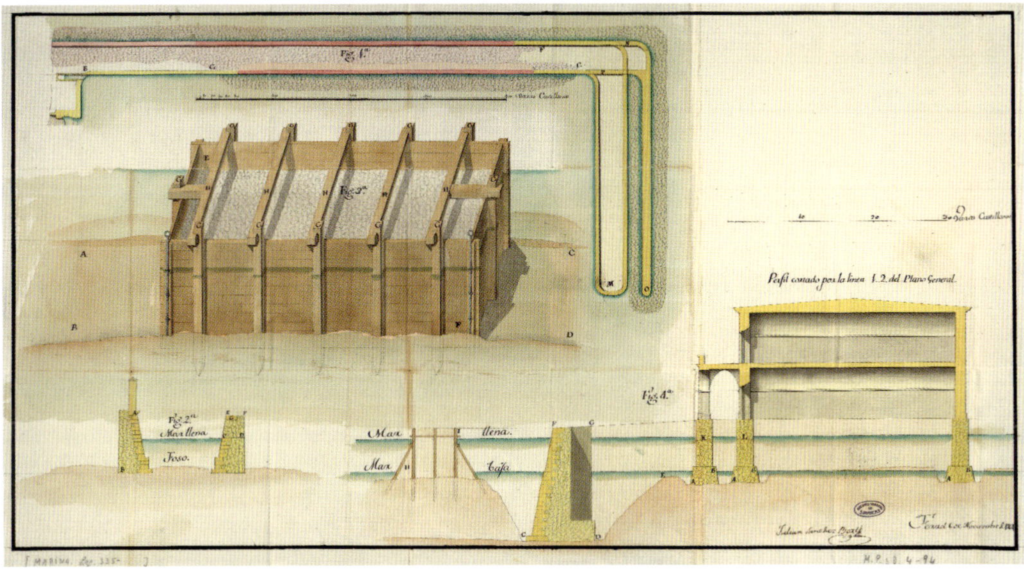

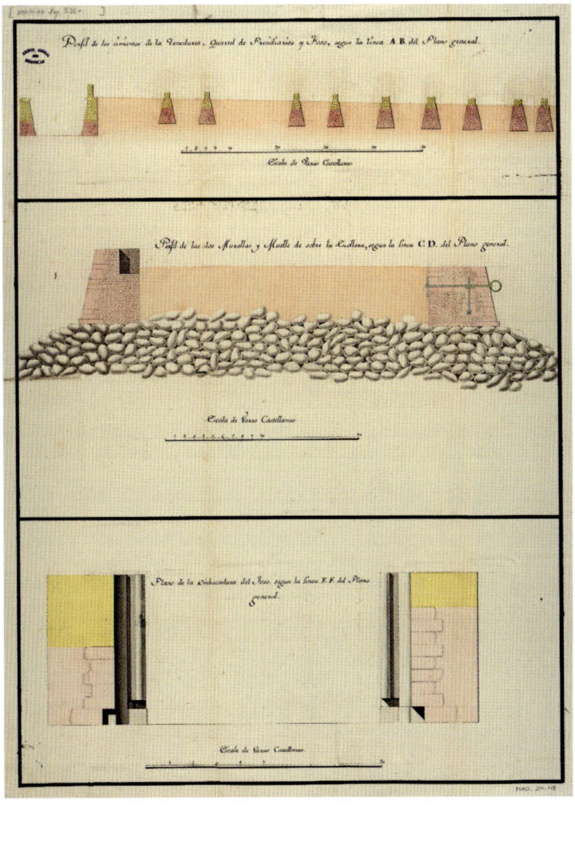

of jetties. While work continues to the dockyards of Cartagena on the Atlantic coast, the other major Spanish dockyard, that of El Ferrol, the engineer Julián Sánchez Bort (1727-85) designs, halfway throught the 1770s, a major expansion using a variation of the caisson system as it is spread in the Mediterranean basin in ports such as Genoa or Civitavecchia. Also taking advantage of the phenomenon of low tide, Sánchez Bort proposes, in 1764, a bottomless wooden caisson, with piles to be driven into the ground and sides equipped with mechanisms such as those imposed with Labelye, in order to make it possible to remove the caissons and reuse them. Thanks to these caissons, or formwork with removable sides, large sectors of concrete wall perfectly matched can be created, as described also by Bossut and Viallet.

Experience gained at the port of Toulon at the end of the 1760s for the construction of new drydocks for ships leads to the premises for the invention, in the next decade, of the most colossal caisson ever built up to this point. After proposals to build the drydock in the nearby port of Seyne, and after the trials of construction of beds with reverse arches, based on the model of the Spanish drydocks built at Cartagena,[19] in 1768 a definitive project has not yet been found, due partially to the difficulty of obtaining trunks of fir or pine of sufficient length for the construction of the cofferdam. The naval officer Maximin De Bompar (1698-1773) and the Commissaire Général de la Marine of Toulon, Félix-Antoine Dasque (1710-79), are ordered to consult the person still reputed to be the greatest expert on hydraulic construction, Milet de Monville,[20] to hear his views on the most suitable location for the drydock. Due to the nature of the terrain, according to Milet de Monville, the drydock should be built at the port of Castineau, near the prisoners' hospital, on a site included in the project for a plaza with drydocks and other port facilities which he has developed himself. From February to June 1768 soundings are made and experiments are conducted for the construction of the caissons, with the use of over 30 pumps to remove the water from the enclosure.[21] But the results convince Milet de Monville and the other members of the commission to declare, on 5 June, that it is impossible to proceed with the casing system.[22] During that same year Pierre-Joseph Laurent (1713-73), engineer and entrepreneur as well as creator of important hydraulic works in Flanders and of canals, is also called in and proposes, in 1769, the construction of "one or several forms suitable to receive first-class vessels," but by 1773 his project has still not been developed.[23]

[19] "Etat de la forme du Port de Toulon," 41AP1, Archives Nationales, Paris.
[20] "Registre des ordres du Roy, et depêches concernant la Marina de Levant, année 1768," Marine, B2, 389, Archives Nationales, Paris.
[21] "Memoire concernant les experiences qui ont été faites pour connoitre la nature du terrein, ainsi que la quantité et l'abondance des sources qui pouvoient se trouver dans l'emplacement aupres du Port Castineau," 4 June 1768, 41AP1, Archives Nationales, Paris.
[22] Maximin De Bompar, Charles-Marin Hurson, François Milet de Monville, Laurent Jean-François Truguet, report, 5 June 1768, *ibid.*
[23] *Cf.* "Propositions du S.r Laurens pour se charger, à forfait et à ses risques, perils et fortune, de la construction d'une ou plusieurs formes à Toulon," 1769, 41AP1, Archives Nationales, Paris. In 1773 Laurent is still developing the project (*cf.* "Copie de la lettre de Mr. Laurent à Mr. Potiev premier commis de la Marine, à Bouchain, le 24 may 1773," 41AP1, Archives Nationales, Paris).
[24] François Milet de Monville, "Plan Profil et Elevation de la Tour projettée sur l'Ecueiul Foucart," F14, 10226-10228, 1 Cherbourg 10, Archives Nationales, Paris. For the German-language treatise published in the 1760s see, for example, [Gaspar Walter], *Brücken-bau oder anweisung wie allerlei arten von brücken sowohl von holz als steinen nach dem desten regeln der zimmerkunst dauerhaft anzelegen sind [...]*, Augsbourg, Beith, 1766.

**300-301**. François Milet de Monville, Foucard rocky coast, Port-de-Bouc, near Marseille (F.14.10226-10228, 1 Cherbourg 10, Achives Nationales, Paris).

In his various applications of concrete for hydraulic works, it is not only for the experience gained at the port of Toulon that Milet de Monville stands out. In order to indicate the so-called shoal of Foucard, dangerous for navigation at Port-de-Bouc, near Marseille, one of the ports of Lyon, towards the end of the 1760s or in the early 1770s he designs a "small tower" or "column" to place on a pedestal of stone and concrete, and to be built in two phases – first the inner core, then the external masonry with its truncated conical form. To make the masonry, Milet de Monville invents an ingenious system of watertight formwork, using canvas barrels: poles and wooden hoops, connected by irons belts and ropes to form an adjustable mechanism, hold in position a "*toile royale à prelart*," or tarpaulin, that is lowered to the shoal bed with the lower edge raised by ropes, to form a space into which to pour the first layer of concrete and stones – the weight of this layer fully spreads the canvas. Iron clamps ensure grip between the two masonry parts.[24]

### The concrete of De Caire for the port of Versoix

The founding, supported even by François-Marie Arouet, known as Voltaire (1694-1778), of a city with a trade port on Lake Léman, at Versoix, to welcome

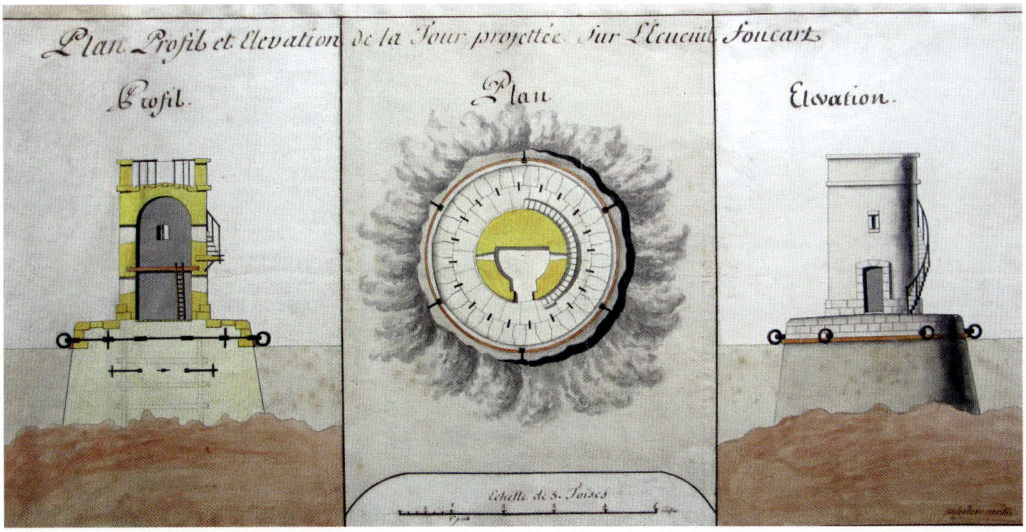


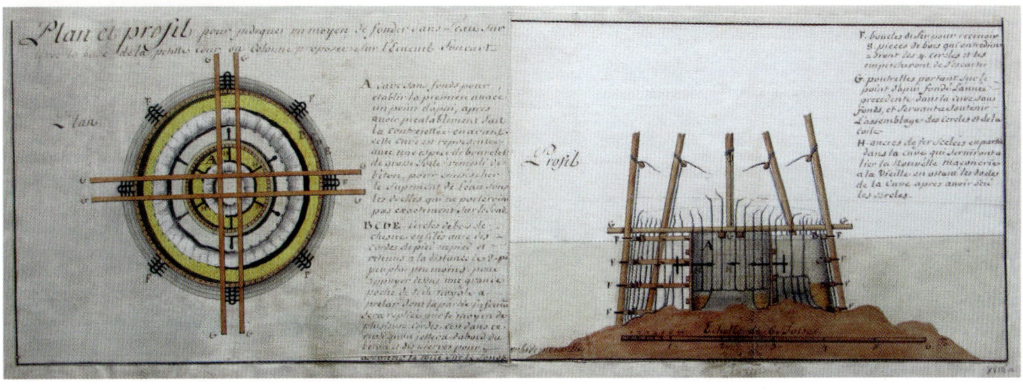

families persecuted in Geneva, becomes an opportunity for further experimentation on concrete for masonry built in water. Undertaken with a project drawn up in 1767 by Pierre-Joseph De Bourcet (1700-80),[25] the construction of the port and the city of Versoix is assigned first to André Clapier, "previously charged with the construction of the fortifications" of the Iles d'Hyères,[26] and then, with a contract dated 23 September 1768, to the architect-entrepreneur Léonard Racle (1736-91), a pupil of De Saint-André, owner of a factory of ceramics, or *argile-marbre*, at Grand Saconnex, near Geneva, and a protegé of Voltaire.[27] The engineer François De Caire (1731-92), Chevalier de l'Ordre Royal et Militaire de Saint-Louis,[28] is in charge of the construction. The work at Versoix is suspended in 1771, when the promoter of the initiative, Étienne-François de Choiseul, comte de Stainville and then duc de Choiseul (1719-85), Chief Minister under Louis XV, falls into disgrace.

The decision to use concrete masonry is taken after the first construction work on the port, when it becomes clear, after about eight months, that it is impossible to build the works at the site initially indicated, which causes problems for the construction of cofferdams due to the bed. It is decided to make the foundation of the port directly in the lake.

The port in the lake is composed of a rectangular enclosure formed by two moles. When the foundations of the moles are being built in the lake, and the system of dewatering again proves to be unfeasible, it is decided to resort, as we read in the documents of the time, "to a type of construction that was not taken into consideration during the cost estimate."[29] The construction of the foundations of the moles happens thanks to wooden formwork into which to pour the concrete – it is the model of the "dammed-up voids" described by Fontana regarding the Roman moles at Anzio, of the *passonata* in Ancona of Vanvitelli, of the wooden molds of Milet de Monville at Toulon, or the coffer-work of Palladio, Ware and Semple. At the port of Versoix, the stones of the moles above water will lay on the concrete foundations. It is significant that the works described in the documents include "filling of the caissons with large gravel."[30] At the same time, certain buildings in the city are also constructed with the use of formwork – reference

[25] *Cf.* "Memoire instructif pour M. de Bourcet sur le projet de former à Versoix une ville de commerce qui puisse dans la suite etre fermée de murs, et qui soit susceptible de fortifications au besoin," (sent 28 June 1767), H173-174, Archives Nationales, Paris.

[26] *Cf.* François De Caire, "Versoix, 22 septembre 1768. Memoire et etat de la dépense faite par economie sous la conduite du S.r André Clapier," H175, Archives Nationales, Paris.

[27] *Cf.* François Chatillon, *Léonard Racle, un architecte du siècle des Lumières*, Centre des Hautes Etudes de Chaillot, 2001.

[28] For the project of the moles of the port see the drawing, François De Caire, "Plan du port projete à Versoix," 9 June 1768, H179, Archives Nationales, Paris. In this phase, De Caire envisions the construction of a grand cofferdam to make the foundations of the moles of the port. In September 1770, the progress of the work of the city is described in a travel journal – large streets without any houses. "Three sides of this square," we read, regarding the port under construction, "are formed by a wall four feet thick and twenty feet high, built out into the Lake upon piles, with an entrance to let vessels in" (Anna Riggs Miller, *Letters from Italy, describing the Manners, Customs, Antiquities, Paintings, &c. of that Country, in the Years MDCCLXX and MDCCLXXI*, London, E. and C. Dilly, 1777, p. 13). In 1859 circa, during excavation for the construction of a house, parts of an aqueduct are found, considered Roman, in "concrete and brick" (entry *Versoix*, in, Marc Lutz, *Dictionnaire géographique-statistique de la Suisse. Nouvelle édition*, Lausanne, F. Blanchard, 1859, vol. I, p. 464). "An irrigation canal was to be drawn from the Versoix river; I oversaw the project and had it executed. It is the only useful outcome of this crazy enterprise," Céard recalls (Nicolas Céard, *Mémoire et observations historiques et critiques sur la route du Simplon et autres objets d'art, Paris*, Goeury, 1820, p. 43). Céard also describes the state of the port: "There exists, nonetheless, a shallow port in ruins that cost almost two million" (*ibid.*, p. 43, note 1). *Cf.* François Walter, *Voltaire et Versoix. La ville incertaine: port fortifié ou cité de la tolérance ?*, in, Erica Deuber-Pauli, Jean-Daniel Candaux, ed., *Voltaire chez lui: Genève et Ferney*, Genève, Skira, 1994, pp. 207-27.

[29] François De Caire, Léonard Racle, "Memoire general. Versoix," [18 March 1771], H175, Archives Nationales, Paris.

[30] François De Caire, Léonard Racle, "Versoix. Toisé général," 20 October 1769, H175, Archives Nationales, Paris. See also, François De Caire, handwritten note, 25 November 1770, H175, Archives Nationales, Paris.

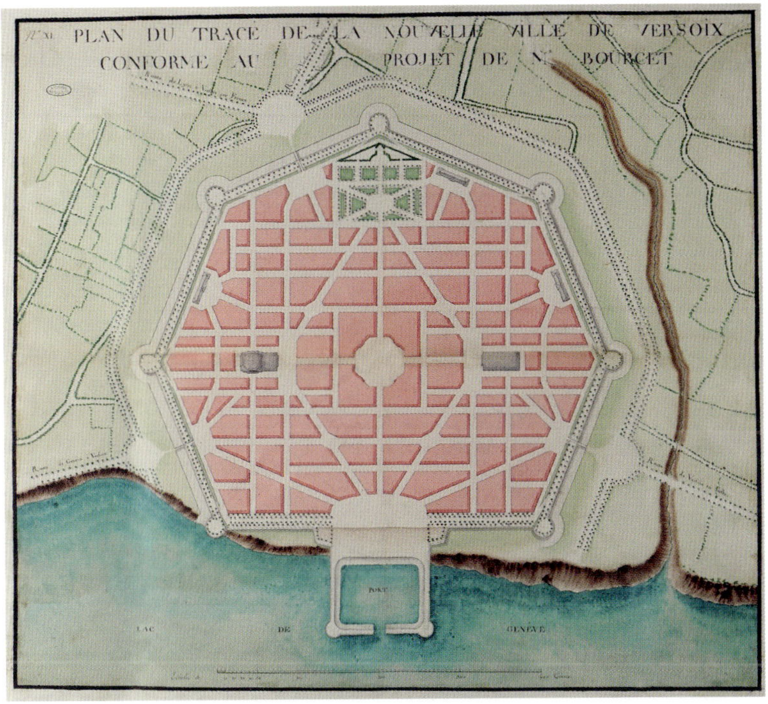


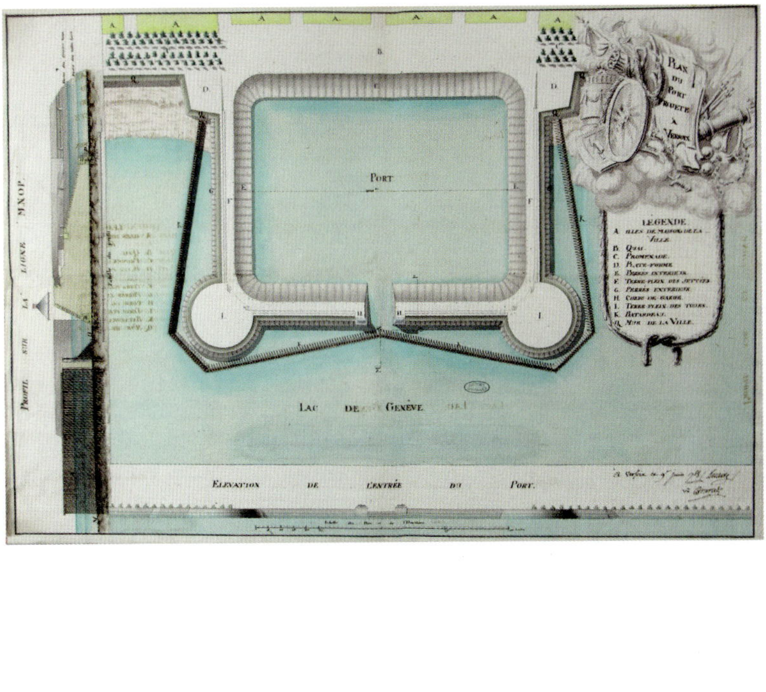


**302**. Pierre-Joseph De Bourcet, plan of the city of Versoix, 1767 (H179, Archives Nationales, Paris).

**303**. François De Caire, project for the port of Versoix, 9 June 1768 (H179, Archives Nationales, Paris).

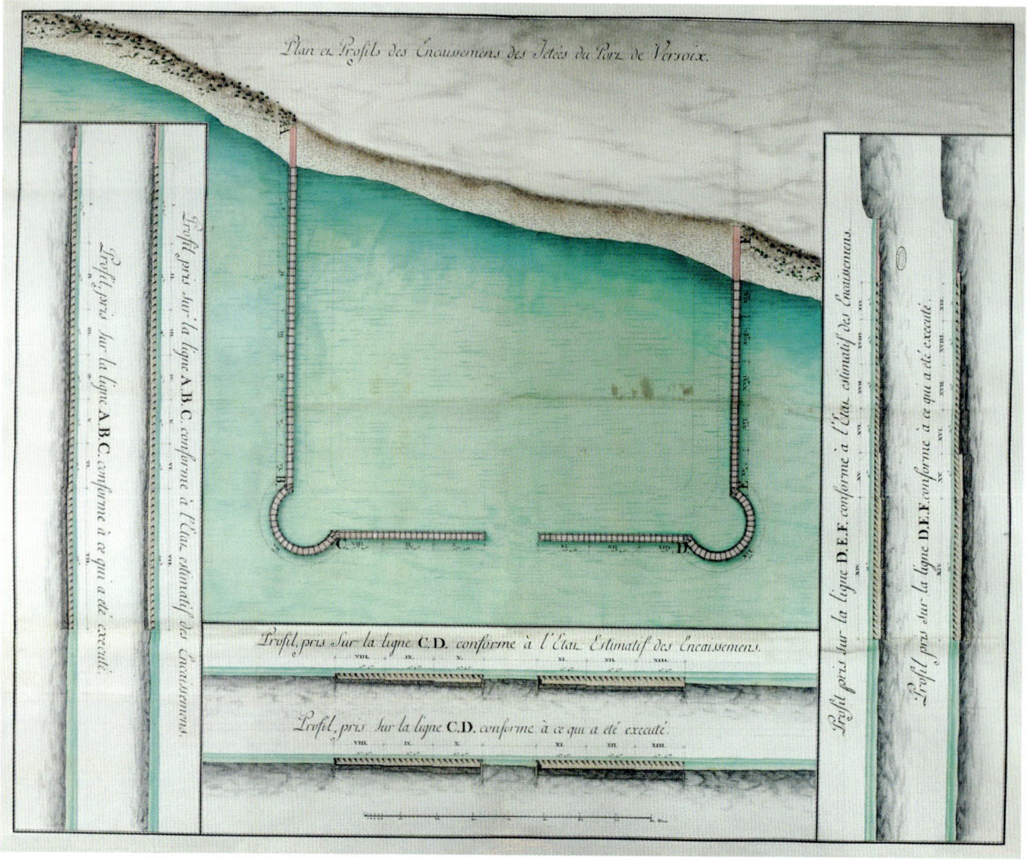

is made to walls "made of earth or rammed earth."[31]

"The jetties, whose foundations we were obliged to make using concrete masonry," we read in a document from 1771, "required wooden coffer [*encaissement*] in oak, whose piles were first rammed with a pile-driver to from 4 to 500 weight, and then damped to refusal [...]."[32]

De Caire is in charge of the works with concrete done at Versoix. He experiments with the fabrication of a concrete "composed of 5 parts lime, 3 parts sand, 3 parts cement and 2 parts rubble or chippings."[33] To test the capacity of the compound to set firmly, De Caire uses the method of the coffers of Milet de Monville described by Bélidor. Probably continuing to follow the indications contained in Bélidor's manuals, De Caire creates a single pouring of about 370 meters, done inside wooden formwork: "one unique block with a length of 190 toises, a width of 14 feet and a height of 15 to 18 feet."[34]

At the end of 1770, when financing runs out for the works at Versoix, have been done excavations "for the foundation and extension masonry for the coffers," the cofferdams in clay, excavations for the "foundations for the

[31] François De Caire, Léonard Racle, "Versoix. Toisé et etat général," 20 October 1769, H175, Archives Nationales, Paris.
[32] "Memoire general. Versoix," [18 March 1771], H175, Archives Nationales, Paris.
[33] Nicolas Céard, "Essai sur les moyens de construire à la mer, soit des forts avancés ou des digues pour en former des ports ou des rades," Versoix, 10 February 1785, Ms 1080, Ecole des Ponts et Chaussées, Paris.
[34] *Ibid.*

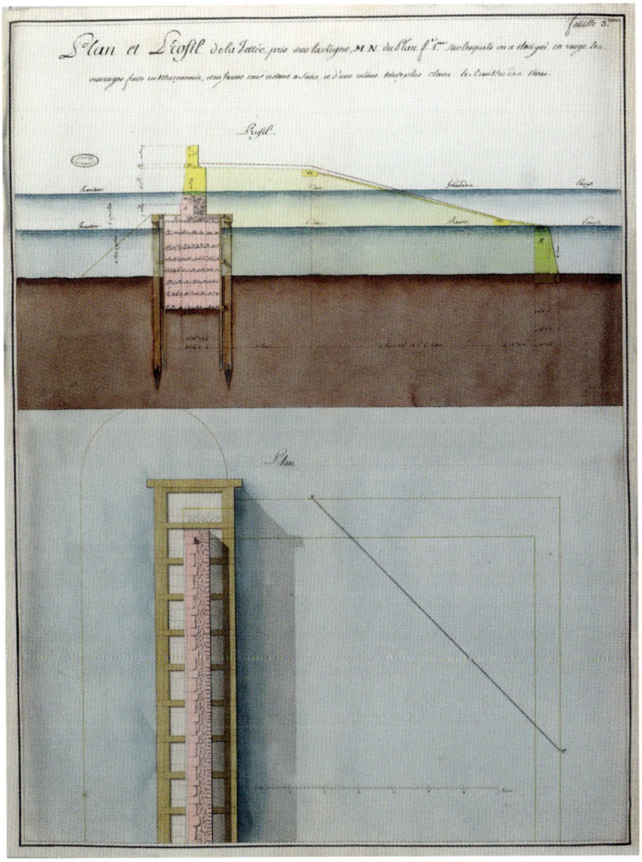

**304-305**. Port of Versoix, plans and sections of the wooden formwork for the jetties (H179, Archives Nationales, Paris).

coffers of the jetties," about "181 toises and 6 inches squared of gravel agglutinated with clay and other natural soils made by a curing machine for the concrete masonry foundation of coffers for the jetty," and "concrete masonry."[35]

One of the most detailed descriptions of the concrete works done in the port is prepared in December 1772 by Nicolas Aubry, an engineer of the Ponts et Chaussées sent to Versoix to assess the possibility of resuming the work interrupted in 1771 – he is assisted by the engineer Jean Querret (1703-76).

"All of the enclosing walls," Aubry writes, "have their foundations, from the low waters to the bank, supported on a solid mass of concrete composed of gravel and lean lime mixed with cement, rubble and brick debris. It is a solid mass enclosed by two parallel walls of sheet piles 9 feet thick in straight sections and 11 to 12 feet when between the two towers. These sheet piles are 4 feet thick on uneven heights that follow the bottom of the lake where they are fixed. They are bordered by 502 piles in two rows that have 11 to 13 inches of rough trimming, and penetrate the ground 4 to 15 feet, depending on the circumstances. Each row is supported by 3 slag bars crowned with walings,

[35] François De Caire, Léonard Racle, "Versoix, 1769 et 1770. Toisé des ouvrages faites pour la construction du port de Versoix jusqu'au 15 decembre 1770 inclusivement exécutés par le Sr Leonard Racle, entrepreneur des dits ouvrages conformement au marché a luy passé le 23 septembre 1768," 15 December 1770, H175, Archives Nationales, Paris.

**306-308**. Nicolas Aubry, Nicolas Céard, (attributed to), survey and project of completion of the mole of the port of Versoix (H179, Archives Nationales, Paris).

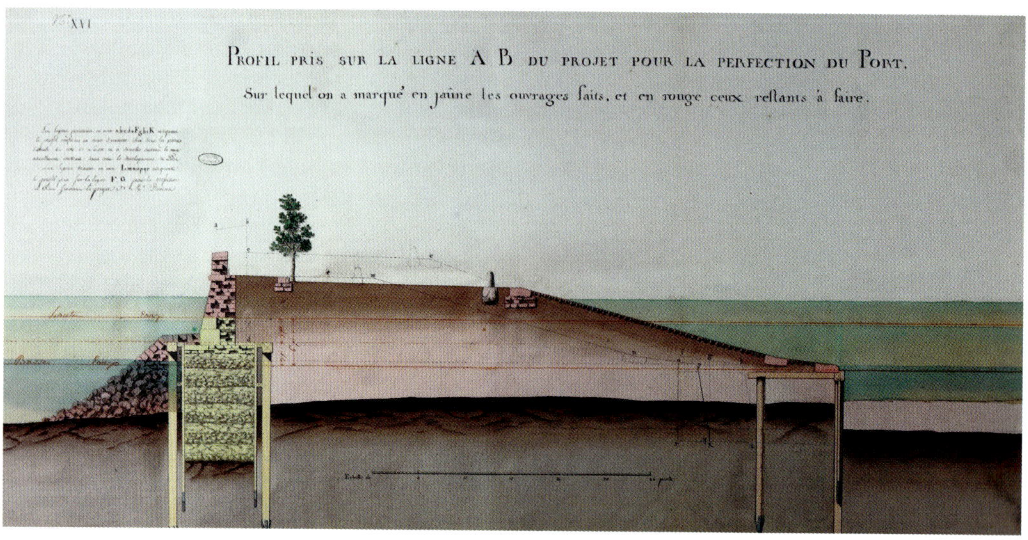

and 6 by 6 foot moises run across the two rows. The top of the foundation is covered, outside the walls and between the wood sections, by a 1 foot high, 2 feet wide coping in cut stone with reduced width in the straight zones, and 3 feet 8 inches in the exterior sections of the towers, independently of the fact that the stones in this coping are embedded 13 inches into the foundation's base. The horizontal surplus for the walls, running from the low waters to the bank, is 3 feet wide and built on natural elements composed of rubble masonry bound by lime and cement."[36]

After the report on the completed works, the engineer Nicolas Céard (1745-1821) is sent to Versoix in the first months of 1773 to assist Aubry and to resume the work on the port and the city. Céard reports on what De Caire has accomplished, starting with the description provided by Bélidor of the tests of Milet de Monville at Toulon. Céard's observations indicate that during the pouring of the compound there have been losses of limewash.

"The author of this treatise," Céard writes, "has seen the same endeavor [of Milet de Monville] at Versoix by Mr. De Caire, Chief Military Engineer, in the lake with the same results as 15 years ago and, at the time I am writing,

[36] Nicolas Aubry, "Devis des ouvrages à faire pour perfectionner le port de Versoix conformément à sa construction actuelle, et aux plans et profil dailleurs cijoints," 15 December 1772, H176, Archives Nationales, Paris. *Cf.* also, Nicolas Aubry, Léonard Racle, "Procès verbal de reconnaissance des ouvrages [...] faites jusqu'à ce jour pour le Port et le projet de la ville de Versoix," 24 November 1772, H176, Archives Nationales, Paris.

**309**. Port of Saint-Jean-de-Luz. Plan of the situation before the 1770s (Ms Fol. 3/2, Ecole des Ponts et Chaussées, Paris).

**310**. Port of Saint-Jean-de-Luz. Plan detail with indication, in dotted lines, of the position of the old securty wall ("Ancien mur de Garantie") and the already built new wall ("Perré de Garantie"), and of the section still to be completed (F.14.10265.5, Saint-Jean-de-Luz, Archives Nationales, Paris).

the breakage in the works are still in the water without change *as well as the entire port* that makes up, as we have said, an enormous block exposed to backwashes in many areas. One can observe that this manner of construction with masonry, submersed while still setting, as Bélidor indicates, to 15 to 20 feet deep has the disadvantage of leaking lime. This can be seen to have happened at Versoix with the sediment of lime around the coffers [...]."[37] "Since the woodwork for the Versoix port," continues Céard, "that is exposed to waves had been dismantled after 7 years of construction, and in order to guarantee the interior of the exposed walls, a never-before seen concrete masonry was needed (see the hydraulic architecture [of Bélidor]) [...]."[38]

Tests are made of the strength of the concrete produced by De Caire at the port of Versoix, which also serve to demonstrate the structural potential of the compound for the first builders of bridges in concrete only: "During the tests done on the foundation of Versoix port, it was recognized that this mortar, submersed in water, became as hard as rock within three months."[39]

Céard himself moves forward with the experience of concrete gained at Versoix, becoming one of the main protagonists of the use of concrete in experiments carried out systematically in the mid-1780s in the ports on the English Channel.

## Concrete works of Picault, Robert and Descolins at the port of Saint-Jean-de-Luz

The rise of concrete in hydraulic works has another chapter, just as fundamental as that of the projects at Versoix for applications of an experimental character, whose protagonists are Jean-Louis-François Picault and Robert de Parades, engineers en chef of the Généralité of Auch, and a young student of Perronet just out of the Ecole des Ponts et Chaussées, Joseph-Michel Descolins. The fact that Picault and Robert are active at Auch is important for their proposals regarding concrete construction of hydraulic works, given the fact that in the Auch region it is current practice to build the walls of houses with caissons in poured concrete.

The site of the new experiences with concrete is the port of Saint-Jean-de-Luz, built in a bay of the Bay of Biscay at the mouth of the Nivelle River, near the Spanish border. Ever since the 1600s, there has been a plan to build piers at the ends of the bay to counter the erosive force of ocean storms that struck the port. The project of Vauban to close the bay with two piers guides the work done since the start of the 1700s for the construction, in approximately 1704-08, on the sandy terrain in front of the port, constantly eroded by the sea, of a wall in blocks of granite, or *mur de garantie*, repeatedly damaged by sea storms.[40]

Following the assignment to work on the bay on the part of the Ponts et Chaussées, Picault, in collaboration with Robert, goes back to Vauban's solution

[37] *Ibid.*

[38] "Observations du Sr Perronet sur le projet envoyé par M. Céard le 7 juin 1784 d'un cone tronqué en charpente et maçonnerie pour l'enceinte & la rade de Cherbourg. Réponses du Sr Céard aux observations de M. Perronet qui lui ont été adressées par M. De la Millière le 26 juin 1784," Jean-Rodolphe Perronet, 21 June 1784, and Nicolas Céard, 1 July 1784, Ms 969, Ecole des Ponts et Chaussées, Paris. For Perronet's memoir see, "Observations du Sr. Perronet sur le projet envoyé par M... [Céard] le 7 juin 1784 d'un cône tronqué en charpente et maçonnerie pour l'enceinte de la Rade de Cherbourg," 21 June 1784, Ms 972, Ecole des Ponts et Chaussées, Paris.

[39] "Mémoire sur la maniere de mouler les ponts en mortier de béton. Expérience sur la force et la tenacité de la maçonnerie faite en mortier de béton," O.1.1294, 182, Archives Nationales, Paris.

[40] On the port see, Jean-Pierre Daguenet, *Ports de Saint-Jean-de-Luz et du Socoa*, in, *Ports maritimes de la France. Tome sixième (2e partie). Des Calonges à Hendaye*, Paris, Imprimerie Nationale, 1887, pp. 997-1035.

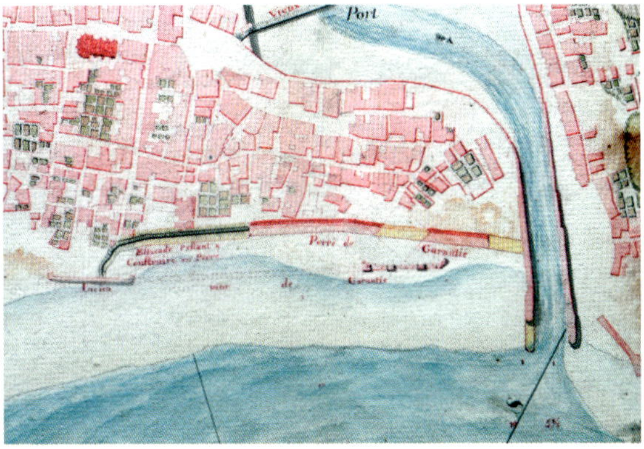


of the two piers. The new project for the extension of the masonry ports for about 80 *toises* of the canal, and for the work on the port, is examined by the Assemblée des Inspecteurs Généraux des Ponts et Chaussées at the meeting of 15 April 1764.[41] Some technical details have to do with the installation of the carpentry for the masonry. Règemorte advises "to ram the walls of the cofferdam to prevent water infiltration, as is done with locks: Mr. Picault has made plans to undergo his own works without these precautions."[42]

The works include the reinforcement of the base of the granite wall and the lengthening of this structure in the ocean, with "a jetty that necessitates a foundation built on moving sand in an area that is the most exposed to furious assaults by the ocean."[43] Two structures, also known as *barres*, will have to form the piers of Socoa and Sainte-Barbe to protect the bay.

Picault and Robert plan to use the concrete described in the treatises of the author who increasingly emerges as the inspiration for French engineers conducting research on concrete, namely Bélidor, often mentioned in the correspondence of Descolins. The use of this material should also be seen in the context of the pursuit of a type of masonry capable of standing up to the corrosion caused by worms that devour the foundations and piles, as well as to the violent action of the ocean currents.[44]

The concrete for the works at Saint-Jean-de-Luz is to be prepared with Italian pozzolana. The supply comes from ships that follow new routes of transport of the pozzolana from Civitavecchia, which precisely in the second half of the 1700s, starting with the works of Saint-Jean-de-Luz, begin to venture more and more as far as the Channel and beyond, after the cargos that reached Brest in the 1740s. Probably the ships that set off in June 1767 from Civitavecchia to bring to Saint-Jean-de-Luz three cargos of 126, 440 and 140

[41] Minutes of the meeting of 15 April 1764, in, "Extrait du Journal des assemblées des Ponts et Chaussées tenues chez Mr. Trudaine, depuis 1748 jusqu'en 1774," Ms 233, vol. XIII, Ecole des Ponts et Chaussées, Paris.

[42] A few days later, on 22 April, the same Assemblée des Inspecteurs Généraux des Ponts et Chaussées discusses Picault's project for the "rebuilding of lock gates, the piers and the port at Cherbourg that were destroyed by the English during the last war" (*ibid.*)

[43] Joseph-Michel Descolins, letter to Jean-Rodolphe Perronet, 23 May 1768, Ms 3085, Ecole des Ponts et Chaussées, Paris. See the report with attached drawings, "Suitte du Memoire abregé sur les ouvrages du Port de S.t Jean de Luz. Contregarde du Mur de Garantie," *ibid.*

[44] Joseph-Michel Descolins, letter to Jean-Rodolphe Perronet, August 1776, *ibid.* Perronet's reply is sent on 7 September 1776, as noted on the letter.

cartloads of pozzolana have been ordered precisely to make the works designed by Picault and Robert (the ships headed for Saint-Jean-de-Luz reach Civitavecchia from Leghorn).[45]

In May 1768 Descolins reaches Saint-Jean-de-Luz with the assignment of assisting Picault, who continues to take part in the work even when the worksite remains entirely under the supervision of Descolins. The composition of the cement with pozzolana planned for the works at Saint-Jean-de-Luz is exceptional in construction along the Atlantic Coast, according to what Descolins reports to Perronet, though pozzolana had previously been used at the port of Brest.

"We are going to use concrete made with pozzolana for the first time in the ocean, hoping that it will harden more than cement mortar," Descolins writes to Perronet as soon as he has reached Saint-Jean-de-Luz. "I will start after tomorrow to ram the piles and drive the cofferdams in order to establish the foundation; we are currently building Mr. Bélidor's raft to prevent the accumulations of sand, a lot of pile-drivers, etc. Then Mr. Picault and Mr. Robert will come spend some time here to put things in order, after which I will remain alone; I can only be very flattered by these men's welcoming, and I am pleased to have made their acquaintances [...]."[46]

The works for the construction of the "stone facing intended to serve as the foundation underpinning for the wall made out of granite blocks" do not begin until July 1769, due to delays in financing and difficulties caused by the constant silting up of the worksite, in spite of the erection of a temporary protective wall or *contregarde*, which cannot be raised further without running the risk of breakage by the waves; furthermore "the laying of the stones is also taking a very long time," since they must be well connected, also with irons.[47] The cutting of the stones is also complex, to make the narrow courses grip together and hold the slabs in position.[48] It is worth noting that certain technical solutions proposed for the connection between the stones are reminders of the construction system of the Venetian *murazzi* developed by Tirali. More in general, the masonry work envisioned at Saint-Jean-de-Luz resembles the system developed for the levees with stones and pozzolana to defend the Venetian Lagoon.

We cannot exclude the possibility that the pozzolana from Civitavecchia that reaches Saint-Jean-de-Luz is used for tests to establish the characteristics of the masonry and to banish those doubts about the concrete reported by Descolins in 1768 immediately after the first encounter with Picault and Robert – "hoping that it will harden more than cement mortar." The technique used for the deployment of the concrete is that of pouring into coffers made with wooden boards and poles, which Picault, Robert and Descolins can find described in the treatises of Bélidor.

[45] "Suitte de l'Etat des batiments françois qu'on fait commerce a Civitavecchia pendant les trois mois d'avril, may, et juin de l'an 1767," in, "Correspondence Consulaire, Civitavecchia, 1744-1774," AE, B1.374, Archives Nationales, Paris.

[46] Joseph-Michel Descolins, letter to Jean-Rodolphe Perronet, 23 May 1768, Ms 3085, Ecole des Ponts et Chaussées, Paris.

[47] Joseph-Michel Descolins, letter to Jean-Rodolphe Perronet, 15 September 1769, *ibid*. See the drawing, "Profil du mur de garantie de St. Jean de Luz et du Perré qui doit lui servir de rempiettement commencé en juillet 1769," *ibid*.

[48] See the drawing by Descolins, "Plan du mur de garantie et du Perré," 1769, *ibid*.

[49] Joseph-Michel Descolins, letter to Jean-Rodolphe Perronet, 15 September 1769, *ibid*.

[50] Joseph-Michel Descolins, letter to Jean-Rodolphe Perronet, August 1776, *ibid*. Perronet's reply is sent on 7 September 1776, as noted on the letter.

[51] *Ibid*.

**311**. Joseph-Michel Descolins, "Counterguard to create a ledge to support the foot of the security wall," port of Saint-Jean-de-Luz (drawing attached to, [J.-M. Descolins], "Suitte du Memoire abregé sur les ouvrages du Port de S.t Jean de Luz. Contregarde du Mur de Garantie," [1768], Ms 3085, Ecole des Ponts et Chaussées, Paris).

The first technical problems, however, arise in the process of pouring into the coffers, due to the seabed that does not permit sinking of the boards. Picault attempts different solutions, as Descolins reports to Perronet in September 1769: "Concerning the work done on the piers on the seabed, almost the entire company has tried their hand with different experiments to form concrete coffers [*encaissements*] up to 12 feet below the surface of the sea at low tide; not able to do so with piles, we have tried to sink 12 foot long caissons, solidly built, and this way form our coffers by [eliminating ?] the sand on the inside by using Bélidor's large shovel and method, by steadily driving them at the angles; this has not been successful and Mr. Picault was finally obliged to decide to return to using sheet piles, to sink and steady them as much as we could and then to cut them back in such a way that the coffers will only be 9 or 10 feet under the water, which will still be sufficient."[49]

After starting with the driving of the piles and sheet piles, the works have to be interrupted in 1769 due to the reduction of financing and the repairs required by the *mur de garantie* – the causes of the damage include sea worms that devour the foundations, and the lack of efficacy of the rubble poured to consolidate the base of the wall.[50] It is perhaps not until 1773 that Descolins manages to return to the construction of a jetty, using the method envisioned by Picault. The quality of the pozzolana cement obtained in the first works in the sea is narrated in the observations of Descolins. "Our concrete works very well," he tells Perronet, "it hardened to such an extent that, even with strong strikes from a pickaxe, it was very difficult to demolish a section that was slightly too elevated for our massive berm to be placed on top and it was impossible to ram it down further."[51]

During the winter of 1773 a fierce storm damages the embankment built to reinforce the base of the *mur de garantie*, while in February 1774 breaches are opened in the masonry. Descolins proposes different solutions to repair the damage, based on the use of a pile structure, or *crêche*, of greater or lesser area, variously inclined and closed by means of wooden boards to form a coffer to fill with concrete masonry; as an alternative, he proposes true isolated caissons, 12 feet in length, again to fill with concrete, like those tested by Picault for the

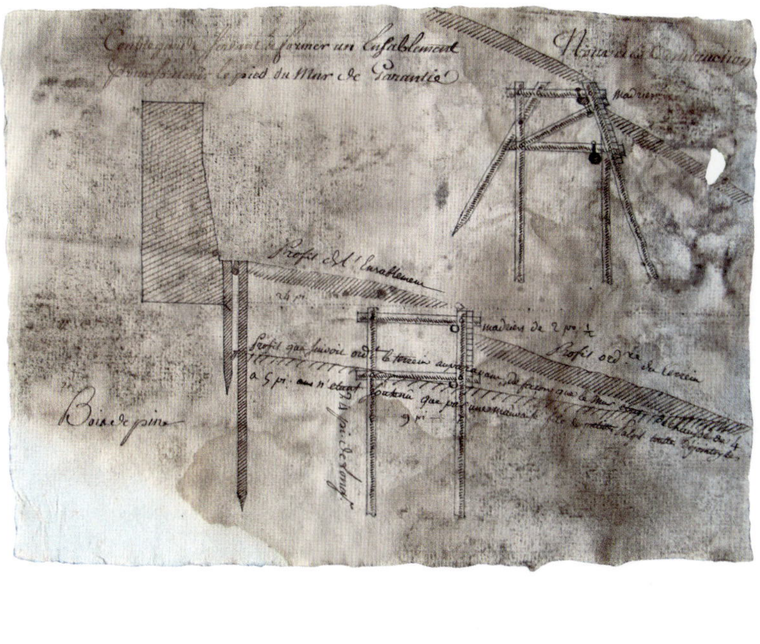

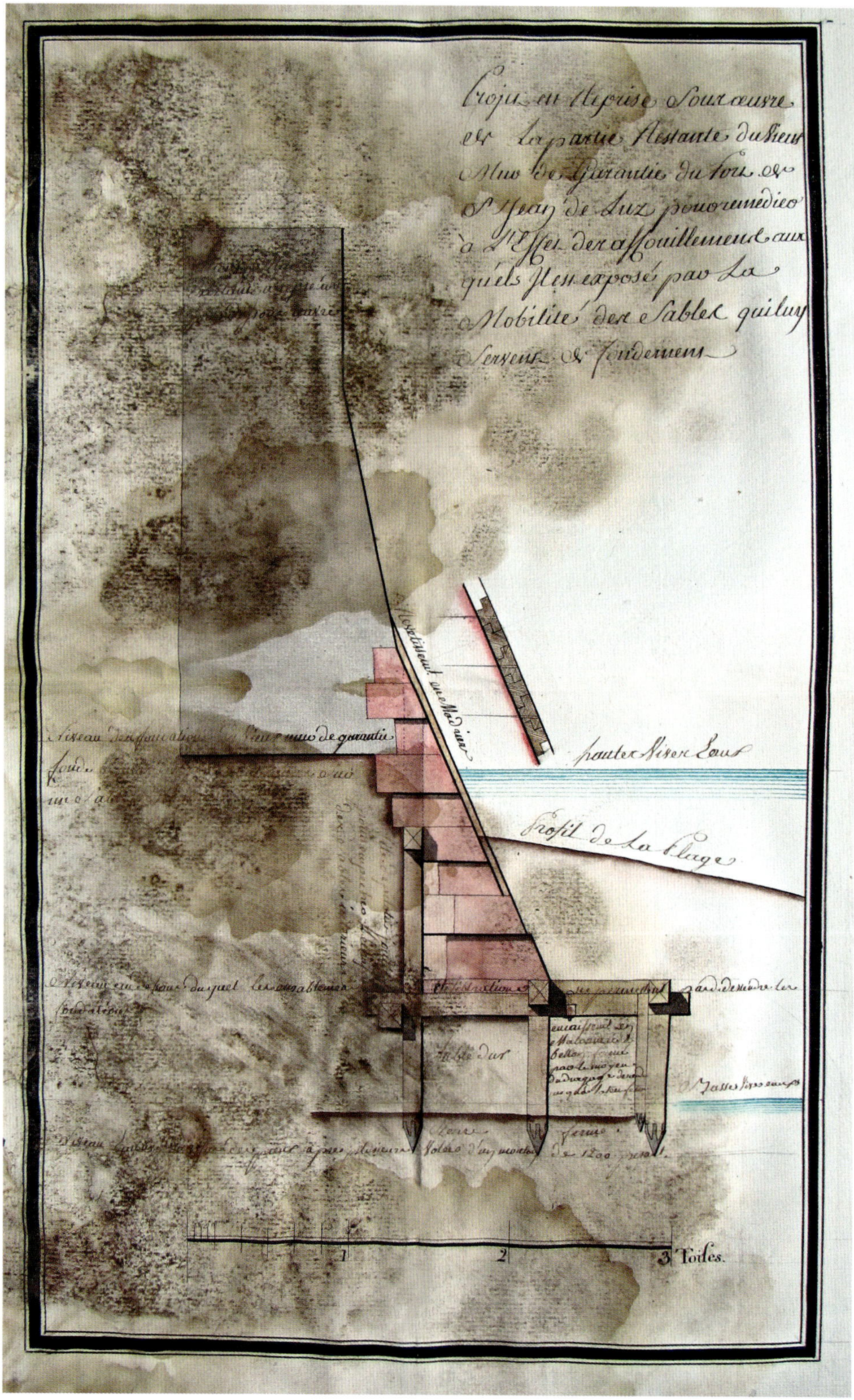
Croquis en Reprise Sous œuvre de la partie Restante du vieux Mur de Garantie du Fort de St Jean de Luz pour remédier à l'Effet des affouillements auxquels il est exposé par la Mobilité des Sables qui luy Servent de Fondemens
Profil de la Plage
1
2
3 Toises.

**312**. Joseph-Michel Descolins, "Recovery project work for the remaining part of the old security wall of the Port St Jean de Luz, to remedy the effect of undermining to which it is exposed caused by the movement of sand which acts as its foundation," 1769 circa. At the foot of the wall an "encaissement" with concrete ("betton") is envisioned (Ms 3085, Ecole des Ponts et Chaussées, Paris).

**313-314**. Joseph-Michel Descolins, "Profile of the security wall of St Jean de Luz and of the riprap of restoration, began in July 1769" and "Plan of the security wall and riprap," port of Saint-Jean-de-Luz (Ms 3085, Ecole des Ponts et Chaussées, Paris).

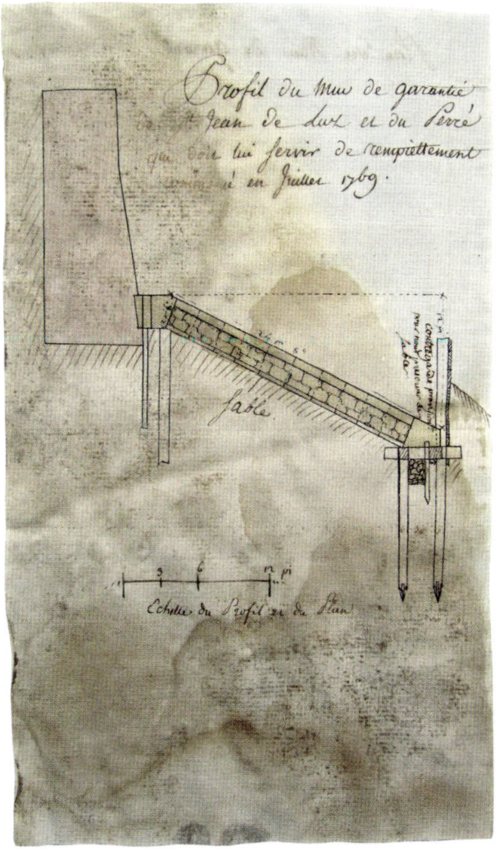


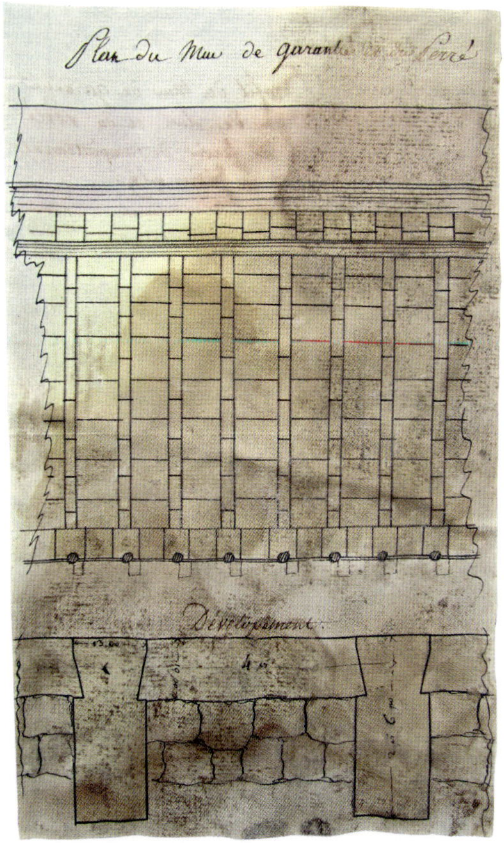

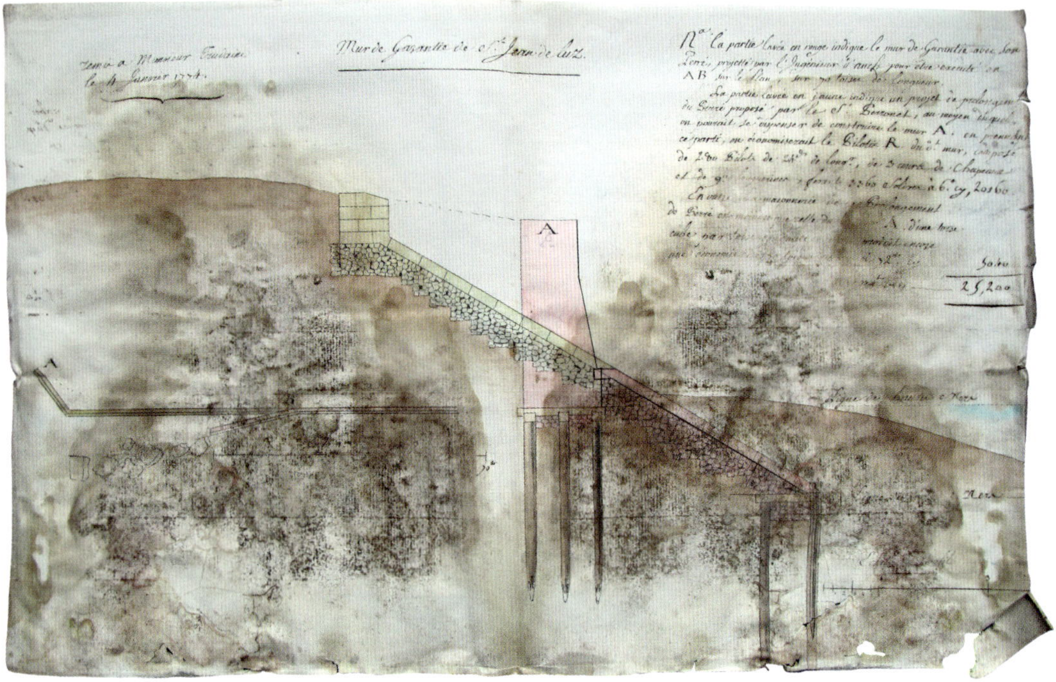
Mur de Garantie de St Jean de luz.
A

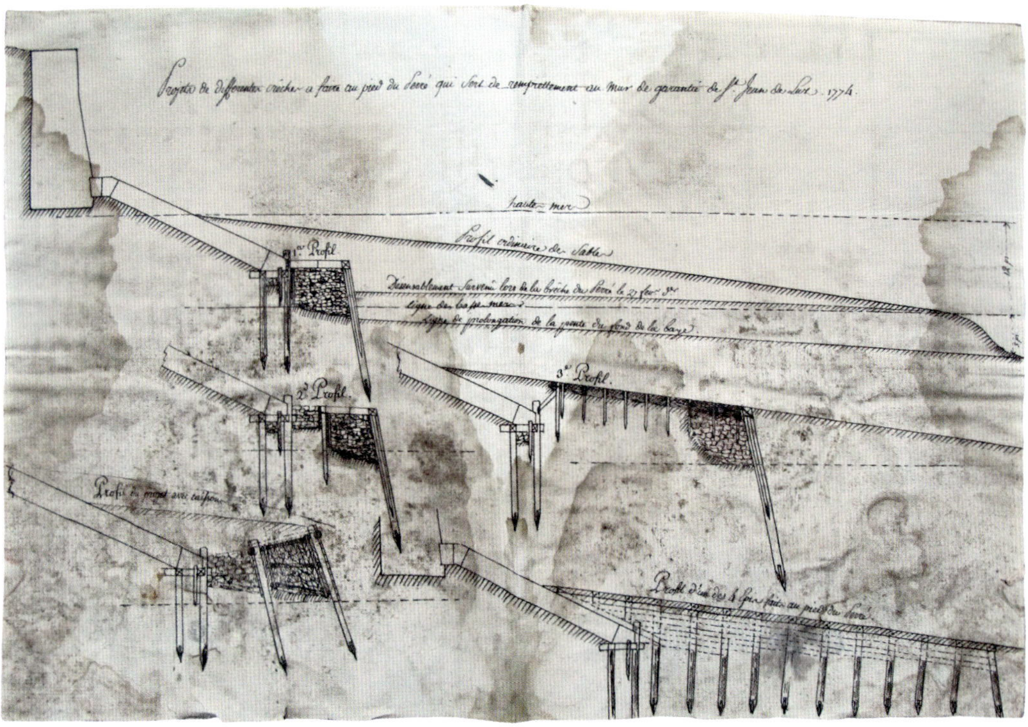
haute mer
Profil ordinaire de Sable
1er Profil
2e Profil
3e Profil

**315**. Joseph-Michel Descolins, "Security wall de St. Jean de Luz. Submitted to Mr. Trudaine on the 4 January 1774." The extension of the riprap up to the top is proposed by Perronet, to avoid raising the wall indicated by the letter A (Ms 3085, Ecole des Ponts et Chaussées, Paris).

**316**. Joseph-Michel Descolins, "Projects for different row of piles [*crêche*] to be made at the foot of the riprap that acts to restore the security wall of St. Jean de Luz, 1774," port of Saint-Jean-de-Luz (Ms 3085, Ecole des Ponts et Chaussées, Paris).

foundation of the jetty. "Then the crib [*crèche*] will be filled with concrete masonry," Descolins explains, "and then thrown in water where it has the property of hardening; and if we observe, during the moments when the bottom sand itself is removed, that the sea is too harsh with its onslaught of shale and pebbles, then we could cover it with rubble which the sea would only wash over."[52]

Other technicians consulted to repair the damages suggest different solutions to protect the mortar from the water during the setting phase. The engineer of the Ponts et Chaussées of Bordeaux, François-Antoine Valframbert, suggests reinforcing the wall, as a subfoundation, with piles and "coffers in concrete masonry,"[53] and the construction of a new embankment in cut stone, protecting the masonry during setting by means of a covering of well interlocked wooden boards, a solution also applied by Smeaton for the base of the Eddystone lighthouse – Valframbert specifies "cement mortar" as the masonry.[54] But the works with pozzolana at the bay of Saint-Jean-de-Luz do not achieve the expected results, as can be seen in a report dated 1777: "We have a stock of 160000 tons of pozzolana, and we do not know what to do with it or how to use it. Coffers have been made at the tip of the quay or jetty that the sea is undermining (because caves have formed, large enough to fit two to three people), we have filled some with stones mixed with pozzolana, that remain as soft as the moment it was first deployed."[55]

When Descolins is working to propose solutions to repair the damages caused by storms on the masonry works, the discovery of the composition of the cement of Loriot is officially made public, thus marking the beginning of European research on the formula of Roman cement without pozzolana.

[52] Joseph-Michel Descolins, "Observations sur la crêche à faire au pied du Perré qui sert de rempiettement au mur de garantie de la ville de St. Jean de Luz," [February 1774], *ibid.* See the drawing, sent by Descolins to Perronet on 14 October 1774, and entitled "Projets de differentes crêches a faire au pied du Perré qui sert de rempiettement au mur de garantie de St. Jean de Luz, 1774," *ibid.*

[53] François-Antoine Valframbert, "Projet en reprise sous œuvre de la partie restante du vieux mur de garantie du port de St Jean de Luz pour remedier à l'effet des affouillements aux qu'els il est exposé par la mobilité des sables qui luy servent de fondement," 22 July 1777, *ibid.*

[54] François-Antoine Valframbert, "Port Maritime de St. Jean de Luz. Memoire," 22 July 1777, *ibid.* In some parts of the defensive levee, Perronet himself proposes lengthening the stone cladding of the embankment, avoiding the erection of a high barrier wall – there are no indications on the nature of the masonry (see the drawing, "Mur de Garantie de St. Jean de Luz," sent to Trudaine on 4 January 1774 and perhaps prepared by Descolins, with the note "The section washed in yellow indicates Perré's prolongation project proposed by Sr. Perronet [...]," *ibid.*).

[55] D'Antin, "Mémoire sur les ouvrages de St. Jean de Luz, sur l'etat actuël du port, et sur les moyens de le retablir," 1777, Marine D2. 54, Archives Nationales, Paris.

# Name index